Calculus

A computer algebra approach

Second Edition

Calculus

A computer algebra approach

Second Edition

by Iris Anshel and Dorian Goldfeld
Columbia University, New York

International Press
www.intlpress.com

Calculus: A computer algebra approach, 2nd edition
by Iris Anshel and Dorian Goldfeld (Columbia University, New York)

ISBN 13: 978-1-57146-222-0
ISBN 10: 1-57146-222-8

Paperback re-issue 2011
This title was previously published in 1996 under ISBN 1-57146-038-1 (casebound).

Printed in the United States of America.

DEDICATION

The authors would like to dedicate this book to their children, Ada and Dahlia, who played quietly while their parents worked on the manuscript.

Acknowledgement

The authors would like to thank Professor Hervé Jacquet, of the Columbia University Mathematics Department, for his motivating idea that we initiate calculus reform at Columbia and for suggesting several elegant and succinct proofs. Professors, Hervé Jacquet, Duong H. Phong, Ramuald Dabrowsky, Patrick X. Gallagher and Cormac O'Sullivan provided indispensable editorial assistance during the preparation of the manuscript, and we thank International Press for making the text available to students at very low cost. The additional exercise appearing at the end of each chapter were contributed by Maia Berkane and Kevin Oden, the authors with to express their gratitude their appreciation for their efforts. We are deeply grateful to Professor Morton Friedman, of the Columbia University School of Engineering, for creating a computer laboratory which has allowed us to present an experimental course based on this material. His constant support and encouragement, as director of the Gateway Project, have been invaluable: without his efforts this project would still be on the drawing board. Finally, we wish to thank the Gateway Project for financial support.

Contents

INTRODUCTION

The evolution of the teaching of calculus is at a critical juncture. For some time there has been an emphasis on the computational aspects of calculus in conjunction with the various applications of the method. It is the advent of highly accessible computer algebra systems (CAS) and various sophisticated calculators which has driven us to reevaluate how calculus should be presented. The use of this book does *not* require a computer laboratory, or even access to an elaborate CAS. A graphing calculator suffices in that the emphasis is not on mass computation or programming, but rather on the understanding of the underlying concepts.

A CAS is a computing device with the following capabilities:

(1) *It is a calculator,* i.e., it can perform arithmetic and compute values of standard functions,

(2) *It has 2–dimensional graphics capabilities,* i.e., it can produce graphic displays of functions of a single variable,

(3) *It can do calculus,* i.e., it can compute derivatives and integrals of functions,

(4) *It can do algebra,* i.e., it can expand and simplify algebraic expressions,

(5) *It can create functions,*

(6) *It has 3–dimensional graphics capabilities,* i.e., it can produce graphic displays of functions of several variables.

In light of the immense capabilities of the CAS we are led us to asking the following questions: (1) what approach should we take to this discipline now that the mechanics of computation have been automated, and (2) how can we incorporate CAS with the classical teaching methods?

It is the authors opinion that the new technology is a moment of opportunity. No one would consider learning, for example, Physics or Biology without working in a laboratory; computation for anyone without a CAS is very limited. Calculus is both an intellectual breakthrough and a powerful tool in research and development (in a variety of fields). To understand it as such (and to apply it as a problem solving tool in modern settings) one must obtain both an understanding of the vital concepts in their abstract form and in a wide range of examples. Our philosophy is to focus on the meaning of the most important definitions and theorems and to *experiment* with them on the CAS.

It should be noted that this text is not a "how–to" manual for a particular CAS. Our aim is to teach the concepts of calculus without getting lost in the quagmire of programming. Almost all the exercises and examples can be done at the level of using the CAS as a calculator. Beyond this, even on the most sophisticated CAS, only a few commands need be learned.

Many students are familiar with introductory calculus upon entering college. With this in mind, we have opted to quickly review the foundational material (such as real numbers, functions and their graphs, lines, and circles). Although the introductory material is covered in relatively little space, almost nothing is assumed on the part of the reader (that is to say, the text is substantially self–contained). When the material becomes less familiar and more complex the sections and chapters become more substantial. Throughout the text we focus on core concepts which are presented from first principles. Once a given concept is introduced we utilize the CAS to rapidly view it in many different contexts. This approach encourages the student to develop *hands–on* experience with Calculus. We have found that such a CAS experience is superior to reading hundreds of worked out examples in a book. Evidence of the benefits of this method were apparent when, in 1993–94, an experimental computer laboratory was set up at Columbia University (in conjunction with a course based on this book).

It is the authors hope that this book is intuitively rigorous and indicates how mathematics is thought about. There is little purpose to memorizing massive numbers of formulae and algorithms (all of which the CAS knows). It is a natural outcome of this pedagogical perspective that solutions of problems are presented analytically, and whenever possible and appropriate, algorithmically. The emphasis is consistently on how to derive the formulae, why the algorithms work, and how the CAS can solve Calculus problems.

This edition begins by covering both differential and integral calculus for functions of one variable, mathematical modeling and optimization, basics of ordinary differential equations, and then moves on to differential calculus for vector valued functions and functions of several variables. Much time is spent on vector geometry, coordinate systems, and two and three dimensional graphical display using the CAS. The latter part of the text includes multiple integration, vector fields and line integrals, surface integrals Stoke's theorem, an overview via differential forms, as well as an introduction to Fourier series including a proof of the Fourier expansion theorem.

This book assumes that students have access to a basic CAS with capabilities (1), (2), and (3) above. Among the many exercises the majority can be solved either by hand or with a basic CAS.

Chapter 0

What is Calculus?

0.1 The Real Numbers

The integers, $\{\dots, -3, -2, -1, 0, 1, 2, 3, \dots\}$ can be visualized as equally spaced points marked off on a line.

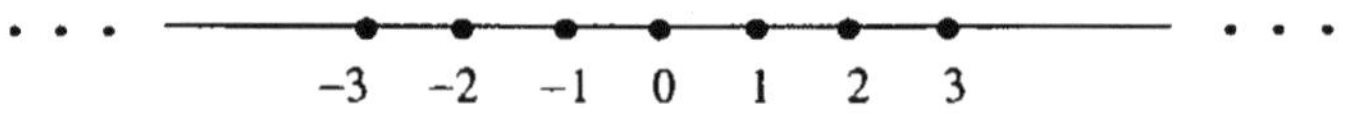

The resulting figure is called the **number line**. When we form all possible ratios of integers, for example, $\frac{1}{2}$, $\frac{3}{4}$, $\frac{13}{4}$, we obtain what the ancient Greeks termed the **rational numbers**. By marking the rational numbers on the number line we are left with a seemingly dense picture.

Consider now the geometric construction of $\sqrt{2}$ on the number line.

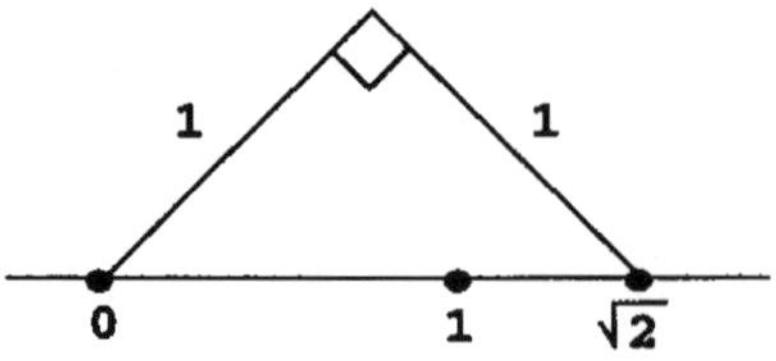

Being that $\sqrt{2}$ is on the line it is natural to ask, is $\sqrt{2}$ rational? This question was investigated by Hippasus, a student of Pythagoras. He proved (as shall we) that $\sqrt{2}$

is not rational. This realization so disturbed his colleagues that he was put to death! (There were, however, others who saw this breakthrough in a positive light — and sacrificed 100 oxen.) The proof that $\sqrt{2}$ is not rational is a classical example of *proof by contradiction* and proceeds as follows. Suppose $\sqrt{2} = \frac{a}{b}$, where a, b are integers. Then $2 = \frac{a^2}{b^2}$ or

$$a^2 = 2b^2.$$

In general any integer can be expressed as the product of a power of 2 and an odd number, e.g., if $a = 40$ then $a = 8 \cdot 5 = 2^3 \cdot 5$. Now expressing a and b in this form we have

$$a = 2^e \cdot a_0$$
$$b = 2^f \cdot b_0,$$

where both a_0 and b_0 are odd numbers. Now plugging these expressions into the formula $a^2 = 2b^2$ gives us

$$2^{2e} \cdot a_0^2 = 2 \cdot 2^{2f} \cdot b_0^2 = 2^{2f+1} \cdot b_0^2.$$

Since both a_0^2 and b_0^2 are odd numbers (by definition) this dictates that the powers of 2 must *agree* and we deduce $2e = 2f + 1$. But $2e$ is an even number and $2f + 1$ is an odd number. Hence the assumption that $\sqrt{2}$ is rational has led us to a contradiction and must be false.

The Greeks termed numbers which are not rational **irrational** (reflecting their attitude). When we combine the integers, the rational numbers, and the irrational numbers we obtain the number system worked with in calculus, the **real numbers**, denoted by $\mathbb{R}$. Every real number can be expressed by an infinite decimal expansion,

$$\sqrt{2} = 1.414\ldots$$
$$\pi = 3.14159\ldots$$

0.2 What is Calculus?

Calculus is a mathematical method which uses an infinite process to solve finite problems. The two basic problems which the theory investigates are:

Problem [1] (The problem of the tangent) How can we find the tangent line to a curve at a point P?

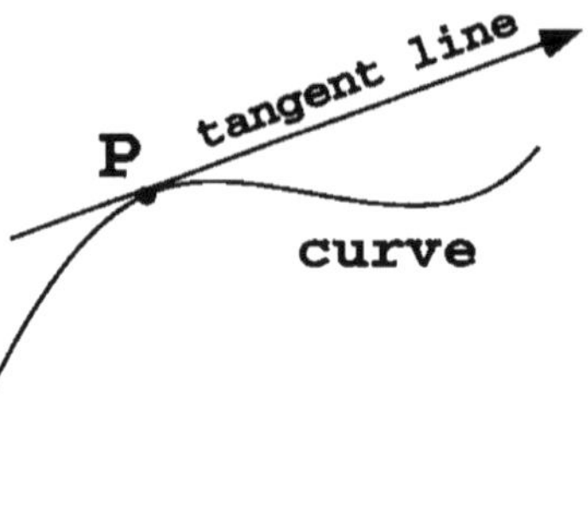

Problem [2] (The problem of area) If a curve bounds a region A, how can we compute the area of A?

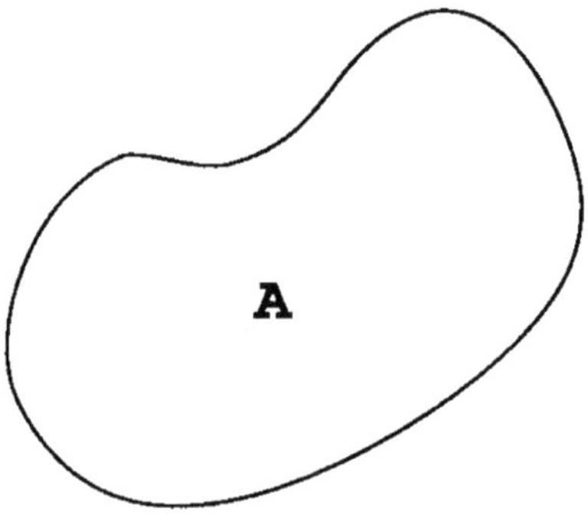

These problems arise naturally in modern context. Imagine, for example, the trajectory of a stealth bomber to be the curve. If a missile is fired, it will travel along the path of the tangent line to the curve. This leads one to want to compute the equation of the tangent line precisely. Consider also an oil spill on the ocean. If we know the currents and wind conditions we can obtain an approximation of the boundary curve for the spill and then, using calculus, we can compute the area of the region covered by oil.

Sir Isaac Newton (and G. Leibnitz independently,) developed differential and integral calculus to solve problems [1] and [2], respectively. It appears, at first sight, that the above problems are unrelated. It is truly remarkable that the solutions to these problems are inverse to each other. Differential calculus provides a sequential process for obtaining the slope of the tangent line to a curve at a point. The computation of the area of the region $\mathcal{R}$ consists of the same sequential process in the reverse order. When this is formally stated we obtain the *Fundamental Theorem of Calculus.* A rigorous derivation of this theorem is given in our text and is the central goal of Part I of our text.

The key concept upon which calculus is based is the limit (which ultimately allows us to describe the real numbers cohesively). An irrational number, such as $\sqrt{2}$

or π, cannot be described in finite terms but can be defined as a *limit of a sequence* of rational numbers. The numbers,

$$1,\ 1.4,\ 1.41,\ 1.414, \ldots \to \sqrt{2}$$
$$3,\ 3.1,\ 3.14,\ 3.141,\ 3.1415, \ldots \to \pi$$

should be thought of as sequences which *approach* $\sqrt{2}$ and π in the limit. This idea already occurs in Greek mathematics with regard to the question of area. By inscribing successive polygons (where the number of sides increases) inside a circle,

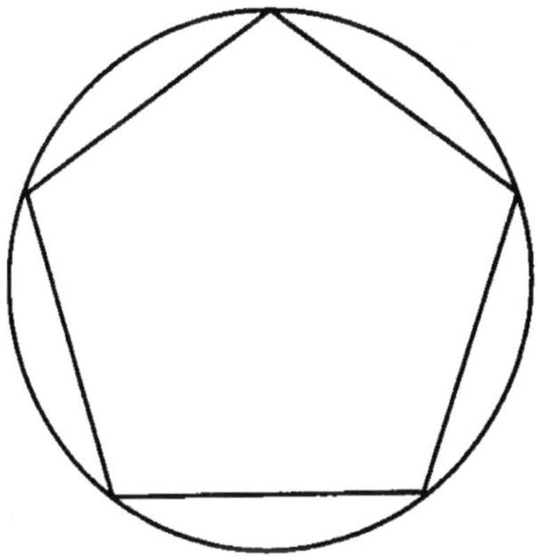

Archimedes was able to create sequences whose entries are better and better approximations to the area of the circle and approach the exact area of the circle in the limit. Similarly, the fundamental idea which ultimately provides the solution to problem [1] is also given by a limiting process.

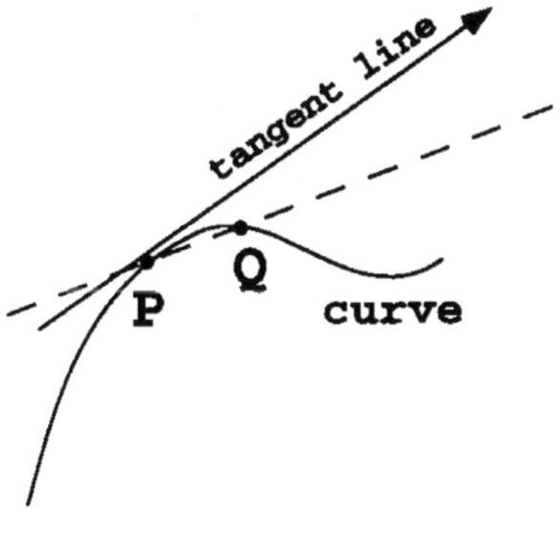

By choosing a point Q near P, the dotted line passing through P and Q will get closer and closer to the tangent line at P as Q approaches P. Newton's breakthrough was to abstractly analyze *the limit* of Q approaching P. This is the essence of differential calculus.

The above descriptions are of infinite processes. The concept of utilizing an infinite process to analyze a finite (and even concrete) problem was in fact revolutionary and it permeated throughout science, engineering, economics, biology, etc. But the ramifications of this breakthrough should not be seen as only technical ones. One should always keep in mind that there have also been sociological and cultural consequences of this way of thinking, the industrial revolution being one striking example.

Chapter I
Functions and their Graphs

1.1 Functions

One of the essential concepts studied in mathematics is the notion of a function between sets (a set being a collection of objects). Thus we begin with,

> **Definition:** *Given two sets A, B, a **function** from A to B is a rule which assigns to each element of the set A a unique element of the set B.*

Functions occur everywhere. For example, if A is the set of vegetables in a grocery store and B is the set of prices, then we have a well–known function (the price function). The rule of this particular function is determined by the store manager to maximize profit. Notice that the price is unique. A bunch of carrots cannot cost 90¢ and 83¢ simultaneously. There is only one price. This is why the general definition requires that the rule assigns to each element of A a *unique* element of B.

The general mathematical definition of function given above is particularly suitable for computers. By identifying the sets A and B and the rule between them we are forced to think in the language of the computer. It is generally necessary to name a function and describe the rule by a notation. Functions are usually represented by a letter such as F (for function). If x is in the set A, denoted $x \in A$, then $F(x)$ will denote the unique element of the set B determined by the rule of the function. We pictorially denote a function by

$$F : A \longrightarrow B$$

$$x \longmapsto F(x).$$

Example [1.1] (The price function) Let

$$A = \{\text{onions, carrots, potatoes, broccoli, cabbage, peas}\}$$

be a set of 6 vegetables. Let

$$B = \{65, \$1.00, \$1.25, \$1.60, \$1.75\}$$

be a set of 5 prices. We define a price function P by the rule

$$\begin{aligned} &P(\text{onions}) = \$1.00 \\ &P(\text{carrots}) = \$1.25 \\ &P(\text{potatoes}) = 65¢ \\ &P(\text{broccoli}) = \$1.25 \\ &P(\text{cabbage}) = 65¢ \\ &P(\text{peas}) = \$1.75 \end{aligned}$$

where, for example, P(broccoli) stands for the price of broccoli (per pound).

Example [1.2] (The squaring function) Let $A = B$ be the set of all real numbers. The squaring function is the function that to each element $x \in A$ assigns the value $x^2 \in B$. If we use the letter S to denote the squaring function then it is defined by the rule $S(x) = x^2$. When we construct this function on the CAS we can quickly verify, for example, $S(1.357) = 1.84145$, $S(256) = 65536$.

Example [1.3] (Loan shark function) Imagine yourself as a loanshark lending people $100 and requiring that they owe you 10% more each week. That is to say after one week they owe you $110. Now 10 percent of $110 is $11, and thus after two weeks they owe you $121. It is easy to see that after x weeks they owe you $(100) \cdot (1.1)^x$ dollars. This defines the loan shark function $L(x) = 100 \cdot (1.1)^x$. When we define this function on the CAS we obtain the values: $L(2) = \$121, L(10) = \$259.37$, $L(52) = \$14,204.29$, $L(104) = \$2,017,619.45$. This is called an **exponential function** and we say that its values **increase exponentially**. In fact, after 2 years you will be owed over 2 million dollars!

Example [1.4] Your CAS already contains a multitude of built in special functions. For example we can immediately compute the following:

$$\sin(\pi/8) = 0.3826834, \quad \tan(2.34) = -1.0329, \quad \log(81) = 4.394449.$$

In the next example we see that not all functions involve numbers exclusively.

Example [1.5] (The interval function) Let A denote the set of all real numbers and let $B = \{\text{True, False}\}$ denote the set of the 2 words True and False. Upon constructing the interval function $i(x)$ which is defined by the rule

$$i(x) = \begin{cases} \text{True}, & \text{if } 1 \le x < 2 \\ \text{False}, & \text{otherwise.} \end{cases}$$

we can verify the values $i(1) = \text{True}$, $i(2) = \text{False}$, $i(1.2331) = \text{True}$, $i(-1) = \text{False}$.

Example [1.6] (Absolute value function) Define the function $A(x) = |x|$ by the rule

$$A(x) = \begin{cases} x, & \text{if } x \ge 0 \\ -x, & \text{if } x < 0. \end{cases}$$

Again we can verify the values $A(-15) = 15$, $A(3) = 3$, $A(-2) = 2$, $A(0) = 0$.

Exercises for §1.1

(1) Input the price function of Example [1.1] on your CAS by simply memorizing every value of the function.

(2) Construct the function $f(x) = 2x^3 - 7x^2 + 1$ on your CAS. Evaluate $f(100)$, $f(5.7)$, and $f(\pi)$ to 10 decimal places.

(3) Construct the function $f(x) = 2\cos(x) - 3x^4 + 2$ on your CAS. Evaluate $f(1)$, $f\left(\frac{\pi}{2}\right)$, and $f(3)$ to 5 decimal places.

(4) Construct the function $f(x) = (\sin(x))^3 + x^2 + 11$ on your CAS. Evaluate $f(1)$, $f(3)$, and $f(5\pi/2)$ to 37 decimal places.

(5) Construct the absolute value function of Example [1.6] on your CAS using the IF–THEN command on your CAS.

(6) Construct the function

$$g(x) = \begin{cases} x^2 & \text{if } x \le 0 \\ 2^x & \text{if } x > 0 \end{cases}$$

on your CAS. Evaluate $g(-3)$, $g(-2)$, $g(1)$, $g(10)$ and evaluate $g(1/2)$ to 10 decimal places.

(7) Construct the function $c(t)$ which given the temperature t in Fahrenheit outputs the temperature in Centigrade. Hint: $c(32) = 0$, $c(212) = 100$. Evaluate $c(50)$, $c(70)$, $c(90)$.

(8) Assume a car is traveling at a constant speed of 50 miles per hour. Construct a function $d(t)$ which outputs the distance traveled by the car after t hours.

(9) Construct the interval function of Example [1.5] on your CAS.

(10) Construct the loanshark function $L(x)$ of Example [1.3] on your CAS. Evaluate $L(52)$, $L(104)$.

(11) A car rental company offers cars at $50 a day plus 20 cents per mile. Construct a function which outputs the total cost of renting a car for one day assuming m miles are traveled.

(12) Construct on your CAS a function $u(n)$ which outputs the word EVEN if n is an even integer and outputs ODD if n is an odd integer.

1.2 The Domain and Range of a Function

Let $F : A \longrightarrow B$ be an arbitrary function.

Definition: *The set A is called the* ***domain*** *of F and the set of values of f, which is denoted $f(A) = \{f(a) \mid a \in A\}$, is called the* ***range*** *of F.*

If we want to program a function on a CAS it is very important to know its domain and range. The programming procedures vary substantially from one type of set to the next. In general, when a function is being studied it is necessary to carefully ascertain the largest possible domain and range for which the function is well defined.

For example, on a CAS it is not possible to divide by zero. Hence the domain of the function

$$F(x) = \frac{1}{x},$$

is the set of all real numbers except zero, as is the range (note that zero cannot be in the range since the equation $\frac{1}{x} = 0$ has no solution in x).

Consider now the more complex example,

$$F(x) = \frac{1}{(x-3)^2(x+1)^2}.$$

This function is well defined except when $x = 3$ or $x = -1$, and hence its domain consists of the set of all real numbers except $3, -1$. What about the range? Since the square of a real number must always be positive, we see that the range of this function can only consist of positive real numbers. Can any number $a > 0$ be in the range? This question is equivalent to asking whether we can, given any $a > 0$, solve the equation

$$\frac{1}{(x-3)^2(x+1)^2} = a.$$

This equation is equivalent to

$$(x+1)(x-3) = x^2 - 2x - 3 = \frac{1}{\sqrt{a}},$$

i.e., $x^2 - 2x - (3 - \frac{1}{\sqrt{a}}) = 0$. Applying the quadratic formula we obtain two solutions of the equation:

$$x = \frac{2 \pm \sqrt{4 + 4(3 + 1/\sqrt{a})}}{2} = 1 \pm \sqrt{4 + 1/\sqrt{a}}.$$

Thus the range of f does in fact consist of *all* positive real numbers. We can concisely present this data using the language of sets as follows:

$$\begin{aligned} \text{Domain}(F) &= \{-\infty < x < \infty \text{ and } x \neq -1, 3\} \\ \text{Range}(F) &= \{0 < y < \infty\}, \end{aligned}$$

where we have put $y = F(x)$. Note that 0 cannot be in the range!

There is one other problem that can arise which needs to be discussed. Consider the square root function $s(x)$ = the *square root of* x. Notice first that every positive real number has 2 square roots, for example

$$\begin{aligned} &s(4) = \pm 2 \\ &s(5) = \pm 2.236068\ldots \\ &s(9) = \pm 3 \\ &\text{etc.} \end{aligned}$$

It is therefore necessary to specify which square root (the positive or the negative one) the function s(x) chooses. Usually one specifies that $s(x)$ is the positive root,

$$\begin{aligned} &s(4) = \sqrt{4} = 2 \\ &s(5) = \sqrt{5} = 2.236068\ldots \\ &s(9) = \sqrt{9} = 3 \\ &\text{etc.} \end{aligned}$$

Example [1.7] (Negative square root function) It is possible to define another square root function, s (by abuse of notation we also call it s) which chooses the negative root. In this case

$$\text{Range}(s) = \{x \leq 0\},$$

i.e., the range of s is the set of all real numbers less than or equal to 0. Note that 0 is in the range since $s(0) = 0$.

What about the domain of s? Since the square root of a negative number cannot be a real number, it is necessary to restrict the domain of s:

$$\text{Domain}(s) = \{x \geq 0\}.$$

Remark: This restriction on the domain can be dropped upon allowing imaginary numbers in our range! In that imaginary numbers are not studied in first year calculus we have restricted ourselves in the above example to domains and ranges consisting of real numbers.

Example [1.8] (Designer square root function) One can construct on the CAS a designer square root function $s(x)$ where $s(x)$ is chosen to be the positive square root if the leftmost digit in the decimal expansion of x is even and $s(x)$ is chosen to be the negative square root if the left most digit is odd. The values of this function change sign frequently, for example $s(1) = -1$, $s(4) = 2$, $s(9) = -3$, $s(16) = -4$, $s(25) = 5$, $s(\pi) = -1.772454\ldots$, etc. What are the domain and range of this function? Clearly the domain consists of all numbers which are greater than or equal to zero. The range of the function cannot be compactly described, but it consists of the output of $s(x)$ where x ranges over numbers greater than or equal to zero.

Example [1.9] (Cube root function) The cube root function

$$c(x) = \sqrt[3]{x},$$

is quite different from the square root function. For example

$$\begin{aligned}
&\sqrt[3]{8} = 2 \ \text{ since } 2 \cdot 2 \cdot 2 = 2^3 = 8 \\
&\sqrt[3]{-8} = -2 \ \text{ since } (-2) \cdot (-2) \cdot (-2) = (-2)^3 = -8 \\
&\sqrt[3]{27} = 3 \\
&\sqrt[3]{-27} = -3.
\end{aligned}$$

It is immediately evident that there is only one possible answer (which is a real number). Hence the ambiguity of signs which complicated matters with the square root function is not present here. We cannot design bizarre cube root functions! There is only one cube root function assuming we continue to restrict ourselves to domains and ranges consisting entirely of real numbers.

What about higher root functions such as $\sqrt[4]{x}$, $\sqrt[5]{x}$, $\sqrt[6]{x}, \ldots$? The even root functions

$$\sqrt[4]{x}, \ \sqrt[6]{x}, \ \sqrt[8]{x}, \ldots$$

will behave exactly the way $\sqrt{x}$ did, while the odd root functions

$$\sqrt[5]{x}, \ \sqrt[7]{x}, \ \sqrt[9]{x}, \ldots$$

will be analogous to $\sqrt[3]{x}$.

Exercises for §1.2

Find the domain and range for the following functions.

(1) $f(x) = 1/(x^2 - x + 2)$. **(2)** $f(x) = 1/(x^2 + 2x + 1)$.

(3) $f(x) = x/x^2 + 2x + 2$. **(4)** $f(x) = 1/x^2 + 2x + 5$.

(5) Construct on your CAS a square root function which chooses the negative root each time. What are the domain and range of this function?

(6) Construct the function $c(x) = \sqrt[3]{x}$ on your CAS. What are the domain and range of c?

(7) Construct the function $c_{19}(x) = \sqrt[19]{x}$ on your CAS. What are the domain and range of this function? Verify the values: $c_{19}(9) = 1.1226\ldots$, $c_{19}(100) = 1.27427\ldots$ Evaluate $c_{19}(10^6)$ to 10 decimal places.

(8) Construct the function $f(x) = x^{2/3} = (\sqrt[3]{x})^2$ on your CAS. What are the domain and range of this function?

(9) What are the domain and range of the absolute value function $|x|$?

(10) Construct the designer square root function of Example [1.7] on your CAS.

(11) Explain why the function

$$f(x) = \begin{cases} 0, & \text{if } x = \dfrac{a}{b} \text{ with integers } a, b \text{ and } b \neq 0 \\ 1, & \text{otherwise} \end{cases}$$

cannot be constructed on your CAS.

1.3 The Graph of a Function.

Consider a function

$$F : \mathbb{R} \longrightarrow \mathbb{R},$$

whose domain consists of the entire set of all real numbers, $\mathbb{R}$, and whose range is some subset of $\mathbb{R}$. As shown in chapter 0, every real number can be considered as a point on the number line:

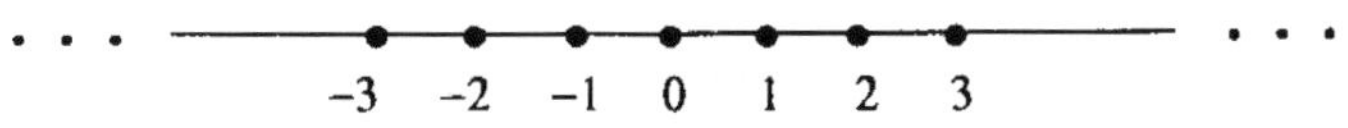

René Descartes had the revolutionary idea of juxtaposing two such number lines (an x–axis and a y–axis) at right angles so that they intersect at the origin, the resulting figure is called the **Cartesian plane**. By convention the x–axis is drawn parallel to the bottom edge of this page. Any line parallel to the x–axis is termed a horizontal line, and any line parallel to the y–axis is termed a vertical line.

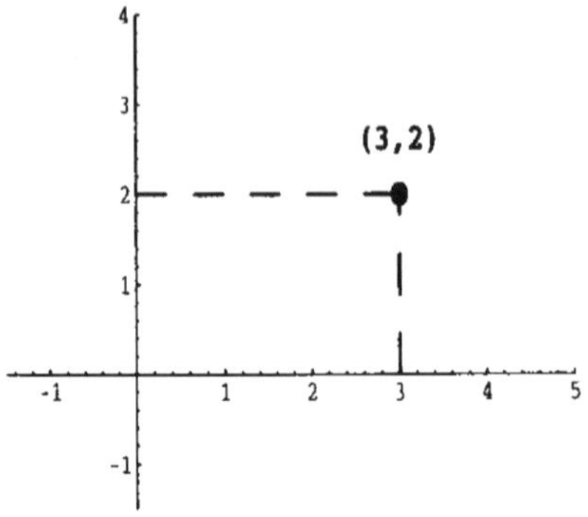

With every pair of real numbers (a, b) there is a corresponding point in the above diagram: this is simply obtained by starting at 0, moving "a" steps in the horizontal direction (i.e., along the x–axis) and then moving "b" steps in the vertical direction from the point "a" on the x–axis. In the diagram above, the point $(3, 2)$ is marked. Conversely if the Cartesian plane is fixed, associated with any point P in the plane, there is a pair of real numbers associated to P. By this simple stroke of genius Descartes had put in place a method of converting *any function* into a 2–dimensional picture or a *graph*.

If, given a function F, we write

$$y = F(x),$$

then this simply means that for every real number x there is a unique real number $y = F(x)$ determined by the rule of the function F. If one plots the points $(x, F(x))$ as x runs over the real numbers, one obtains the **graph** of F. One terms $y = F(x)$ the **equation** of the graph.

Example [1.11] Consider the cubing function $y = x^3$. Let's plot the following 5 points corresponding to $x = 0, 1, 2, -1, -2$.

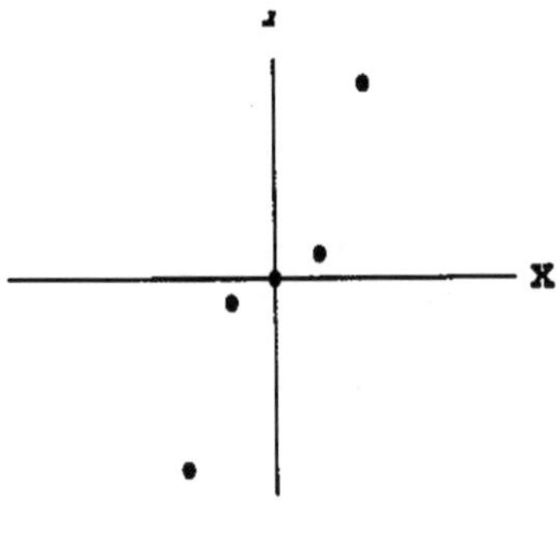

By plotting many more points we fill in the dots and obtain the graph:

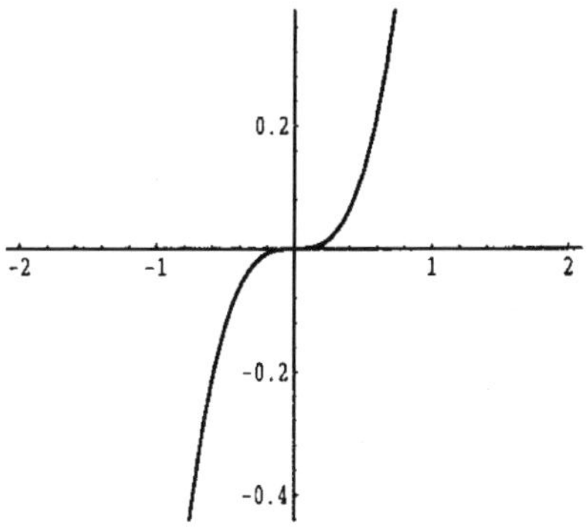

Remark: Every function has a graph. It is not true, however, that every graph comes from a function. For example, the graph

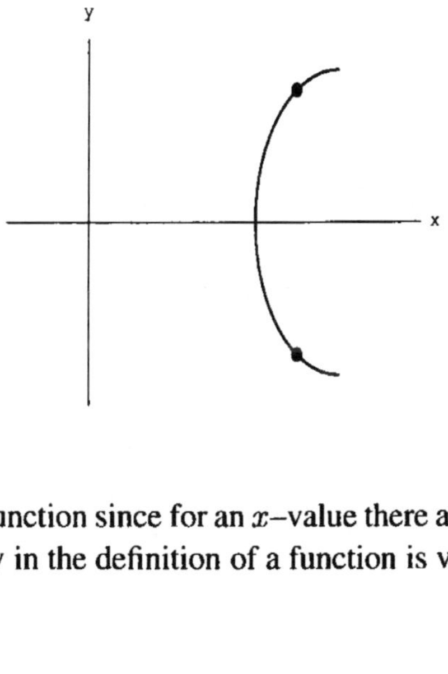

cannot correspond to a function since for an x–value there are usually two y–values. The uniqueness property in the definition of a function is violated.

Exercises for §1.3

(1–9) Using the plotting function on your CAS, output the graphs of all the functions in exercises 1–9 of §1.1.

(10) Construct $f(x) = x^2$ on your CAS. Graph the functions $y = f(x)$, $y = f(x+1)$, $y = f(x+10)$, and $y = f(x-10)$. What happens?

(11) Construct $f(x) = x^3$ on your CAS. Graph the functions $y = f(x)$, $y = f(x+1)$, $y = f(x+10)$, and $y = f(x-10)$. What happens?

(12) Construct $f(x) = x^2$ on your CAS. Graph $y = f(x)$, $y = f(x) + 1$, $y = f(x) + 10$, and $y = f(x) - 10$. What happens?

(13) Construct $f(x) = x^3$ on your CAS. Graph $y = f(x)$, $y = f(x) + 1$, $y = f(x) + 10$, and $y = f(x) - 10$. What happens?

1.4 The Trigonometric and Exponential Functions

In this section we briefly review the formal definitions of the trigonometric functions $\sin(\theta)$, $\cos(\theta)$, and $\tan(\theta)$, as well as the exponential function, a^x.

Consider a circle of radius 1 centered at 0.

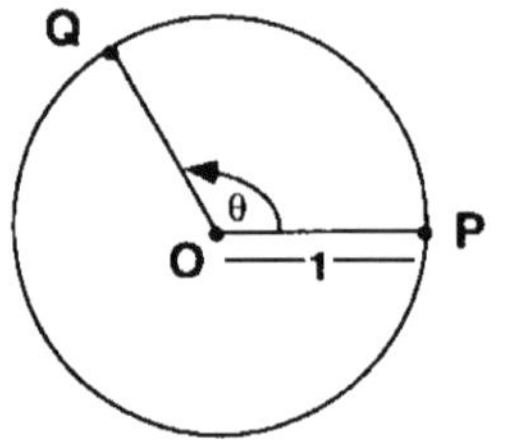

Let OP be a fixed line of reference which, in the figure above, is depicted horizontally. By definition, the angle θ is defined to have measure ℓ **radians** if the length of the arc PQ is ℓ. Since the entire circumference of the circle has length 2π, the angle corresponding to a full rotation (360 degrees) has measure 2π radians. Similarly, a right angle (90 degrees) has measure $\pi/2$ radians. We also allow angles of arbitrarily large measure: for example, an angle of 6π radians corresponds to 3 full rotations.

Let $0 \leq \theta \leq \pi/2$. Consider a right triangle

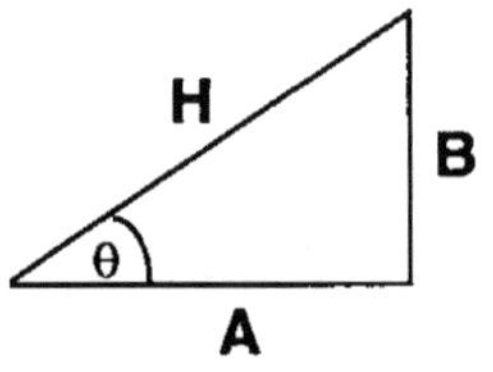

whose hypotenuse has length H and whose sides have length B and A (where we assume the side of length A is adjacent to θ). With this notation in place we define the following **trigonometric functions**:

$$\sin(\theta) = \frac{B}{H}, \quad \cos(\theta) = \frac{A}{H}, \quad \tan(\theta) = \frac{B}{A}.$$

(In a complete presentation of trigonometry one discusses the functions

$$\cot(\theta) = \frac{1}{\tan(\theta)}, \quad \sec(\theta) = \frac{1}{\cos(\theta)}, \quad \csc(\theta) = \frac{1}{\sin(\theta)}.$$

In that this section is intended only as a brief review, the properties of the latter set of functions are not detailed in this section.) Notice first that these definitions depend only on the angle θ: if the angle appears in two distinct right triangles

then the triangles are necessarily *similar*, and thus we have equality of the various ratios i.e.,

$$\frac{A}{H} = \frac{A'}{H'}, \quad \frac{B}{H} = \frac{B'}{H'}, \quad \text{and} \quad \frac{B}{A} = \frac{B'}{A'}.$$

Another important observation is that we may translate the Pythagorean theorem, $A^2 + B^2 = H^2$, into the identity

$$\sin^2(\theta) + \cos^2(\theta) = 1.$$

Example [1.12] $\sin\left(\frac{\pi}{4}\right) = \cos\left(\frac{\pi}{4}\right) = \frac{\sqrt{2}}{2}$, and $\tan\left(\frac{\pi}{4}\right) = 1$.

These computations follow from the diagram below.

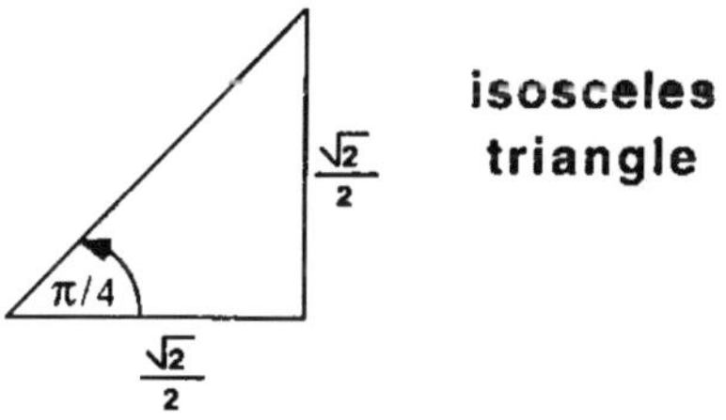

Example [1.13] $\sin\left(\frac{\pi}{3}\right) = \frac{\sqrt{3}}{2}$, $\cos\left(\frac{\pi}{3}\right) = \frac{1}{2}$, and $\tan\left(\frac{\pi}{3}\right) = \sqrt{3}$.

Again the identities follow from a diagram.

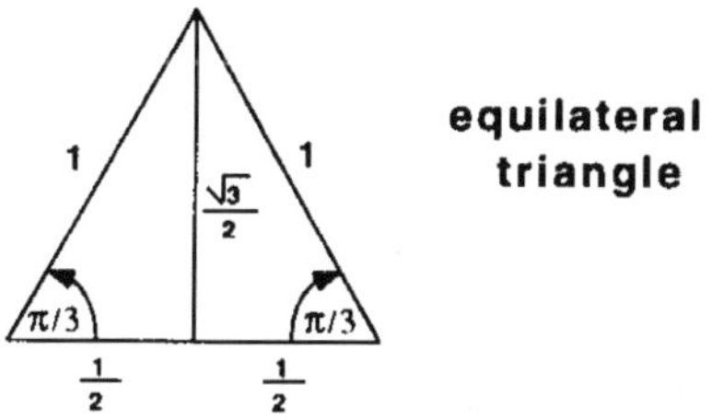

Example [1.14] $\cos(0) = 1$, $\sin(0) = 0$, $\cos\left(\frac{\pi}{2}\right) = 0$, and $\sin\left(\frac{\pi}{2}\right) = 1$.

We shall discuss the identities $\cos(0) = 1$ and $\sin(0) = 0$ leaving the remainder of the claims in the example to the reader. Let θ be a very small angle.

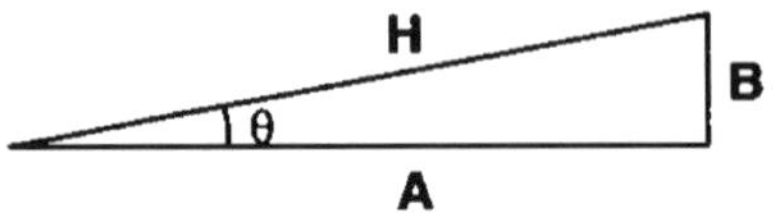

As θ gets smaller and smaller, the lengths of A and H get closer to each other and the length of the side opposite to θ gets smaller and smaller. Hence $\cos(\theta) = \frac{A}{H}$ must approach 1 as θ approaches 0. Similarly $\sin(\theta) = \frac{B}{H}$ approaches 0 as θ approaches 0. A precise formulation of this argument requires the notion of a limit which will be discussed at length in §4.1.

To complete our abbreviated discussion of the trigonometric functions we must enlarge the domain on which our functions are defined to include angles $\theta > \pi/2$. Thus consider θ on the Cartesian plane subsumed by the x–axis and a line segment of length 1 which is centered at the origin.

Let (x, y) be the coordinates of the endpoints of the line segment.

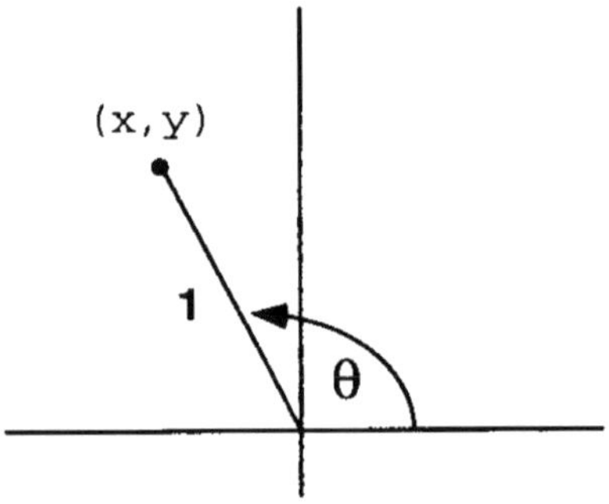

We define

$$\sin(\theta) = y, \quad \cos(\theta) = x, \quad \tan(\theta) = \frac{y}{x}.$$

Remark: This definition agrees with our previous one when $0 \leq \theta \leq \pi/2$ as can be seen in the diagram below.

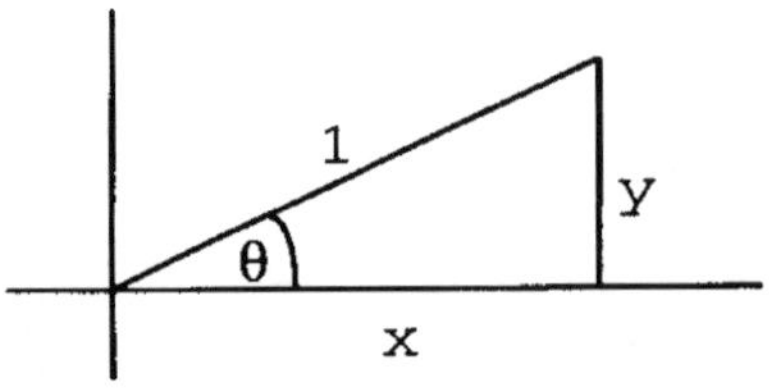

Example [1.15] $\sin\left(\frac{2\pi}{3}\right) = \frac{\sqrt{3}}{2}$, $\cos\left(\frac{2\pi}{3}\right) = -\frac{1}{2}$.

This final identity is derived from the following second quadrant diagram.

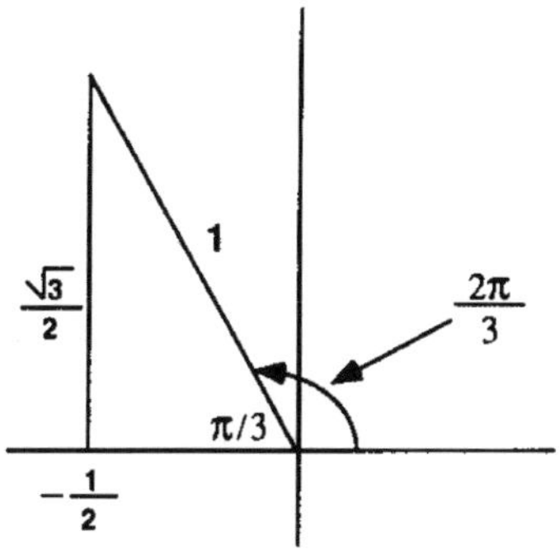

Example [1.16] The graph of the function $y = \sin(x)$ is depicted below.

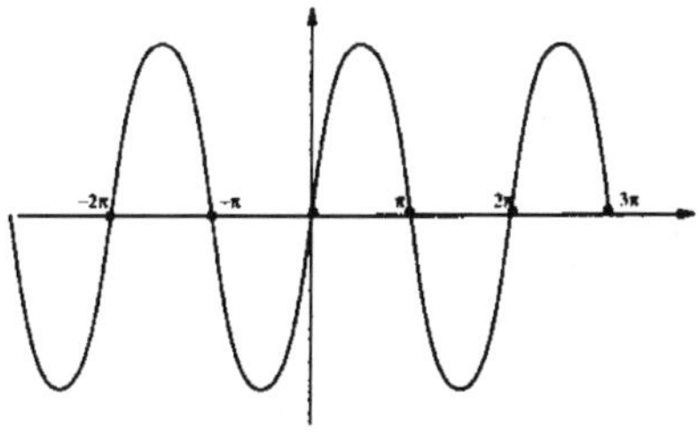

It is clear that after every full rotation (i.e., after an angle of 2π radians) that the graph must repeat itself.

We conclude this chapter by introducing the very important **exponential function**. Fix a real positive number a and consider the function on the set of rational numbers (denoted $\mathbb{Q}$),

$$x \mapsto a^x,$$

where $x \in \mathbb{Q}$. When x is a positive integer the meaning of a^x is simply a multiplied by itself x times. For example $a^5 = a \cdot a \cdot a \cdot a \cdot a$. We define $a^{-1} = \frac{1}{a}$, and more generally, given a positive integer x,

$$a^{-x} = \frac{1}{\underbrace{a \cdot a \cdots a}_{x\text{–times}}}.$$

When we set $a^0 = 1$, we observe the remarkable identity for the exponential function: for any integers x and y,

$$a^x \cdot a^y = a^{x+y}. \tag{1.1}$$

Question: How should we define a^x when x is not an integer?

We would like to enlarge the definition while maintaining the identity (1.1). In fact, this identity dictates how we should proceed. For example, if have defined $a^{\frac{1}{5}}$, then necessarily $a^{\frac{1}{5}}$, when multiplied by itself 5 times, gives us a, i.e.,

$$\begin{aligned} a^{\frac{1}{5}} \cdot a^{\frac{1}{5}} \cdot a^{\frac{1}{5}} \cdot a^{\frac{1}{5}} \cdot a^{\frac{1}{5}} &= a^{\left(\frac{1}{5}+\frac{1}{5}+\frac{1}{5}+\frac{1}{5}+\frac{1}{5}\right)} \\ &= a^1 \\ &= a. \end{aligned}$$

Thus $a^{\frac{1}{5}}$ is the fifth root of a, i.e., $a^{\frac{1}{5}} = \sqrt[5]{a}$. Similarly, given an arbitrary m

$$\begin{aligned} \underbrace{a^{\frac{m}{n}} \cdots a^{\frac{m}{n}}}_{n\text{-times}} &= a^{\left(\frac{m}{n}+\cdots+\frac{m}{n}\right)} \\ &= a^m, \end{aligned}$$

and we conclude that in general

$$a^{\frac{m}{n}} = \sqrt[n]{a^m},$$

where we recall that, by convention, we choose the positive real n^{th}– root when n is even.

Exercises for §1.4

Compute the following values. Check your answers with the CAS.

(1)	$\cos(3\pi/2)$.	**(2)**	$\sin(6\pi)$.	**(3)**	$\tan(2\pi)$.
(4)	$\cos(4\pi/3)$.	**(5)**	$\sin(4\pi/3)$.	**(6)**	$\tan(5\pi/3)$.
(7)	$\sin(5\pi/3)$.	**(8)**	$\cos(5\pi/3)$.	**(9)**	$\sin(3\pi/4)$.
(10)	$\cos(3\pi/4)$.	**(11)**	$\tan(3\pi/4)$.	**(12)**	$\cos(5\pi/4)$.
(13)	$\sin(5\pi/4)$.	**(14)**	$\tan(7\pi/4)$.	**(15)**	$\cos(7\pi/4)$.

(16) Evaluate $10^{3/4}$, $15^{11/13}$, $100^{6/7}$ to 10 digit accuracy on your CAS.

(17) Evaluate $(10^{3/4})^4$, $(15^{11/13})^{13}$, $(100^{6/7})^7$ on your CAS.

(18) Show that $(x^\alpha)^\beta = x^{\alpha\beta}$.

(19) Graph the function $y = \cos(x)$ with your CAS. Discuss the symmetries of the graph and the similarities to the graph of the function $\sin(x)$.

(20) Graph the function $y = \tan(x)$ with your CAS. What are the symmetries of the graph.

(21) Prove that $\cos(\theta) = \sin(\pi/2 - \theta)$.

(22) Prove that $\cos(\theta + \pi) = -\cos(\theta)$ and $\cos(\theta + 2\pi) = \cos(\theta)$.

Additional exercises for Chapter I

---

(1) Express the area of a square as a function of its sides of length x.

(2) Express the volume of the cube as a function of the length x of its sides.

(3) Express the area of the circle as a function of its radius and its circumference. Use your CAS to graph the two functions.

(4) Suppose your credit card has an outstanding balance of \$1025 and you do not plan to make any more purchases with it. The credit card company charges 1% interest per month and requires a minimum payment of \$50 each month. Express the amount owed as a function of the number of months elapsed. Construct the function on your CAS and determine how many months it will take to completely pay off the balance.

Find the domain and range for the following functions. Graph each function with your CAS.

(5) $f(x) = \frac{x}{|x|}$.

(6) $f(x) = \sqrt{\frac{x-1}{x+2}}$.

(7) $f(x) = \sqrt{\frac{x^2-16}{x+16}}$.

(8) $f(x) = \frac{1}{1-3\sin(x)}$.

(9) (a) Let $f(x) = x + 6$ and $g(x) = x^2$. What are the range and domain of f and g?

(b) Determine $g(f(x))$ and $f(g(x))$. What are their respective ranges and domain? Do f and g commute?

In the following, determine whether the equation defines y as a function of x, x as a function of y, both or neither. It may help to graph the solution set of the equation with your CAS.

(10) $3x - 6y = 10$.

(11) $x^2 + y^2 = 25$.

(12) $xy^2 = 2$.

(13) $\frac{2-2x}{1+y} = 1$.

(14) $x^2y^3 + 4 = 8$.

(15) $\frac{xy}{2-xy} = 1$.

(16) What is the domain and range of $\cos(\theta)$? $\sin(\theta)$? $\tan(\theta)$? With your CAS contruct the function $f_\beta(\theta) = \cos(\theta + \beta)$ for $0 \le \theta \le 2\pi$ and β fixed. Graph this function when $\beta = \frac{\pi}{6}$.

(17) Can you derive a formula for $\cos(2\theta)$, $(\sin(2\theta))$ in terms of $\cos(\theta)$ and $\sin(\theta)$? **(Hint:** Rotate the axes through an angle θ.) You may derive the formulas for $\cos(\theta + \beta)$, and $\sin(\theta + \beta)$ by the same method. Verify this formula with your CAS by taking $\beta = \frac{\pi}{6}, \frac{\pi}{3}, \frac{\pi}{4}$ and θ not necessarily equal to β.

(18) With your CAS plot the function $f(\theta) = \sin(\beta) + \sin(\theta)$ for several fixed β.

(19) With your CAS graph $f(\theta) = 1 + \tan^2(\theta)$. Graph $\frac{1}{\cos^2(\theta)}$. What do you notice? What identity can you derive?

(20) Graph the function $f(x) = 2^x$ on your CAS. Further, use your CAS to evaluate

$$\frac{2^h - 1}{h}$$

for $h = .1, .025, .01, .005, .0025, .001$. What do you notice?

Chapter II
The Algebra of Functions

2.1 An Informal Introduction to Algebras

In that the precise definition of an *algebra* is lengthy and technical we shall confine ourselves to an informal discussion which is sufficient for our needs. Briefly, an algebra consists of a set $\mathcal{A}$ together with a set of *operations* on $\mathcal{A}$: an operation on $\mathcal{A}$ allows one to take any two elements in $\mathcal{A}$ and generate a (unique) third element in $\mathcal{A}$. The algebra everyone is familiar with is the algebra of real numbers $\mathbb{R}$. The operations are addition and multiplication, together with the inverses of these operations, subtraction and division. While $\mathbb{R}$ is well–known (and seemingly elementary from this point of view) there exist in nature algebras with complex and unfamiliar forms with operations far removed from those found in $\mathbb{R}$.

Example [2.1] One of the simplest algebras is the **binary algebra** of two symbols 0 and 1,

$$\mathcal{A} = \{0, 1\}.$$

We define the operations as follows:

$$\begin{aligned} 0+0=0,\ \ 0+1=1,\ \ 1+0=1,\ \ 1+1=0 \\ 0-0=0,\ \ 0-1=1,\ \ 1-0=1,\ \ 1-1=0 \\ 0*0=0,\ \ 0*1=0,\ \ 1*0=0,\ \ 1*1=1. \end{aligned}$$

Then $\mathcal{A}$ is an algebra with the three operations $+$, $-$, $*$. The only surprise here is the fact that $1 + 1 = 0$. This can be explained if we identify 0 and 1 as positions of a switch and the operation $+1$ as flipping the switch and $+0$ as doing nothing.

Clearly flipping twice amounts to returning to the original position i.e., $1 + 1 = 0$. This algebra is the basis for computer circuitry.

Exercises for §2.1

(1) Construct an addition function A on your CAS with

$$\text{Domain}(A) = \{(0,0),\ (0,1),\ (1,0),\ (1,1)\} \text{ and } \text{Range}(A) = \{0,1\}$$

which satisfies

$$A((0,0)) = 0,\ \ A((0,1)) = 1,\ \ A((1,0)) = 1,\ \ A((1,1)) = 0.$$

(2) Evaluate $1+1+1$ and $1+1+1+1$ with the additive function you constructed in the previous exercise.

2.2 The Algebra of Functions

Given two functions $f(x)$, $g(x)$ we may add (or subtract) them to obtain a new function

$$f(x) \pm g(x).$$

Similarly, we may multiply them to obtain

$$f(x) \cdot g(x).$$

The only requirement necessary for these operations to be well defined is that the domain of f, g should coincide and that the operations of addition, subtraction, and multiplication are valid in the ranges of f, and g. It is clear that everything works beautifully if we assume that

$$\begin{aligned} &\text{Domain}(f) = \text{Domain}(g) = \mathbb{R} \\ &\text{Range}(f),\ \text{Range}(g) \subseteq \mathbb{R}. \end{aligned}$$

Can we divide functions? The answer is yes provided we don't divide by 0. Ultimately, if we plug in a real number for the variable x the ratio

$$f(x)/g(x)$$

can be computed provided $g(x) \neq 0$.

Question: *What are the simplest functions?*

The simplest functions are the constant functions

$$f(x) = c$$

where c is a fixed real number. Here $f(x)$ takes the same value c for any x. The graph of this function is a horizontal line which intersects the y–axis at the point $(0, c)$.

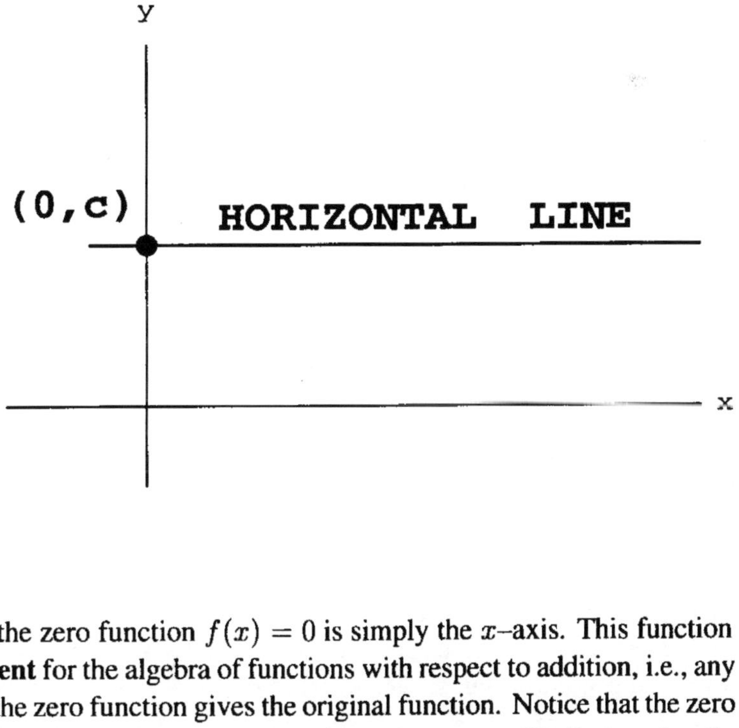

The graph of the zero function $f(x) = 0$ is simply the x–axis. This function is the **identity element** for the algebra of functions with respect to addition, i.e., any function added to the zero function gives the original function. Notice that the zero function is the only such function with this property. Similarly, the function $f(x) = 1$ is the identity element of the algebra of functions with respect to multiplication. Notice that $f(x) = 1$ behaves like the number 1 in the usual algebra of real numbers.

It would appear that the algebra of functions is very similar in structure to the algebra of real numbers. A surprise awaits us, however. The algebra of functions contains two more operations! Nature has endowed this algebra with a much richer structure than the algebra of real numbers, and as we shall see, this leads to some intriguing developments.

Definition *Given two functions f, g such that Range(g) is contained in Domain(f), we may form the* **composition** *of the functions, denoted $f \circ g$, which is a new function defined by the rule*

$$\Big(f \circ g\Big)(x) = f\Big(g(x)\Big),$$

i.e., the function of the function.

Example [2.2] Let $f(x) = x^2$, and $g(x) = x + 1$. Then $f\Big(g(x)\Big) = g(x)^2 = x^2 + 2x + 1$, $g\Big(f(x)\Big) = f(x) + 1 = x^2 + 1$.

This example shows that in general

$$f \circ g \neq g \circ f,$$

which is a phenomena that does not occur for either addition or multiplication in the algebra of real numbers.

Example [2.3] Let

$$f(x) = \frac{x^3 - 7x}{x^2 + 1}, \qquad g(x) = x^4 + 1.$$

Then using our CAS we see that $f + g = x^4 + 1 + \frac{x^3-7x}{x^2+1}$, $g - f = x^4 + 1 - \frac{x^3-7x}{x^2+1}$, $f/g = \frac{x^3-7x}{(x^2+1)(x^4+1)}$, $f \circ g = \frac{(x^4+1)^3-7(x^4+1)}{(x^4+1)^2+1}$, $g \circ f = \left(\frac{x^3-7x}{x^2+1}\right)^4 + 1$.

Example [2.4] (Population Function) This example (due to the physicist M. Feigenbaum) indicates a system which behaves quite normally and smoothly for a wide range of initial conditions but suddenly becomes **chaotic** at certain critical values.

Fix a real number $0 \le r \le 4$, and let $P_r(x) = r \cdot x(1 - x)$ denote a population function where if x denotes the population of a certain species in a given year (for simplicity we take x to be a number between 0 and 1, the true population being one million times x), then $P_r(x)$ denotes the population in the following year. What happens to the population over time? It is easily seen that the population in the nth year is given by

$$P_r \underbrace{\circ \cdots \circ}_{n\text{–times}} P_r,$$

the composition of P_r with itself n–times.

Exercises for §2.2

(1) Let $f(x) = x^3 + 2x + 1$ and $g(x) = x^2 - 3$. Evaluate $f + g,\ g - f,\ f/g,\ f \cdot g,\ f \circ g$, and $g \circ f$.

(2) Let $f(x) = x^4 - 1$ and $g(x) = 3x^2 - 2x + 1$. Evaluate $f + g,\ g - f,\ f/g,\ f \cdot g,\ f \circ g$, and $g \circ f$.

(3) Let $f(x) = x^2 + 2$ and $g(x) = 3^x$. Evaluate $f + g,\ g - f,\ f/g,\ f \cdot g,\ f \circ g$, and $g \circ f$.

(4) Let $f(x) = 2^{1-x}$ and $g(x) = 2^{4x-3}$. Evaluate $f \circ g$, and $g \circ f$.

(5) Let $f(x) = \sqrt{x}$ and $g(x) = x^2$. Evaluate $f \circ g$, and $g \circ f$.

(6) Let $f(x) = \sqrt[3]{x}$ and $g(x) = x^3$. Evaluate $f \circ g$, and $g \circ f$.

(7) Let $f(x) = \cos(x - 1)$ and $g(x) = x^4 + 2$. Evaluate $f \circ g$, and $g \circ f$.

(8) Let $f(x) = \sqrt{x}$. Evaluate $f \circ f \circ f \circ f$.

(9) In example [2.4] (the population function), let $r = 2$. Show that the population eventually stabilizes at 500,000, i.e., pick a value of x such as $x = 0.2$, compute $P_2(0.2)$, $(P_2 \circ P_2)(0.2)$, $(P_2 \circ P_2 \circ P_2)(0.2), \ldots$ and show that these numbers converge to 0.5. Let $r = 1$. Show that the population eventually becomes extinct, i.e., it stabilizes at 0.

(10) Let $r = 3.2$. Show that in this case the population eventually alternates between the two values 0.5 and 0.8. Try a slightly larger value of r which is less than 3.57. What happens? Now try $r = 3.57$ and show that the population grows to infinity in a chaotic manner. What happens when r increases further?

2.3 The Identity and the Inverse Function

Recall, we observed that the functions $f(x) = 0$ and $f(x) = 1$ are the identity elements with respect to the operations of addition and multiplication, respectively (these are analogous to the numbers 0 and 1 in the algebra of real numbers). We are thus led to the following

Question: *What is the identity element for the operation of composition of functions?*

The answer is the identity function

$$I(x) = x.$$

This follows immediately from the identities,

$$(f \circ I)(x) = f\Big(I(x)\Big) = f(x)$$
$$(I \circ f)(x) = I\Big(f(x)\Big) = f(x)$$

for every function f, i.e.,

$$(f \circ I) = (I \circ f) = f.$$

Example [2.5] When we define $I(x) = x$, $f(x) = \cos(x^3 - 1)$ on the CAS we can easily verify that $I \circ f = f \circ I$.

As we observed before, the inverse of the operation of multiplication of functions (when it exists) is the operation of division i.e., f multiplied by $\frac{1}{f}$ yields the multiplicative identity:

$$f \cdot \frac{1}{f} = \frac{1}{f} \cdot f = 1.$$

Similarly, for the operation of addition, the inverse operation is subtraction i.e., f added to $-f$ yields the additive identity:

$$f + (-f) = (-f) + f = 0.$$

Question: *What is the inverse operation associated to composition of functions?*

Definition: *Given a function f, the* ***inverse of*** f*, denoted f^{-1}, is a function whose domain is the set Range(f), whose range is the set Domain(f) and which is defined by the rule*

$$f \circ f^{-1} = f^{-1} \circ f = I,$$

where I is the identity function for composition. If it is not possible to define a function f^{-1} satisfying the above rule, then we say that f^{-1} does not exist.

Warning: It is *critical* to realize that, in general,

$$f^{-1} \neq \frac{1}{f}.$$

When does the inverse of a function exist? To answer this question we require an important concept.

Definition: *A function $f : A \to B$ is termed* ***one–to–one*** *if for each element b in* Range(f) *there is a* ***unique*** *element a in the domain A such that $f(a) = b$.*

We can now identify which functions are invertible i.e., have an inverse.

Proposition [2.6] *A function has an inverse if and only if it is one–to–one.*

Proof: Left to the reader in exercise 5 of §2.3.

Example [2.7] The function $h(x) = \frac{1}{2}x$ is the inverse of the doubling function $d(x) = 2x$ (which is clearly one–to–one) since

$$d \circ h(x) = d\,(h(x)) = d\left(\frac{1}{2}x\right) = x,$$

and

$$h\,(d(x)) = h(2x) = x.$$

Example [2.8] Let a, b be any fixed real numbers with $a \neq 0$. What is the inverse of $f(x) = ax + b$?

Before searching for the inverse function we must first check that $f(x)$ is one–to–one. In fact our hypothesis that $a \neq 0$ insures that f is one–to–one: if x_1 and x_2 are distinct numbers, then $ax_1 + b \neq ax_2 + b$ provided $a \neq 0$. Returning now to finding the inverse of f, a function $g(x)$ is the inverse of $f(x)$ provided

$$f(g(x)) = x.$$

In this case we obtain the identity

$$ag(x) + b = x,$$

i.e.,

$$f^{-1}(x) = g(x) = (x - b)/a.$$

Example [2.9] What is the inverse of $f(x) = x^2 + 5$ on the infinite interval $x \geq 0$?

Observing that on the infinite interval $x \geq 0$, $f(x)$ is one–to–one, we are seeking a function $g(x)$ such that $f(g(x)) = x$. Since $f(x) = x^2 + 5$ we have

$$f(g(x)) = g(x)^2 + 5 = x,$$

and upon solving for $g(x)$ we arrive at the inverse of $f(x)$,

$$f^{-1}(x) = g(x) = \sqrt{x - 5}.$$

Remark: Having found the inverse of various functions we note that whenever f^{-1} exists, it is unique. This can be seen as follows. Suppose that the function g is another inverse of the function f. Then, by definition, $g \circ f = f \circ g = I$, and we conclude that

$$f^{-1} = (g \circ f) \circ f^{-1} = g \circ (f^{-1} \circ f) = g.$$

Exercises for §2.3

(1) Show that the squaring function and the square root function form a pair of inverse functions.

(2) Show that the cubing function and the cube root function form a pair of inverse functions.

(3) Verify with the CAS that the log function and the exp function form a pair of inverse functions.

(4) Show that the inverse of the function $f(x) = x^N$ is $f^{-1}(x) = x^{1/N}$.

(5) Verify Proposition [2.6].

(6) Given a function f which is one–to–one, show that the inverse of f^{-1} is the original function f.

(7) Let $y = mx + b$ be the function of a straight line. What is its inverse function? Is it also a line?

(8) Find $f^{-1}(x)$ for the function $f(x) = x^N + a$.

(9) Find a function $f(x)$ which is its own inverse, i.e., $f(x) = f^{-1}(x)$.

(10) The function $\cos(x)$ is one–to–one on the interval $0 \leq x \leq \pi$. Hence it has a well defined inverse which is denoted $\arccos(x)$ whose domain is $-1 \leq x \leq 1$. Evaluate $\arccos(1)$, $\arccos(-1)$, $\arccos\left(\frac{1}{2}\right)$, $\arccos\left(-\frac{1}{2}\right)$, $\arccos\left(\frac{\sqrt{3}}{2}\right)$, and $\arccos\left(-\frac{\sqrt{3}}{2}\right)$.

(11) The function $\sin(x)$ is one–to–one on the interval $-\frac{\pi}{2} \leq x \leq \frac{\pi}{2}$. Hence it has a well defined inverse which is denoted $\arcsin(x)$ whose domain is $-1 \leq x \leq 1$. Evaluate $\arcsin(1)$, $\arcsin(-1)$, $\arcsin\left(\frac{1}{2}\right)$, $\arcsin\left(-\frac{1}{2}\right)$, $\arcsin\left(\frac{\sqrt{3}}{2}\right)$, and $\arcsin\left(-\frac{\sqrt{3}}{2}\right)$.

(12) Show that $\cos(\arcsin(x)) = \sqrt{1 - x^2}$ for $-1 \leq x \leq 1$. **Hint:** Recall that

$$\cos(u) = \sqrt{1 - \sin^2(u)}$$

for $-\frac{\pi}{2} \leq u \leq \frac{\pi}{2}$ since $\cos(u)$ is positive on this interval.

Additional exercises for Chapter II

(1) Express $h(x) = \tan(x^5)$ as the composition of two functions f and g.

(2) Express $h(x) = \frac{2}{6+\cos(x)}$ as the composition of two functions f and g.

(3) Express $h(x) = \frac{2\sin^2(x)}{6+\cos^2(x))}$ as the composition of two functions f and g.

(4) On your CAS graph the function

$$f(x) = \begin{cases} \frac{1}{3}x - 1, & \text{if } x \leq 3 \\ x^2 - 6x + 9, & \text{if } x > 3. \end{cases}$$

Does f have an inverse? If so find it. (**Hint:** $x^2 - 6x + 9 = (x-3)^2$).

(5) With your CAS, graph e^x and $\log x$. Then graph the function $f(x) = x$. What do you notice?

(6) Let

$$f(x) = \frac{1}{(x-2)(x-3)}.$$

What is the domain of f? What is the range of f? Does the inverse of f exist? Find the largest domain where f and f^{-1} are both defined. Use your CAS to graph f^{-1} on that domain.

(7) In general, is $f \circ g = g \circ f$?

(8) Is it ever true that $f \circ g = g \circ f$?

(9) Find $f(x)$ if $f(3x+1) = \frac{x}{x^2+1}$ (**Hint:** put $z = 3x + 1$ and find $f(z)$).

(10) Find $f(x)$ if $f(x+1) = x^2 - 6x + 10$.

(11) Find functions f and g so that $f(g(x)) = \sin(\sqrt{x^2+1})$.

(12) Let $f(x)$ be such that $f(x) = 0$ only for $x = 0$ and let $g(x) = x^3 - 6$. For what values of x is $f \circ g(x) = 0$?

(13) A function is said to be **even** if $f(x) = f(-x)$ for x in the domain of f, and is said to be **odd** if $f(x) = -f(-x)$. With your CAS graph the function $f(x) = x^2 - 1$. Is $f(x)$ even or odd? What kind of geometric property does $f(x)$ have?

Classify the following functions as even, odd, or neither (the definition is given in the problem (13) above) and graph each with your CAS.

(14) $f(x) = |x|$.

(15) $f(x) = \sin(x)$.

(16) $f(x) = |x - 2| + 3$.

(17)

$$f(x) = \begin{cases} -x - 2, & \text{if } x \leq -2 \\ -x^2 + 4, & \text{if } -2 < x < 2 \\ x - 2, & \text{if } x \geq 2. \end{cases}$$

(18) If $f(x)$ is odd, what conditions should $g(x)$ satisfy for $g \circ f(x)$ to be even? odd?

(19) Let $f(x) = (x - a)^2$. With your CAS graph $f(x)$ for some fixed value of a. What geometric property does $f(x)$ have? Now suppose $g(x) = e^x$, does $g \circ f(x)$ have the same geometric property as $f(x)$? Express the property algebraically.

(20) Let a be a constant and suppose that f is a function satisfying

$$f(a - x) = f(a + x)$$

for all x. What geometric property must the graph of f have?

Chapter III

Lines, Circles, and Curves: a Review

3.1 Lines

The Cartesian coordinate system introduced in §1.3 allows us to view functions pictorially as graphs. Most functions which appear in nature (i.e., in physics, biology, economics, etc.) look like curves. For example the graph of the loanshark function (see Example [1.8] of §1.2) looks like

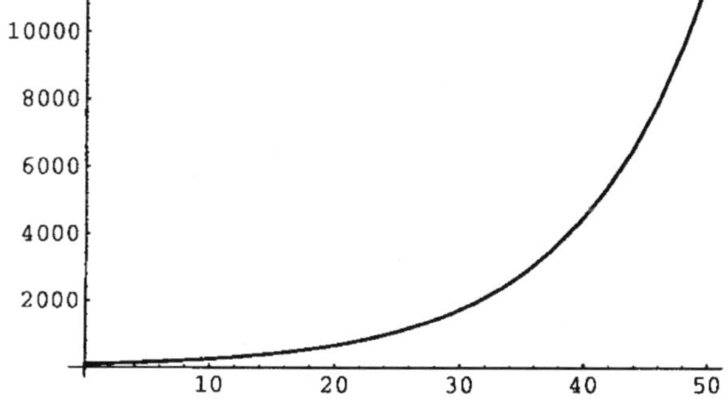

The more complex function, the designer square root function (cf. Example [1.8], §1.2), does not visualize as a smooth curve.

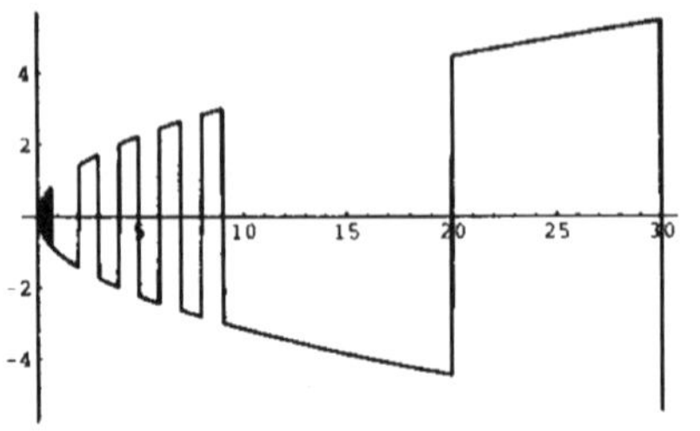

A powerful method of studying functions is to use information derived from the geometry of the associated graph. Functions whose graphs are well understood geometric figures, such as lines and semicircles, can be shown to have particularly nice properties. We single out these special functions.

Problem [3.1] *Determine all possible functions whose graph is a straight line.*

The solution to this problem relies on a basic principle of classical Euclidean geometry: *An arbitrary line L intersects any horizontal line at the same angle.*

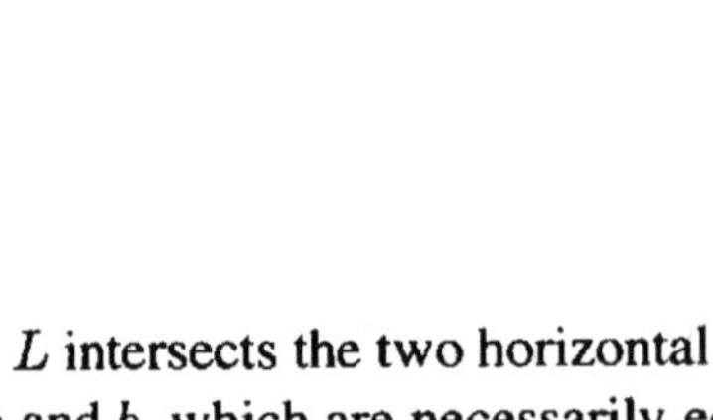

In the figure above, the line L intersects the two horizontal lines (at points P and Q say) and forms the angles a and b, which are necessarily equal.

It follows that when we form two right triangles (along the horizontal lines depicted above)

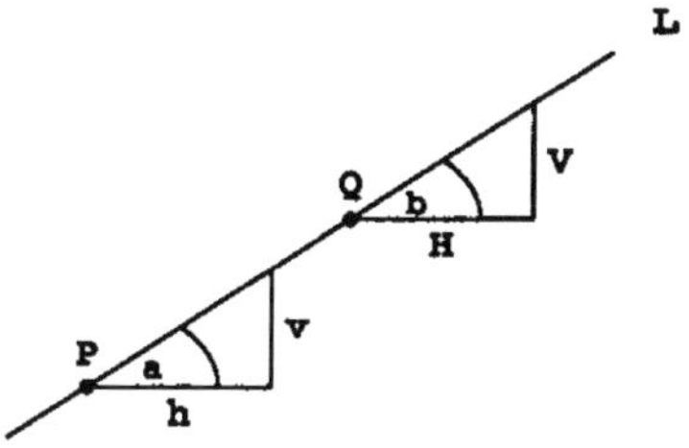

since the angles a and b are equal, the tangent of a, $\tan(a) = \frac{v}{h}$ must be equal to $\tan(b) = \frac{V}{H}$:

$$\frac{v}{h} = \frac{V}{H}.$$

The ratio $\frac{v}{h}$ (of vertical to horizontal distance) is called the **slope** of the line. By the above discussion, the slope of the line L is well defined (i.e., is independent of the point P). It is interesting to note that this concept appears over two thousand years ago in ancient Egyptian papyri in reference to pyramid building.

If a line is drawn in the Cartesian plane

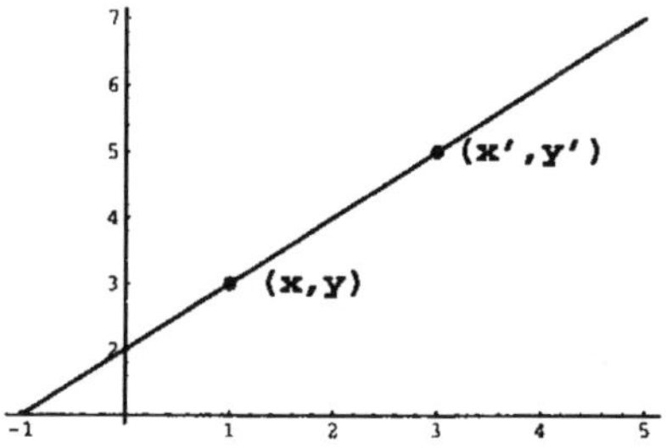

and (x, y) , (x', y') are any two points on the line, then, without reference to any angles, we have the alternate:

Definition: *The **slope** of the line which goes through the distinct points (x, y), (x', y') is defined to be*

$$\frac{y' - y}{x' - x}.$$

Warning: The definition is well defined provided the line L is not vertical.

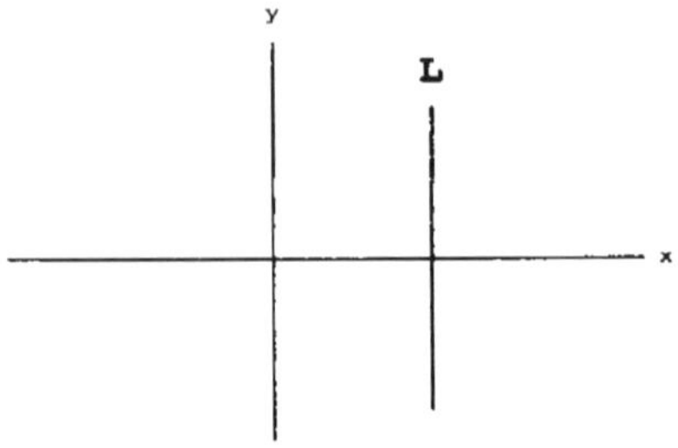

In this case the x–coordinates of every point on L are equal, and in the computation of the slope we would be dividing by zero. Heuristically, a vertical line is said to have infinite slope (note that a vertical line is never the graph of a function).

We now focus on an arbitrary line L with finite slope m. Since L is not vertical, it must intersect the y–axis at some point $(0, b)$.

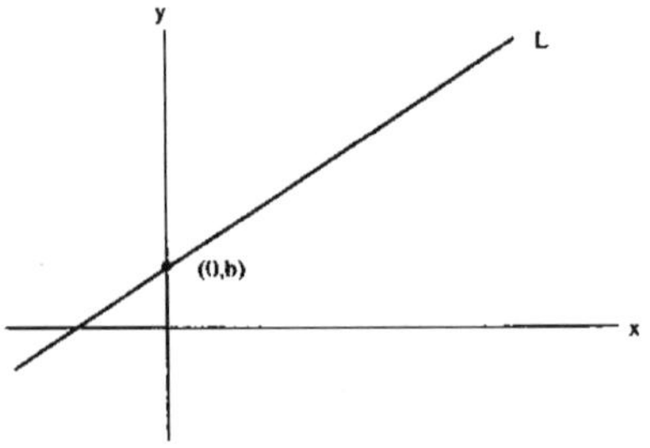

If (x, y) is any other point on L, then we have

$$m = \frac{y - b}{x - 0}.$$

Simplifying this we obtain the well–known equation

$$y = mx + b.$$

In summary we have shown:

Proposition [3.2] *Every function F whose graph is a straight line must be of the form*

$$F(x) = mx + b$$

where m is the slope of the line and $(0, b)$ is the point of intersection of the line with the y–axis.

Remark: A very important example is the horizontal line. Here the slope $m = 0$ and the function is of the form

$$F(x) = b,$$

where $(0, b)$ is the point where the line intersects with the y–axis.

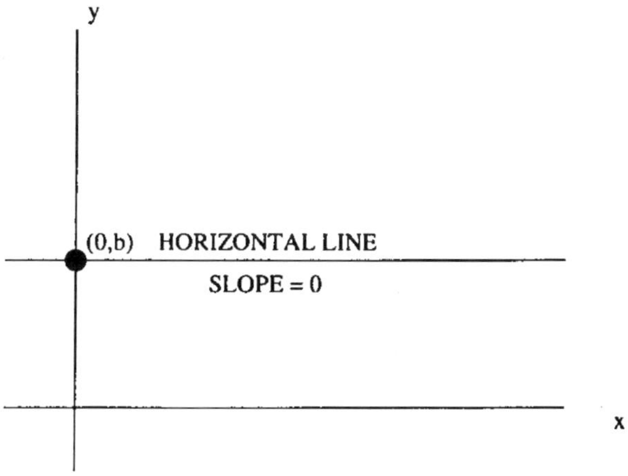

Example [3.3] Find the slope and the y–intercept of the line $2x - 3y = 5$.

Beginning with the original equation, we employ a little elementary algebra to rewrite the equation as $3y = 2x - 5$, or $y = \frac{2}{3}x - \frac{5}{3}$. Now we see that the slope of the line is $\frac{2}{3}$ and the y–intercept is $b = -\frac{5}{3}$.

Example [3.4] Consider the two lines $y = 3x + 2$ and $y = 2x - 1$. Where do these lines intersect?

The point of intersection is the point (x, y) which lies on both lines simultaneously. To find this point we must solve two equations in two unknowns:

$$\begin{aligned} y - 3x &= 2 \\ y - 2x &= -1. \end{aligned}$$

Subtracting these equations yields $-x = 3$, and by plugging this into one of the equations of the lines we see that $y = -7$.

Example [3.5] Find the equation of the line going through the points $(1, 3)$ and $(2, 4)$.

The slope of the line is $m = \frac{4-3}{2-1} = 1$. Hence the equation takes the form $y = x + b$. Since the point $(1, 3)$ is on the line we can plug in $x = 1$ and $y = 3$

into the equation $y = x + b$ yielding $3 = 1 + b$, i.e., $b = 2$. We conclude that the equation of the line is given by $y = x + 2$.

Exercises for §3.1

(1) Find the slope and y–intercept of the line $6x + 2y = 11$. Graph this line with your CAS.

(2) Find the slope and y–intercept of the line $x - 3y = \pi$. Graph this line with your CAS.

(3) Find the point of intersection of the two lines $3x - 2y = 1$ and $2x + y = 5$. Graph these lines with your CAS.

(4) Consider two arbitrary lines

$$\begin{aligned} L_1 : a_1 x + b_1 y + c_1 &= 0 \\ L_2 : a_2 x + b_2 y + c_2 &= 0 \end{aligned}$$

Where do these lines intersect? Show that L_1 intersects L_2 if $a_1 b_2 - a_2 b_1 \neq 0$. Otherwise they are parallel.

(5) Find the equation of the line going through the two points $(-2, 3)$ and $(1, -2)$. Plot these points and the line through them on your CAS.

(6) Repeat problem (5) with the points $(1, 4)$ and $(2, -3)$.

(7) Graph the two lines

$$\begin{aligned} L_1 : 2x - 3y &= 13 \\ L_2 : 3x + 2y &= 7 \end{aligned}$$

with your CAS. Show that they are perpendicular. **Hint:** Use the fact that two lines are perpendicular if the product of their slopes is -1.

(8) Two lines are said to be **parallel** if they never intersect. Show that any two lines with the same slope are parallel.

(9) Verify the fact that any two lines with slopes m_1 and m_2 are perpendicular if $m_1 \cdot m_2 = -1$.

(10) Determine whether or not the three points $(1, 3)$, $(-1, 5)$, $(9, -4)$ lie on a straight line.

(11) Find the equation of the line which passes through $(-1, 2)$ and is perpendicular to the line $2x - 3y = 4$.

(12) Find the equation of the line which is the perpendicular bisector of the line segment whose endpoints are $(1, 3)$ and $(4, -1)$. Graph the line segment and its perpendicular bisector with your CAS.

3.2 Circles

Problem [3.2] *Determine the functions whose graphs are semi-circles.*

We attack this problem by recalling the Pythagorean theorem which states that in a right triangle of hypotenuse of length c and sides of length a, b,

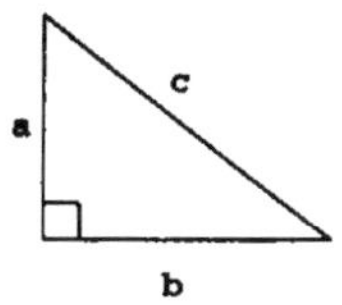

we have

$$a^2 + b^2 = c^2.$$

The proof of the Pythagorean theorem is given in the following diagram:

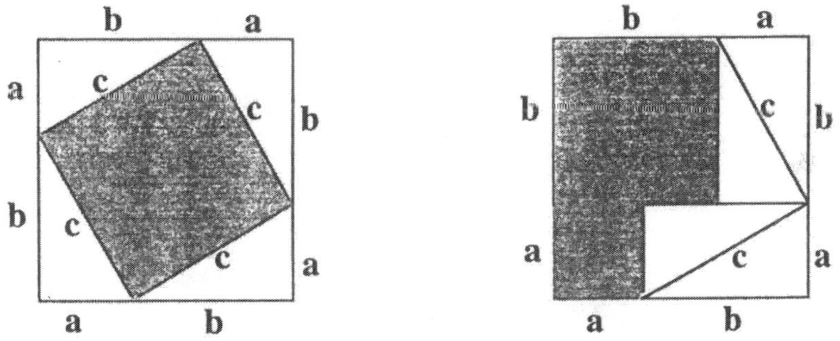

This ancient formula allows us to find the distance d between any two points (x, y) and (x', y') in the Cartesian plane

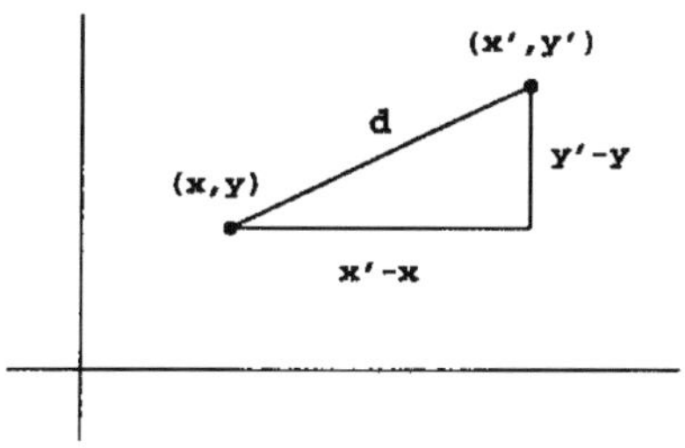

by the formula

$$d^2 = (x' - x)^2 + (y' - y)^2,$$

or

$$d = \sqrt{(x' - x)^2 + (y' - y)^2}.$$

Now the circle of radius r centered at the point (c, d) consists of the locus (collection) of points (x, y) whose distance from (c, d) is r.

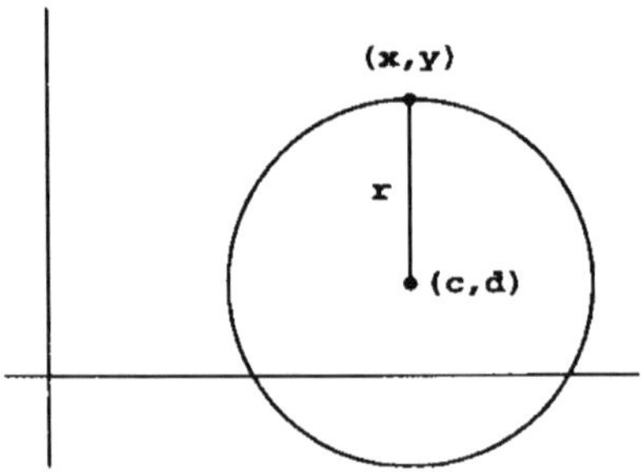

It follows that

$$r^2 = (x - c)^2 + (y - d)^2,$$

and hence

$$y - d = \sqrt{r^2 - (x - c)^2}$$
$$y = \sqrt{r^2 - (x - c)^2} + d.$$

Following the standard convention, we use the positive square root function above to obtain:

Proposition [3.7] *Every function whose graph is a semicircle of radius r centered at* (c, d) *must be either of the form*

$$F(x) = \sqrt{r^2 - (x - c)^2} + d, \quad \text{or} \quad F(x) = -\sqrt{r^2 - (x - c)^2} + d.$$

In summation, we have arrived at the remarkable conclusion that if a function F is not of the form $F(x) = mx + b$ then its graph must be curved! It cannot be a straight line. Furthermore, if F is not of the form

$$F(x) = \pm\sqrt{r^2 - (x - c)^2} + d$$

then its graph cannot be circular.

Example [3.8] Find the equation of a circle of radius 5 centered at $(1,3)$. Construct a function whose graph is a semicircle of radius 5 centered at $(1,3)$.

The equation of the circle is simply $(x-1)^2+(y-3)^2=25$. The function whose graph is the semicircle is given by

$$F(x)=\sqrt{25-(x-1)^2}+3.$$

Example [3.9] Compute the distance between the points $(1,1)$ and $(5,-2)$.

The distance d is given by

$$\begin{aligned} d &= \sqrt{(5-1)^2+(-2-1)^2} \\ &= \sqrt{25}=5. \end{aligned}$$

Example [3.10] Find the shortest distance between the point $(4,1)$ and the line $y=3x-1$.

To solvc this problem we must first find the equation of the line which passes through $(4,1)$ and is perpendicular to the line $y = 3x - 1$. Once this is done we must find the point P where the two lines intersect, and then the distance between P and $(4,1)$ will give us our final answer. By Exercise (9) of §3.1, we know that any line which is perpendicular to $y=3x-1$ must have slope $-\frac{1}{3}$, and thus takes the form $y=-\frac{1}{3}x+b$. Since $(4,1)$ is on the line we find that the line we seek is given by $y=-\frac{x}{3}+\frac{7}{3}$. The intersection of the two lines is computed by letting $3x-1=-\frac{x}{3}+\frac{7}{3}$. This yields $x=1$ and the point P is found to be $P=(1,2)$. To complete the exercise we compute the distance d between $(1,2)$ and $(4,1)$: $d=\sqrt{(4-1)^2+(2-1)^2}=\sqrt{10}$. In conclusion $\sqrt{10}$ is the shortest distance between $(4,1)$ and the line $y=3x-1$.

Exercises for §3.2

In the following five exercises find the equation of the circle which is centered at P and whose radius is R. Graph the upper and lower semicircle with your CAS.

(1) $R=1,\ P=(1,3)$.

(2) $R=3,\ P=(-1,5)$.

(3) $R=10,\ P=(-3,-2)$.

(4) $R=5,\ P=(1.3,-2.1)$.

(5) $R = 2,\ P = (\pi, -1.03)$.

(6) Compute the distance between the points P_1 and P_2 for $P_1 = (1,3), P_2 = (3,-1)$, $P_1 = (0,1), P_2 = (-1,-3)$, and $P_1 = (2,3), P_2 = (-2,11)$.

(7) Find the shortest distance between the point $(1,3)$ and the line $2x - 4y = -1$.

(8) Find the shortest distance between the point $(-1,2)$ and the line $y = 5x - 2$.

(9) (Approximation to π) Estimate the the value of π the way Pythagoras did.

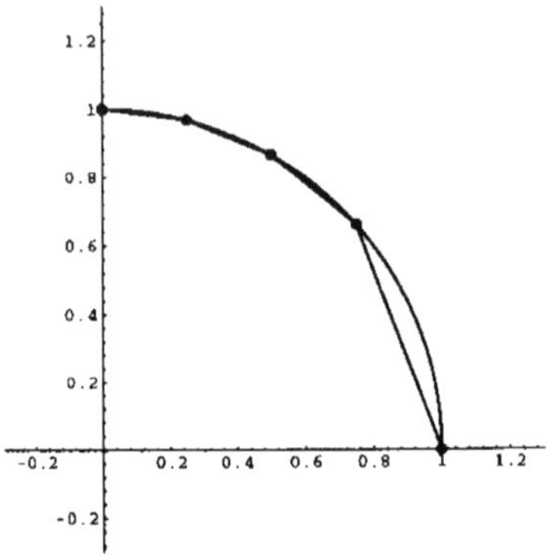

Choose the following four points on the quarter circle:

$$(0,1),\quad \left(\frac{1}{4}, \sqrt{1-\frac{1}{16}}\right),\quad \left(\frac{1}{2}, \sqrt{1-\frac{1}{4}}\right),\quad \left(\frac{3}{4}, \sqrt{1-\frac{9}{16}}\right).$$

These points are equally spaced with respect to the x-axis. Add the distance between these points. This approximates $\frac{1}{4}$ of the circumference (i.e., $\frac{1}{2}\pi$). Thus when we multiply the sum by 2 we obtain an approximation for π.

(10) Repeat Exercise (9) above with eight points on the quarter circle to obtain another approximation of π. How much better is this approximation that the one obtained previously.

Additional exercises for Chapter III

(1) Find the general equation of the line passing through the points $P_1 = (x_1, y_1)$ and $P_2 = (x_2, y_2)$.

In the problems 2-4, graph the line joining P_1 and P_2 with your CAS, and find the midpoint of the line segment joining P_1 and P_2.

(2) $P_1 = (2,0), P_2 = (-4,6)$

(3) $P_1 = (-1,7), P_2 = (-4,6)$

(4) $P_1 = (\sqrt{3},1), P_2 = (-1,1)$

(5) Can you find a formula to compute the midpoint of the line segment joining two arbitrary points P_1 and P_2.

(6) Find k, so that $(3, k)$ is equidistant from $(4, 6)$ and $(9, 1)$.

(7) Find the equation of the line parallel to $y = 6x - 1$ whose y-intercept is 6.

(8) Find the equation of the line which is the perpendicular bisector of the line segment joining $P_1 = (2, 6)$ and $P_2 = (3, 11)$. Graph the line and the line segment with your CAS.

(9) Find the equation of the line which is the perpendicular bisector of the line segment joining $P_1 = (5, -1)$ and $P_2 = (4, 8)$. Graph the line and the line segment with your CAS.

(10) Graph the following parabola with your CAS and determine any lines of symmetry: $y = -(x - 2)^2 - 3$.

(11) Find the equation of a parabola which has line of symmetry $x = 1$, faces down and has maximum value y=10. (**Hint:** look at problem 10 and graph several versions by changing the 2 or 3).

(12) Find the relationship between the slope of a line and its angle θ with the x-axis. (assume $\theta \leq \pi$).

(13) Let L_1 and L_2 be two lines with slope m_1 and m_2, intersecting at a point p. Let θ be the angle formed by these two lines at p. Let β_1 and β_2 be the angles between L_1 and the x-axis, L_2 and the x-axis, respectively. Assume $\beta_2 > \beta_1$. Can you calulate $\tan(\theta)$ in terms of β_1, β_2.

(14) Consider the line segment which joins two arbitrary points a, b. Let A be a point so that the angle α between the line segments Aa and Ab is a right angle. What is the geometric figure described by the point A as it moves, keeping the angle α equal to $\frac{\pi}{2}$?

(15) On your CAS graph $(x - 1)^2 + (y - 1)^2 - 1 = 0$.

- **(a)** Replace $(x - 1)^2$ by $\frac{(x-1)^2}{4}$, and graph the resulting equation. What can you say?
- **(b)** Replace $(y - 1)^2$ by $\frac{(y-1)^2}{4}$, and graph the resulting equation. What do you notice? Is there a center?

(16) Graph $\frac{(x-1)^2}{4} - (y - 1)^2 = 1$ with your CAS. Can you find symmetries? A center? The figure obtained is called a left-right hyperbola. The equation

$$(y - 1)^2 - \frac{(x - 1)^2}{4} = 1$$

is the equation of an "up-down" hyperbola. Graph it identifying symmetries and centers. These equations represent some of the fundamental conic section equations.

(17) Consider again the equation

$$\frac{(x-1)^2}{4} - (y-1)^2 = 1.$$

What happens if the right hand side is changed to zero? What happens if it is replaced by -2?

(18) Consider again the equation in the previous problem. On your CAS investigate what happens if the right hand side is replaced by 0. Graph the figures representing the solutions of

$$\frac{(x-1)^2}{4} - (y-1)^2 = 1.$$

(19) Identify the following graph by putting the equation in a form similar to the one in problem (18):

$$x^2 - y^2 + 4x + 2y + 2 = 0.$$

(20) Identify the following graph by putting the equation in a form similar to the one in problem (18):

$$2x^2 + y^2 - 4x + 6y + 9 = 0.$$

Chapter IV

Limits and Continuity

4.1 Limit of a Sequence

Consider a sequence of real numbers such as

$$4.01,\ 4.001,\ 4.00001,\ 4.000000001, \ldots$$

If we assume this sequence continues indefinitely (subject to the rule that we have the number 4, a decimal point, a block of zeros whose length keeps doubling and then a 1) it is clear that the numbers in this sequence get closer and closer to 4. We say that the sequence **converges** to 4, or that the **limit** of the sequence is 4. Formally we have the following definition:

Definition: *An infinite sequence of real numbers*

$$a_1,\ a_2,\ a_3,\ a_4, \ldots$$

***converges** to a, or approaches the **limit** a, if*

$$|a - a_n|$$

becomes arbitrarily small for all $n > N$ with $N \to \infty$. We concisely denote this data as

$$\lim_{n\to\infty} a_n = a,$$

or

$$a_1,\ a_2,\ a_3,\ \ldots \to a.$$

Example [4.1] In the sequence above, setting $a_1 = 4.01, a_2 = 4.001, a_3 = 4.00001$, etc. we have that

$$|4 - a_n| = 10^{-2^n} = .\underbrace{00\ldots00}_{2^n \text{ times}}1.$$

It follows that for any large N if $n > N$ then $|4 - a_n| < 10^{-2^N}$. This proves that for this example, $a_1, a_2, a_3, \ldots \to 4$.

4.2 The Limit of a Function

The concept of the limit of a function is one of the essential ideas upon which calculus is based. From an intuitive point of view the matter is simple. The difficulty arises in making the intuition precise. Many students gloss over the precise definition and take a major loss in their understanding of calculus. In an attempt to remedy this situation we begin with a simple:

Paradox: Define the function

$$f(x) = \frac{x-1}{x^2-1}$$

on your CAS. If you ask for the value of $f(1)$ the CAS will respond in turn that it cannot divide by zero. Nevertheless, a simple application of elementary algebra dictates the identity

$$f(x) = \frac{(x-1)}{(x+1)(x-1)} = \frac{1}{x+1},$$

which leads us to the opinion that $f(1)$ should equal $1/2$.

What is going on here? The problem lies in that the above function is not well defined at $x = 1$ and hence the computer does not properly understand the rule for the function at this point. If your CAS has a SIMPLIFY command, then if this command is invoked, it will cancel out the factor $(x-1)$ in both the numerator and denominator of the function and arrive at the conclusion that

$$f(x) = \frac{1}{x+1}.$$

Unfortunately, SIMPLIFY, will not always do the trick. To see this, consider for example, the function

$$g(x) = \frac{\sin(x)}{x}.$$

We can construct $g(x)$ on your CAS and easily verify the values $g(\pi/2) = 0.63662\ldots$, $g(1.5) = 0.665\ldots$. Were we to try to evaluate $g(0)$, however, the computer would respond with error messages. Even if we try to reconstruct $g(x)$ using SIMPLIFY, upon asking for $g(0)$ we would *still* be faced with error messages.

Example [4.2] Based on CAS computation, what value should $g(0)$ have?

If we consider the sequence of numbers (which approach 0),

$$1, 0.1, 0.01, 0.001, 0.0001, \ldots$$

and evaluate $g(x)$ at each point of the sequence we obtain $g(1) = 0.84147\ldots$, $g(0.1) = 0.99833\ldots, g(0.01) = 0.999999833\ldots, g(0.0001) = 0.999999998\ldots$. This leads us to expect that $g(0) = 1$. A heuristic proof of this can be obtained by examining the diagram:

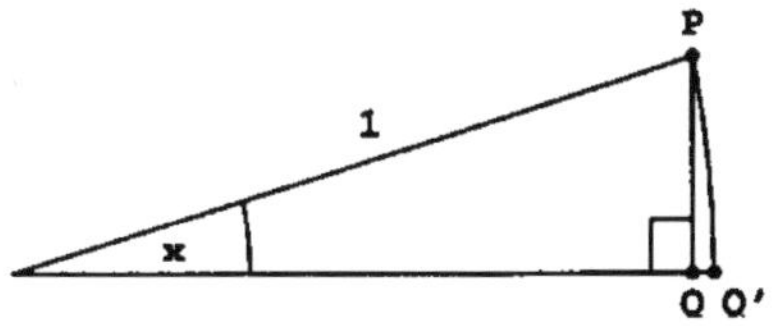

In that the hypotenuse of the triangle is 1 we see that

$$\sin(x) = \frac{\text{OPPOSITE}}{\text{HYPOTENUSE}} = \text{LENGTH}(PQ).$$

But the angle x in radians is by definition the arc length of the arc PQ'. As the angle x approaches 0, the arc approaches the line PQ and thus the ratio of their lengths approaches 1.

Example [4.3] Let $f(x) = \frac{x-1}{x^2-1}$. Using a sequence of computations, what should the value of $f(1)$ be?

In this example, when we examine, for example, the sequence $f(0.9) = 0.52631579$, $f(0.99) = 0.50251256$, $f(.0.999) = 0.5002501$, and $f(0.9999) = 0.500025$, we are led to suspect that $f(1) = \frac{1}{2}$.

This informal discussion leads us to the conclusion that, for the two functions $f(x) = \frac{x-1}{x^2-1}$ and $g(x) = \frac{\sin(x)}{x}$, while the SIMPLIFY rule may not always work, these functions can still be defined at the points of bad behavior $x = 1$, $x = 0$, respectively, by a limiting procedure: viz.

$$f(0.9) = 0.52631579, \;\; f(0.99) = 0.50251256, \;\; f(0.999) = 0.5002501, \ldots$$

$$g(0.1) = 0.99833, \;\; g(.01) = 0.999999833, \;\; g(.001) = 0.999999998.$$

In both the above cases the limit of the sequence gives the expected value of the function.

Definition: *Let $F(x)$ be a function. Assume there exists a real number b such that*

$$\lim_{n \to \infty} F(a_n) = b$$

for any sequence of points

$$a_1, a_2, a_3, \ldots \to a.$$

Then we say

$$\lim_{x \to a} F(x) = b.$$

Remarks: The notation $\lim\limits_{x \to a} f(x) = b$ is used mainly when $f(a)$ is not well defined. If $f(a)$ is well defined, and the above limit exists, then $\lim\limits_{x \to a} f(x) = f(a)$.

Example [4.4] Consider $f(x) = \frac{x-1}{x^2-1}$. Since $f(x)$ is well defined at $x = 3$ it is not necessary to go through a limiting process to evaluate the limit of $f(x)$ as $x \to 3$. We need simply evaluate $f(3)$ to compute the limit:

$$\lim_{x \to 3} \frac{x-1}{x^2-1} = \frac{3-1}{9-1} = \frac{1}{4}.$$

In contrast the limit as $x \to 1$ (the point where $f(x)$ is not well defined) must be evaluated by considering the sequence

$$f(a_1),\ f(a_2),\ f(a_3),\ \ldots$$

where $a_1, a_2, a_3, \ldots \to 1$ is any sequence approaching 1. Since each a_i in the sequence is not equal to 1 we can simplify the fractions:

$$f(a_i) = \frac{a_i - 1}{a_i^2 - 1} = \frac{1}{a_i + 1}.$$

We are then left with a sequence approaching $\frac{1}{2}$,

$$\frac{1}{a_1+1},\ \frac{1}{a_2+1},\ \frac{1}{a_3+1},\ \ldots \to \frac{1}{2},$$

and can conclude that

$$\lim_{x \to 1} \frac{x-1}{x^2-1} = \frac{1}{2}.$$

This can be rewritten as

$$\lim_{x \to 1} \frac{x-1}{x^2-1} = \lim_{x \to 1} \frac{1}{x+1} = \frac{1}{2}.$$

Example [4.5] Evaluate the limit

$$\lim_{x \to 2} \frac{x-2}{x^2-x-2}.$$

This example is analogous to the previous Example where we turn out to be able to simplify the terms in the limit sequence. Upon doing so we see that

$$\lim_{x \to 2} \frac{x-2}{x^2-x-2} = \lim_{x \to 2} \frac{1}{x+1} = \frac{1}{3}.$$

Example [4.6] Evaluate the limit

$$\lim_{x \to 0} \frac{\sin(3x)}{x}.$$

Here we begin by rewriting the limit as

$$\lim_{x \to 0} \frac{\sin(3x)}{x} = \lim_{x \to 0} 3 \cdot \frac{\sin(3x)}{3x}.$$

If we set $y = 3x$, then as $x \to 0$ we see that $y \to 0$. Since $\lim_{y\to 0} \frac{\sin(y)}{y} = 1$, we deduce that

$$\lim_{x\to 0} \frac{\sin(3x)}{x} = 3.$$

Exercises for §4.2

In problems(1)–(8), evaluate the following limits

(1) $\lim_{x\to 0} \frac{3x}{x^2-2x}$.

(2) $\lim_{x\to 2} \frac{x^2+1}{2x-7}$.

(3) $\lim_{x\to 1} \frac{(x-1)^2}{x^3-2x^2+x}$.

(4) $\lim_{x\to 2} \frac{2x-4}{x^2+x-6}$.

(5) $\lim_{x\to 0} \frac{\sin(5x)}{x}$.

(6) $\lim_{x\to 0} \frac{\sin(2x)}{x\cos(x)}$.

(7) $\lim_{x\to 0} \frac{(\cos(x))^2\cdot\sin(x)}{x^2+x}$.

(8) $\lim_{x\to 0} \frac{\sin(\pi x)}{x^2+2x}$.

(9) Let $f(x) = \frac{3x}{x^2-2x}$. Compute $f(0.1)$, $f(0.01)$, $f(0.001)$, $f(-0.1)$, $f(-0.01)$, $f(-0.001)$. Compare your answer with the result obtained in Exercise (1).

(10) Let $f(x) = \frac{(\cos(x))^2\cdot\sin(x)}{x^2+x}$. Compute the values of $f(0.1)$, $f(0.01)$, $f(0.001)$, $f(-0.1)$, $f(-0.01)$, and $f(-0.001)$. Compare your answer with the result obtained in Exercise (7).

(11) Using the identity

$$|\sqrt{x} - \sqrt{a}| = \left|\frac{x-a}{\sqrt{x}+\sqrt{a}}\right|,$$

prove that $\lim_{x\to a} \sqrt{x} = \sqrt{a}$ for $a > 0$.

(12) Define the function $g(x) = x^x$ on your CAS. Evaluate $g(0.1)$, $g(0.01)$, $g(0.001)$, $g(0.0001)$. Can you guess the limit of $g(x)$ as $x \to 0$ with $x > 0$?

(13) Generalize Exercises (9) and (10) from §3.2 by dividing the quartercircle into 4, 8, 16, 32, ... etc pieces. In each case by computing the sum and multiplying by 2 we obtain an approximation of π. Show that this gives a sequence of numbers converging to π.

4.3 Continuous Functions

Intuitively, a function is continuous provided its graph is a curve without any breaks in it, i.e., it can be drawn in a continuous manner without lifting the pen from the paper.

Example [4.7] The following graphs are continuous:

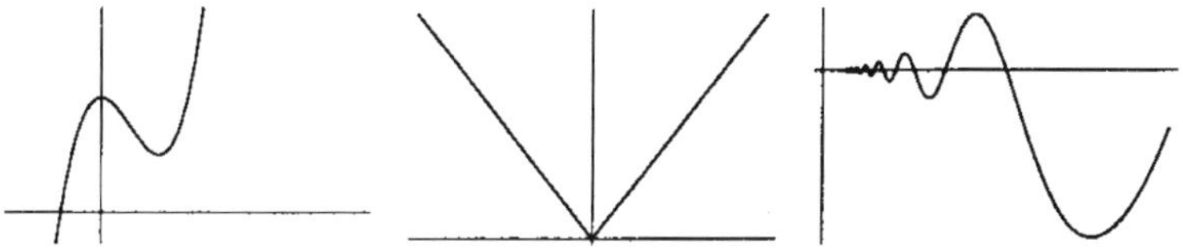

while the following are not:

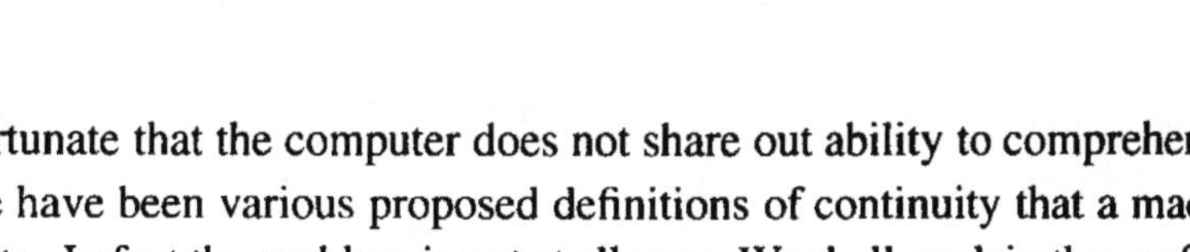

It is unfortunate that the computer does not share out ability to comprehend this.

There have been various proposed definitions of continuity that a machine can incorporate. In fact the problem is not at all easy. We shall explain the profound and elegant solution developed over the last one hundred years.

Definition: *A function $F(x)$ is said to be* ***continuous on an interval*** $c \leq x \leq d$ *if*

$$\lim_{x \to a} F(x) = F(a)$$

for all points a in the interval $c \leq a \leq d$. A function $F(x)$ is termed ***continuous*** *if it is continuous on every interval.*

Example [4.8] Graph the function

$$F(x) = \begin{cases} x, & \text{if } \ 0 \le x < 1 \\ x^2 - 3, & \text{if } \ 1 \le x \le 2 \end{cases}$$

on the interval $0 \le x \le 2$. Is it continuous?

By examining the graph of $F(x)$

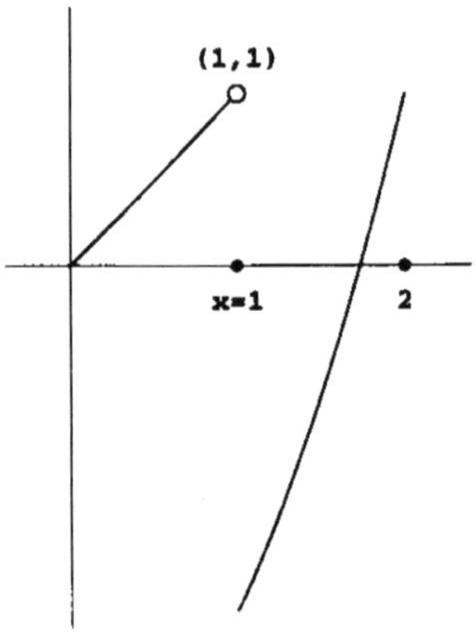

we immediately see a discontinuity at $x = 1$. The small circle indicates that the point $(1, 1)$ is not part of the graph. To see that the limit does not exist we compute

$$F(0.9) = 0.91 \ \ F(0.99) = 0.9801, \ \ F(0.999) = 0.998001, \ldots$$
$$F(1.1) = -1.79, \ \ F(1.01) = -1.9899, \ \ F(1.001) = -1.997999, \ldots$$

Clearly there are two possible outcomes depending on whether you approach 1 from the left, for example with the sequence $0.9, \ 0.99, \ 0.999, \ldots \to 1$, or from the right, for example with the sequence, $1.1, 1.01, 1.001, \ldots$ Hence the limit does not exist and $x = 1$ is a point of discontinuity.

Example [4.9] Graph the function

$$F(x) = \begin{cases} \frac{x-1}{x^2-1}, & \text{if } \ x \ne 1 \\ 13, & \text{if } \ x = 1 \end{cases}$$

on the interval $-\frac{1}{2} \le x \le 2$. Is it continuous?

The graph is given by

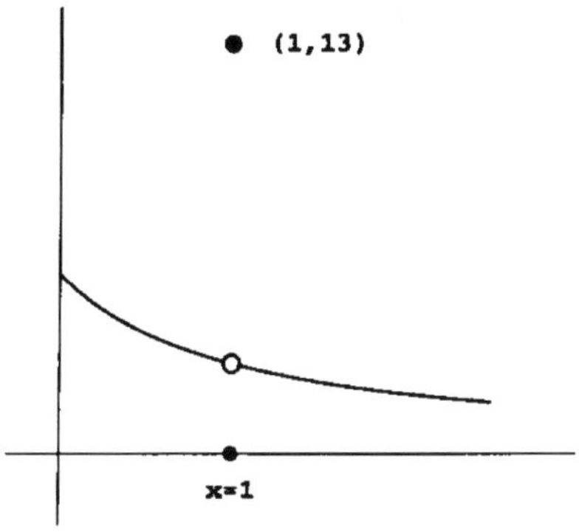

where the point $(1, 1/2)$ is not part of the graph while the point $(1, 13)$ is. Clearly there is a discontinuity at $x = 1$. In this example we have (via example [4.4])

$$\lim_{x \to 1} F(x) = \frac{1}{2}.$$

But $F(1) = 13$; the criterion for continuity is obviously violated.

Example [**4.10**] Explain why the function

$$F(x) = \begin{cases} \dfrac{x-1}{x^2-1} & \text{if } x \neq 1 \\ \dfrac{1}{2} & \text{if } x = 1 \end{cases}$$

is continuous. In this example, since we know (via Example [4.4]) that

$$\begin{aligned} \lim_{x \to 1} \frac{x-1}{x^2-1} &= \frac{1}{2} \\ &= F(1), \end{aligned}$$

we can deduce that the function does not have a discontinuity at $x = 1$ and the graph is continuous.

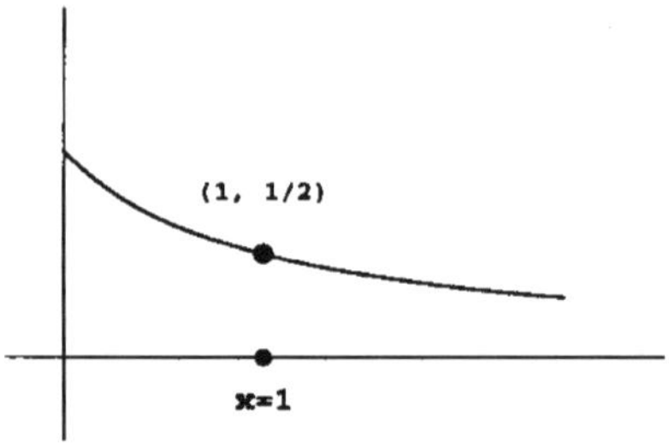

Exercises for §4.3

Use your CAS to graph the following functions. Find all points of discontinuity.

(1) $$f(x) = \begin{cases} 3x, & \text{if } 0 \le x < 1, \\ 2x - 1, & \text{if } 1 \le x \le 2. \end{cases}$$

(2) $$f(x) = \begin{cases} 1 - x^2, & \text{if } -1 \le x \le 2, \\ -3, & \text{if } 2 < x \le 4. \end{cases}$$

(3) $$f(x) = \begin{cases} 2x - 1, & \text{if } -2 \le x < -1, \\ x^2, & \text{if } -1 \le x < 1. \\ x^3 - 2x + 1, & \text{if } 1 \le x < 2. \end{cases}$$

(4) $$f(x) = \begin{cases} \dfrac{\sin(x)}{x}, & \text{if } -1 \le x < 0, \\ 2x^2 + 1, & \text{if } 0 < x \le 1. \end{cases}$$

(5) $$f(x) = \begin{cases} \dfrac{\sin(x)}{x}, & \text{if } -\pi \le x < \pi \text{ and } x \neq 0, \\ 5, & \text{if } x = 0. \end{cases}$$

(6) $$f(x) = \begin{cases} \dfrac{2x - 2}{x^2 + x - 2}, & \text{if } 0 \le x < 1, \\ x + 1, & \text{if } 1 \le x \le 2. \end{cases}$$

4.4 The Algebra of Limits

Consider the set S, of all convergent sequences

$$a_1, a_2, \ldots \to a$$

For example, the sequence given in §4.1

$$4.01, 4.001, 4.00001, 4.000000001.... \to 4$$

is in S. The set S inherits all the algebraic properties of the real numbers.

Proposition [4.11] *Let*

$$a_1, a_2, a_3 \ldots \to a$$

$$b_1, b_2, b_3 \ldots \to b$$

be any two convergent sequences in S. Then the sequences

$$a_1 \pm b_1, a_2 \pm b_2, a_3 \pm b_3, \ldots \to a \pm b,$$

$$a_1 \cdot b_1, a_2 \cdot b_2, a_2 \cdot b_2, \ldots \to a \cdot b,$$

$$a_1/b_1, a_2/b_2, a_3/b_3, \ldots \to a/b,$$

are also in S.

The proofs of the above statements follow easily: for example the second assertion is proved by considering

$$\begin{aligned} |a_n b_n - ab| &= |a_n b_n + \overbrace{-a_n b + a_n b}^{\text{The Trick: Add 0}} - ab| \\ &\le |a_n b_n - a_n b| + |a_n b - ab| \\ &= |a_n| \cdot |b_n - b| + b \cdot |a_n - a|, \end{aligned}$$

which necessarily becomes arbitrarily small for n sufficiently large since both $|b_n - b|$ and $|a_n - a|$ become arbitrarily small.

The algebra of convergent sequences further extends to form the algebra of limits of functions.

Proposition [4.12] *The following identities hold:*

$$\lim_{x\to a}\Big(f(x)\pm g(x)\Big) = \lim_{x\to a} f(x) \pm \lim_{x\to a} g(x)$$
$$\lim_{x\to a}\Big(f(x)\cdot g(x)\Big) = \Big(\lim_{x\to a} f(x)\Big)\cdot\Big(\lim_{x\to a} g(x)\Big)$$
$$\lim_{x\to a}\frac{f(x)}{g(x)} = \frac{\lim\limits_{x\to a} f(x)}{\lim\limits_{x\to a} g(x)},$$

where the last statement holds provided $\lim\limits_{x\to a} g(x) \neq 0$.

Additional exercises for Chapter IV

---

(1) Let $f(x) = \frac{x-2}{x^2-4}$. On your CAS compute $f(1.5), f(1.6), f(1.7), \cdots f(1.9), f(1.99), f(1.999)$. Do these values seem to approach a finite value as x gets closer to 2? Is $f(x)$ continuous?

(2) Compute $\lim\limits_{x\to 2} f(x)$.

Evaluate the following limits. Graph the functions with your CAS.

(3) $\lim\limits_{x\to 1} \frac{x^4-1}{x-1}$.

(4) $\lim\limits_{x\to 9} \frac{x-9}{\sqrt{x}-3}$.

(5) $\lim\limits_{y\to 4} \frac{4-y}{2-\sqrt{y}}$.

(6) Let $f(x) = \begin{cases} \dfrac{x^2-16}{x+4}, & \text{if } x \neq -4 \\ k, & \text{if } x = -4. \end{cases}$

Find k so that $f(x)$ is continuous.

(7) Let $f(x) = \begin{cases} \dfrac{1}{4}x - 2, & \text{if } x \leq k \\ x^2 - 16x + 64, & \text{if } x > k. \end{cases}$

Find k so that $f(x)$ is continuous.

(8) Let $f(x) = \frac{x}{\sin(x)}$. Does the limit of $f(x)$, as x approaches 0 exist? On your CAS compute $f(x)$ for $x = 1, 0.1, 0.01, 0.001, 0.0001, \cdots$. Can you find a different method to compute the limit, based on the knowledge $\lim\limits_{x\to 0} \frac{\sin(x)}{x} = 1$.

(9) Let $f(x) = \frac{\sin(x)}{x} + \frac{x^2}{x(x+1)}$. What is $\lim_{x \to 0} f(x)$?

(10) Let

$$f(x) = \frac{x-3}{(x^2-9)(x-2)} - \frac{x-1}{x^3-1}.$$

Use your CAS to compute $\lim_{x \to 3} f(x)$ and $\lim_{x \to 1} f(x)$.

(11) Let

$$f(x) = \frac{x-3}{(x^2-9)(x-2)} \cdot \frac{x-1}{x^3-1}.$$

What is $\lim_{x \to 3} f(x)$ and $\lim_{x \to 1} f(x)$?

(12) Show that if f and g are functions which are continuous at b, then

(a) $f+g,\ f-g,\ f \cdot g$ are all continuous at b.

(b) f/g is continuous at b if $g(b) \neq 0$.

(13) If g is continuous at b and f is continuous at $g(b)$, show that $f \circ g$ is continuous at b.

(14) Suppose f is such that $f(-1) = 2.5,\ f(0) = 2,\ f(1) = 3,\ f(3) = 1,\ \lim_{x \to 1^-} f(x) = 2,$ and $\lim_{x \to 1^+} f(x) = 0$, where $\lim_{x \to 1^-} f(x)$ is defined to be the limit of f as x approaches 1 from the left and $\lim_{x \to 1^+} f(x)$ is defined to be the limit of f as x approaches 1 from the right. Can you sketch a graph of f? Is $f(x)$ continuous at $x = 1$?

(15) On your CAS, use the appropriate command function to factor the rational function

$$R(x) = \frac{x^2+6x+8}{x^2-4}.$$

Find $\lim_{x \to -2} \frac{x^2+6x+8}{x^2-4}$ and graph $R(x)$.

We will often want to understand how a function $f(x)$ behaves as x becomes very large. The techniques for computing $\lim_{x \to \infty} f(x)$ are quite similar to those for a finite limit.

Evaluate the following limits and graph the functions with your CAS.

(16) $\lim_{x \to \infty} \frac{3x^2+1}{x^2+3}$ (**Hint:** Factor out x^2 from the numerator and the denominator and cancel, Then use the fact that $\lim_{x \to \infty} \frac{1}{x^2} = 0$).

(17) $\lim_{x \to \infty} \frac{5x^2+x-11}{6x^2+3}$.

(18) $\lim_{x \to \infty} \frac{\sin(x)}{x}$.

(19) $\lim_{x \to \infty} \frac{x^2+x-2}{3x^3+2x}$.

(20) $\lim_{x \to +\infty} \frac{x}{|x|}$.

(21) $\lim_{x \to -\infty} \frac{x}{x^2+3}$.

Chapter V

The Derivative

5.1 Tangent Lines

Let $y = f(x)$ be an arbitrary function which is continuous on an interval which contains a point P.

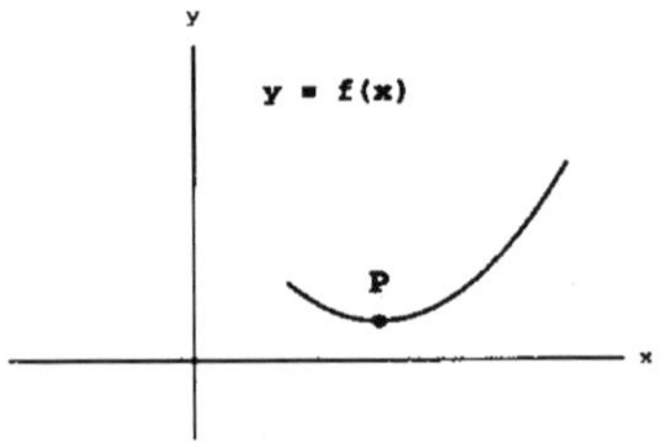

Let Q be a point on the curve which is close to P. Consider the line L_{PQ} which passes through both points P and Q.

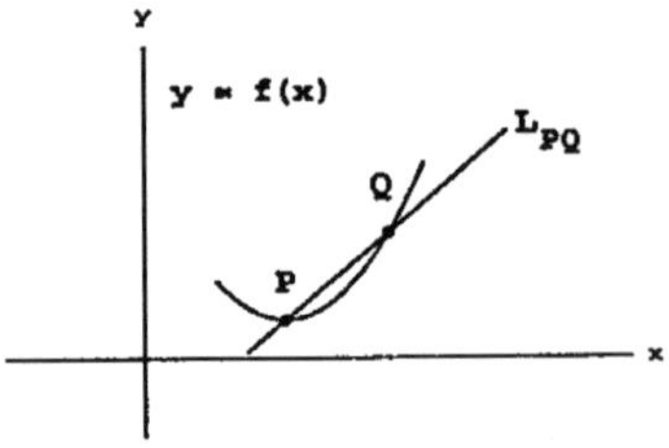

If we let Q move closer to P and form a new line the picture changes.

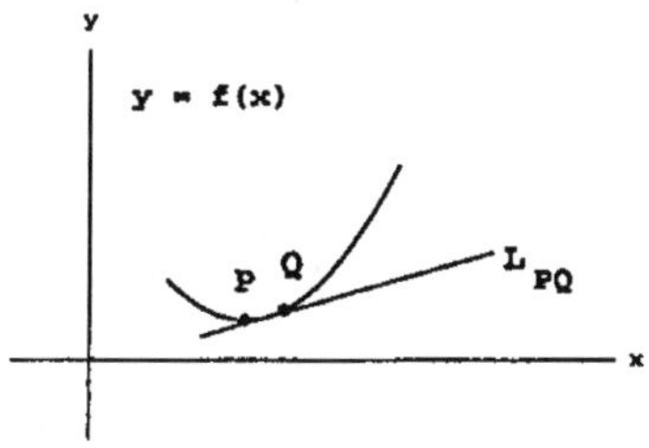

Definition: *The* ***tangent*** *line to the curve* $y = f(x)$ *at the point* P *is the limit of the lines* L_{PQ} *as* $Q \to P$. *If the limit does not exist, then there is no tangent line.*

Remarks:
(1) The precise definition of limit of lines is omitted for the sake of brevity — we rely on the readers intuition.
(2) Assuming the graph of $f(x)$ is not itself a line, locally (in a small neighborhood of P) the tangent line defined above (if it exists) touches the curve only at P, viz.

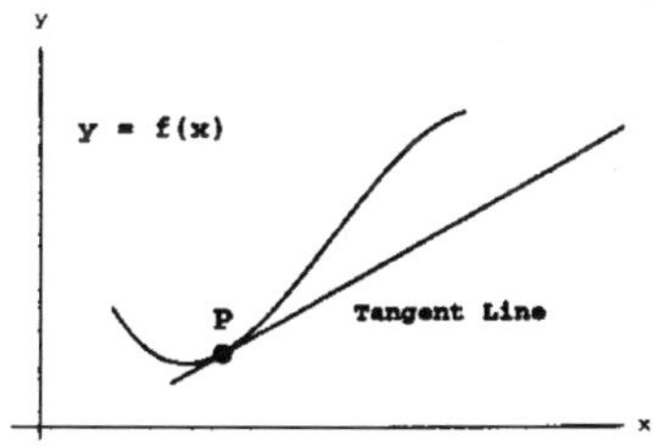

It may, however, intersect the curve at some point $Q \neq P$.

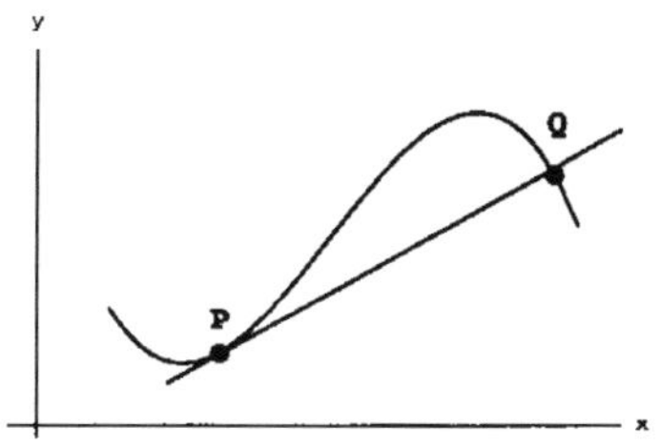

(3) When the graph of $f(x)$ is in fact a line, and thus $f(x) = mx + b$, the tangent line must coincide with the graph itself.

Warning: There exist continuous curves where the tangent line at a point does not exist!

Example [5.1] The absolute value function, $y = |x|$ has the graph:

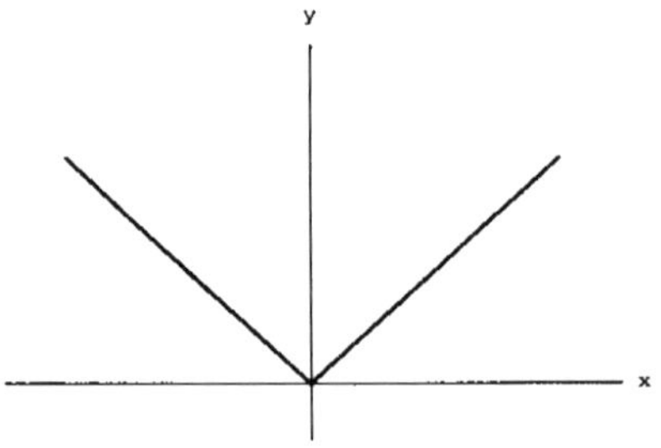

Visually we can see the two possible tangent lines at $(0, 0)$. Since the limit condition of the definition is violated in this example, there is no tangent line at the origin.

Example [5.2] Graph the function $y = x^{2/3} = (\sqrt[3]{x})^2$ using your CAS. Show that visually it has 2 possible tangent lines at the origin.

Here again we visually see two possible tangent lines at the origin.

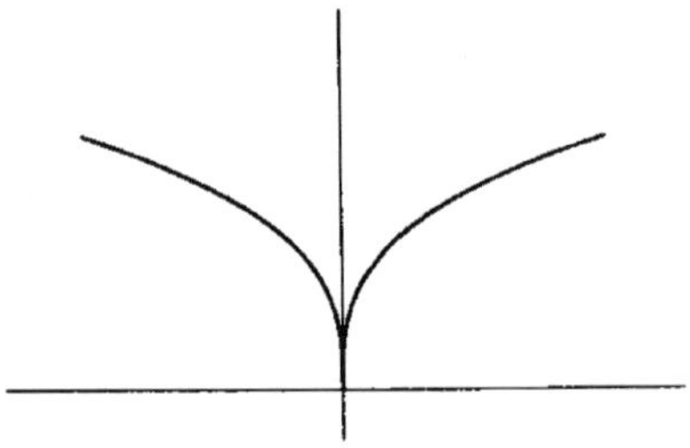

5.2 The Derivative of a Function

The concept of the derivative is the fundamental building block of calculus. Despite its importance, however, it is surprisingly simple to define.

We assume the function $y = f(x)$ is continuous on an interval containing a and has a unique tangent line at this point.

> **Definition:** *The* ***derivative*** *of the function $f(x)$ at the point $x = a$ is defined to be the slope of the line L, the tangent line at the point $(a, f(a))$.*

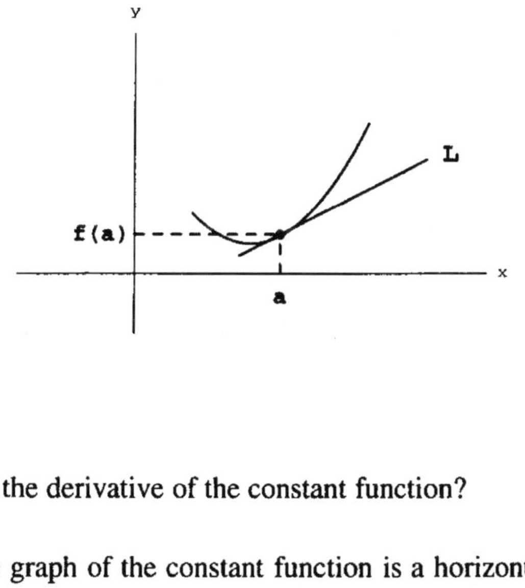

Question: What is the derivative of the constant function?

Answer: Since the graph of the constant function is a horizontal line the tangent line at any point coincides with the graph itself. Hence the slope of the tangent line is 0, i.e., the derivative of the constant function is 0.

Question: What is the derivative of the function $f(x) = 3x + 2$ at the point $(1, 5)$?

Answer: Here again the graph of the function is a straight line and hence the tangent line coincides with the graph, i.e., the tangent line has the equation $y = 3x + 2$. We then see that the derivative $= 3 =$ slope of the tangent line.

Question: What is the derivative of the function $f(x) = -8x + 7$ at the point $(0, 7)$.

Answer: -8.

Question: What is the derivative of the function $f(x) = x^2$ at the point $(0, 0)$.

Answer: Graphing the function $y = x^2$,

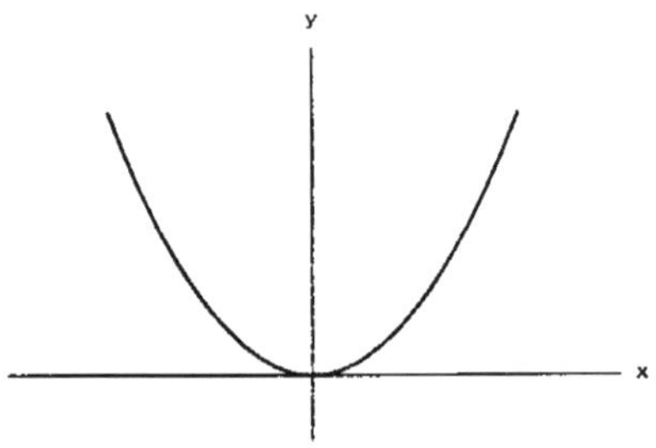

we see that the tangent line at the origin $(0, 0)$ is simply the x–axis whose slope is zero. Hence the derivative $= 0$.

There are various standard notations for the derivative of a function $f(x)$ at $x = a$. The variety is due mainly to the fact that calculus was developed in parallel in both England and Germany. The most prevalent notations are

$$f'(a), \quad \left.\frac{df}{dx}\right]_{x=a}, \quad \frac{df}{dx}.$$

Example [5.3] Let $f(x) = 18x - 91$ and $g(x) = x^2$. Compute $f'(3)$ and $g'(0)$.

From the discussion above, we see that $f'(3) = 18$ and $g'(0) = 0$. In the alternative notation, $\left.\frac{df}{dx}\right]_{x=3} = 18$, $\left.\frac{dg}{dx}\right]_{x=0} = 0$.

5.3 Computing the Derivative using Limits

The derivative of a function may be computed using limits. Before exploring the general case, consider the following example:

Example [5.4] Let $f(x) = x^2$. Compute $f'(2)$, the slope of the tangent line to this curve at the point $(2, 4)$.

First we graph the function and label the point $P = (2, 4)$.

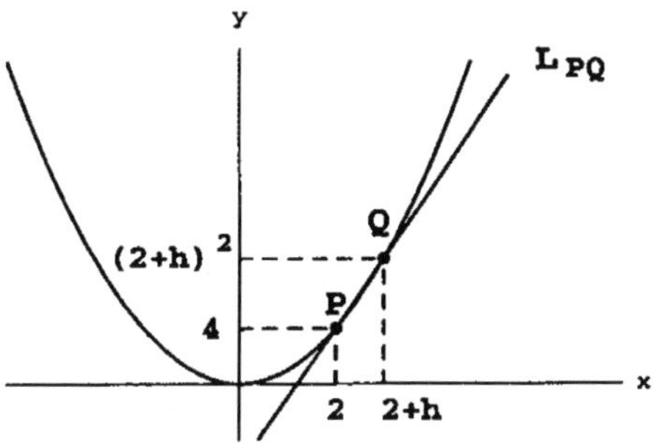

Next we consider a point Q which is near P with coordinates $(2+h, (2+h)^2)$. By definition the slope of the line L_{PQ} (the line which passes through P, Q) is given by

$$\begin{aligned} \text{slope}(L_{PQ}) &= \frac{(2+h)^2 - 4}{h} \\ &= \frac{4h + h^2}{h} \\ &= 4 + h. \end{aligned}$$

Now as $h \to 0$, the point $P \to Q$ and L_{PQ} approaches the tangent line. The slope of the tangent line must therefore be 4, i.e., $f'(2) = 4$.

We can now follow the above model to compute the derivative of f at an arbitrary fixed point a.

Example [5.5] Find the derivative of $y = x^2$ at an arbitrary point a.

Let $P = (a, a^2)$, and $Q = (a+h, (a+h)^2)$.

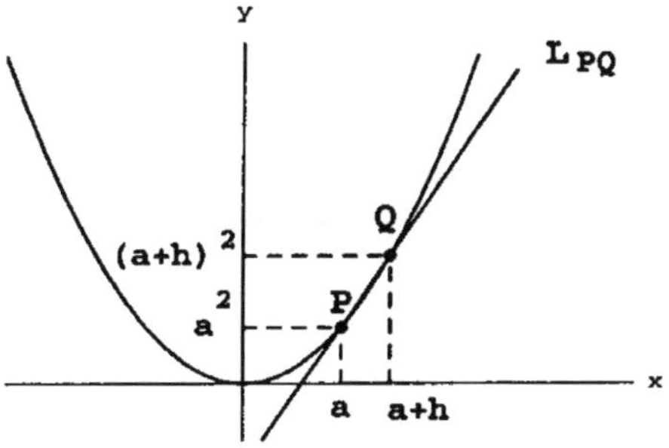

Again we compute the slope of L_{PQ}:

$$\begin{aligned}\text{slope}(L_{PQ}) &= \frac{(a+h)^2 - a^2}{h} \\ &= \frac{2ah + h^2}{h} \\ &= 2a + h.\end{aligned}$$

As $h \to 0$ the line L_{PQ} approaches the tangent line, and we see that

$$f'(a) = 2a.$$

Finally, we observe that the above procedure can be applied to an arbitrary function $y = f(x)$. Let $P = (x, f(x))$ be a fixed point on the curve and let $Q = (x+h, f(x+h))$ be a point close to P.

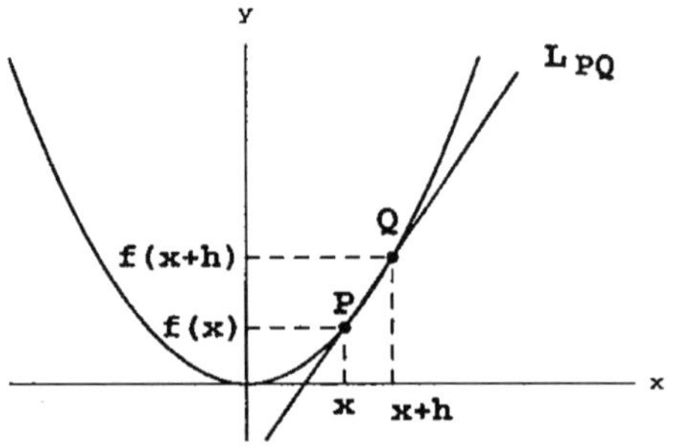

In this general case when we compute the slope of L_{PQ} we obtain:

$$\text{slope}(L_{PQ}) = \frac{f(x+h) - f(x)}{h}.$$

Letting $h \to 0$ we obtain the slope of the tangent line. This leads to an alternative definition of the derivative.

Definition: *Let $f(x)$ be a function which is continuous on an interval which contains x. Then the* ***derivative*** *of f is given by*

$$f'(x) = \lim_{h \to 0} \frac{f(x+h) - f(x)}{h}.$$

If the above limit does not exist, then the derivative does not exist. We say $f(x)$ is ***differentiable*** *if, for every real number x, $f'(x)$ exists.*

Example [5.6] Evaluate the derivative of $f(x) = x^3$ using limits.

Beginning with the definition, we see that

$$\begin{aligned}\frac{f(x+h)-f(x)}{h} &= \frac{(x+h)^3 - x^3}{h} \\ &= \frac{3x^2h + 3xh^2 + h^3}{h} \\ &= 3x^2 + 3xh + h^2.\end{aligned}$$

When we let $h \to 0$ we conclude that $f'(x) = 3x^2$.

Example [5.7] Evaluate the derivative of $f(x) = \frac{1}{x}$ (for $x \neq 0$) using limits.

In this case we have

$$\begin{aligned}\frac{f(x+h)-f(x)}{h} &= \frac{\frac{1}{x+h} - \frac{1}{x}}{h} \\ &= \frac{x - (x+h)}{hx(x+h)} \\ &= -\frac{1}{x(x+h)}.\end{aligned}$$

Upon letting $h \to 0$, we arrive at the conclusion,

$$f'(x) = \lim_{h \to 0} -\frac{1}{x(x+h)} = -\frac{1}{x^2}.$$

Warning: If the function $f(x)$ does not have a unique tangent line at a point x, then $f'(x)$ will not exist! For example, the absolute value function $y = |x|$ does not have a derivative at $x = 0$.

Remark: If $f(x)$ is a differentiable function then $f'(x)$, since it is unique for every x, will also be a function. This is the reason for choosing the notation $f'(x)$ over the notation $\frac{df}{dx}$ which is seen in older texts.

Exercises for §5.3

Use the EXPAND and SIMPLIFY capability of your CAS to evaluate $\frac{f(x+h)-f(x)}{h}$ for the functions in Exercise (1)–(10). Complete the computation of $f'(x)$ by mentally computing the limit as $h \to 0$.

(1) $f(x) = x^4$.

(2) $f(x) = x^{19}$.

(3) $f(x) = x^{50}$.

(4) $f(x) = x^2 + 3x - 1$.

(5) $100x^{18} + 3x^2 - 11$.

(6) $f(x) = \frac{1}{2x+1}$.

(7) $f(x) = \frac{x-1}{x+1}$.

(8) $f(x) = \sqrt{x}$. **Hint:** Use the identity

$$(\sqrt{x+h} - \sqrt{x}) = (\sqrt{x+h} - \sqrt{x}) \cdot \overbrace{\frac{\sqrt{x+h} + \sqrt{x}}{\sqrt{x+h} + \sqrt{x}}}^{\text{the trick}}.$$

(9) $f(x) = \sqrt{x + 11}$.

(10) $f(x) = \frac{1}{\sqrt{2x-1}}$.

5.4 Finding the Equation of the Tangent Line

Let $y = f(x)$ be a continuous function which possesses a unique tangent line at every point x.

Question: What is the equation of the tangent line L to the curve $y = f(x)$ at the point $x = a$?

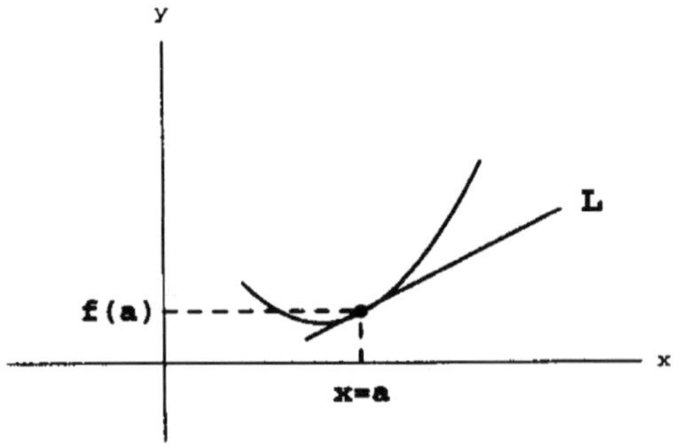

Previously we have shown that the equation of any line takes the form

$$y = mx + b$$

where m is the slope of the line and b is the y–intercept of the line. By definition, since L is the tangent line, we have that

$$\text{slope}(L) = f'(a).$$

Thus the equation for L must be of the form

$$y = f'(a)x + b.$$

How do we compute b? There are various methods. Perhaps the simplest is to use the fact that L passes through the point $(a, f(a))$. When we substitute $x = a$ and $y = f(a)$ into the equation of L we obtain,

$$f(a) = f'(a) \cdot a + b.$$

Solving for b yields

$$b = f(a) - f'(a) \cdot a.$$

So the equation of the tangent line now looks like

$$\begin{aligned} y = f'(a)x + b &= f'(a)x + \Big(f(a) - f'(a) \cdot a\Big) \\ &= f'(a)\,(x - a) + f(a). \end{aligned}$$

Example [5.8] Find the equation of the tangent line to the curve $y = x^3$ at the point $(2, 8)$.

This problem can be attacked as follows:

Step 1. Construct $f(x) = x^3$ on your CAS.

Step 2. Compute $f(2) = 8$, $f'(2) = 12$.

Step 3. Plug in $a = 2$ in the tangent line equation $y = f'(a)\,(x - a) + f(a)$ to obtain

$$\begin{aligned} y &= 12(x - 2) + 8 \\ &= 12x - 16. \end{aligned}$$

Example [5.9] Find the equation of the tangent line to the curve $y = \frac{1}{x}$ at the point $x = 5$.

Since $f(x) = \frac{1}{x}$ and $f'(x) = -\frac{1}{x^2}$ it follows that $f(5) = \frac{1}{5}$ and $f'(5) = -\frac{1}{25}$. Plugging these values into the tangent line equation $y = f'(a)(x - a) + f(a)$ we get the equation of the tangent line at $x = 5$:

$$y = -\frac{1}{25}(x - 5) + \frac{1}{5}$$

$$= -\frac{x}{25} + \frac{2}{5}.$$

Exercises for §5.4

In exercises (1)–(6), find the equation of the tangent line to the curve $y = f(x)$ at the point P. Using your CAS, graph the curve, the point P, and the tangent line at P.

(1) $f(x) = 2x^3 - x + 4$, $P = (0, 4)$.

(2) $f(x) = x^5 - 3x^4 + 2x + 9$, $P = (1, 9)$.

(3) $f(x) = 2/(x + 1)$, $P = (2, 2/3)$.

(4) $f(x) = x + \frac{1}{x}$, $P = (2, 5/2)$.

(5) $f(x) = \sqrt{x + 1}$, $P = (1, \sqrt{2})$.

(6) $f(x) = \sqrt{x + 1}$, $P = (3, 2)$.

(7) Find two tangent lines to the curve $y = x^2 - 2x + 4$ which pass through the origin. Graph the curve with its two tangent lines.

(8) Show that $f(x) = 3x^2$ and $g(x) = 2x^3 + 1$ have the same tangent line at $(1, 3)$. Graph the curves and the tangent line.

(9) Does the curve $y = x^3 + 17x - 6$ have a tangent line which is parallel to the line ℓ given by the equation, $y = 4x + 37$. Graph the curve, the line ℓ, and the tangent (if it exists).

(10) Does the curve $y = x^3 + 7x - 14$ have a tangent line which is parallel to the line $y = 19x + 8$? Graph the curve, the line ℓ, and the tangent (if it exists).

Given a curve $y = f(x)$ the **normal line** at a point P is defined to be the line which passes through P and is perpendicular to the tangent line to the curve at P. In the following problems graph the curve, the point P, the tangent line at P, and the normal line at P.

(11) $f(x) = x^3 - 8$, $P = (2, 0)$.

(12) $f(x) = \frac{1}{x}$, $P = (2, 1/2)$.

(13) $f(x) = 2x^5 - 6x + 4$, $P = (0, 4)$.

(14) $f(x) = x + x^2 + x^3$, $P = (0, 0)$.

5.5 Higher Derivatives

Given a differentiable function $f(x)$, its derivative, $f'(x)$ will itself be a new function which may also be differentiable. If this is the case, upon taking the derivative of $f'(x)$ we obtain

$$f''(x)$$

which is termed the **second derivative** of $f(x)$. Various notations are used for the second derivative:

$$\frac{d^2}{dx^2}f(x) = \frac{d}{dx}\left(\frac{df}{dx}\right) = \frac{d}{dx}(f'(x)) = f''(x).$$

If $f(x)$ is a differentiable function and its derivative $f'(x)$ is also differentiable, then we term $f(x)$ a **twice differentiable** function.

Example [5.10] Evaluate the second derivative of the function $f(x) = x^3$.

From our computations in Example [5.6] and Example [5.5] we see that $f'(x) = 3x^2$ and $f''(x) = 6x$.

Example [5.11] For $x \neq 0$ evaluate the second derivative of $f(x) = \frac{1}{x}$.

Having seen in Example [5.7] that $f'(x) = -\frac{1}{x^2}$, to compute the second derivative we need to evaluate the limit

$$\begin{aligned}
\lim_{h\to 0} \frac{\left(\frac{-1}{(x+h)^2}\right) - \left(\frac{-1}{x^2}\right)}{h} &= \lim_{h\to 0} \frac{-x^2 + (x+h)^2}{h \cdot x^2 \cdot (x+h)^2} \\
&= \lim_{h\to 0} \frac{2hx + h^2}{h \cdot x^2 \cdot (x+h)^2} \\
&= \lim_{h\to 0} \frac{2x + h}{x^2(x+h)^2} \\
&= \frac{2}{x^3}.
\end{aligned}$$

We conclude that $f''(x) = \frac{2}{x^3}$.

Having defined the second derivative we can similarly define higher derivatives. A function $f(x)$ is said to be n**–times differentiable** if

$$f(x),\ \frac{df}{dx},\ \frac{d}{dx}\left(\frac{df}{dx}\right) = \frac{d^2f}{dx^2},\ \frac{d}{dx}\left(\frac{d^2f}{dx^2}\right) = \frac{d^3f}{dx^3},\ \ldots,\ \frac{d^nf}{dx^n},$$

are all well defined functions.

Example [5.12] Find the third derivative of the function $f(x) = x^3$.

As we have seen, $f'(x) = 3x^2$ and $f''(x) = 6x$. Since the graph of $f(x) = 6x$ is a line whose slope is 6 we conclude that $f'''(x) = 6$.

Example [5.13] Find the nth derivative of the function $f(x) = x^4$.

The first derivative of $f(x) = x^4$ was computed in Exercise (1) of §5.3. It is $f'(x) = 4x^3$. The second derivative is

$$\frac{d^2 f}{dx^2} = 12x^2,$$

the third derivative is

$$\frac{d^3 f}{dx^3} = 24x,$$

while the fourth derivative is

$$\frac{d^4 f}{dx^4} = 24.$$

All higher derivatives (i.e., when $n > 4$) are thus seen to be 0.

Exercises for §5.5

Evaluate the first, second, and third derivatives of the following functions. Check your answers with your CAS.

(1) $f(x) = 2x^2$.

(2) $f(x) = 3x^3$.

(3) $f(x) = 7x^4$.

(4) $f(x) = \frac{2}{x}$.

(5) $f(x) = \sqrt{x}$.

(6) $f(x) = 3x^2 - 2x + 1$.

(7) $f(x) = 7x^3 - 2x + 3$.

(8) $f(x) = 4x - 3$.

(9) $f(x) = \frac{1}{2x+1}$.

(10) $f(x) = 7x^9 - 2x^3 + 3x - 14$.

(11) Explain why the polynomial $13x^5 - 5x^4 + 2x - 12$ has its nth–derivative $\frac{d^n f}{dx^n} = 0$ for $n > 5$.

(12) Give an example of a function $y = f(x)$ which satisfies $f(1) > 0$, $f'(1) < 0$, and $f''(1) > 0$.

Additional exercises for Chapter V

In the first two problems, find the equation of the tangent line to the curve, and the normal line to the curve $y = f(x)$ at the point P, and graph all of the aforementioned on your CAS.

(1) $f(x) = -x^2 + 12x - 3, \quad P = (1, 8).$

(2) $f(x) = \frac{1}{\sqrt{2x-1}}, \quad P = (1, 1).$

(3) On your CAS, compute the derivative of $x, x^2, x^3, \cdots, x^{12}$ using the limit definition of derivative. Do you notice any regular patern? Can you infer the expression for the derivative of x^n, where n is a positive integer. Verify your expression using the limit definition.

(4) Let a, b, c, d be constants satisfying $ad - bc = 1$. Compute the derivative of $\frac{ax+b}{cx+d}$ using the limit definition of the derivative.

(5) Find the points in the curve $y = x^3 - 2x + 1$ where the tangent is horizontal.

(6) Show that the curve $y = x^3 + 8x - 11$ has tangents of slope greater than or equal to 8.

(7) Graph the following function on your CAS.

$$f(x) = \begin{cases} x^2 + 4x + 4 & \text{if } x \leq 0 \\ 4x + 4 & \text{if } x > 0. \end{cases}$$

Show that $f(x)$ is continuous. Show that $f(x)$ has a derivative everywhere. What is the derivative at $x = 0$?

(8) Graph the following function on your CAS.

$$f(x) = \begin{cases} x^3 - 9x^2 + 27x - 27 & \text{if } x \leq 4 \\ 3x - 11 & \text{if } x > 4. \end{cases}$$

Show that $f(x)$ is continuous. Show that $f(x)$ has a derivative everywhere. What is the derivative at $x = 0, x = 4, x = 5$?

(9) Let $f(x) = \frac{1}{x}$. Using the limit definition and your CAS, derive the derivative of $f(x)$. Now let $f(x) = \frac{1}{x^2}$, then $\frac{1}{x^3}, \cdots$. What are the derivatives? Is there a pattern?

(10) Let $f(x) = \sin(x)$. Using the limit definition of the derivative show that $f'(0) = 1$.

(11) Graph the function $f(x) = \cos(x)$, with your CAS. By observing the graph, determine if $f'\left(\frac{5\pi}{7}\right)$ is positive or negative. **Hint:** Try to imagine if the tangent line has positive or negative slope.

(12) Graph the function $f(x) = 2^x$ with your CAS. By observing the graph, determine if $f'(3)$ is positive or negative.

(13) On your CAS, graph the function $f(x) = x^{1/3} + 1$. Use the limit definition of the derivative to check whether or not $f(x)$ is differentiable at $x = 0$. Is it continuous at $x = 0$?

(14) Suppose that the function f is differentiable at $x = 2$ and $\lim\limits_{h \to 0} \frac{f(2+h)}{h} = 6$. Find $f(2)$ and $f'(2)$.

(15) Show that any two tangent lines to the parabola $y = ax^2$, $a \neq 0$, intersect at a point that is on the vertical line halfway between the points of tangency.

Chapter VI
Basic Applications of the Derivative

6.1 Velocity

Imagine a car traveling at the constant speed of 55 mph (miles per hour). Then after $\frac{1}{2}$ hour the car will will have traveled $22\frac{1}{2}$ miles, while after 3 hours it will have traversed 165 miles. These are special cases of the equation

$$s(t) = 55t,$$

where $s(t)$ denotes the distance traveled (in miles) after t hours. We have expressed the distance traveled as a function $s(t)$ of time (for the sake of clarity the letter s has become the standard notation for distance rather than the letter d which is usually used in the derivative notation). What if the car is accelerating i.e., going faster and faster. In this case the distance function might be of the form

$$s(t) = 10t^2,$$

which indicates that more distance is traveled in a given amount of time than was traveled when the speed or **velocity** was constant. In this case $s(10) = 1000$ which is almost twice as much distance covered than by going at the constant speed of 55 mph.

Example [6.1] (Drunk Driver) Let us assume the function measuring the distance from the origin for a drunk driver is given by the equation

$$s(t) = t \cdot \left(t - \frac{1}{6}\right) \cdot \left(t - \frac{1}{10}\right).$$

Then $s(1) = \frac{5}{6} \cdot \frac{9}{10} = \frac{45}{60} = \frac{3}{4}$, i.e., the drunk driver travels $\frac{3}{4}$ of a mile in 1 hour. But $s\left(\frac{1}{6}\right) = 0$ which indicates the drunk driver is back to his starting point after $\frac{1}{6}$hour = 10 minutes. He has gone forwards and backwards!

Remark: The distance function $s(t)$ depends on how an object moves. Very often it is found by experimentation.

Problem: Assume we know the distance function $s(t)$ for a moving object. How do we find the speed or velocity of the object at any time t?

Since

$$\text{average velocity} = \frac{\text{distance travelled}}{\text{time elapsed}},$$

we see that the average velocity between times t and $t + h$ is given by

$$\frac{s(t+h) - s(t)}{h}.$$

Hence, the instantaneous velocity at time t is obtained as a limit:

$$\lim_{h \to 0} \frac{s(t+h) - s(t)}{h} = s'(t),$$

the derivative of the distance function.

Example [6.2] How fast is the drunk driver going after 10 minutes? How fast is he going after 30 minutes?

Expanding the function (with the CAS) and taking the derivative gives us

$$s(t) = t^3 - \frac{4}{15}t^2 + \frac{t}{60}$$

$$s'(t) = 3t^2 - \frac{8}{15}t + \frac{1}{60}.$$

Hence $s'(1/6) = .011111$mph, and $s'(1/2) = 1/2$ mph.

Example [6.3] (Falling Apple) This is the example that, by legend, inspired Sir Isaac Newton to develop calculus. We are led to believe that while sitting under an apple tree Sir Isaac had the opportunity to observe an apple falling — according to myth, he was hit on the head by an apple at the moment of his inspiration!

After much experimentation physicists have arrived at a good approximation for the distance traveled (in feet) by an object falling to the ground;

$$s(t) = 16t^2.$$

Notice that this formula is independent of the size and weight of the object. This is the result of a famous experiment of Galileo Galilei in 1589 which was performed at the leaning tower of Pisa. A small stone and a cannon ball were dropped simultaneously from the top of the tower, and were observed to hit the ground at the same time.

Example [6.4] If an apple falls from a height of 13 feet, how long does it take to hit the ground. What will its velocity be when it hits the ground?

The amount of time it takes the apple to hit the ground is obtained by solving the equation $s(t) = 16t^2 = 13$. The solution is clearly $t = \sqrt{13/16} = \frac{1}{4}\sqrt{13}$. The velocity on impact will be $s'(\sqrt{13}/4) = 32\sqrt{13}/4 = 8\sqrt{13}$ feet per second.

Remark: In Example [6.3] people sometimes intuitively believe that the velocity of the apple when it hits the ground = 0. This is simply incorrect: consider why one doesn't jump out of a window.

Example [6.5] (Ball thrown in the air) Assume a ball is thrown straight up with an initial velocity v_0 from a height s_0 at time $t = 0$.

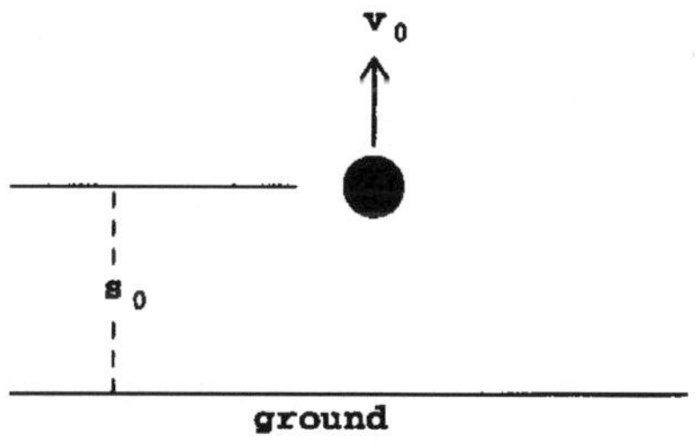

Let $s(t)$ denote the height of the ball (measured from the ground) at time t. Clearly $s(0) = s_0$. Furthermore, we must have $s'(0) = v_0$. In view of the known distance function for the falling apple (see Example [6.3]) we infer that

$$s(t) = -16t^2 + v_0t + s_0.$$

The minus sign in front of the $16t^2$ is due to the fact that in this model the force of gravity is pulling the ball down, i.e., $s(t)$ is not distance traveled, but rather the height function.

Example [6.6] A ball is thrown straight up with an initial velocity of 32 feet per second from a height of 48 feet. How long will it take for the ball to hit the ground and what will its velocity be on impact?

In this example we are told that $v_0 = 32$ and $s_0 = 48$. Hence the position of the ball is given by

$$s(t) = -16t^2 + 32t + 48,$$

and the velocity is given by

$$s'(t) = -32t + 32.$$

The ball will hit the ground at the time t when $s(t) = 0$, i.e., when

$$-16t^2 + 32t + 48 = 8(-2t^2 + 4t + 6) = 0.$$

Hence the ball will hit the ground in 3 seconds and the velocity on impact will be $s'(3) = -64$ feet per second. That the velocity is negative indicates that the ball is traveling down.

Example [6.7] In example [6.6] what is the maximum height the ball attains.

The ball will reach its peak when its velocity is zero (i.e., before it heads for the ground) and thus we must solve the equation

$$s'(t) = -32t + 32 = 0.$$

The solution is clearly $t = 1$ second, and the height after one second will be exactly $s(1) = -16 + 32 + 48 = 64$ feet.

Exercises for §6.1

(1) A stone is dropped from a tall building which is 900 feet high. How long will it take the stone to hit the ground and what will its velocity be on impact?

(2) A ball is thrown straight up with an initial velocity of 8 feet per second from a height of 16 feet. What is the maximum height the ball attains?

(3) A bean bag is thrown up with an initial velocity of 8 feet per second from a height of 8 feet. How long will it take the bean bag to hit the ground and what will its velocity be on impact?

(4) On the surface of the distant planet Xenon, a rock thrown straight up into the air with initial velocity v_0 from a height s_0 travels the distance $s(t)$ where $s(t)$ is given by

$$s(t) = -100t^3 + v_0 t + s_0.$$

If $v_0 = 1000$ feet per second and $s_0 = 50$ feet, determine the maximum height the rock attains. What will be the velocity of the ball after 1 second?

(5) A spaceship is about to land on Earth. Its height $h(t)$ (measured in miles) at a time t (measured in seconds) is given by the function

$$h(t) = 120 - 120t + 30t^2.$$

How long will it take the space ship to land and what will its velocity be when it touches the ground?

6.2 Newton's Method

Let $f(x)$ be an arbitrary differentiable function whose graph crosses the x–axis.

Problem: Find a solution in x to the equation $f(x) = 0$.

Newton developed a beautiful and cunning method to attack this problem. We illustrate his idea — it was one of the early triumphs of calculus.

Since the graph of $y = f(x)$ crosses the x–axis it must look like

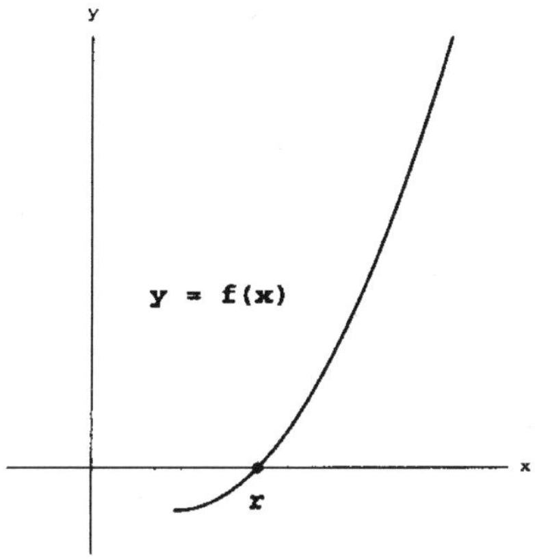

near the point of crossing $x = r$. Clearly $x = r$ is by definition the solution to our problem. But how do we find r?

Newton's idea is to begin with a guess for the value of r, let's call it r_1. If r_1 is close to r and we draw the tangent line L to the curve $y = f(x)$ at the point $(r_1, f(r_1))$, we obtain

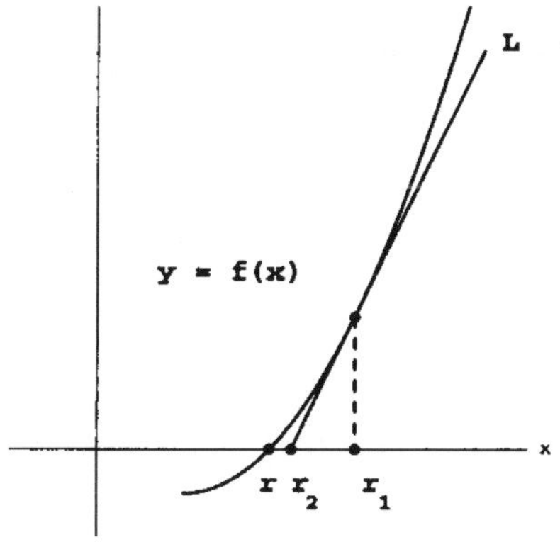

We label a new point (where L intersects the x–axis) by r_2. The point r_2 is now much closer to r. We can calculate r_2 as follows. The equation of the line L is

given by (see §5.4)

$$y = f'(r_1) \cdot x + \Big(f(r_1) - f'(r_1) \cdot r_1\Big).$$

Since this line passes through the point $(r_2, 0)$ we see that

$$0 = f'(r_1) \cdot r_2 + f(r_1) - f'(r_1) r_1,$$

and we can now solve for r_2:

$$r_2 = r_1 - \frac{f(r_1)}{f'(r_1)}.$$

Repeating this procedure we obtain a point

$$r_3 = r_2 - \frac{f(r_2)}{f'(r_2)}.$$

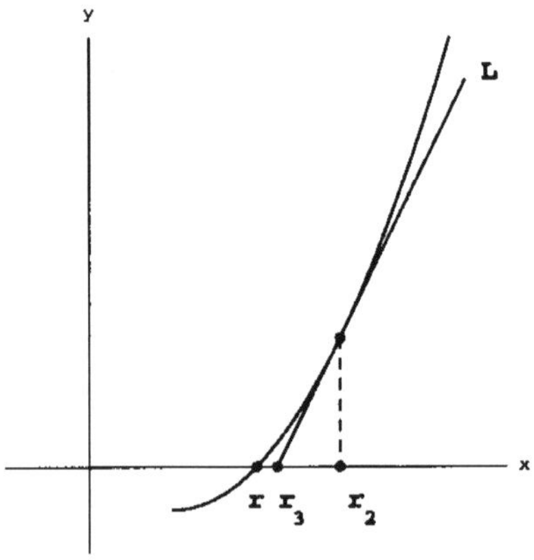

Upon iterating this process, we obtain a sequence of points

$$r_1, r_2, r_3, \ldots \to r$$

where the n-th term is given by

$$r_n = r_{n-1} - \frac{f(r_{n-1})}{f'(r_{n-1})},$$

(for $n = 2, 3, 4, \ldots$) which converges to the real root r.

Warning: If the first approximation r_1 is not close enough to r then Newton's method may break down and the sequence $r_1, r_2, r_3, \ldots$ may not converge. In this case a better initial approximation r_1 should be tried.

Example [6.8] Find an approximate solution to the equation $x^2 - 5 = 0$ using Newton's method.

The derivative of $f(x) = x^2 - 5$ is the linear function $f'(x) = 2x$. To begin Newton's method we must begin with a sensible choice for r_1, in this case let's try $r_1 = 2$. Then

$$r_2 = 2 - \frac{f(2)}{f'(2)} = 2 - \frac{-1}{4} = 2.25,$$

and, continuing with the method,

$$r_3 = 2.25 - \frac{f(2.25)}{f'(2.25)} = 2.25 - \frac{(2.25)^2 - 5}{4.5} = 2.236\ldots$$

Considering the minimal amount of computation we did, we have arrived at a very nice approximation, $r \approx 2.236$. Note that $2.236^2 = 4.9997\ldots$ Were we to continue with the method we would, of course, arrive at an even finer approximation.

Exercises for §6.2

Use the CAS to assist in the computations below.

(1) Starting with $r_1 = 2$, find $9^{1/3}$ to 5 decimal places. **Hint:** Choose $f(x) = x^3 - 9$.

(2) Starting with $r_1 = 1$, use Newton's method to approximate (to 5 decimal places) a solution to the quadratic equation $x^2 + x - 3 = 0$. Find another starting value r_1 where Newton's method breaks down.

(3) Use Newton's method to approximate a real root of the equation $x^5 - 3x^2 + 2 = 0$.

(4) Use Newton's method to approximate a real root of the equation $x^7 - 3x^4 + 2 = 0$.

(5) Show that Newton's method must fail for the equation $x^2 + 1 = 0$.

Additional exercises for Chapter VI

(1) A plane travelling at a speed of 500 mph and an altitude of 2000 miles drops an object. How much time will it take the object to reach the ground? What is the velocity on impact? On your CAS graph the function representing the trajectory.

(2) Suppose the average velocity of a vehicle moving along a straight road for t minutes is given by $t^2 + t$ for all $t \geq 0$. What is the instantaneous velocity at $t = 0$?

(3) Use Newton's method to approximate the root of $x^3 - 3x^2 + 1 = 0$ which lies in the interval $0 \le x \le 1$. Use your CAS and choose a starting value $x_1 = .5$.

(4) Use Newton's method to approximate the root of $2x^3 - 4x + 1 = 0$ which lies in the interval $0 \le x \le 1$. Use your CAS and choose a starting value $x_1 = .5$. Graph the function.

Use Newton's method to approximate the coordinates of the point of intersection of the following curves. Graph each with your CAS.

(5) $y = \frac{1}{x^2}$ and $y = x^3 - 3 + \frac{2}{x^2}$.

(6) $y = x$ and $y = \cos(x)$.

(7) $y = x^2$ and $y = x^7 + 3x^5 - 4$.

(8) $y = 3x^5 + 2x^3 - 2$ and $y = 2x^5 - 17$.

(9) $xy = 1$ and $y = x^3 - 6x + 9$.

Solve using Newton's method; verify and graph the curves with your CAS.

(10) $x^5 + 9x + 1 = 0$.

(11) $\frac{4}{3}\pi x^3 + 2x - 3 = 0$.

(12) $x^4 + x^3 + x^2 = 1$.

(13) $x^5 + x^3 + 2x = 5$.

(14) $x^4 - 4x^2 - 4x = 8$.

(15) $x^5 + 3x^3 = 17$.

(16) $x^7 - 3x^5 + 9x^3 - 14x + 1 = 0$.

(17) A drunk driver's distance from the bar (he either travels north or south and the bar is the origin) is given by $s(t) = 3t^3 - 200t + 1$. How fast is the drunk driver going after 5 minutes? After 10 minutes?

(18) The distance of a projectile above the ground is given by $s(t) = at^2 + bt + c$ where $a < 0$ and t is time in seconds. What is the maximum height of the projectile? At what time does it reach its maximum height?

(19) A driver's distance, north or south of home (the origin) is given by $s(t) = at^3 + bt^2 + ct + d$ where $a \neq 0$ and t is time in hours. What is the driver's velocity? How often does the driver return home?

Chapter VII
The Rules of Calculus

7.1 The Primary Rules

Before we can proceed to develop further applications of the derivative we must find some efficient method to compute derivatives of a wide range of functions without resorting to the original definition. We begin with four primary rules: let c be a constant, $f(x)$ and $g(x)$ be arbitrary differentiable functions, and let N be a real number, then

$$\text{Rule 1.}\quad \frac{d}{dx}c = 0,$$

$$\text{Rule 2.}\quad \frac{d}{dx}x^N = Nx^{N-1},$$

$$\text{Rule 3.}\quad \frac{d}{dx}cf(x) = cf'(x),$$

$$\text{Rule 4.}\quad \frac{d}{dx}(f(x) + g(x)) = f'(x) + g'(x).$$

Before verifying these initial rules observe how useful they are.

Example [7.1] Let $f(x) = 3x^{92} - 4x^{13} + 6x - 3$. Our rules allow us to compute $f'(x)$ visually (with very little effort):

$$f'(x) = 3 \cdot 92x^{91} - 4 \cdot 13x^{12} + 6.$$

The example is completed by working out the arithmetic,

$$f'(x) = 276x^{91} - 52x^{12} + 6.$$

Example [7.2] The derivative of $f(x) = \pi x^{5.03} - x^{-2} + 111$ is simply $f'(x) = (5.03)\pi x^{4.03} + 2x^{-3}$.

Verification of Rules (1)–(4)

Rule 1. That the derivative of the constant function is zero was demonstrated in §5.2.

Rule 2. We examined the case of $N = 0, 1, 2, 3$ in §5.3. We defer the general proof to §17.1 at which point we will be in a position to utilize logarithms in the verification.

Rule 3. This rule follows easily from the definition of the derivative and the basic properties of limits:

$$\begin{aligned}\frac{d(cf(x))}{dx} &= \lim_{h\to 0} \frac{cf(x+h) - cf(x)}{h} \\ &= c \cdot \lim_{h\to 0} \frac{f(x+h) - f(x)}{h} \\ &= cf'(x).\end{aligned}$$

Rule 4. Here again the original definition of the derivative and the basic properties of limits come into play:

$$\begin{aligned}\frac{d(f(x) + g(x))}{dx} &= \lim_{h\to 0} \frac{[f(x+h) + g(x+h)] - [f(x) + g(x)]}{h} \\ &= \lim_{h\to 0} \frac{[f(x+h) - f(x)] + [g(x+h) - g(x)]}{h} \\ &= \lim_{h\to 0} \frac{f(x+h) - f(x)}{h} + \lim_{h\to 0} \frac{g(x+h) - g(x)}{h} \\ &= f'(x) + g'(x).\end{aligned}$$

That the *converse* of Rule (1) holds is an important result which we shall need later on.

Proposition [7.3] *If $f(x)$ is a differentiable function and $f'(x) = 0$ for all x, then $f(x) = c$ for some constant c, i.e., f is the constant function.*

Although the details of the proof are omitted, the intuition behind the proposition is simple. Since the derivative of f is 0 at all points, the tangent lines to the curve must all be horizontal. There is only one way this can happen, the curve must be a horizontal line itself, i.e., $f(x) = c$ for some c.

Exercises for §7.1

Using the primary rules compute the derivatives of the following functions visually. Check your answers with your CAS.

(1) $3x^5 - 2x^2 + 1$.

(2) $17x^2 + 2$.

(3) $x^{100} + 57$.

(4) $x^{15} + 3x^{-7} - 11$.

(5) $2x^{-5} + 3x^{-1} + 5$.

(6) $x^{-100} + 8$.

(7) $-2x^{1.1} + 5x^{7.009}$.

(8) $2x^{-7.81} + \pi x^5$.

(9) $(3.1)x^{-0.001} + 11$.

(10) $-(1.2)x^{-0.001} + 1590$.

(11) $\frac{1}{x^2} + 5x + 3$.

(12) $\frac{7}{x^{100}} - \frac{2}{x^4} + 1.3$.

(13) $x^{1/3}$.

(14) $\frac{1}{\sqrt{x}} + x^{3/4}$.

(15) $2x^{-5/4} + \frac{3}{\sqrt[3]{x}}$.

7.2 The Product and the Quotient Rules

We now present three rules which are essential to computing the derivatives of more complex functions. While the form these rules take is not entirely intuitive, we do begin to sense the nature of the operation of differentiation. In Rule 6 and Rule 7 we assume that $g(x) \neq 0$.

Rule 5. (Product Rule) $\frac{d}{dx}\Big(f(x)\cdot g(x)\Big) = f'(x)\cdot g(x) + f(x)\cdot g'(x),$

Rule 6. $\frac{d}{dx}\left(\frac{1}{g(x)}\right) = -\frac{g'(x)}{g(x)^2},$

Rule 7. (Quotient Rule) $\frac{d}{dx}\left(\frac{f(x)}{g(x)}\right) = \frac{f'(x)\cdot g(x) - f(x)\cdot g'(x)}{g(x)^2}.$

These rules allow us to compute the derivatives of a large class of functions. We begin with a few examples.

Example [7.4] The derivative of $(x^3+2)\cdot(x^{19}-3x+1)$ can be quickly seen to be

$$3x^2(x^{19}-3x+1)+(x^3+2)(19x^{18}-3).$$

Example [7.5] The derivative of $1/(x^4+2)$ is given by

$$-\frac{4x^3}{(x^4+2)^2}.$$

Example [7.6] The derivative of the quotient $\frac{x^2}{x^4+2}$ is

$$\frac{2x(x^4+2)-x^2(4x^3)}{(x^4+2)^2}.$$

Verification of Rules (5)–(7)

Rule 5. To prove the product rule we will need the help of a little trick along with the basic techniques we used when proving the previous rules.

$$\frac{d\Big(f(x)\cdot g(x)\Big)}{dx} = \lim_{h\to 0}\frac{f(x+h)\cdot g(x+h) - f(x)g(x)}{h}$$

$$= \lim_{h\to 0}\frac{f(x+h)\cdot g(x+h) - \overbrace{f(x)\cdot g(x+h) + f(x)\cdot g(x+h)}^{\text{The Trick: Add Zero}} - f(x)g(x)}{h}$$

$$= \lim_{h\to 0}\frac{f(x+h)\cdot g(x+h) - f(x)g(x+h)}{h} + \lim_{h\to 0}\frac{f(x)g(x+h) - f(x)g(x)}{h}$$

$$= \lim_{h\to 0} g(x+h)\cdot\frac{f(x+h)-f(x)}{h} + \lim_{h\to 0} f(x)\cdot\frac{g(x+h)-g(x)}{h}$$

$$= g(x)\cdot f'(x) + f(x)\cdot g'(x).$$

Rule 6. We leave this proof to the reader.

Rule 7. Since $\frac{f(x)}{g(x)} = f(x)\cdot\frac{1}{g(x)}$ we can verify this rule by combining rules 5 and 6.

Exercises for §7.2

For each of the following pairs of functions $f(x)$ and $g(x)$, compute the derivative of $f(x)\cdot g(x)$, $g(x)^{-1}$, and $f(x)/g(x)$. Verify your answers with your CAS.

(1) $f(x) = x^7 + 3,\ g(x) = 2x^3 + 1.$

(2) $f(x) = 3x^2 - 5,\ g(x) = 3x^6 - x.$

(3) $f(x) = \frac{2}{x^4} + 1,\ g(x) = 11x^3 + 2x + 1.$

(4) $f(x) = 10x^{0.1} + 3,\ g(x) = x^{1.9} - 2.$

Compute the derivatives of the following functions visually.

(5) $\frac{\sqrt{x}+3}{x^5-1}.$

(6) $\frac{1}{\sqrt[3]{x}+2x^4}.$

(7) $\frac{x^{2/3}-11}{x^{2/3}+11}.$

(8) $\frac{3\sqrt{x}+x^{17}}{4x^5-2x+3}.$

(9) $\frac{1/x-3/x^2}{x^2+1}.$

(10) $\frac{x^{-0.3}+4x^{1.9}}{\sqrt{x}+1}.$

(11) Expressing x^2 as $x^2 = x\cdot x$, use the product rule to compute the derivative of x^2.

(12) Expressing x^3 as $x^2 = x^2 \cdot x$, use the product rule to compute the derivative of x^3.

(13) Assuming the identity

$$\frac{d}{dx}x^N = Nx^{N-1},$$

for some integer $N > 1$, use the product rule to prove that

$$\frac{d}{dx}x^{N+1} = (N+1)x^N.$$

Observe that exercises (11)–(13) yield a proof by induction of Rule 2 for all positive integers N.

(14) Verify the identity $(f \cdot g \cdot h)' = f'gh + fg'h + fgh'$.

(15) Use the identity in exercise (14) to compute the derivative of $(x^{10} + 18) \cdot \sqrt{x-1} \cdot (1/x^4 + 6x - 2)$.

(16) Use the identity in exercise (14), along with the quotient rule, to compute the derivative of the function $1/(1-x)(1-x^2)(1-x^3)$.

7.3 The Chain Rule

Having obtained formulae for the derivatives of sums and products, the derivative of the composite of two function remains to be tackled. Since the composition of functions has no analog (i.e., in the algebra of real numbers) the rules which follow are not at all obvious. The verifications are cumbersome and we omit them.

$$\text{Rule 8. (Chain Rule)} \quad (f \circ g)'(x) = f'(g(x)) \cdot g'(x),$$

$$\text{Rule 9. (Alternate Form)} \quad \frac{d(f \circ g)}{dx} = \frac{df}{dg(x)} \cdot \frac{dg}{dx}.$$

Remark: In general, recall that the notation $\frac{df}{dx}$ simply means the derivative of f at the point x, i.e., $f'(x)$. We can of course consider $\frac{df}{dw}$ where w is some other point of interest. Thus the notation which appears in the alternate form of the chain rule, $\frac{df}{dg(x)}$, is by definition the derivative of f at the point $g(x)$.

Example [7.7] Let $f(x) = x^{25} + 2x^{10} - 1$, and $g(x) = x^{10} - 11$. Compute $(f \circ g)'(x)$.

Observing that $f'(x) = 25x^{24} + 20x^9$ and $g'(x) = 10x^9$, the chain rule allows us to easily compute the desired derivative:

$$\begin{aligned}(f \circ g)'(x) &= f'(g(x)) \cdot g'(x) \\ &= (25g(x)^{24} + 20g(x)^9) \cdot 10x^9 \\ &= (25(x^{10} - 11)^{24} + 20(x^{10} - 11)^9) \cdot 10x^9.\end{aligned}$$

If we consider the effort it would take to differentiate $(f \circ g)(x) = (x^{10} - 11)^{25} + 2(x^{10} - 11)^{10} - 1$ directly, the power and utility of the chain rule becomes obvious.

Example [7.8] Let $f(x) = x^{\frac{1}{3}} = \sqrt[3]{x}, \ \ g(x) = x^{43} - 3x^{11} + 2$. Compute $(f \circ g)'(x)$.

Since $f'(x) = \frac{1}{3}x^{-\frac{2}{3}}$ and

$$f'\Big(g(x)\Big) = \frac{1}{3}\Big(g(x)\Big)^{-\frac{2}{3}} = \frac{1}{3}\left(x^{43} - 3x^{11} + 2\right)^{-\frac{2}{3}},$$

the chain rule tells us that the derivative in question is

$$\frac{d}{dx}\left((x^{43} - 3x^{11} + 2)^{\frac{1}{3}}\right) = \frac{1}{3}\left(x^{43} - 3x^{11} + 2\right)^{-\frac{2}{3}} \cdot \left(43x^{42} - 33x^{10}\right).$$

The following proposition is perhaps the most frequently used consequence of the chain rule.

Proposition [7.9] *Let $g(x)$ be a differentiable function and let N be a real number such that $g(x)^N$ is well defined. Then*

$$\frac{d}{dx}g(x)^N = Ng(x)^{N-1} \cdot g'(x).$$

Proof: Letting $f(x) = x^N$ we see that $f(g(x)) = g(x)^N$. Since

$$f'(x) = Nx^{N-1},$$

we can now apply the chain rule to obtain the desired formula:

$$\begin{aligned}\frac{d}{dx}g(x)^N &= (f \circ g)'(x) \\ &= f'(g(x)) \cdot g'(x) = Ng(x)^{N-1} \cdot g'(x).\end{aligned}$$

Example [7.10] What is the derivative of $f(x) = (x^3 + 2x + 1)^{1000}$?

In this example we can apply Proposition [7.9] to compute the derivative with little effort:

$$1000(x^3 + 2x + 1)^{999} \cdot (3x^2 + 2).$$

(It is worth considering how horrific this computation would be without the chain rule!)

Example [7.11] Compute the derivative of $f(x) = \sqrt{x^5 - 3}$.

To apply Proposition [7.9] in this example we first rewrite $f(x)$ in the form $f(x) = (x^5 - 3)^{\frac{1}{2}}$. Having done this we see that

$$f'(x) = \frac{1}{2}(x^5 - 3)^{-\frac{1}{2}} \cdot 5x^4$$
$$= \frac{5x^4}{2\sqrt{x^5 - 3}}.$$

Exercises for §7.3

Compute the derivatives of the functions in exercises (1)–(15) without simplifying your answers. Check your results with the CAS.

(1) $(2x^{19} - 2x^2 + 1)^{10}$.

(2) $(8x^4 - 2)^5$.

(3) $(2x^3 + 1)^{-10}$.

(4) $\sqrt{x^{11} - 3x + 7}$.

(5) $(x^2 + 1)^{-2/3}$.

(6) $\sqrt[4]{\frac{x-1}{x+1}}$.

(7) $(2x^{40} - 3.01)^{-0.23}$.

(8) $(x^\pi + 1)^\pi$.

(9) $\left(\frac{x^2+1}{x}\right)^{21}$.

(10) $\frac{1}{\sqrt{x^3-2}}$.

(11) $\frac{5x+1}{(x^4-3x+1)^{10}}$.

(12) $(3 + \sqrt[3]{x})^{500}$.

(13) $\sqrt{1 + \sqrt{\sqrt{x}}}$.

(14) $\sqrt[3]{1 + \sqrt[3]{1 + \sqrt[3]{x}}}$.

(15) $\sqrt{2 + \sqrt{\sqrt{1 + \sqrt{x}}}}$.

Calculate $(f \circ g)'(x)$ for the following pairs of functions. Verify your answers with the CAS.

(16) $f(x) = 2x^3 - 3x + 1, \quad g(x) = 5x^2 + x - 10.$

(17) $f(x) = (x + 1)^5, \quad g(x) = (\sqrt{x} + 1).$

(18) $f(x) = \sqrt{x^2 - 3}, \quad g(x) = x^3 - 2.$

(19) $f(x) = x^6 - x^5 + x^4 - x^3, \quad g(x) = x^{2/3} - 3.$

(20) $f(x) = (x^3 + 1)^{100}, \quad g(x) = \frac{1}{\sqrt{x-2}}.$

(21) $f(x) = (x^3 + 1)^{-100}, \quad g(x) = x^{1.1} + 3.$

7.4 Derivatives of Trigonometric Functions

There are two basic rules which govern differentiation of trigonometric functions.

$$\text{Rule 10.}\quad \frac{d}{dx}\Big(\sin(x)\Big) = \cos(x),$$

$$\text{Rule 11.}\quad \frac{d}{dx}\Big(\cos(x)\Big) = -\sin(x).$$

Before verifying these final identities, let's observe that, by combining these two rules with our previous ones, it is relatively easy to differentiate almost any trigonometric function.

Example [7.12] Show that $\frac{d}{dx}\Big(\tan(x)\Big) = \sec^2(x)$.

Since $\tan(x) = \frac{\sin(x)}{\cos(x)}$ we can evaluate the derivative of $\tan(x)$ using the quotient rule together with rules 10 an 11:

$$\begin{aligned}\frac{d}{dx}\left(\frac{\sin(x)}{\cos(x)}\right) &= \frac{\cos(x)\cdot\cos(x) - \sin(x)\cdot(-\sin(x))}{(\cos(x))^2} \\ &= \frac{\cos^2(x) + \sin^2(x)}{\cos^2(x)} \\ &= \frac{1}{\cos^2(x)} = \sec^2(x).\end{aligned}$$

Example [7.13] Evaluate the derivative of $f(x) = \cos(x^4 + 3x - 1)$.

In order to use the chain rule here we set $f(x) = \cos(x)$ and $g(x) = x^4 + 3x + 1$. The derivatives of f and g are $f'(x) = -\sin(x)$ and $g'(x) = 4x^3 + 3$, and finally

$$\frac{d(f\circ g)}{dx} = -\sin(x^4 + 3x - 1)\cdot(4x^3 + 3).$$

In order to prove our trigonometric differentiation rules we will require an important lemma.

Lemma [7.14] *The following limits hold;*

$$\lim_{h\to 0}\frac{\sin(h)}{h}=1,\quad \lim_{h\to 0}\frac{\cos(h)-1}{h}=0.$$

Proof: That $\lim_{h\to 0}\frac{\sin(h)}{h}=1$ was demonstrated in §4.2. The second limit requires a small amount of maneuvering:

$$\lim_{h\to 0}\frac{\cos(h)-1}{h} = \lim_{h\to 0}\frac{\cos(h)-1}{h}\cdot\overbrace{\frac{\cos(h)+1}{\cos(h)+1}}^{\substack{\text{The Trick:}\\ \text{Mult. by 1}}} = \lim_{h\to 0}\frac{\cos^2(h)-1}{h\cdot(\cos(h)+1)}$$

$$= \lim_{h\to 0}\frac{-\sin^2(h)}{h\cdot(\cos(h)+1)}$$

$$= \left(\lim_{h\to 0}\frac{-\sin(h)}{h}\right)\cdot\left(\lim_{h\to 0}\frac{\sin(h)}{\cos(h)+1}\right)$$

$$= -1\cdot\left(\frac{0}{1+1}\right)=0.$$

Verification of Rules (10) and (11)

Notice first that Rule 11 follows from Rule 10 and the formulae

$$\cos(x)=\sin(x+\frac{\pi}{2}),\quad \cos(x+\frac{\pi}{2})=-\sin(x),$$

since these identities allow us to deduce

$$\frac{d}{dx}\cos(x)=\frac{d}{dx}\sin(x+\frac{\pi}{2})=\cos(x+\frac{\pi}{2})=-\sin(x).$$

When we recall the classical trigonometric formula

$$\sin(x+h)=\sin(x)\cos(x)+\cos(x)\sin(h),$$

the proof of Rule 10 proceeds as follows:

$$\begin{aligned}
\frac{d}{dx}\Big(\sin(x)\Big) &= \lim_{h\to 0} \frac{\sin(x+h)-\sin(x)}{h} \\
&= \lim_{h\to 0} \frac{[\sin(x)\cos(h)+\cos(x)\sin(h)]-\sin(x)}{h} \\
&= \lim_{h\to 0} \frac{\sin(x)\cos(h)-\sin(x)}{h} + \lim_{h\to 0} \frac{\cos(x)\sin(h)}{h} \\
&= \sin(x)\lim_{h\to 0} \frac{\cos(h)-1}{h} + \cos(x)\lim_{h\to 0} \frac{\sin(h)}{h} \\
&= 0 + (\cos(x))\cdot 1 \\
&= \cos(x).
\end{aligned}$$

Exercises for §7.4

Evaluate the derivatives of the following functions. Check your answers with the CAS.

(1) $\cos(x^4+3x^2-2)$.

(2) $\sin(2x^{19}-3x^5-2x+11)$.

(3) $\tan(x^3+3x^2-6)$.

(4) $\cot(x^2+2)$.

(5) $\cos\left(\frac{x-1}{x+1}\right)$.

(6) $\sin\left(\frac{x-1}{x+1}\right)$.

(7) $\cos(\cos(x))$.

(8) $\cos(\cos(\cos(x)))$.

(9) $\sin(\sin(x^2+2))$.

(10) $\sqrt{(\sin(x))^3+3x}$.

(11) $\left(\sin(x^4+2)\right)^{3/4}$.

(12) $\cos(\cos(x^{10}-3x^2+2))$.

(13) $\tan(\sqrt{x^3+2})$.

(14) $\cot(\sqrt[3]{x^3-3x+2})$.

(15) $\frac{\sin(x)}{\cos(x)-\sin^2(x)}$.

(16) $\frac{\cos(x)+3}{\sqrt{x+2}}$.

(17) $\sin(x)\cdot(\cos(x))^3$.

(18) $(1+3\cos(x))^{20}$.

(19) $\left(\cos(x^2+1)\right)^3$.

(20) $\frac{1-\cos(x)}{1+\cos(x)}$.

Additional exercises for Chapter VII

---

(1) Using the rules of derivation, obtain the derivative of $f(x) = a_n x^n + a_{n-1}x^{n-1} + \cdots + a_0$ where $a_n, a_{n-1}, \cdots, a_0$ are constants independent of x. Verify your answer with your CAS.

(2) Let $f(x) = 6x^4 - x^2$. Find the values of x for which $f'(x) = 0$. Use your CAS to graph $f'(x)$.

(3) Let $f(x) = x^4 - 1$.

(a) What geometric property does this function have? What does this property indicate about the graph of f?

(b) If you have derived the derivative of f at the point x can you deduce the derivative at the point $-x$ without having to recompute it?

(c) Use your CAS to graph $f'(x)$. What geometric property does $f'(x)$ have?

(d) Can you generalize the answers of the questions above using the limit definition of derivative?

(4) Let $f(x) = x^5 + 2x^3 - x$. Answer the same three questions for this function as you did for the problem (3).

(5) Prove the rule of differentiation of a composition of functions which is $(f \circ g)' = f'(g(x)).g'(x)$ by use of the limit definition of derivative. Obtain $(f \circ g)'$ for $f(x) = 3x+4, g(x) = x^2+4x$. Verify your result with your CAS.

(6) Suppose $f(x)$ and $g(x)$ are functions such that for all x in their common domain, $f'(x) = g'(x)$. Can you conclude that $f(x) = g(x)$? (**Hint:** take $h(x) = f(x) - g(x)$ and use the rules of differentiation.

(7) On your CAS obtain $f'(x)$ for $f(x) = \frac{1}{x^2-a}$ where a is a positive constant. The second derivative $f^{(2)}(x)$ is defined as the derivative of $f'(x)$ and so on, the nth derivative of f is $f^{(n)}(x)$. Obtain the first 6 derivatives of f given above, using your CAS. Can you observe a pattern? Can you deduce a formula for $f^n(x)$?

(8) Let $f(x) = (x-1)^2(x^3 - 4x - 2)$. The value $x = 1$ is said to be a double root of f. On your CAS, graph $f(x)$ and $f'(x)$. What do you notice about $f(1)$ and $f'(1)$? Obtain $f^{(2)}(x)$ and then $f^{(2)}(1)$.

Calculate $(f \circ g)'$, for the following functions.

(9) $f(x) = \cos(x^2 + 3), g(x) = \frac{1}{(x-3)}$.

(10) $f(x) = x^3 - 9x^2 + 27x - 27, \; g(x) = x^{1/3} + 3$.

(11) $f(x) = \frac{3x-12}{x-2}, \; g(x) = \frac{2x-12}{x-3}$.

(12) Suppose $(f \circ g)' = 1$. Can we conclude that f and g are inverses?

(13) Suppose $(f \circ g)' = 0$. Can we conclude that either f or g is constant?

In the following problems, find the equation of the tangent line to the given curve at the given point.

(14) $y = \cos(\pi(x^2 - 1)), \; \left(\frac{\sqrt{2}}{2}, 0\right)$

(15) $y = \tan(\sqrt{x}), \; \left(\frac{\pi^2}{9}, \sqrt{3}\right)$

(16) $y = \frac{3-x}{2+x}, \; \left(1, \frac{2}{3}\right)$

(17) $y = \sqrt{1 + \sqrt{x}}, \; (4, \sqrt{3})$

(18) $y = \left(\sin\left(\frac{\pi}{3}(x^3 + 2x + 1)\right)\right)^{10}, \; \left(0, \left(\frac{\sqrt{3}}{2}\right)^{10}\right)$.

Chapter VIII

Implicitly Defined Functions and their Derivatives

8.1 Implicit Functions

Suppose we are given an equation involving x and y. If it is possible to solve for y in terms of x, the we have defined y as a function of x. The function so obtained is said to be **implicitly** defined by the equation.

Example [8.1] Consider the equation

$$-x^2 + 2x + y^2 = 3.$$

Solving for y in terms of x yields:

$$y = \sqrt{x^2 - 2x + 3}.$$

Remark: As we saw in §1.2, Example [1.8], there are many possible square root functions. Hence the equation $-x^2 + 2x + y^2 = 3$ implicitly defines many possible functions. A simple convention which is often used is to always take the positive square root function.

8.2 Implicit Differentiation

Implicit differentiation is a method to quickly compute the derivative of an implicitly defined function.

Example [8.2] What is the derivative of the implicitly defined function in Example [8.1]?

One way to solve this problem is to simply differentiate $y = \sqrt{x^2 - 2x + 3}$ and obtain

$$\frac{1}{2}(x^2 - 2x + 3)^{-\frac{1}{2}} \cdot (2x - 2).$$

We now indicate another method of solving this problem which is called the method of **implicit differentiation**. Begin by considering the original equation

$$-x^2 + 2x + y^2 = 3.$$

When we solve for y in terms of x we obtain $y = f(x)$ for some function $f(x)$ which in this case is $f(x) = \sqrt{x^2 - 2x + 3}$. Remarkably, the four step method of implicit differentiation (detailed below) does not require knowing what the function f is. All we need to know is that some function f exists.

Step 1: Replace y by $f(x)$ in the original equation. This yields:

$$-x^2 + 2x + f(x)^2 = 3.$$

Step 2: Differentiate both sides of this equation with respect to x using the rules for calculus (since both sides of the above equation are themselves functions of x this is valid). We obtain

$$-2x + 2 + 2f(x) \cdot f'(x) = 0.$$

Step 3: Solve for $f'(x)$. This requires nothing more than simple algebra:

$$\begin{aligned} 2f(x) \cdot f'(x) &= 2x - 2 \\ f'(x) &= \frac{2x - 2}{2f(x)} = \frac{x - 1}{f(x)}. \end{aligned}$$

Step 4: Replace $f(x)$ by y to obtain:

$$y' = \frac{x - 1}{y}.$$

Remark: This agrees with the previously found solution $\frac{1}{2}(x^2-2x+3)^{-\frac{1}{2}}\cdot(2x-2)$ after we replace y by $\sqrt{x^2-2x+3}$.

Example [8.3] Compute the derivative y' (as an expression involving both x and y) of the implicitly defined function

$$y^7 - 8x^4 \cdot y^3 + 2y + x^3 = 3.$$

Step 1: Replace y by $f(x)$ to obtain

$$f(x)^7 - 8x^4 \cdot f(x)^3 + 2f(x) + x^3 = 3.$$

Step 2: Differentiate both sides of the equation with respect to x:

$$7f(x)^6 \cdot f'(x) - 32x^3 \cdot f(x)^3 - 24x^4 \cdot f(x)^2 \cdot f'(x) + 2f'(x) + 3x^2 = 0.$$

Step 3: Solve for $f'(x)$:

$$7f(x)^6 \cdot f'(x) - 24x^4 \cdot f(x)^2 \cdot f'(x) + 2f'(x) = 32x^3 \cdot f(x)^3 - 3x^2$$

$$\left[7f(x)^6 - 24x^4 \cdot f(x)^2 + 2\right] \cdot f'(x) = 32x^3 \cdot f(x)^3 - 3x^2,$$

and hence

$$f'(x) = \frac{32x^3 \cdot f(x)^3 - 3x^2}{7f(x)^6 - 24x^4 \cdot f(x)^2 + 2}.$$

Step 4: Replace $f(x)$ by y to obtain the solution:

$$y' = \frac{32x^3 \cdot y^3 - 3x^2}{7y^6 - 24x^4 \cdot y^2 + 2}.$$

Remark: In this problem it is not possible to find $f(x)$. The equation cannot be solved!

Question: What is the derivative of the implicitly defined function of example [8.3] at the point $(1,1)$?

Answer: By simply plugging in the given values for $x = 1$ and $y = 1$ into the solution obtained in step 4 above we obtain the value of the derivative at the point $(1,1)$:

$$y' = \frac{32-3}{7-24+2} = -\frac{29}{15}.$$

Exercises for §8.2

In exercises (1)–(9), compute y' for the following functions in two different ways. First, by solving for y as a function of x, and secondly by the 4 step method of implicit differentiation.

(1)	$x + y^2 = 1$	**(4)**	$xy^2 = 1$	**(7)**	$y^2 + 2xy + x^2 = 1$
(2)	$x^2 + y^2 = 1$	**(5)**	$xy^2 = 2$	**(8)**	$x^2y^3 = x - 2$
(3)	$xy = 1$	**(6)**	$xy^4 = 3$	**(9)**	$y^2 - 3x^2y - 2 = 0$

In exercises (10)–(15) use the 4 step algorithm to compute y' (as an expression involving both x and y) for the equations:

(10)	$x \cdot \cos(x) + y \cdot \cos(x) = 13$	**(13)**	$e^{xy^2} + 2\cos(x^3y) = 4$
(11)	$y^7 - 2x^3y^4 + 3xy^2 + x^5 = 3$	**(14)**	$\ln\left(\frac{x-y}{x+y}\right) = y^3$
(12)	$x^2y + y\tan(x) = y^3$	**(15)**	$2^{x^2+y^2} = 17$

(16) Compute y' at the point $\left(\frac{1}{\sqrt[3]{2}}, \frac{1}{\sqrt[3]{2}}\right)$ for the function defined implicitly by the equation $x^3 + y^3 = 1$.

(17) Find the equation of the tangent line to the curve $(\cos(\pi x))^2 y + y^3 = x + 10$ at the point $x = 0,\ y = 2$.

(18) Find the equation of the tangent line to the curve $(\sin(2x)) \cdot xy + y^4 = x + 1$ at the point $x = 0,\ y = 1$.

(19) Show that the circles given by the equations $x^2 + y^2 - 2x - 4y - 20 = 0$, and $x^2 + y^2 - 20x - 28y + 196 = 0$ are tangent at the point $(4, 6)$.

8.3 The Exponential, the Natural Logarithm, and the Hyperbolic Functions

The exponential function a^x was defined in §1.4 in the case x is a rational number. We can extend the definition to the case of x being irrational by setting

$$a^x = \lim_{x' \to x} a^{x'},$$

where the limit is over rational numbers approaching x. For example, if $x = \pi = 3.14159\ldots$ we may let

$$x' = 3,\ 3.1,\ 3.14,\ 3.141,\ 3.1415,\ 3.14159, \ldots \longrightarrow \pi$$

run through rational values approaching π.

Question: What is the derivative of a^x?

By definition,

$$\begin{aligned}\frac{d}{dx}a^x &= \lim_{h\to 0}\frac{a^{x+h}-a^x}{h}\\ &= \lim_{h\to 0}\frac{a^x\cdot a^h - a^x}{h}\\ &= a^x\cdot\lim_{h\to 0}\frac{a^h-1}{h}.\end{aligned} \tag{8.1}$$

We have thus reduced the problem of computing the derivative of a^x to the problem of computing $\lim\limits_{h\to 0}\frac{a^h-1}{h}$.

Definition: *The number e is defined to be the unique real number with the property that*

$$\lim_{h\to 0}\frac{e^h-1}{h} = 1. \tag{8.2}$$

The proof of the existence and uniqueness of e is deferred to §17.1. Nevertheless, it may be noted that $e > 1$, because otherwise $\frac{e^h-1}{h}$ would tend to a nonpositive number. Further it can be shown that $e \approx 2.718\ldots$. The graph of $f(x) = e^x$ is depicted below.

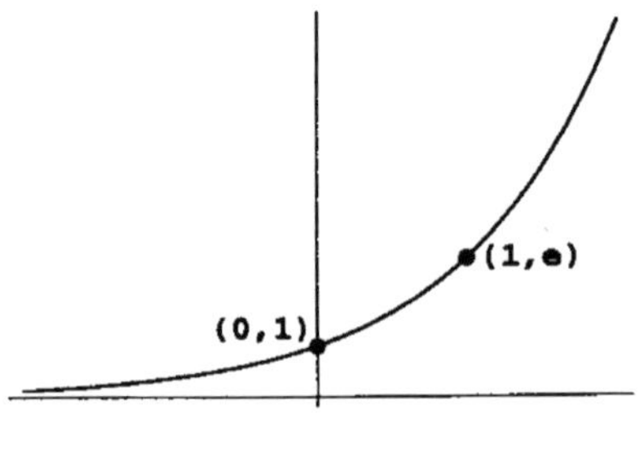

Combining equations (8.1) with (8.2) allows us to deduce the following proposition.

Proposition [8.4]

$$\frac{d}{dx}e^x = e^x.$$

Example [8.5] Evaluate $\frac{d}{dx}e^{(x^4-2x+1)}$.

This can easily be accomplished with the chain rule:

$$\frac{d}{dx}e^{(x^4-2x+1)} = (4x^3 - 2)e^{(x^4-2x+1)}.$$

We next consider the inverse of the exponential function which is another striking and essential example.

> **Definition:** *The inverse function of e^x is defined to be $\ln(x)$,* ***the natural logarithm function****. It satisfies the identities*
>
> $$\ln(e^x) = x$$
>
> $$e^{\ln(x)} = x.$$

The graph of the natural logarithm function is pictured below.

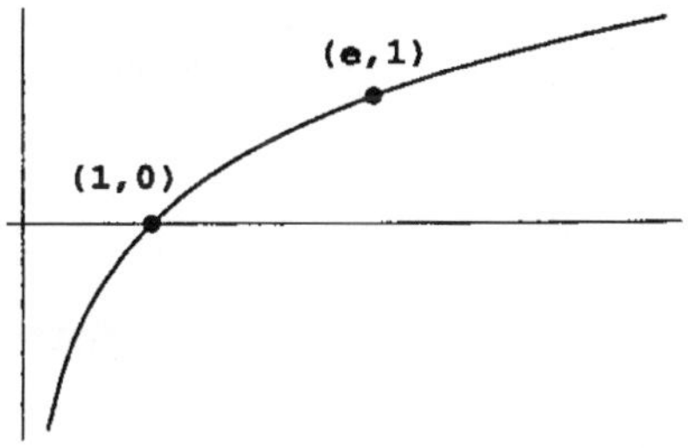

The equation

$$e^{\ln(x)} = x,$$

actually defines $\ln(x)$ implicitly as a function of x. When we differentiate both sides of this equation with respect to x (using the chain rule and Proposition [8.4]), we obtain

$$\frac{d}{dx}e^{\ln(x)} = e^{\ln(x)} \cdot \left(\frac{d}{dx}\ln(x)\right) = x \cdot \left(\frac{d}{dx}\ln(x)\right) = 1,$$

and we have proved:

Proposition [8.6]

$$\frac{d}{dx}\ln(x) = \frac{1}{e^{\ln(x)}} = \frac{1}{x}.$$

An alternate description of e is given in the following proposition.

Proposition [8.7] (Alternate definition of the number e) *The number e may be expressed as the limit*

$$e = \lim_{h\to 0}(1+h)^{\frac{1}{h}}.$$

Proof: Since $(1+h)^{\frac{1}{h}} = e^{\frac{\ln(1+h)}{h}}$, it suffices to show that $\lim_{h\to 0}\frac{\ln(1+h)}{h} = 1$. But since $\ln(1) = 0$, we see that

$$\begin{aligned}\lim_{h\to 0}\frac{\ln(1+h)}{h} &= \lim_{h\to 0}\frac{\ln(1+h)-\ln(1)}{h}\\ &= \frac{d}{dx}\ln(x)\Big]_{x=1} = 1.\end{aligned}$$

We are now in a position to differentiate the general exponential function $f(x) = a^x$.

Proposition [8.8] *Fix a positive real number a. Then*

$$\frac{d}{dx}a^x = a^x \cdot (\ln a).$$

Proof: Since $a = e^{\ln(a)}$ and thus $a^x = e^{\ln(a)\cdot x}$, we may derive the desired result by differentiating both sides of the latter identity:

$$\begin{aligned}\frac{d}{dx}a^x &= \frac{d}{dx}e^{\ln(a)\cdot x}\\ &= \ln(a)\cdot e^{\ln(a)\cdot x}\\ &= \ln(a)\cdot a^x.\end{aligned}$$

Example [8.9] Evaluate $\frac{d}{dx}x^x$.

In order to transform x^x into a form we can manipulate we begin with the fact $x = e^{\ln(x)}$, and then raising both sides to the x power we obtain the identity $x^x = e^{\ln(x)\cdot x}$. We can then differentiate our function using the chain rule:

$$\begin{aligned}\frac{d}{dx}x^x &= \frac{d}{dx}e^{\ln(x)\cdot x}\\ &= \left(\frac{d}{dx}\Big(\ln(x)\cdot x\Big)\right)\cdot e^{\ln(x)\cdot x}\\ &= \left(\frac{1}{x}\cdot x+\ln(x)\cdot 1\right)x^x\\ &= \Big(1+\ln(x)\Big)x^x.\end{aligned}$$

Example [8.10] Evaluate $\frac{d}{dx}(\ln(x)+x^4)^{10}$.

With the chain rule we compute

$$\frac{d}{dx}(\ln(x)+x^4)^{10} = 10(\ln(x)+x^4)^9\cdot\left(\frac{1}{x}+4x^3\right).$$

Example [8.11] Evaluate $\frac{d}{dx}(x^2+1)^x$.

Since $e^{\ln(Y)} = Y$ for any Y, we may write $x^2+1 = e^{\ln(x^2+1)}$ and

$$(x^2+1)^x = e^{\left(\ln(x^2+1)\right)\cdot x}.$$

Differentiating with the chain rule yields:

$$\begin{aligned}\frac{d}{dx}(x^2+1)^x &= \frac{d}{dx}e^{\left(\ln(x^2+1)\right)\cdot x}\\ &= \left(\frac{2x}{(x^2+1)}+\ln\left(x^2+1\right)\right)e^{\left(\ln(x^2+1)\right)\cdot x}\\ &= \left(\frac{2x}{(x^2+1)}+\ln\left(x^2+1\right)\right)(x^2+1)^x.\end{aligned}$$

Remark: It should be observed that, in general, to differentiate $f(x) = g(x)^x$ we first express $f(x)$ as

$$f(x) = e^{\ln(g(x))\cdot x},$$

and then employ the chain rule.

Hyperbolic Functions

It has been found that certain combinations of exponential functions behave very much like the classical trigonometric functions. The names of these functions are, for convenience, obtained by putting the letter h (which stands for hyperbolic) at the end of the name of a trigonometric function.

Definition: *The basic **hyperbolic functions** are defined as follows:*

$$\begin{aligned} \sinh(x) &= \frac{e^x - e^{-x}}{2} \\ \cosh(x) &= \frac{e^x + e^{-x}}{2} \\ \tanh(x) &= \frac{\sinh(x)}{\cosh(x)}. \end{aligned}$$

It can be immediately verified that

$$\frac{d}{dx}\sinh(x) = \cosh(x), \quad \frac{d}{dx}\cosh(x) = \sinh(x),$$

and

$$\frac{d}{dx}\tanh(x) = \frac{1}{(\cosh(x))^2}.$$

To see further analogies between the hyperbolic functions and the classical trigonometric functions see exercises (13)–(16) of this section.

Exercises for §8.3

Evaluate the derivatives of the following functions. Check your answers with your CAS.

(1)	$y = e^{x^3}$.	**(5)**	$y = x^{x^2}$.	**(9)**	$y = \left(x^2 + \cos(x)\right)^x$.
(2)	$y = e^{5x^2+3}$.	**(6)**	$y = \ln(3x^2 + 2)$.	**(10)**	$y = \ln(x)/\sqrt{e^x}$.
(3)	$y = 2^{x^2}$.	**(7)**	$y = \left(\ln(2x^4 - 1)\right)^2$.	**(11)**	$y = \cos(\ln(x^2 + 1))$.
(4)	$y = x^{2x}$.	**(8)**	$y = (\ln(x))^x$.	**(12)**	$y = \ln\left(\sin(e^x)\right)$.

(13) Prove that $(\cosh(x))^2 - (\sinh(x))^2 = 1$.

(14) Prove that $1-(\tanh(x))^2 = \frac{1}{(\cosh(x))^2}$.

(15) Establish the addition formulae:

$$\sinh(x+y) = \sinh(x)\cosh(y) + \cosh(x)\sinh(y)$$

$$\cosh(x+y) = \cosh(x)\cosh(y) + \sinh(x)\sinh(y)$$

(16) Evaluate the derivatives

$$\frac{d}{dx}\sinh(x^2), \quad \frac{d}{dx}\tanh(\ln(x)), \quad \frac{d}{dx}\cosh(x^{-3}).$$

(17) Approximate e with your CAS by evaluating $(1+h)^{\frac{1}{h}}$ for $h = 1/10,\ 1/20,\ 1/50,\ 1/100$. Check to see if $\frac{e^h-1}{h} \approx 1$ for these values.

8.4 The Derivative of the Inverse Function

Let f, f^{-1} be a pair of inverse functions as in the definition of section 2.3. Then necessarily

$$f(f^{-1}(x)) = f^{-1}(f(x)) = x.$$

Notice that these equations define the function $f^{-1}(x)$ implicitly.

Since the inverse function is defined implicitly we can differentiate implicitly when searching for the derivative of the inverse. That is to say beginning again with the equation

$$f(f^{-1}(x)) = x,$$

we can differentiate both sides with respect to x to obtain

$$\begin{aligned}\frac{d}{dx}f(f^{-1}(x)) &= f'(f^{-1}(x))\cdot(f^{-1})'(x)\\ &= 1.\end{aligned}$$

Solving this last equation gives us a final proposition.

Proposition [8.12] *Let $f^{-1}(x)$ be the inverse of a differentiable one–to–one function $f(x)$. If $f^{-1}(x)$ is differentiable then*

$$\frac{d}{dx}f^{-1}(x) = \frac{1}{f'(f^{-1}(x))}.$$

Remark: In order that the derivative of f^{-1} be defined at the point x, it is necessary that $f'\left(f^{-1}(x)\right) \neq 0$.

Example [8.13] Find the derivative of $f^{-1}(x)$ where $f(x) = x^2$.

There are two ways to attack this problem. The elementary approach is to observe that $f^{-1}(x) = \sqrt{x}$ and thus $\frac{d}{dx} f^{-1}(x) = \frac{1}{2} x^{-\frac{1}{2}}$. However, we can also apply Proposition [8.12]: since $f'(x) = 2x$, we can determine that

$$\frac{d}{dx} f^{-1}(x) = \frac{1}{2f^{-1}(x)} = \frac{1}{2\sqrt{x}}.$$

Example [8.14] Find the derivative of $f^{-1}(x)$ where $f(x) = x^3$.

Here we simply apply our proposition: we have $f^{-1}(x) = x^{\frac{1}{3}}$, $f'(x) = 3x^2$, and thus

$$\frac{d}{dx} f^{-1}(x) = \frac{1}{3(f^{-1}(x))^2} = \frac{1}{3x^{2/3}}.$$

Observe that $y = \sin(x)$ is a one–to–one function from the interval $-\pi/2 \leq x \leq \pi/2$ to the interval $-1 \leq x \leq 1$. Hence its inverse, $\arcsin(x)$, is a well defined function whose domain is the interval $-1 \leq x \leq 1$, and whose range is the interval $-\pi/2 \leq x \leq \pi/2$.

Example [8.15] Evaluate

$$\frac{d}{dx} \arcsin(x).$$

To obtain the derivative of $\arcsin(x)$ we again apply our proposition to see that on the interval $-1 \leq x \leq 1$,

$$\frac{d}{dx} \arcsin(x) = \frac{1}{\cos\left(\arcsin(x)\right)}.$$

In this case we can simplify the solution tremendously. Since $\cos(\theta) \geq 0$ on the interval $-\pi/2 \leq \theta \leq \pi/2$, we may express

$$\cos(\theta) = \sqrt{1 - \sin^2(\theta)}$$

(where here we are taking the positive square root). It follows that

$$\begin{aligned} \frac{d}{dx} \arcsin(x) &= \frac{1}{\sqrt{1 - \left(\sin\left(\arcsin(x)\right)\right)^2}} \\ &= \frac{1}{\sqrt{1 - x^2}}, \end{aligned}$$

which is valid in the range $-1 < x < 1$. Clearly, the derivative of $\arcsin(x)$ is not defined at $x = \pm 1$.

Exercises for §8.4

For each of the following functions $f(x)$, find the inverse function, $f^{-1}(x)$, explicitly determine its domain and range, and compute the derivative of $f^{-1}(x)$ using proposition [8.12].

(1) $y = x^4$.

(2) $y = x^N$.

(3) $y = \frac{1}{x}$.

(4) $y = e^{x^2}$.

(5) $y = \cos(x)$.

(6) $y = \tan(x)$.

(7) $y = \sqrt{\sin(x)}$.

(8) $y = a^x$.

(9) $y = \cot(x)$.

(10) $y = \sinh(x)$.

(11) $y = \cosh(x)$.

(12) $y = \tanh(x)$.

Additional exercises for Chapter VIII

Find the derivative of y with respect to x by use of the implicit differentiation method for problems 1 to 4.

(1) $y^3 + 2y = x^4$.

(2) $6x^2 + 6y = 20$.

(3) $6x^3 + 6y = 6xy$.

(4) $5xy = (x - y)^2$.

Find the angle of intersection (the smallest of two possible angles) between the following pairs of curves and use your CAS to graph these curves.

(5)

$$2xy + y = 3$$

$$2y^3 - (x + 1) = 0.$$

(6)

$$x^2 + y^2 = 4$$

$$y - x = 0.$$

(7)

$$x^2 + y^2 = x + y$$

$$y - x^2 = 1/2.$$

(8)

$$x^2 + y^2 - 2xy = x$$

$$y - x = 0.$$

By using the implicit differentiation method or the chain rule, obtain the derivative of y with respect to x for problems 9 and 10.

(9) $y = \frac{1}{u^2+1}, u = \frac{1}{x-1}$.

(10) $x = u^3 - u^2, y = u^3 - 2u$.

(11) Suppose that $y = xf(x)$. On your CAS, compute $\frac{dy}{dx}, \frac{d^2y}{dx^2}, \cdots, \frac{d^6y}{dx^6}$. Do you observe any pattern? Can you find a formula for $\frac{d^ny}{dx^n}$?

(12) Use implicit differentiation to find the equation of the tangent to the curve $y^2 - 4y + x^2 - 3 = 0$ at the point $(0, 1)$.

(13) On your CAS, graph the function defined implicitly by $2(x^2 + y^2)^2 - 25(x^2 - y^2) = 0$. Use implicit differentiation to find the equation of the tangent to the curve at $(3, 1)$.

(14) On your CAS, graph the function defined by $2x^{4/3} + 2y^{4/3} = 4$. Use implicit differentiation to find the equation of the tangent to the curve at $(-1, 1)$. Does the derivative at $(0, 0)$ exist?

(15) Find $\frac{dy}{dx}$ given that $4y^2t^2 - (t + 1)y = 3$, and t is a differentiable function satisfying $\frac{dt}{dx} = \frac{1}{\tan(t)}$.

Chapter IX

The Maxima and Minima of Functions

9.1 Maximum and Minimum Values

Consider a differentiable function $y = f(x)$. The graph of $f(x)$ will exhibit various hills and valleys as in the diagram below.

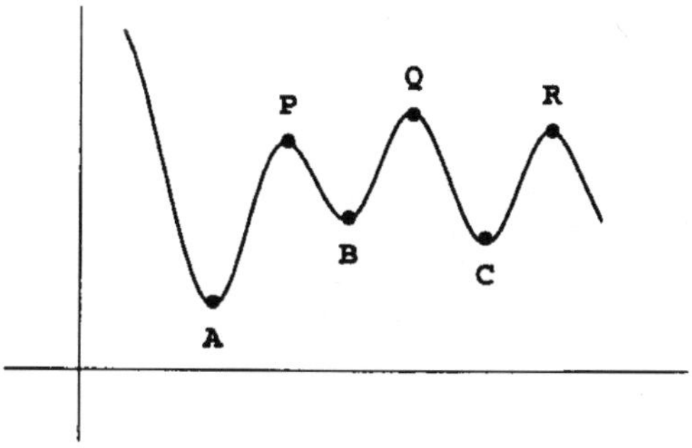

The points P, Q, and R will be maximum values of f corresponding to summits of the hills, while the points A, B, and C will be minimum values of f corresponding to the bottoms of the valleys.

If the function $y = f(x)$ is the (all–important) profit function for a business then the points P, Q, and R will denote times when profits are at their peak while the points A, B, and C denote the times when the profits bottom out. A key feature of the mathematical analysis of profit functions is to try to predict future peaks. Many Wall Street analysts spend their lives in just such an endeavor. Their chief tool is calculus: it provides a powerful technique for quickly computing the maximum and minimum of functions. We shall now develop this theory.

In approaching this problem it is important to realize that the occurrence of maxima and minima depend entirely on the domain of x–values we consider. For example, for $-1 \leq x \leq 1$, the function $y = 3x^4 - 16x^3 + 18x^2$ has a minimum at $x = 0$.

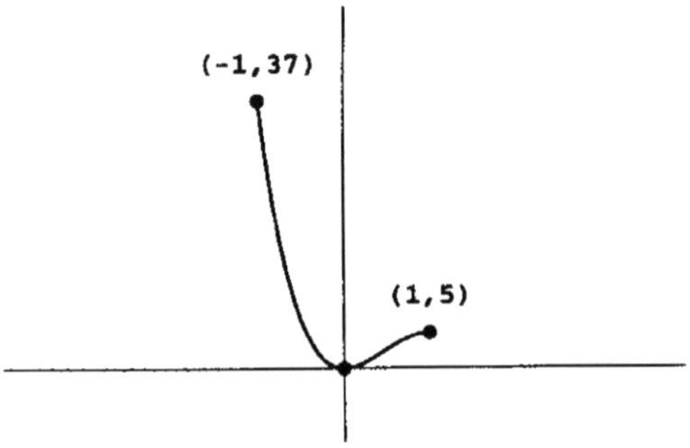

Enlarging the domain to $-1 \leq x \leq 4$, we see a point on the graph even lower at $(3, -27)$.

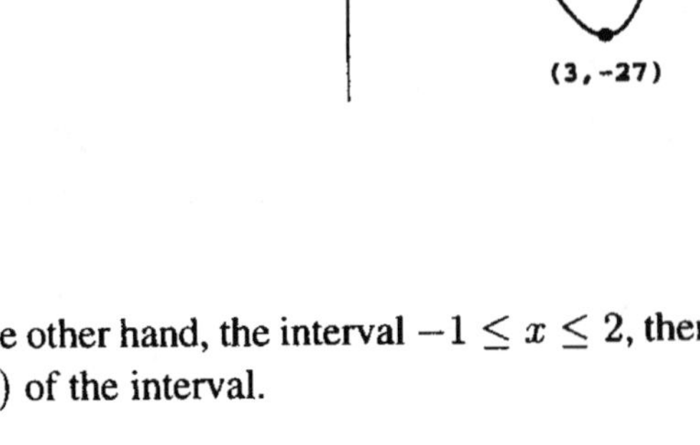

If we consider, on the other hand, the interval $-1 \leq x \leq 2$, then the minimum point is the endpoint $(2, 8)$ of the interval.

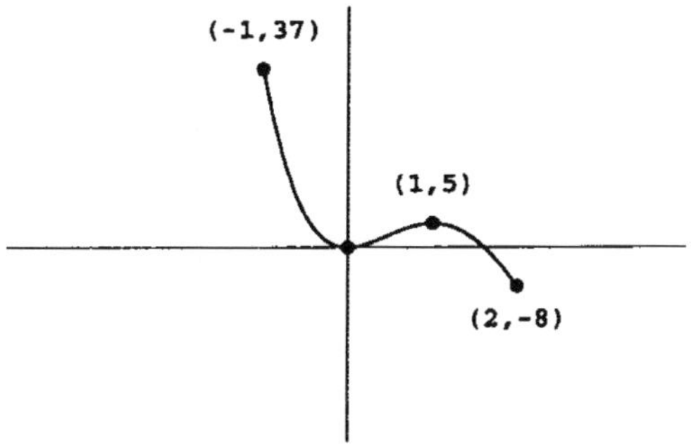

It is such behavior that forces us to specify the interval $a \leq x \leq b$ before we can begin to discuss maximum and minimum values of a function. Any change in the interval (i.e., changing a, b) above may result in drastically different maxima and minima.

Definition: *Fix an interval $a \leq x \leq b$. A function $F(x)$, defined for $a \leq x \leq b$, has a **maximum** at a point c (where $a \leq c \leq b$) if*

$$F(c) \geq F(x)$$

for every x in the interval $a \leq x \leq b$.

*The function F has a **minimum** at a point c in the interval $a \leq c \leq b$ provided*

$$F(c) \leq F(x)$$

for all x in the interval $a \leq x \leq b$.

Returning now to the earlier example, $y = 3x^4 - 16x^3 + 18x^2$

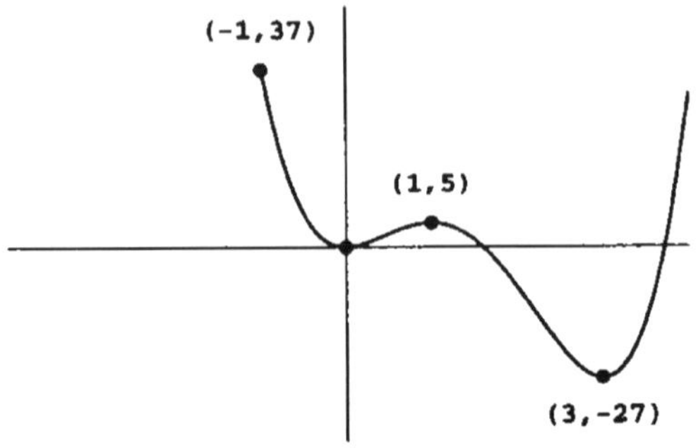

we visually see that $(3, -27)$ is a minimum for the interval $-1 \leq x \leq 4$. On the subinterval $-1 \leq x \leq 1$, however, the point $(0, 0)$ is a minimum. It is clear that each of the points $(-1, 37)$, $(0, 0)$, $(1, 5)$, $(3, -27)$ can be a maximum or a minimum value of the function if we choose a suitable subinterval. To isolate this concept we require a basic definition.

Definition: *Given an interval $a \leq x \leq b$, a point c is termed an* ***interior point*** *provided $a < c < b$. The points a, b are called* ***end points****.*

The above discussion of the function $y = 3x^4 - 16x^3 + 18x^2$ motivates the important:

Definition: *Let $F(x)$ be a function defined on some interval. A point c is called a* ***local maximum*** *(or* ***minimum****) if there exists a subinterval which includes the point c as an interior point, and $x = c$ is a maximum of F (respectively minimum) for this subinterval.*

Question: Given a differentiable function $y = f(x)$, how do we compute the local maxima and minima?

Answer: This question provides another opportunity to apply the concepts of calculus. Intuitively, when one walks up a hill the slope might be, say, .5 (45° incline). As one approaches the summit, the slope decreases, and finally at the peak it is 0. Upon descent the slope will, of course, become negative. To formalize this description, suppose $x = c$ is a local maximum or minimum of the differentiable function

$f(x)$, and consider the tangent line at $f(c)$. This tangent line must be horizontal, i.e., it must have slope = 0. This condition, in the language of calculus, is given by

$$f'(c) = 0.$$

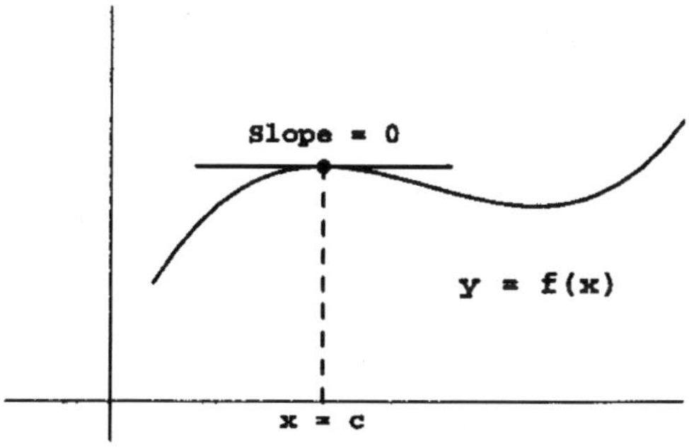

We present this as a formal proposition:

Proposition [9.1] *Let $f(x)$ be a differentiable function on some interval $a \leq x \leq b$. If $x = c$ (with $a < c < b$) is a local maximum or minimum of $f(x)$, then $f'(c) = 0$.*

Proof: We present the proof for a local maximum — the case of a local minimum is essentially the same. For sufficiently small $h \geq 0$, we have

$$f(c+h) \leq f(c),$$

since $f(c)$ is a local maximum. Hence $f(c+h) - f(c) \leq 0$, and dividing by $h > 0$ we see that

$$\frac{f(c+h) - f(c)}{h} \leq 0.$$

Recalling that the limit is independent of the sequence chosen we conclude that

$$\lim_{\substack{h \to 0 \\ h > 0}} \frac{f(c+h) - f(c)}{h} \leq 0.$$

If we now choose $h < 0$ (and $|h|$ is sufficiently small) we still must have the inequality

$$f(c+h) \leq f(c),$$

and thus $f(c+h) - f(c) \leq 0$. But dividing by $h < 0$ reverses the inequality and we have

$$\frac{f(c+h) - f(c)}{h} \geq 0.$$

Hence

$$\lim_{\substack{h \to 0 \\ h<0}} \frac{f(c+h) - f(c)}{h} \geq 0$$

and, since the the limit is unique, we conclude

$$f'(c) = \lim_{h \to 0} \frac{f(c+h) - f(c)}{h} = 0.$$

Warning: If $f(x)$ has a local maximum or minimum at $x = c$ then necessarily $f'(c) = 0$. The reverse may not be true: the function $f(x) = x^3$ satisfies $f'(x) = 3x^2$ and thus $f'(0) = 0$. But $x = 0$ is *not* a local maximum or minimum.

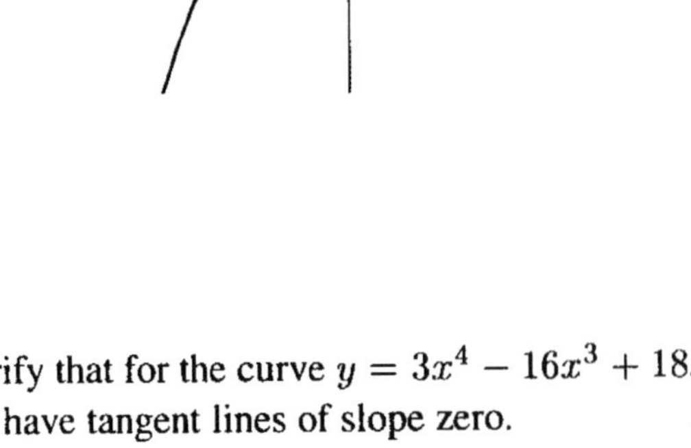

Example [9.2] Verify that for the curve $y = 3x^4 - 16x^3 + 18x^2$, the points (0, 0), (1, 5), and (3, -27) have tangent lines of slope zero.

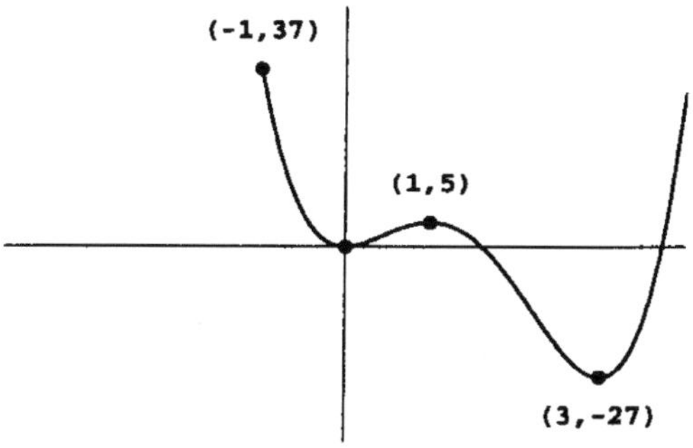

Let $f(x) = 3x^4 - 16x^3 + 18x^2$. We compute

$$\begin{aligned} f'(x) &= 12x^3 - 48x^2 + 36x \\ &= 12x \cdot (x^2 - 4x + 3) \\ &= 12x \cdot (x-1) \cdot (x-3), \end{aligned}$$

from which it immediately follows that $f'(0) = f'(1) = f'(3) = 0$. This confirms that the tangent lines at these points are horizontal, which allows us to conclude that the point (1, 5) is a local maximum and the points (0, 0) and (3, -27) are local minima.

Exercises for §9.1

In the following problems find all local maxima and minima of $f(x)$ in the stated interval. You may use your CAS to quickly compute derivatives and plot functions. Finding the maxima and minima is up to you.

(1) $f(x) = x^3 - 9x^2 + 30x, -4 \le x \le 6.$

(2) $f(x) = x^4 - 2x^3 - 2x^2, \ -1 \le x \le 3.$

(3) $f(x) = x^5, \ -2 \le x \le 2.$

(4) $f(x) = x^4 + x^3, \ -3 \le x \le 3.$

(5) $f(x) = \cos(x) + x, \ -2\pi \le x \le 2\pi.$

(6) $f(x) = \sin(2x) - x, \ -2\pi \le x \le 2\pi.$

(7) $f(x) = \ln(x) - x^3, \ 0 < x \le 3\pi.$

(8) $f(x) = 2^x - x, \ -5 \le x \le 5\pi.$

9.2 The First Derivative Test

Let $f(x)$ be a differentiable function defined on the interval $a \leq x \leq b$. Suppose that for some c in the interval, $f'(c) = 0$. There are three possibilities: Either $x = c$ is a local maximum, a local minimum, or it is neither. In order to obtain criteria to distinguish these three cases, we require the following proposition which will lead naturally to the first derivative test.

Proposition [9.3] *If $f'(x) > 0$ for all points $a \leq x \leq b$ then $f(x)$ is increasing on this interval, i.e.,*

$$f(x_1) < f(x_2) \qquad \textit{for all } \; a \leq x_1 \leq x_2 \leq b.$$

If $f'(x) < 0$ for all points $a \leq x \leq b$ then $f(x)$ is decreasing on this interval, i.e.,

$$f(x_1) > f(x_2) \qquad \textit{for all } \; a \leq x_1 \leq x_2 \leq b.$$

Pictorially, an increasing function looks like

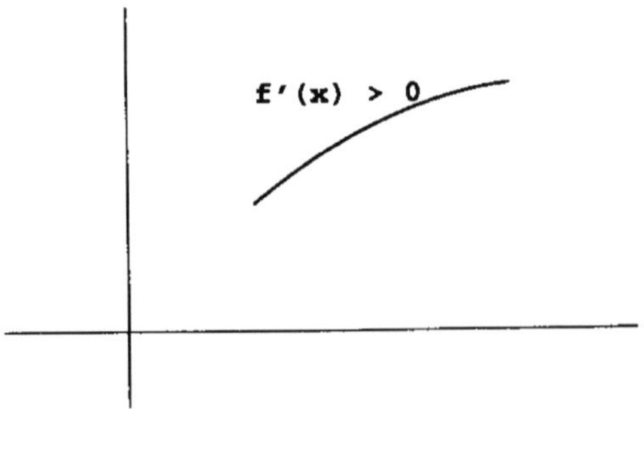

while a decreasing function looks like

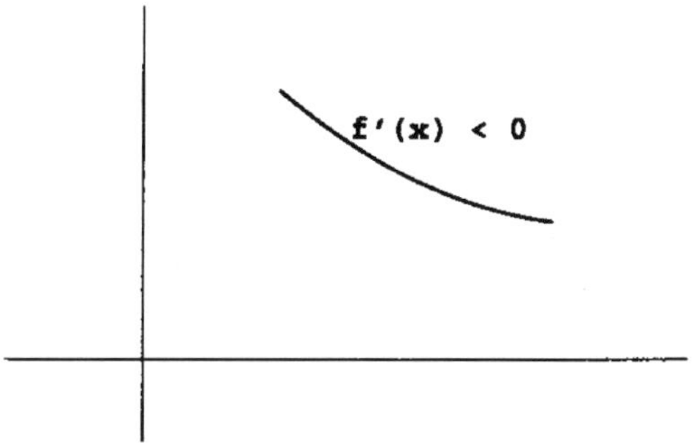

Proof: We prove that $f'(x) > 0$ implies that $f(x)$ is increasing — the case of $f'(x) < 0$ is similar. Since by definition

$$f'(x) = \lim_{h \to 0} \frac{f(x+h) - f(x)}{h}$$

the condition $f'(x) > 0$ implies that for all sufficiently small $|h|$ we must have

$$\frac{f(x+h) - f(x)}{h} > 0.$$

For $h > 0$, we see that $f(x+h) - f(x) > 0$, or

$$f(x+h) > f(x).$$

For $h < 0$, the signs reverse and we obtain $f(x+h) - f(x) < 0$, or $f(x+h) < f(x)$. Thus f is increasing in a sufficiently small enough interval containing x. Putting these intervals together we conclude that $f(x)$ is increasing on the whole interval $a \le x \le b$.

Proposition [9.4] (The first derivative test) *Suppose $f'(c) = 0$, and $f'(x) \neq 0$ for every other x in a small interval around c. Then in this interval we have three cases.*

Case 1: *If $f'(x) > 0$ for $x < c$ and $f'(x) < 0$ for $x > c$, then f has a local maximum at $x = c$.*

Case 2: *If $f'(x) < 0$ for $x < c$ and $f'(x) > 0$ for $x > c$, then f has a local minimum at $x = c$.*

Case 3: *If $f'(x)$ does not change sign at c then f has neither a local maximum or local minimum at $x = c$.*

The three cases can be viewed pictorially as hills, valleys, or inflection points.

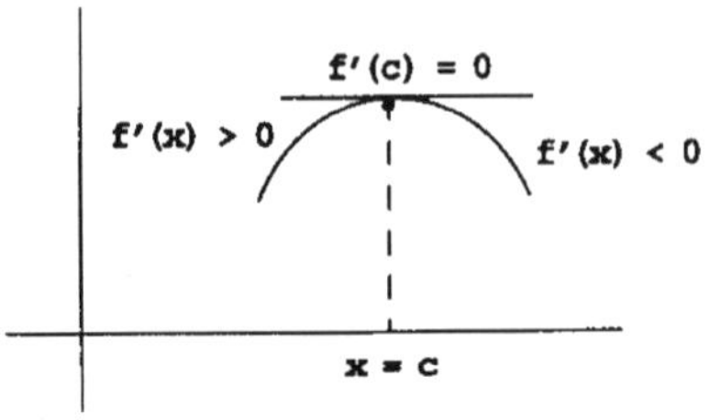

Case 1

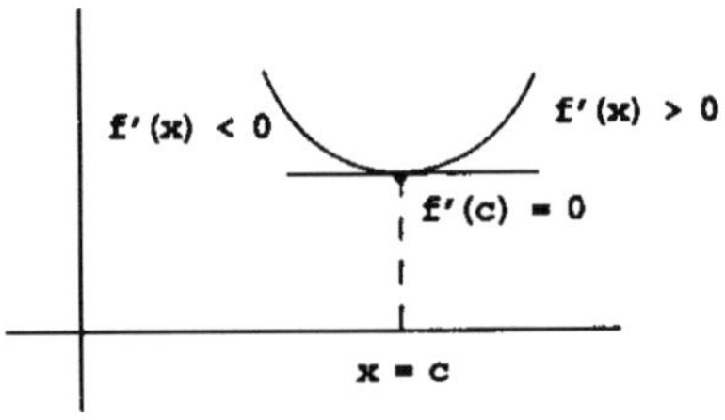

Case 2

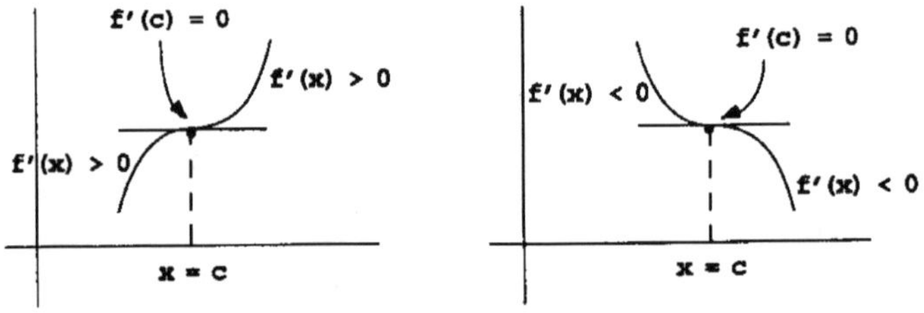

Case 3

For example, in case(1), for $x < c$, $f'(x) > 0$ so the slope of the tangent line is positive at points on the curve approaching $f(c)$ from the left. At the peak $x = c$ the slope is zero, and for $x > c$, $f'(x) < 0$ and the curve is falling.

Example [9.5] Let $f(x) = 2 - x^2$. Using the first derivative test determine where this function is increasing and where this function is decreasing. Find all local maxima and minima.

First of all, $f'(x) = -2x$. Thus there can only be one possible point which is a maximum or minimum. For $x < 0$, $f'(x) = -2x > 0$ so the function is increasing.

For $x > 0,\ f'(x) = -2x < 0$ so the function is decreasing. The point $x = 0$ must be a maximum. In fact, $f(0) = 2$ must be the largest value of the function.

Example [9.6] Let $f(x) = \frac{x^6 - 24x^4}{256}$. Using the first derivative test determine where this function is increasing and where this function is decreasing. Find all local maxima and minima.

Differentiating, we obtain $f'(x) = \frac{6x^5 - 96x^3}{256} = \frac{6x^3(x^2-16)}{256}$, and thus, $f'(0) = 0$, $f'(4) = 0$, and $f'(-4) = 0$. We now apply the first derivative test to classify each of these points. For $x < -4$, $f'(x) = \frac{6x^3(x^2-16)}{256} < 0$ and $f(x)$ is decreasing. For $-4 < x < 0$, $f'(x) > 0$ so $f(x)$ is now increasing. Hence $x = -4$ is a local minimum and the graph in a neighborhood of $x = -4$ looks like:

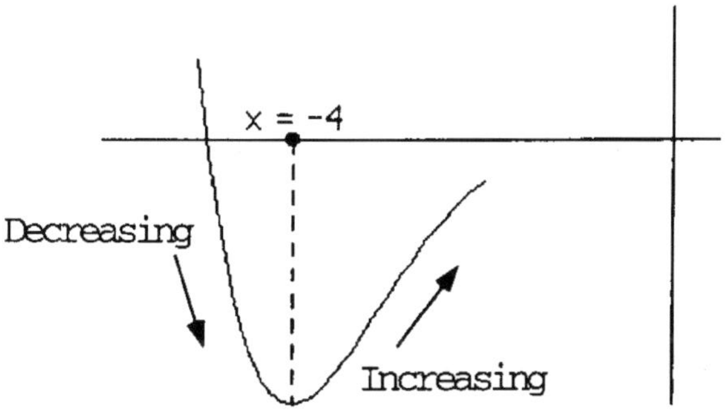

For $0 < x < 4$, $f'(x) = \frac{6x^3(x^2-16)}{256} < 0$, and thus the first derivative test tells us that $x = 0$ is a local maximum

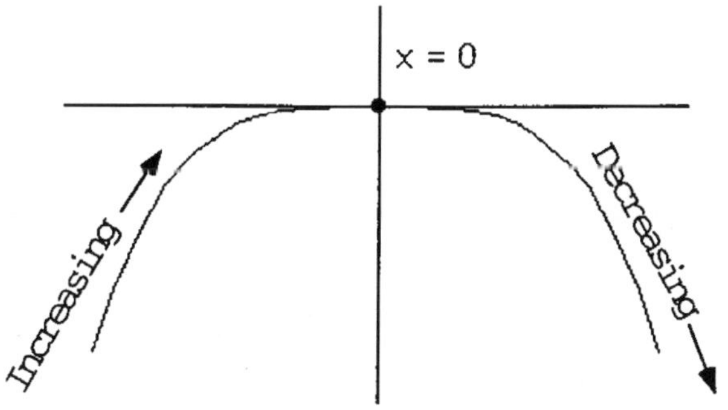

We conclude our analysis by analyzing the point $x = 4$. Notice that for $x > 4$, $f'(x) = \frac{6x^3(x^2-16)}{256} > 0$, applying the first derivative test one last time we find that $x = 4$ is a local minimum.

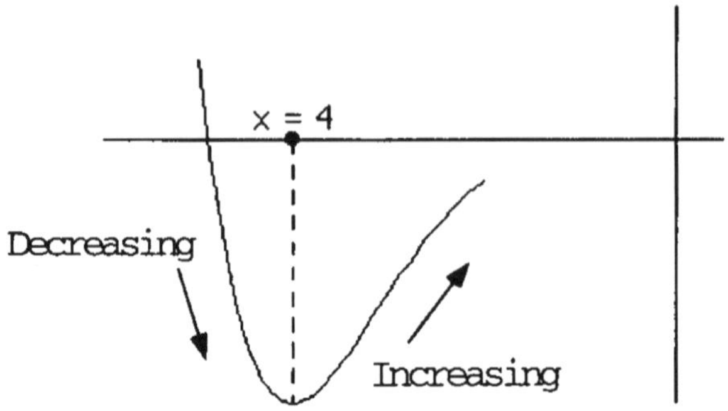

The graph of $y = \frac{x^6 - 24x^4}{256}$,

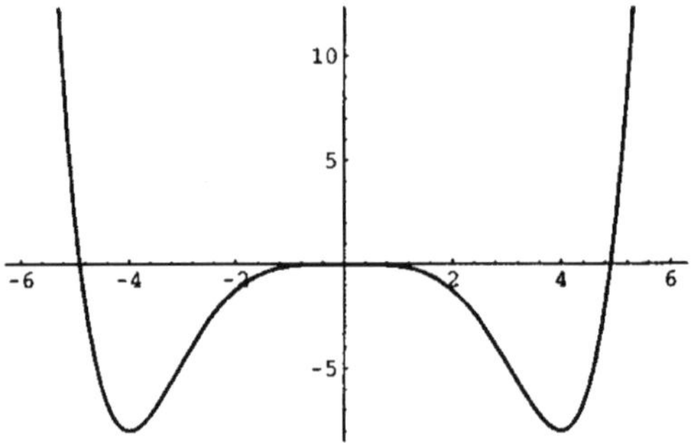

exhibits the two local minima at $x = \pm 4$ and the one local maximum at $x = 0$.

Question: Do there exist functions which do not have a local maxima or local minima?

Answer: There are many such functions, a simple example of which is the function $f(x) = 10^x$. To see that there are no local maxima or local minima in the following graph, simply note that $f'(x) = \ln(10) \cdot 10^x$ and thus $f'(x) > 0$ for *any* x.

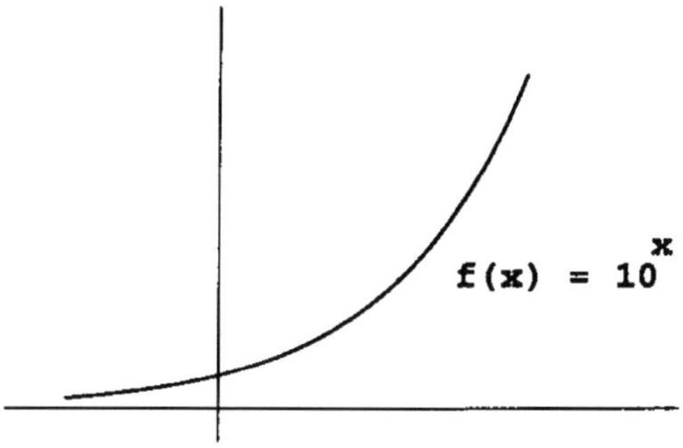

Exercises for §9.2

In the following problems find and classify all local maxima and minima of $f(x)$ with the first derivative test. Use your CAS to quickly compute derivatives. Check your answers by graphing $f(x)$ with your CAS.

(1) $f(x) = 5 + 3x^2.$

(2) $f(x) = 3 - 7x^2.$

(3) $f(x) = x^3 - 9x^2 + 30x.$

(4) $f(x) = x^4 - 2x^3 - 2x^2.$

(5) $f(x) = x^4 + x^3$

(6) $f(x) = x^5.$

(7) $f(x) = 3x^4 - 20x^3 + 48x^2 - 48x + 3.$

(8) $f(x) = 5x^6 - 6x^5 - 45x^4 + 8.$

(9) $f(x) = \cos(x) - x.$

(10) $f(x) = \sin(3x) + x.$

In the following problems determine in which intervals $a \leq x \leq b$ the function $f(x)$ is increasing (or decreasing). Use your CAS to quickly compute derivatives and check your answers by plotting the function.

(11) $f(x) = 2x^2 + x + 1.$

(12) $f(x) = 3 - 4x^2.$

(13) $f(x) = x^3 - 9x^2 + 30x.$

(14) $f(x) = x^4 - 2x^3 - 2x^2.$

(15) $f(x) = x^7.$

(16) $f(x) = x^7 - 11.$

(17) $f(x) = e^x.$

(18) $f(x) = \ln(x), \quad (0 < x).$

(19) $f(x) = \sin(x), \quad (0 \leq x \leq 2\pi).$

(20) $f(x) = 2\cos(x) + x.$

9.3 The Second Derivative Test

Let $f(x)$ be a twice differentiable function and let $x = c$ be a point where $f'(c) = 0$. A quick way to determine if $x = c$ is a local maximum or minimum is the second derivative test. Unfortunately, while this test is rapid, it may not always yield conclusive information.

Proposition [9.7] (The second derivative test) *Assume that*

$$f'(c) = 0.$$

If $f''(c) > 0$ then $x = c$ is a local minimum. If $f''(c) < 0$ then $x = c$ is a local maximum. If $f''(c) = 0$ we obtain no information from this test.

Proof: The condition $f'(c) = 0$ implies that

$$\frac{f'(x)}{x-c} = \frac{f'(x) - f'(c)}{x-c} \longrightarrow f''(c)$$

as $x \to c$. If $f''(c) > 0$, for x very close to c we have

$$\frac{f'(x)}{x-c} > 0.$$

Hence $f'(x) > 0$ if $x - c > 0$ and $f'(x) < 0$ if $x - c < 0$. Applying the first derivative test we conclude that $x = c$ is a local minimum. The proof is similar if $f''(c) < 0$; all signs are reversed.

Warning: If $f''(c) = 0$ the second derivative test fails and *anything* can happen. The function $f(x) = x^3$ has an **inflection** point at $x = 0$ despite the fact that $f'(0) = f''(0) = 0$. The function $g(x) = x^4$ has a local minimum at $x = 0$ and here $g'(0) = g''(0) = 0$. The function $r(x) = 1 - x^4$ has a local maximum at $x = 0$ and again $r'(0) = r''(0) = 0$. Clearly if the second derivative test fails there is no alternative but to go back to the first derivative test which, happily, never fails.

Example [9.8] Using the second derivative test find and classify the local maxima and minima of $f(x) = 2 - 3x^2$.

Since the derivative of $f(x)$ is $f'(x) = -6x$ we need to determine the nature of the point $x = 0$. Taking the second derivative we see that $f''(x) = -6$ for all x, and in particular, $f''(0) = -6$. We conclude, by the second derivative test, that $x = 0$ is a local maximum. The graph is depicted below.

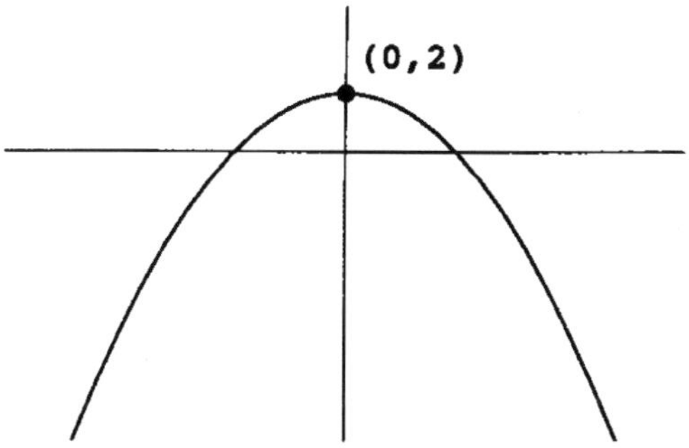

Example [9.9] Apply the second derivative test to find and classify the local maxima and minima of $f(x) = 2\cos(x) + x$ where $0 \leq x \leq 2\pi$.

In this case $f'(x) = -2\sin(x) + 1$, and thus $f'(x) = 0$ when $\sin(x) = \frac{1}{2}$. Since we are working on the interval $0 \leq x \leq 2\pi$, the only possible solutions are $x = \frac{\pi}{6}$ and $\frac{5\pi}{6}$. Now we need to evaluate the second derivative of f, which is $f''(x) = -2\cos(x)$, at the points $x = \frac{\pi}{6}$ and $\frac{5\pi}{6}$:

$$f''\left(\frac{\pi}{6}\right) = -2\cos\left(\frac{\pi}{6}\right) = -\sqrt{3},$$

$$f''\left(\frac{5\pi}{6}\right) = -2\cos\left(\frac{5\pi}{6}\right) = +\sqrt{3}.$$

Hence $x = \frac{\pi}{6}$ is a local maxima, and $x = \frac{5\pi}{6}$ is a local minima. The graph of our function is sketched below.

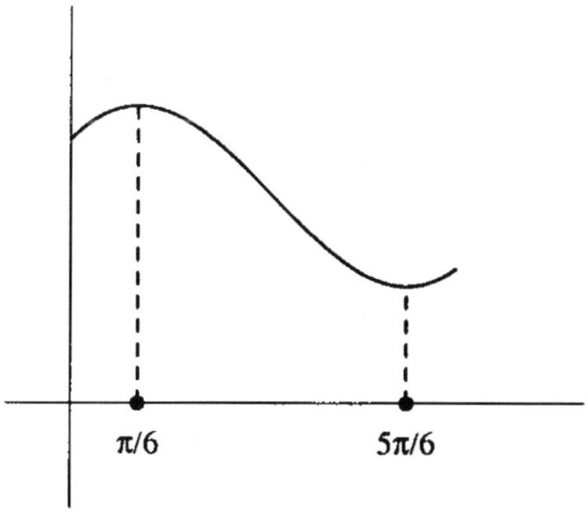

Example [9.10] Analyze the local maxima and minima of the function

$$f(x) = 15x^3 - \frac{3}{4}x^2 + \frac{x}{90} + 1$$

on the interval $-2 \le x \le 2$.

When we graph this function on the interval $-2 \le x \le 2$ we obtain

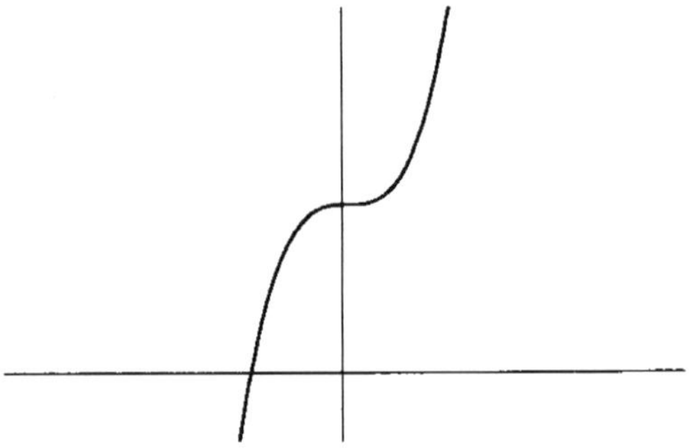

and it is difficult to see that there are any local maxima and minima. Nevertheless, the CAS can factor $f'(x) = 45x^2 - \frac{3}{2}x + \frac{1}{90}$,

$$f'(x) = \frac{1}{90}(90x - 1)(45x - 1).$$

This factorization tells us that there are points where the derivative vanishes: $x = \frac{1}{45}$ and $\frac{1}{90}$. Now $f''(x) = 90x - \frac{3}{2}$, and evaluating at the $x = \frac{1}{45}$ and $\frac{1}{90}$ we obtain

$$f''\left(\frac{1}{45}\right) = \frac{1}{2}, \quad f''\left(\frac{1}{90}\right) = -\frac{1}{2}.$$

Thus $x = \frac{1}{45}$ is a local minimum and $x = \frac{1}{90}$ is a local maximum. It is somewhat disconcerting to discover maxima and minima which were not immediately visible. The fact that we did not see these points in our first diagram was due to our choice of scale on the axes: when we plot $f(x)$ on the interval $0 \le x \le 1/10$ we can in fact see what we discovered with the derivative tests.

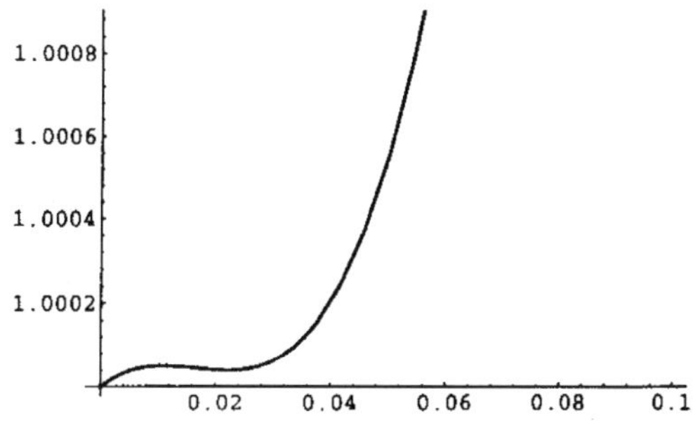

Exercises for §9.3

In the following problems find and classify all local maxima and minima of $f(x)$ using the second derivative test. Use your CAS to quickly compute the derivatives and factor polynomials. Check your answers by graphing $f(x)$.

(1) $f(x) = 4x^2 - 11.$

(2) $f(x) = x^2 + 6x - 3.$

(3) $f(x) = 5x^2 - 3x + 11.$

(4) $f(x) = 3x^4 + 16x^3 + 6x^2 - 72x.$

(5) $f(x) = x^4 + 4x^3 - 8x^2 + 48x.$

(6) $f(x) = x^3 + 2x^2 + 3.$

(7) $f(x) = 2\sin(x) + x - 2.$

(8) $f(x) = \cos(x) + \sqrt{3}x - 1.$

(9) $f(x) = \ln(x) + x^2, (0 < x).$

(10) $f(x) = e^x - 5x + 3.$

(11) $f(x) = 2^x - 3x + 1.$

(12) $f(x) = 3\ln(x) - \frac{1}{x} + 2, (0 < x).$

(13) $f(x) = \tan(x) + \frac{x}{2} + 3.$

(14) $f(x) = \tan(x) + \frac{4}{3}x.$

In problems (15)–(17) the points where $f'(x) = 0$ are extremely close together and hard to see on a graph. Find the local maxima and minima and find an appropriate interval to graph the function $f(x)$ so that these extremities are visible.

(15) $f(x) = 20x^3 - 181x^2 + \frac{73712}{135}x.$

(16) $f(x) = x^2(2.2048 - 2.8x + x^2).$

(17) $f(x) = 3750x^4 - 30200x^3 + 91203x^2 - 122412x + 11111.$

Additional exercises for Chapter IX

(1) Let $f(x) = x^2 + px + q$. Find values of p and q such that $f(1) = 3$. Is there a local maxima or minima of f on the interval $0 \leq x \leq 3$? Is this value a maximum? A minimum?

(2) Graph $f(x) = 3\cos\left(\frac{x}{3}\right) + 4\cos\left(\frac{x}{2}\right)$ on your CAS, then find the the maximum and minimum values. (**Hint:** f is periodic).

(3) Graph

$$f(x) = \begin{cases} 4x - 2, & \text{if } x < 1, \\ -(x-2)(x+1), & \text{otherwise,} \end{cases}$$

on your CAS.

(a) Determine the maximum of f on the interval $0 \leq x \leq 6$ by inspection.

(b) Does the answer in part (a) contradict the first derivative test or the second derivative test? Explain.

(c) Does this example yield another criterion for finding potential extrema?

(4) Graph

$$f(x) = \begin{cases} 3x - 1, & \text{if } x < 2 \\ x^2 - 6x + 13, & \text{if } 2 \leq x < 4 \\ -x + 9, & \text{otherwise} \end{cases}$$

on your CAS. Using problem 3 above, determine all possible maxima and minima on the interval $0 \leq x \leq 9$.

(5) Show that $F(x) = 25\sec(x)$ has a minimum value for $-\frac{\pi}{2} < x < \frac{\pi}{2}$ but no maximum. Graph it with your CAS.

(6) Use your CAS to graph $f(x) = \sin(x^2) \cdot \sec x$ on the interval $-\frac{\pi}{2} \leq x \leq \frac{\pi}{2}$.

(a) Try to determine local maxima and minima for the graph.

(b) Verify your results by the first and second derivative test. (Use your CAS to verify)

(7) Show that if f is continuous on $a \leq x \leq b$ and differentiable on (a, b) and $f(a) = f(b)$ then there is a value $a < x < b$ such that $f'(x) = 0$. This result is called **Rolle's theorem.**

(8) Prove the mean value theorem. If f is continuous and differentiable on $a < x < b$ then there is a value $a < x < b$ such that $f'(x) = \frac{f(b)-f(a)}{b-a}$. **Hint:** Use Rolle's theorem.

(9) Prove the following consequence of the mean value theorem: If f is defined on an interval $a \leq x \leq b$ and $f'(x) = 0$ for all $a < x < b$, then f is a constant on the interval.

In problems (10)-(14), find and classify all local maxima and minima of $f(x)$. Graph the functions with your CAS.

(10) $$f(x) = \begin{cases} 2x - 7 & \text{if } x < 4 \\ x^2 - 6x + 9 & \text{if } x \geq 4. \end{cases}$$

(11) $$f(x) = \begin{cases} x^3 - 3x + 2 & \text{if } x < 2 \\ 9x - 14 & \text{if } x \geq 2. \end{cases}$$

(12) $f(x) = e^{-x} \sin(x)$, $(x \geq 0)$.

(13) $f(x) = e^{-x^2} \cos(x)$ **Hint:** one critical point is easy for the rest you may consider $\tan(x)$ or resort to your CAS.

(14) $f(x) = \frac{e^{-x^2}-1}{x^2}$.

(15) Can you find a function with absolute minimum 0 at the origin and $f'' < 0$ everywhere? Give an example or explain why not.

Chapter X

Classical Optimization Theory

10.1 A Three Step Method for Finding Maxima and Minima

Let $f(x)$ be a differentiable function defined on an interval $a \leq x \leq b$. We have seen in Chapter 9 that the maximum or minimum value of $f(x)$ in this interval must occur either at one of the endpoints $x = a$, b, or at a point where the derivative of f vanishes. This immediately gives us a three step method which can be readily performed by a CAS.

Step 1: Find all points $a \leq c_1, c_2, \ldots, c_m \leq b$ where $f'(c_i) = 0$ $(i = 1, 2, \ldots, m)$.

Step 2: Compute $\{f(a), f(b), f(c_1), f(c_2), \ldots, f(c_m)\}$.

Step 3: Find the largest and smallest numbers in the set of values $\{f(a), f(b), f(c_1), f(c_2), \ldots, f(c_m)\}$. These will correspond to the maximum and minimum values of the function.

Example [10.1] Find the maximum and minimum values of the function $f(x) = x^3 - 3x + 1$ on the interval $0 \leq x \leq 3$.

Step 1: $f'(x) = 3x^2 - 3$, so $f'(1) = f'(-1) = 0$.

Step 2: $f(0) = 1$, $f(3) = 19$, $f(-1) = 3$, and $f(1) = -1$.

Step 3: The maximum value of f occurs at the endpoint $x = 3$, while the minimum value of f occurs at $x = 1$.

Example [10.2] Find the maximum and minimum values of the function $f(x) = \sin^2(x)$ on the interval $0 \leq x \leq 4\pi/3$.

Step 1. $f'(x) = 2\sin(x) \cdot \cos(x)$, thus $f'(0) = f'(\pi/2) = f'(\pi) = 0$.

Step 2. $f(0) = 0$, $f(\pi/2) = 1$, $f(\pi) = 0$, and $f(4\pi/3) = 3/4$.

Step 3. The maximum value of f occurs at $x = \pi/2$, while the minimum value occurs at $x = 0$.

Example [10.3] Find the maximum and minimum values of the function $f(x) = x \cdot e^{-x}$ on the interval $-2 \leq x \leq 2$.

Step 1. $f'(x) = e^{-x} - xe^{-x} = (1 - x)e^{-x}$, and thus $f'(1) = 0$.

Step 2. $f(-2) = -2e^2$, $f(1) = 1/e$, $f(2) = 2/e^2$.

Step 3. Since $1/e > 2/e^2$ we see that the maximum value of f occurs at $x = 1$, while the minimum value occurs at $x = -2$.

Exercises for §10.1

Using the three step algorithm of §10.1 find the maximum and values of $f(x)$ in the stated interval for the following functions. Quickly plot these functions with your CAS to substantiate your calculations.

(1) $f(x) = x^4 - 2x^2 + 3$ for $-5 \leq x \leq 4$.

(2) $f(x) = x^4 - 32x + 148$ for $0 \leq x \leq 3$.

(3) $f(x) = 2x^3 + 3x^2 - 36x + 11$ for $-2 \leq x \leq 2$.

(4) $f(x) = x^7 - 5x^5 + 13$ for $-3 \leq x \leq 3$.

(5) $f(x) = \cos^2(x)$ for $-\pi \leq x \leq \pi$.

(6) $f(x) = \tan(x)$ for $-\pi/4 \leq x \leq \pi/4$.

(7) $f(x) = e^x + e^{-x}$ for $0 \le x \le 6$.

(8) $f(x) = x^2 - \ln(1 + x^2)$ for $-1 \le x \le 1$.

(9) $f(x) = \cos(2x) + 2\cos(x)$ for $0 \le x \le 2\pi$.

(10) $f(x) = \sin(2x) + 2\sin(x)$ for $-2\pi \le x \le 2\pi$.

(11) Suppose $y = f(x)$ is defined implicitly by the equation $e^x = e^y \sin(x)$. Find the maximum and minimum of $y = f(x)$ on the interval $\frac{\pi}{4} \le x \le \frac{3\pi}{4}$.

10.2 Mathematical Modeling

Mathematical Modeling is a process which translates real world problems into mathematical problems which have algorithmic solutions. As an introduction to this vast and vital theory we shall focus on problems which, when translated, are solved by the algorithm presented in §10.1.

Example [10.4] Show that the square has the greatest area amongst all rectangles of a given perimeter.

Solution: First draw a picture of a rectangle.

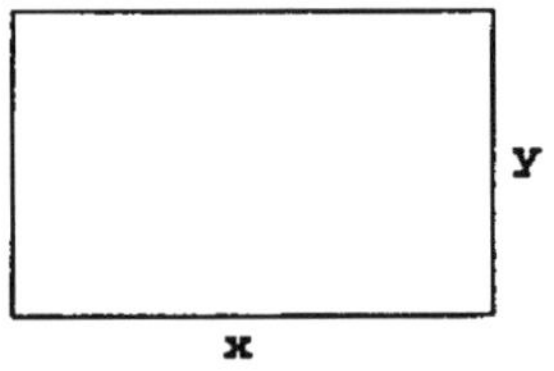

Since the length and width are unknown, we label them x and y. Now let $P =$ the perimeter which was given and is constant. By definition

$$2x + 2y = P,$$

and thus we can express y in terms of x, $y = \frac{P-2x}{2}$. We now define the area function (of the rectangle)

$$A(x) = x \cdot y = x \cdot \frac{P - 2x}{2},$$

which is a function of a single variable, x. This function is well defined provided $0 \le x \le P/2$: if $x < 0$ we would have a negative length (which cannot exist), and if $x > P/2$ then the width $y = \frac{P-2x}{2} < 0$ would again be negative (and not allowable). We can now apply the three step method of §10.1.

Step 1: $A'(x) = \frac{P}{2} - 2x$. Hence $A'\left(\frac{P}{4}\right) = 0$.

Step 2: $A(0) = 0$, $A\left(\frac{P}{2}\right) = 0$, $A\left(\frac{P}{4}\right) = \frac{P^2}{16}$.

Step 3: The maximum area occurs when $x = y = p/4$, i.e., the rectangle is a square.

Remarks: The most important part of this exercise is constructing the area function $A(x)$ and the interval it is defined on, $0 \leq x \leq p/2$. At present no artificial intelligence systems are sufficiently advanced to enable a computer to solve the problem alone. You can program the three step method on your CAS (and then use the computer to help you solve a wide variety of problems), but extracting the function (which *is* the translation of the original problem into a mathematical problem) must be done by you.

Example [10.5] A clothing company is producing cotton T–shirts with small valentines ironed on the front. Each shirt must have the same number of valentines on it, and the shirts are produced in the following manner: (1) all the plain T–shirts are produced at once after which they are sent to the art department, (2) the art department irons on the valentines (in an attractive manner) to each shirt. Assume the art department has only 10,000 valentines and they want to use them all. Assume further that it takes 25 minutes to produce one shirt and it takes one minute to iron on one valentine.

A special customer put in a rush order for 100 completed shirts. How many shirts should the company produce altogether in order to minimize the time it takes to produce the 100 completed shirts.

To begin our attack on this problem let

$$x = \text{number of shirts produced altogether}$$

$$y = \text{number of valentines on each shirt.}$$

Now the art department's desire to use 10,000 valentines translates to the equation

$$xy = 10,000,$$

and the function which determines the time utilized in producing the customers order of 100 shirts is

$$f(x) = 25x + 100y = 25x + \frac{10^6}{x},$$

where $100 \leq x \leq 10^4$. At this point we use the three step algorithm of §10.1.

Step 1. $f'(x) = 25 - \frac{10^6}{x^2}$, and thus $f'(200) = 0$.

Step 2. $f(100) = 25 \cdot 100 + 10^5$, $f(200) = 5000 + 5000 = 10,000$, and $f(10^4) = 25 \cdot 10^4 + 100$.

Step 3. The minimum appears at $x = 200$.

Remarking that when $x = 200$, $y = 50$ we conclude that the company should produce 200 shirts with 50 valentines on each.

When looked at in perspective, a pattern emerges from the above examples. In general when we are given a problem we will be asked to maximize or minimize some quantity which is given as a function of *two variables*. In addition to a description of this function we are given some other piece of information which, when translated, allows us to derive another identity which involves both of the variables. With this identity we can solve for one of

the variables in terms of the other. The function we derived at the beginning of the problem can then be expressed as a function of a single variable and we are reduced to employing the algorithm of §10.1.

Exercises for §10.2

(1) Find two positive integers whose sum is 1000 and whose product is minimized (maximized).

(2) Find two positive numbers whose product is 1000 and whose sum is minimized.

(3) An architect wants to design a rectangular cedarwood closet (completely from cedar) that would be built against a long 7 foot high wall. Assuming that the architect has exactly 700 square feet of cedarwood available to build the closet (with 7 foot walls) what should the dimensions be to maximize the floor area of the closet.

(4) Find the point on the line $y = 4x + 1$ which is closest to the point (0, 18).

(5) Given a line $\ell : y = mx + b$ and a point $P = (u, v)$ find the shortest distance from the point P to the line, i.e. minimize $\sqrt{(u - x)^2 + (v - mx - b)^2}$. Show that this agrees with the purely geometric solution obtained by finding the line ℓ' passing through P and normal to ℓ and measuring the distance from P to the intersection of ℓ and ℓ'.

(6) Find the point on the curve $y = 1 - x^2$ which is closest to the point (3, 1).

(7) Fix a point (a, b) in the first quadrant. Find the equation of the line which goes through the point (a, b) and minimizes the area of the triangle formed with the positive x and y axes.

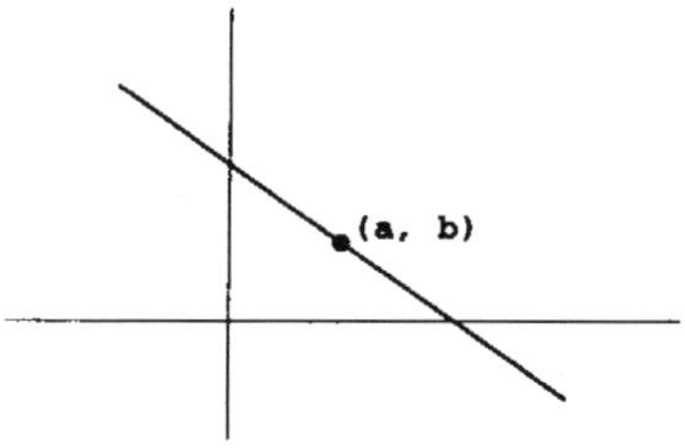

(8) What is the area of the largest rectangle which can be placed inside an equilateral triangle in such a way that the base of the triangle contains one of the sides of the rectangle.

(9) Country $\mathcal{A}$ has been flying reconnaissance missions over country $\mathcal{B}$. At height h thousand feet, a plane can scan an area of

$$A(h) = h^2 \cdot e^{\frac{-h}{2000}}$$

square feet. Country $\mathcal{B}$ has installed satellite and surface to air missile defense systems. After some hard experience country $\mathcal{A}$ has learned the following: flying above $30,000$ feet leaves a plane vulnerable to satellite detection, and flying below $7,000$ feet is too close to the surface to air missile defense system. At what height h does country $\mathcal{A}$ obtain the most data.

(10) A roofing contractor would like to minimize costs when building a roof for a structure of fixed depth 40 feet. When looked at from the front, the structure has the shape of a rectangle with an isosceles triangle on top of it. The triangle has a fixed area 1000 square feet. What proportions should the V–shaped roof have to minimize its total surface area.

(11) A swimmer is on an island, P, 3 miles from the shore. She is able to swim 1.5 m.p.h. and walk 4 m.p.h. Her plan is to swim to some point R on the shore and then walk to a café, C, which is exactly 5 miles from Q. To what point R should she swim to minimize her total time.

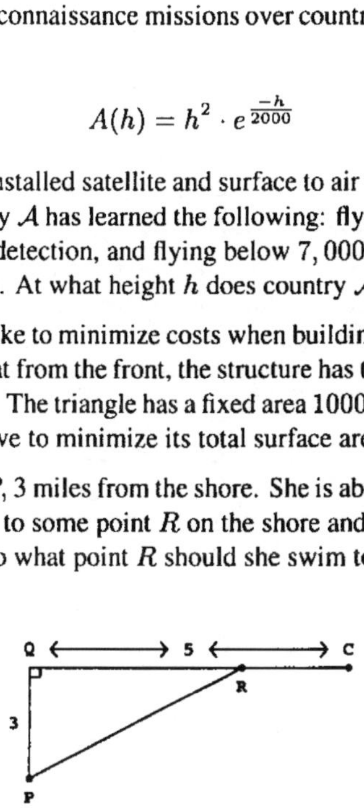

(12) A tall warehouse is surrounded be 8 foot electric fencing. The fence is 20 feet from the warehouse. In order to survey the goings on, the FBI wishes to install a straight wire originating on the ground outside the fencing to the building. The wire must clear the fence by 5 inches.

(a) What is the shortest length of wire they can use?

(b) If the warehouse is only 20 feet high, what is the shortest wire length which can be used?

Hint: Use your CAS to solve the quartic equation that appears in this problem.

(13) A two dimensional simulation of a mountain is given by the plot of the function $f(x) = 50 - x^3 - 3x^2 + 2x$.

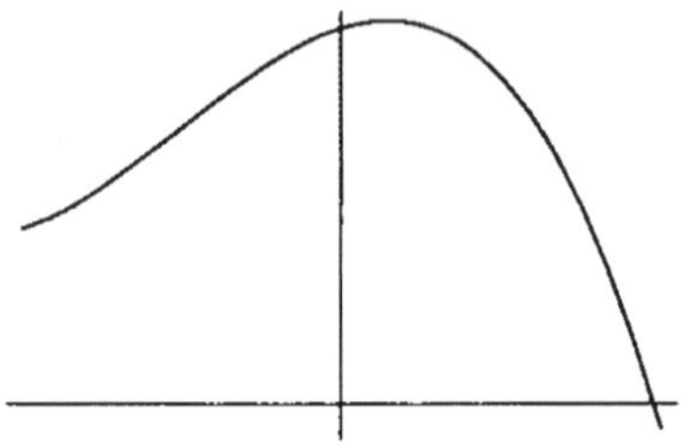

If a hiker climbs this mountain, find the point x on the graph where the hike is the most difficult. **Hint:** The hike is most difficult when the slope of the tangent line is as large as possible.

(14) A posynomial is a function $p(x)$ of the form:

$$p(x) = a_0 + a_1 x^{r_1} + a_2 x^{r_2} + \cdots + a_n x^{r_n}$$

where $a_i > 0$ (for $i = 0, 1, 2, \ldots n$) and where r_i (for $i = 0, 1, \ldots, n$) are arbitrary real numbers. A basic problem of geometric programming is to find the value $x \geq 0$ which minimizes $p(x)$. Solve this problem for posynomials with $n = 2$.

(15) An onyx pendant is to have the shape of a rectangle surmounted by a semicircle (we assume that the diameter of the semicircle is equal to the width of the rectangle). The perimeter of the pendant is to be embedded with a small pearls. There are only enough pearls to cover a 4 inch perimeter. What should the proportions of the pendant be in order to maximize the surface area (and allow for engraving on the back).

(16) A 10 centimeter piece of platinum thread is to be cut into two pieces. One piece is shaped into a square and the second piece is to form a circle. What should the dimensions of the square and circle be in order to maximize the sum of the area of the figures.

(17) Let A, B be any two points on the same side of the x–axis. Let P be a point on the x–axis such that the sum of the distances AP plus BP is minimized. Show that in this case the angle α must equal the angle β in the figure below.

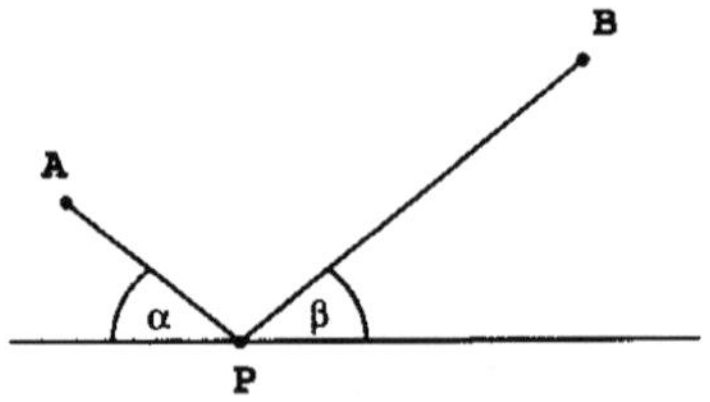

(18) Let $a_1, a_2, \ldots, a_n$ be arbitrary real numbers. Find the real number x which minimizes the sum $(x - a_1)^2 + \cdots + (x - a_n)^2$.

10.3 Surface Area and Volume Problems

Many of the mathematical modeling problems we will come across involve surface area and volume computations, so we briefly review the important concepts and results from this part of geometry. The most basic volume we can compute is a 3–dimensional box

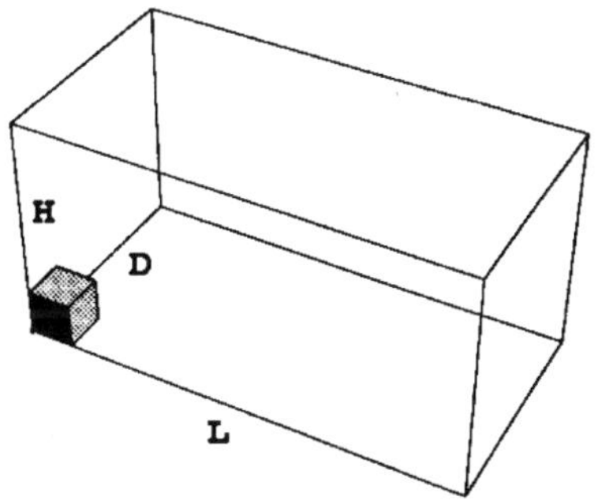

of length L, height H, and depth D whose volume is $L \cdot H \cdot D$. This formula is derived by filling the box with $L \cdot H \cdot D$ unit cubes

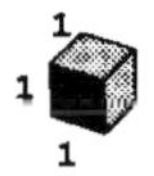

In the course of solving the various optimization problems we will require a knowledge of a variety of surface area and volume formulae. The verifications of the following results will be obtained once we delve into the theory of integration.

1. Area and perimeter of a circle: The area of a circle a radius R is πR^2. It has perimeter $2\pi R$.

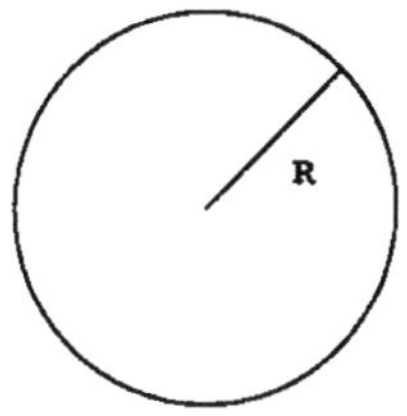

2. Volume of a cylinder: The volume of a cylinder

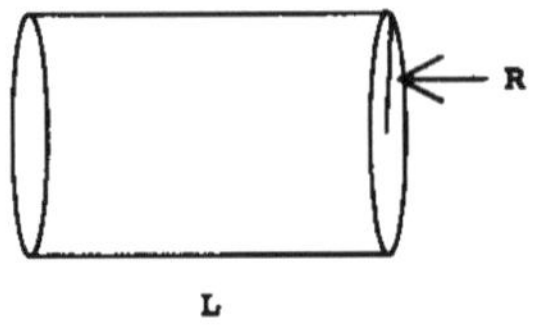

of length L and radius R is $\pi R^2 \cdot L$.

3. Surface area of an open–ended cylinder: The surface area of the above (open–ended) cylinder is $2\pi R \cdot L$. This can be seen by unrolling the cylinder:

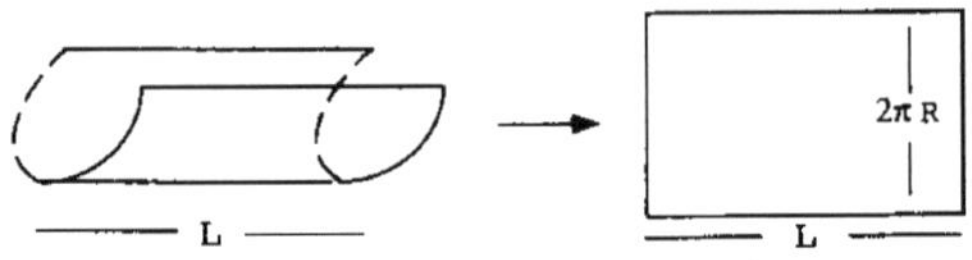

4. Surface area of a closed cylinder: The surface area of a closed cylinder is $2\pi R \cdot L + 2\pi R^2$. We must add on the area of the top and bottom circles.

5. Volume of a circular cone: The volume of a cone of height h and whose base is a circle of radius R is $\frac{1}{3}\pi R^2 h$.

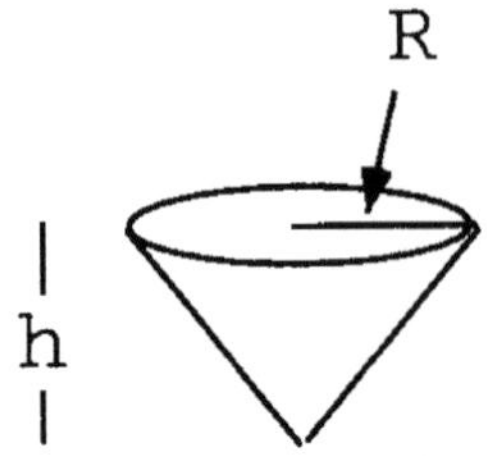

6. Volume and surface area of a sphere: The volume of a sphere of radius R is $\frac{4}{3}\pi R^3$ and its surface area is $4\pi R^2$.

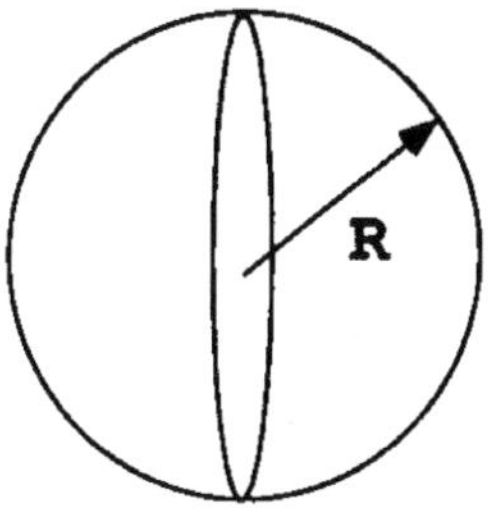

Example [10.6] (Box Folding) A cardboard manufacturing company produces rectangular pieces of cardboard 8 feet wide by 12 feet long. Identical squares of area x^2 are cut out at each corner and the remaining piece of cardboard is then folded (and taped) into a box without a top.

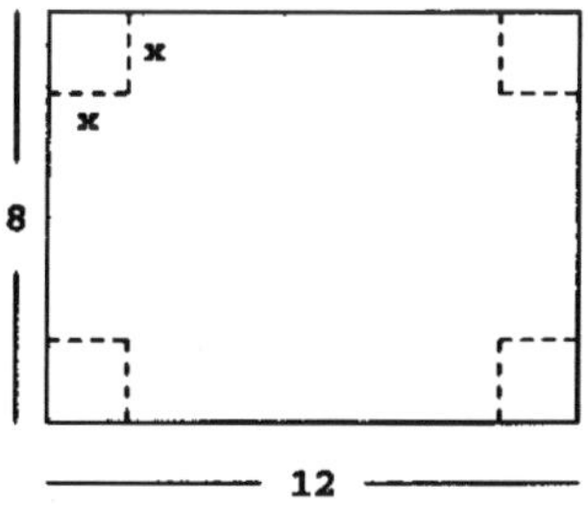

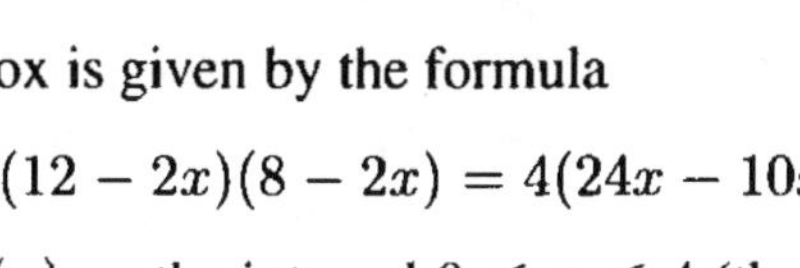

How large should the cut out squares be to maximize the volume of the box?

The volume of the box is given by the formula

$$V(x) = x(12 - 2x)(8 - 2x) = 4(24x - 10x^2 + x^3).$$

In order to maximize $V(x)$ on the interval $0 \leq x \leq 4$ (these endpoints arise naturally in that we do not want the sides of the box to have negative length), we apply the three step algorithm.

Step 1. $V'(x) = 4(24 - 20x + 3x^2) = 0$ implies that

$$x = \frac{20 \pm \sqrt{400 - 4 \cdot 3(24)}}{6} = \frac{20 \pm \sqrt{112}}{6},$$

and thus $x = \frac{20-\sqrt{112}}{6}$ since $x = \frac{20+\sqrt{112}}{6} > 4$ is out of the range of consideration.

Step 2. $V(0) = V(4) = 0$, and $V\left(\frac{20-\sqrt{112}}{6}\right) = 67.6$.

Step 3. We conclude that the volume is maximized when $x = \frac{20-\sqrt{112}}{6}$.

Example [10.7] (Flashlight Production) A company produces cylindrical flashlights.

The end of the flashlight labeled B in the figure is to be made of glass. The rest of the surface (including the opposite end labeled A) is made of yellow plastic. Assume that 12π square inches of yellow plastic are available for each flashlight. What should the dimensions of the flashlight be to maximize the volume of the flashlight and thus allow an array of different batteries to be used inside.

Let r denote the radius of the circular end A and let h be the length of the flashlight. Then by hypothesis the sum of the area of the cylindrical part of the flashlight together with the area of the bottom must be 12π, i.e.,

$$12\pi = \pi r^2 + 2\pi r h, \tag{10.1}$$

and hence $0 \leq r \leq \sqrt{12}$. The volume we wish to maximize in this case is simply $\pi r^2 \cdot h$. In order to solve this problem we must transform this volume into an equation of one variable. This is accomplished by first solving for h in equation (10.1), yielding

$$h = \frac{12\pi - \pi r^2}{2\pi r} = \frac{6}{r} - \frac{r}{2},$$

and then plug this expression into the volume formula to obtain

$$V(r) = 6\pi r - \frac{\pi}{2} r^3.$$

Step 1. $V'(r) = 6\pi - \frac{3\pi}{2} r^2 = 0$ implies $r = 2$.

Step 2. $V(0) = 0$, $V(\sqrt{12}) = 0$, and $V(2) = 8\pi$.

Step 3. The maximum volume occurs when $r = 2$ and $h = 2$.

Exercises for §10.3

(1) A cardboard box is to be designed to have a square base and an open lid. If 32 square feet of cardboard is available to make the box, how should the box be designed to maximize its volume?

(2) A cylindrical can is being designed to have a fixed volume V. Find the dimensions of the can which will minimize the amount of metal needed to make the can.

(3) A jeweler wishes to carve a circular cylinder from a spherical jade bead whose radius is r. What is the largest possible surface area the cylinder can have?

(4) An amateur film maker is showing a film to a large group of friends. She wishes to serve popcorn in large conical containers, and (in order to economize) plans to make the cones out of a circular piece of paper whose radius is R by cutting out a sector and then joining the edges with tape.

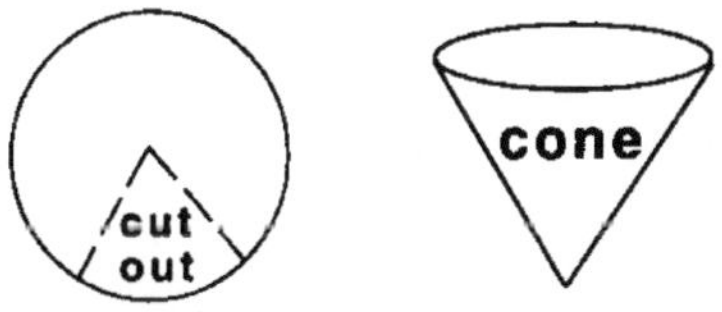

What is the largest possible volume the cone can have?

(5) A gardening supply company is planning to produce a plastic seed starter in the shape of a rectangular open box with square base which has 9 equal partitions.

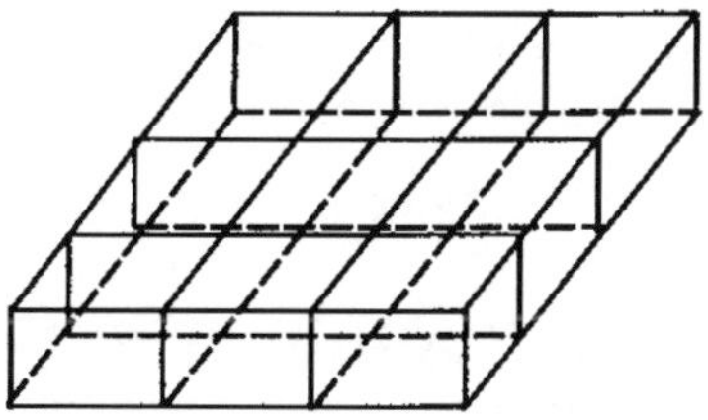

Since the company already produces small 50 cubic inch bags of topsoil they would like the seed starter to have a 50 cubic inch volume. What should the dimensions of the seed starter be to minimize the total amount of plastic necessary for construction.

(6) A nuclear waste disposal company needs to build silos in which to store nuclear waste. The silos are to be closed cylindrical structures with an interior volume of 120 cubic feet and have a thickness of 4 feet. The silos are to be coated with an interior coating which costs \$20 per square foot and an exterior coating which costs \$35 per square foot. What should the dimensions of the silo be so that the cost of the coatings is minimized.

(7) A house is situated on a site with a high water level, the owner wishes to install an outdoor drainage system to prevent possible basement flooding. A duct is to be made from long aluminum sheets which are 18 inches wide by partitioning the sheets into three equal rectangular sections (each of width 6 inches) and folding up the outer sections as indicated in the figure below. Through what angle θ should each of the sides be turned up to maximize the volume of the duct?

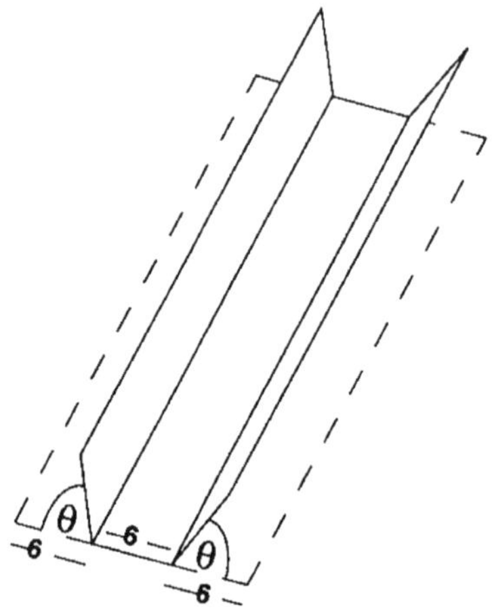

(8) A company is designing a recycling bin. The bin is in the shape of a cylinder with a hemisphere on top.

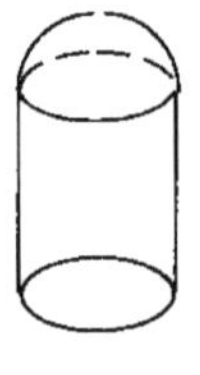

It is required that the volume of the bin be 64 cubic feet. If the material for the hemispherical top costs three times as much as the material for the cylindrical bottom, what should the dimensions of the recycling bin be to minimize the production cost.

10.4 A Simple Mathematical Model in Economics

As a final illustration of mathematical modeling and optimization we shall consider a well known and vital question arising in market research. Assume a company will produce x units of a certain product (such as computers) and plans to put them on the market for sale. The question of what price each unit should be sold for so that all the units can be sold in a given period of time arises immediately.

Definition: *The* ***demand function*** $p(x)$ *is the highest price per unit that the company can charge in order to sell all* x *units in a given period of time.*

Finding the demand function $p(x)$ is not at all easy. One possible approach to simplifying this problem is to assume that the function $p(x)$ is linear and has the form $p(x) = ax + b$ for certain constants a and b. These constants can then be computed by price testing.

Example [10.8] (Price Testing) A computer company has just come out with a new PC. In order to determine the appropriate price for each PC the company does some price testing. At $\$1500$ per PC it is found that $10,000$ units are sold in one week. The following week the price is lowered to $\$1200$ per PC and $15,000$ units are sold. Assuming the demand function $p(x)$ is linear, find $p(x)$.

When we plug in the given data $p(10,000) = \$1500$, and $p(15,000) = \$1200$, into the equation $p(x) = ax + b$ we can easily solve for a and b:

$$1500 = 10,000a + b$$

$$1200 = 15,000a + b,$$

implies

$$a = \frac{-3}{50}, \qquad b = 2100.$$

Thus the demand equation is

$$p(x) = \frac{-3x}{50} + 2100,$$

and we can, for example, obtain the appropriate price the company should charge in order to sell $20,000$ PC's; $p(20,000) = \$900$.

The **total revenue** obtained from selling x units at a price $p(x)$ per unit is given by $x \cdot p(x)$. It is usually desirable to maximize the total revenue (taxation not being a factor in our analysis).

Example [10.9] Referring to Example [10.8], what should the company charge for each PC in order to maximize the total revenue,

The function to be maximized is

$$r(x) = x \cdot p(x) = x \cdot \left(\frac{-3x}{50} + 2100 \right),$$

subject to the bounds $0 \leq x \leq 35,000$ (since we certainly do not want the total revenue to be negative).

Step 1. $r'(x) = \frac{-6x}{50} + 2100 = 0$ when $x = 700 \cdot 25$.

Step 2. $r(0) = r(35,000) = 0$, and $r(700 \cdot 25) = 700 \cdot 25 \cdot 1050$.

Step 3. The total revenue is maximized when $x = 17,500$, and in this case the price per PC would be $p(17,500) = \$1050$.

While maximizing the total revenue is certainly desirable, it is the profit which is often the most important issue. Let $c(x)$ denote the **total cost** of producing all x units. The **profit**, $P(x)$, obtained from selling all x units at the price $p(x)$ per unit is given by

$$P(x) = x \cdot p(x) - c(x).$$

Example [10.10] Assume that the cost of producing one PC in Example [10.8] is \$300. What should the price per PC be to maximize the profit.

In this case the cost function is $300x$ and thus the profit function is

$$P(x) = x \cdot \left(\frac{-3x}{50} + 2100 \right) - 300x = 1800x - \frac{3x^2}{50}.$$

Taking the derivative yields $P'(x) = \frac{-6x}{50} + 1800$ which vanishes at $x = 15,000$. The maximum profit is obtained by calculating P at this point:

$$P(15,000) = 15,000\left(1800 - \frac{3 \cdot 15,000}{50} \right) = \$13,500,000.$$

Exercises for §10.4

(1) A manufacturer of computer chips produces 10 megabyte memory cards. When the cards were sold for \$350 a piece only 6000 chips were sold. When the price was lowered to \$300 in the following month the company managed to sell 8000 chips.

(a) Assuming the demand function $p(x)$ is linear, find $p(x)$.

(b) If the cost of producing one memory card is \$50, what price should the company charge to maximize its profits.

(2) An automobile company is introducing a new model. When they first offered the car at the sticker price of \$25,000 they found that 1000 vehicles were sold in a week. When the price was raised to \$28,000 on;y 800 were sold.

(a) Assuming the demand function $p(x)$ is linear, find $p(x)$.

(b) If the cost of producing one new car is actually \$12,000, what price should the company charge to maximize its profits.

(3) A university ice cream shop estimated their daily demand function to be

$$p(x) = 5 - \frac{x}{40},$$

where x is measured in gallons and $p(x)$ is measured in dollars. The shop has fixed costs (rent, utility bills, wages, etc.) of \$110 per day and it costs them \$1.00 per gallon to purchase the ice cream so that the daily cost function is $110 + x$.

(a) What price should the shop charge per gallon of ice cream in order to maximize profits? How many gallons will it sell per day at this price? What will be the daily profit?

(b) If the landlord raises the rent by \$5 per day, how will the answers in part (a) change?

(4) A gourmet muffin shop has established a daily demand function of

$$p(x) = \frac{-19}{6000}x + \frac{43}{20}.$$

The shop has fixed daily costs of \$50 (this covers rent, utilities, wages, etc.). Assuming it costs the shop 25 cents to produce a muffin, what would the shop charge per muffin in order to maximize profits? What will be the daily profits?

(5) A startup company has received a government loan of \$100,000 to develop and market a new software product. Market research has shown that the monthly demand function for the software should be $p(x) = 80 - \frac{x}{25}$ dollars for $1000 \leq x \leq 3000$. The company has fixed monthly costs of \$10,000 (for rent wages, etc.) and the cost of producing the software is \$5 per unit.

(a) What price should the company charge for its software to maximize profits?

(b) If the price derived in part (a) is utilized, how long will it take to recover the original amount the government loaned the company?

(6) A company is producing cellular telephones. Market research has shown that the demand for the cellular phone is given by $p(x) = \frac{-x}{20} + 200$. Furthermore, the cost of producing x telephones is $2000\sqrt{x}$.

(a) Explain why the average cost of producing a telephone decreases as the number of telephones produced increases.

(b) Show that the profit is maximized when the number of telephones produced, x, satisfies the equation

$$\frac{-x}{10} + 200 - 1000x^{-\frac{1}{2}} = 0.$$

(c) Use your CAS to find the closest integer solution x to the equation given in (b) above.

Additional exercises for Chapter X

Use the three step algorithm of 10.1 to find the maximum and minimum values of f(x), if any, in the stated interval. Graph the functions on your CAS to verify computations.

(1) $f(x) = \frac{x}{\sqrt{2x+1}}$ for $0 \leq x \leq 10$.

(2) $f(x) = \frac{\sin(x)}{2+\cos(x)}$ for $0 \leq x \leq 2\pi$.

(3) $f(x) = 2\sec(x) - \tan(x)$ for $0 \leq x \leq \frac{\pi}{4}$.

(4) $f(x) = \begin{cases} 4x - 2 & \text{if } x < 1 \\ x^2 - 5x + 6 & \text{if } x \geq 1 \end{cases}$ for $\frac{1}{2} \leq x \leq 4$.

(5) $f(x) = \begin{cases} -x - 2 & \text{if } x \leq -2 \\ -x^2 + 4 & \text{if } -2 < x < 2 \\ x - 2 & \text{if } x \geq 2 \end{cases}$ for $-4 \leq x \leq 4$.

(6) Considering problems (4) and (5), what modifications would you make to the three step algorithm.

(7) Find the number in the closed interval $\frac{1}{4} \leq x \leq \frac{3}{4}$ such that the sum of the number and its reciprocal is:

(a) as small as possible;

(b) as large as possible.

(8) A triangle is inscribed in a semicircle of radius 10 so that one side is along the diameter. Find dimensions of the triangle with maximum area.

(9) Show that the right-circular cylinder of greatest volume that can be inscribed in a right circular cone has volume that is $\frac{4}{9}$ the volume of the cone.

(10) A church wishes to install a Norman window (a rectangle topped with a semi- circle). The window is to have perimeter p. Find the radius of the semicircle if the area of the window is to be maximum.

(11) A square sheet of cardboard of side L is used to make an open box by cutting out squares of equal size from the four corners and bending up the sides. What size squares should be cut from the corners to obtain a box with largest possible volume?

(12) A farmer estimates that if he digs his potatoes now he will have 120 bushels, which he can sell at $1.75 per bushel. If he expects his crop to increase 8 bushels per week, but the price to drop 5 cents per bushel per week, in how many weeks should he sell to realize the maximum amount for his crop?

(13) If $C(x)$ is a cost function, then the marginal cost function is $C'(x)$. Why is $C'(x)$ called the marginal cost? A graph of several cost functions may help to answer this question.

(14) Suppose a certain company has cost function $C(x) = 6000 + 42x - 0.004x^2$. Find the marginal cost of producing 300 items.

(15) The average cost per unit is given by $\frac{C(x)}{x}$. With your CAS graph the cost function and the average cost functions of problem 14. Determine where average cost is minimized. Where is cost minimized?

(16) A pipe of negligible diameter is to be carried horizontally around a corner from a hallway 8ft wide into a hallway 4 ft wide. What is the maximum length that the pipe can have?

(17) Find the points on the ellipse $x^2 - 2xy + 4y^2 = 12$ where the abscissa x has its greatest and least values. Graph the ellipse with your CAS to verify your calculation.

(18) The strength of a rectangular beam is proportional to the product of its breadth and the square of its depth. Find the dimensions of the strongest rectangular beam that can be cut from a circular log of diameter 18 inches.

(19) Find the dimensions of the largest rectangle that can be inscribed in the curve $x^{2/3} + y^{2/5} = 8$.

Chapter XI

Graphing Functions

11.1 Graphing with the First and Second Derivative Test

Let $y = f(x)$ be a differentiable function on an interval $a \leq x \leq b$. We now describe a very rapid method for graphing $y = f(x)$ which can easily be executed by hand for a large class of functions. It is presented as a four step algorithm.

Step 1: Graph the two end points $(a, f(a))$, and $(b, f(b))$.

Step 2: Find all real numbers $a \leq x_1 \leq x_2 \leq \ldots \leq x_n \leq b$ such that the condition $f'(x_i) = 0 \; (i = 1, 2, \ldots, n)$ holds. These are termed **critical values**. The corresponding points $(x_i, f(x_i)$ on the graph are termed **critical points**. Graph the critical points:
$(x_1, f(x_1)), \;\; (x_2, f(x_2)), \; \ldots, (x_n, f(x_n))$.

Step 3: Classify each of the critical values $x_1, x_2, \ldots, x_n$ using the first or second derivative tests. Draw in a local maximum, minimum, or inflection at each $(x_i, f(x_i))$, $(i = 1, 2, \ldots, n)$.

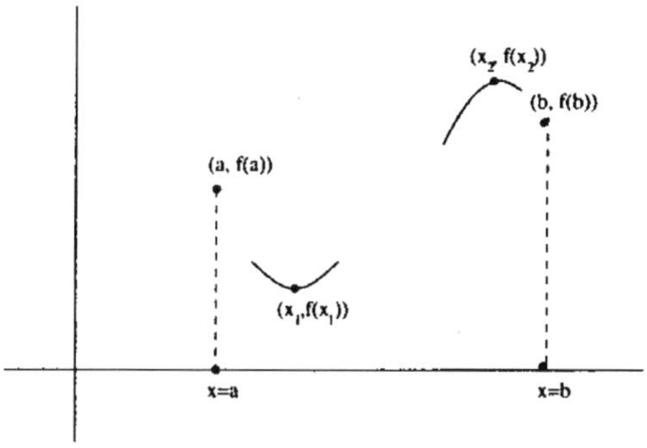

Step 4: Finally, connect the dots to complete the graph.

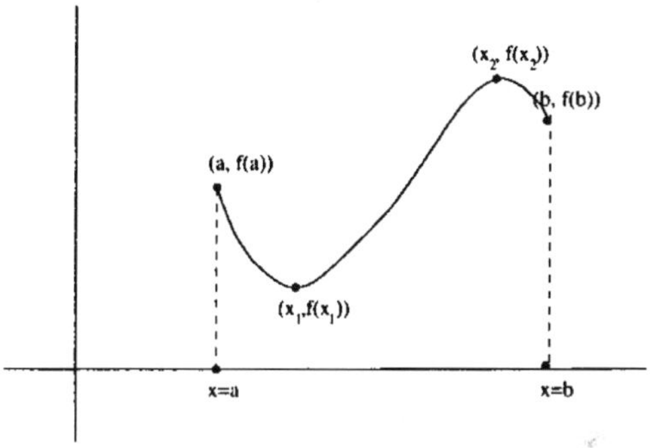

Example [11.1] Let $f(x) = 3x^4 - 8x^3 + 20$ on the interval $-1 \le x \le 3$.

Step 1: $f(-1) = 31$, $f(3) = 47$. We first graph the two endpoints $(-1, 31)$ and $(3, 47)$.

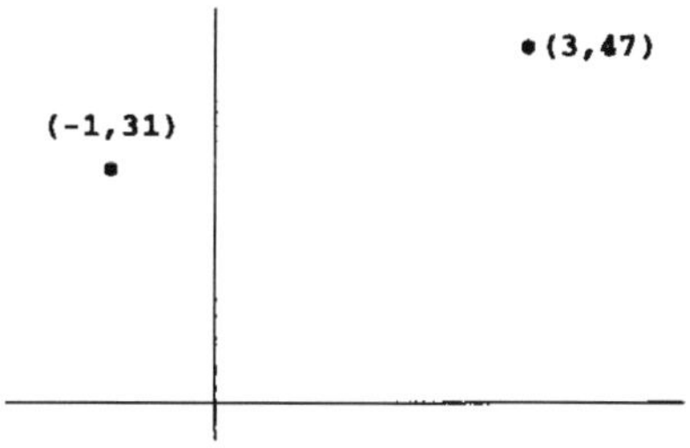

Step 2: $f'(x) = 12x^3 - 24x^2 = 12x^2(x-2)$. When we set $f'(x) = 0$ we see that $f'(0) = f'(2) = 0$, and the critical values are located at $x = 0$ and $x = 2$. We graph the critical points $(0, 20)$ and $(2, 4)$.

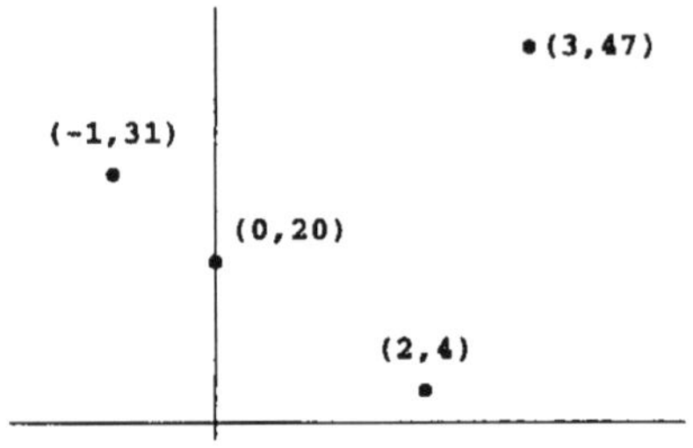

Step 3: $f'(x) = 12x^2(x-2) < 0$ when $x < 0$ and when $0 \le x < 2$. Hence the point $(0, 20)$ is an inflection point with $f(x)$ decreasing on this interval. On the other hand, $f''(2) = 48 > 0$ which tells us that $(2, 4)$ is a local minimum.

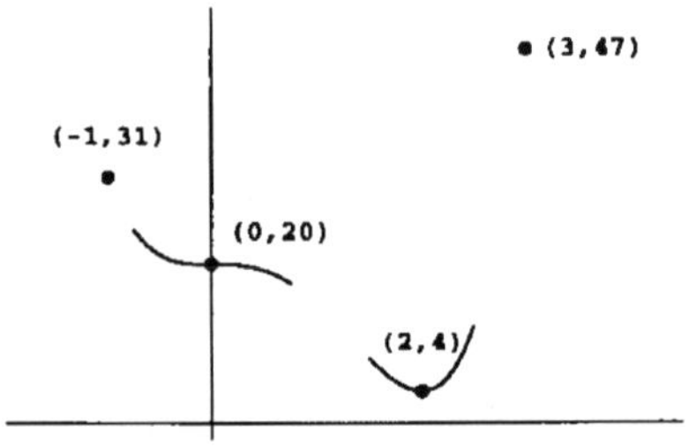

Step 4: We connect the dots to complete the graph.

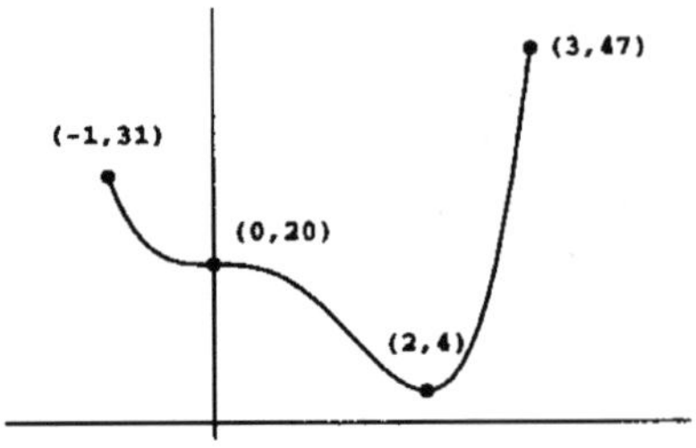

Exercises for §11.1

Use the four step algorithm which was presented in §11.1 to graph each of the following functions on the indicated interval. Use your CAS to quickly compute the first and second derivatives at the various critical values. After sketching the graph, compare your drawing with the graph the CAS plots.

(1) $y = x^2 - 4x - 21, (-10 \le x \le 10)$.

(2) $y = 11 - x^5, (-2 \le x \le 2)$.

(3) $y = \ln(x) - x^3, (\frac{1}{4} \le x \le 2)$.

(4) $y = \cos(x^2), (\frac{1}{2} \le x \le 10)$.

(5) $y = xe^x, (-2 \le x \le 2)$.

(6) $y = e^{x^2}, (-2 \le x \le 2)$.

(7) $y = x^7 - 3x^6 + 2x^5, (-1 \le x \le 2)$.

(8) $y = \sin(x) + \cos(x), (0 \le x \le 2\pi)$.

(9) $y = 3x^4 + 16x^3 - 6x^2 - 48x - 200, (-3 \le x \le 3)$.

(10) $y = x + \frac{1}{x}, (-2 \le x \le -\frac{1}{2})$.

(11) $y = 3x^8 - 52x^6 + 216x^4 - 500, (-4 \le x \le 4)$.

(12) $y = x^x, (\frac{1}{4} \le x \le 1)$.

11.2 Graphing with Cusps

If we graph a function such as $y = x^{\frac{2}{3}}$ a cusp appears at $x = 0$.

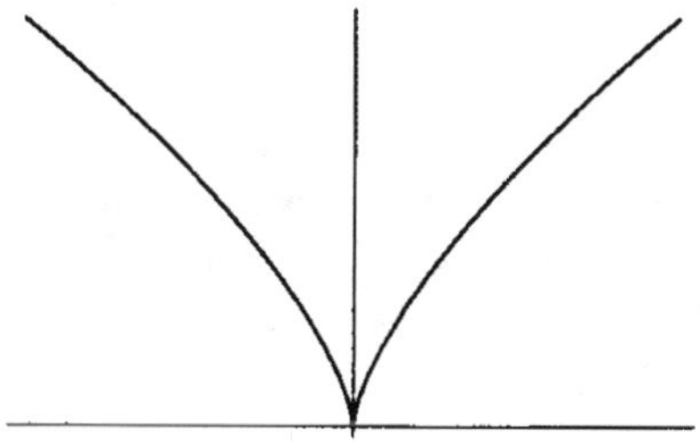

Definition: *If a function $y = f(x)$ is differentiable in a neighborhood of a point $x = c$ except at the point $x = c$ itself,*

$$\lim_{x \to c} f'(x) = \pm\infty,$$

and $f'(x)$ changes sign as x passes through c, then we term the point $x = c$ a ***cusp***.

Remark: Since the derivative becomes infinite as we approach $x = c$ we deduce that the tangent line becomes vertical as we approach the cusp.

For example, in the case $f(x) = x^{\frac{2}{3}}$ we have that $f'(x) = \frac{2}{3}x^{-\frac{1}{3}}$, and, therefore, $f'(0) = \infty$. Since $f'(x) < 0$ for $x < 0$ and $f'(x) > 0$ for $x > 0$ we infer that f is decreasing for $x < 0$ and increasing for $x > 0$.

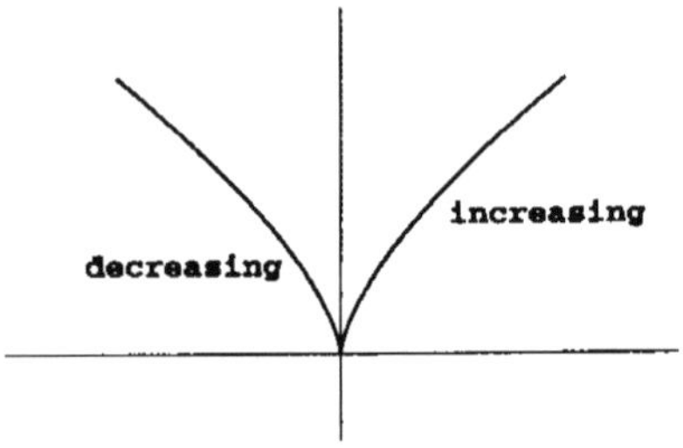

If we graph the function $y = |x - 1| + 1$ on the interval $\frac{1}{2} \leq x \leq \frac{3}{2}$ we find a corner at $x = 1$.

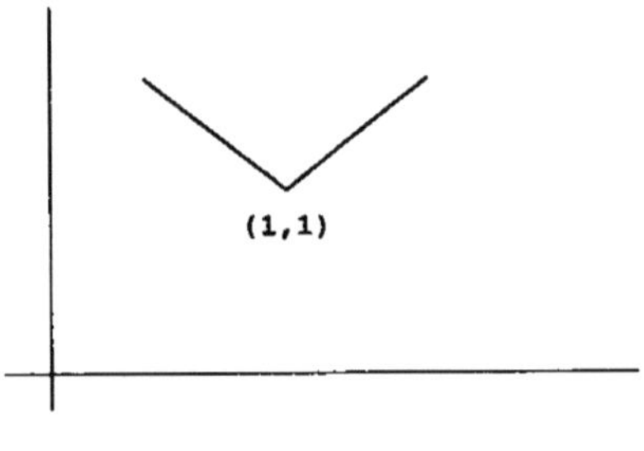

At the point $x = 1$ this curve does not have a derivative — in fact here there are two possible tangent lines which leads to the corner.

To graph all the possible cusps and corners, it will be necessary to add an additional **Step 3′** to the four step method of §11.1.

> **Step 3′:** Find all points $x = c_1, c_2, \dots, c_m$ where $f'(x)$ is infinite or does not exists. Graph the points $(c_i, f(c_i))$ for $i = 1, 2, \dots, m$. At each c_i determine on both sides whether $f(x)$ is increasing or decreasing. Draw in the cusps or corners.

The presence of cusps in the above examples motivates the extension of the definition of a critical value.

Definition: *A **critical value** of a function $f(x)$ is a real number x where either $f'(x) = 0$, $f'(x) = \pm\infty$, or $f'(x)$ does not exist.*

Exercises for §11.2

Use the four step method of §11.2 to graph the following functions on the indicated interval. As before, use your CAS to evaluate the first and second derivatives of the various functions at their critical values. Be sure to include all cusps and corners. Compare your drawing with the graph the CAS plots.

(1) $y = x^{\frac{4}{5}}, (-10 \le x \le 10)$.

(2) $y = \sqrt{4 - x^2}, (-2 \le x \le 2)$.

(3) $y = x^2 - |2 - x|, (-1 \le x \le 4)$.

(4) $y = (x - 5)^{\frac{2}{7}} + 3, (0 \le x \le 10)$.

(5) $y = x^{\frac{2}{7}}(1 - x), (-1 \le x \le 1)$.

(6) $y = x^{\frac{2}{7}}(1 - x) + |2 - x|, (-1 \le x \le 3)$.

(7) $y = |\cos(x)|, (0 \le x \le \pi)$.

(8) $y = x^x + |x - \frac{1}{2}|, (\frac{1}{4} \le x \le 1)$.

11.3 Concavity

Consider a function which is increasing on an interval $a \le x \le b$. There are basically three possible scenarios for this. The function may increase at a constant rate,

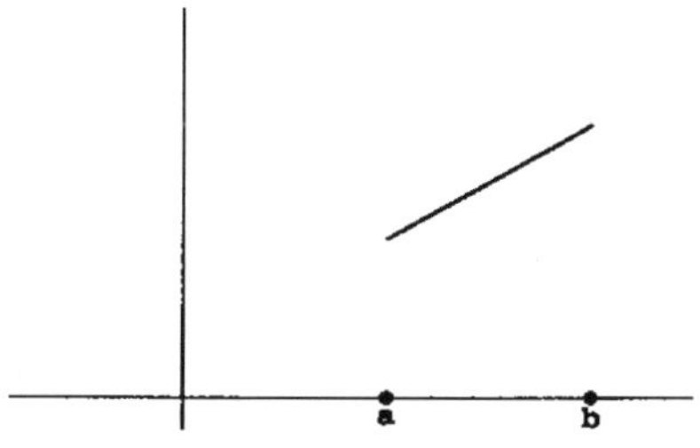

in a concave up manner,

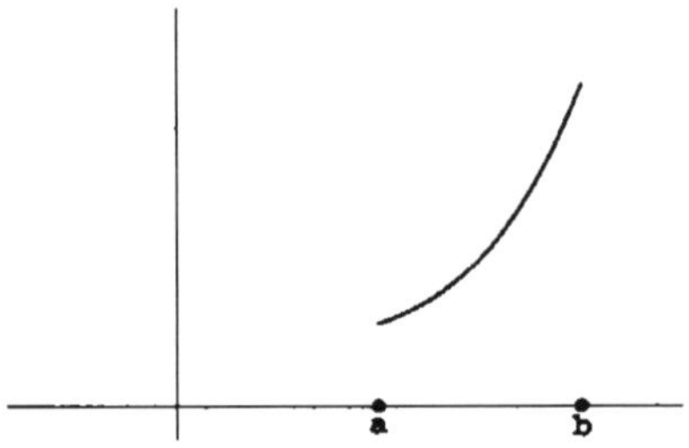

or in a concave down manner.

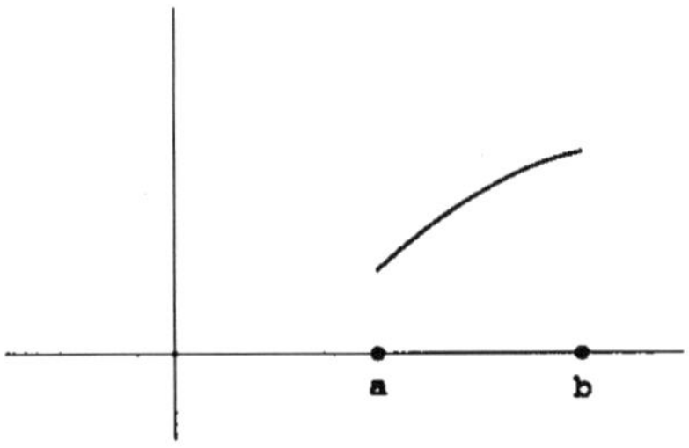

Note that in the concave up situation, the curve lies below the dotted line (whose slope is constant and positive),

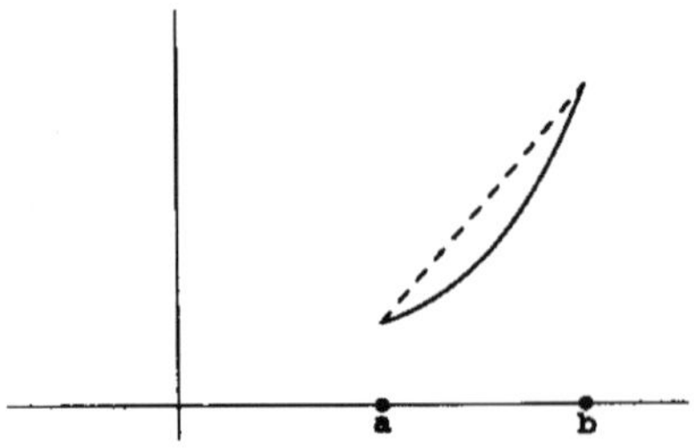

while in the concave down situation it lies above the dotted line.

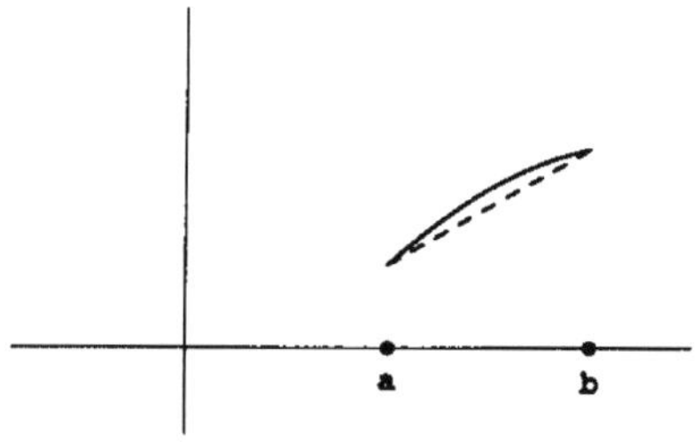

The case of a decreasing function is completely analogous and as before there are three cases.

It is sometimes necessary to know precisely how a function is increasing or decreasing, i.e., to specify the concavity. Here we utilize the second derivative to obtain a simple criterion.

Consider a function which is concave up and increasing.

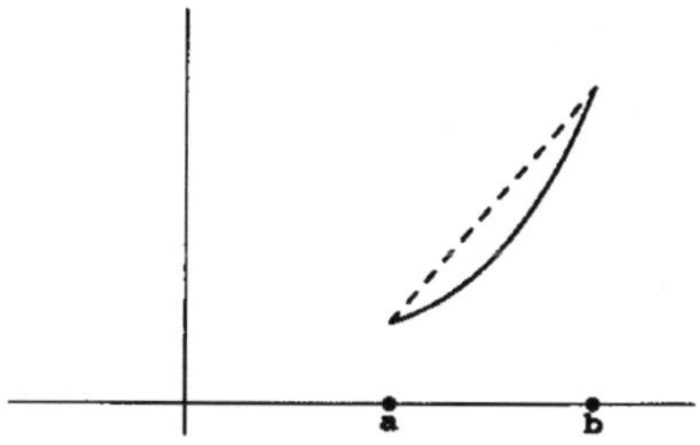

The dotted line (whose slope is constant and positive) is a function of the form $ax+b$ whose second derivative is zero. The slope of the tangent line to the concave up curve must be increasing, i.e., $f'(x)$ must be increasing as x goes from a to b. But this implies $f''(x) > 0$ (here we are using the first derivative test applied to the function $F(x) = f'(x)$ which is known to be increasing). This reasoning allows us to give a mathematically precise definition of concavity.

Definition: *If a function $f(x)$ is twice differentiable on an interval $a \le x \le b$ and $f''(x) \ge 0$ for all $a \le x \le b$ then we say that f is **concave up** on this interval.*

Definition: *If a function $f(x)$ is twice differentiable on an interval $a \le x \le b$ and $f''(x) \le 0$ for all $a \le x \le b$ then we say that f is **concave down** on this interval.*

Example [11.2] Consider the function $f(x) = (1-x)^3$ in the region $0 \le x \le \frac{1}{2}$. Since $f''(x) = 6(1-x)$ we see that $f''(x) > 0$ for all $0 \le x \le \frac{1}{2}$. Hence f must be concave up in this region.

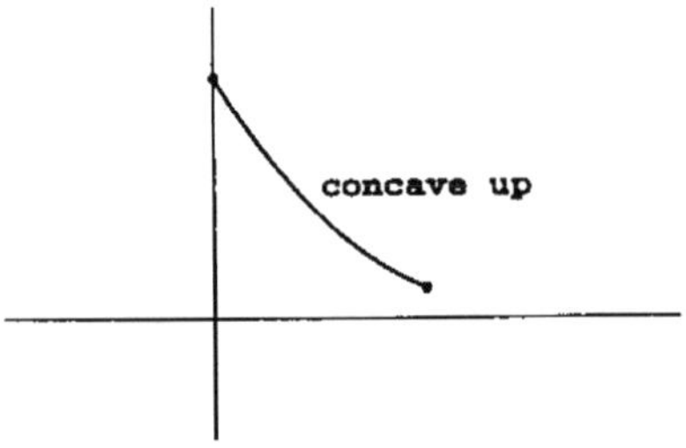

Example [11.3] Consider again the function $f(x) = (1-x)^3$ for $0 \le x \le 2$. Now since $f''(x) = 6(1-x)$ we see that $f''(x) > 0$ for $0 \le x < 1$, $f''(1) = 0$, and $f''(x) < 0$ for $1 < x \le 2$. Hence the concavity changes at the inflection point $x = 1$. This motivates the following definitions.

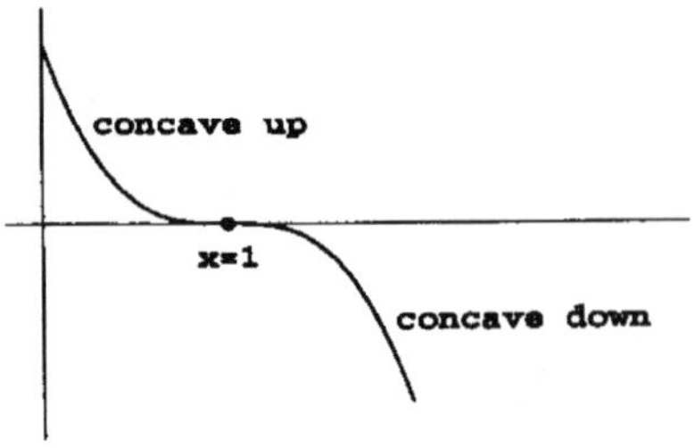

Definition: *If $f''(x) > 0$ for $a \leq x < c$, $f''(c) = 0$, and $f''(x) < 0$ for $c < x \leq b$ then we term $x = c$ an* ***inflection point*** *(or a point where the concavity changes).*

Definition: *If $f''(x) < 0$ for $a \leq x < c$, $f''(c) = 0$, and $f''(x) > 0$ for $c < x \leq b$ then again we term $x = c$ an* ***inflection point*** *(or a point where the concavity changes).*

Warning: Just because $f''(c) = 0$ does not insure that $x = c$ is an inflection point. Consider for example $f(x) = (1-x)^4$. Here $f''(x) = 12(1-x)^2$, thus $f''(1) = 0$. But $x = 1$ is not an inflection point: $f''(x)$ is positive for all $x \neq 1$ and $f(x) = (1-x)^4$ is always concave up.

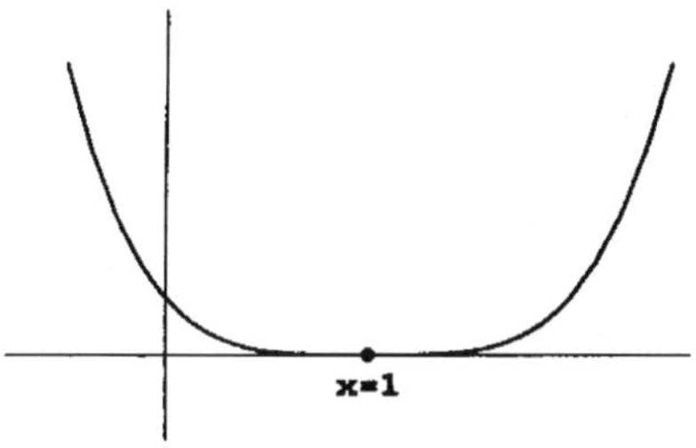

Exercises for §11.3

Determine the concave structure and all inflection points for the graphs of the following functions on the indicated intervals. Use your CAS to quickly compute derivatives and check your answers by plotting the graphs with the CAS.

(1) $y = (x-2)^5$, $(0 \leq x \leq 4)$.

(2) $y = 4x^3 - 3x^2 - 6x - 10$, $(0 \leq x \leq 2)$.

(3) $y = x^2 - 7x + 2$, $(-\infty \leq x \leq \infty)$.

(4) $y = \ln(x)$, $(\frac{1}{2} \leq x \leq 10)$.

(5) $y = (2-x)^6$, $(0 \leq x \leq 4)$.

(6) $y = \sin(x)$, $(-\infty \leq x \leq \infty)$.

(7) $y = e^x$, $(-\infty \leq x \leq \infty)$.

(8) $y = \sin(x) + \cos(x)$, $(0 \leq x \leq 2\pi)$.

(9) $y = x^{\frac{2}{3}}$, $(-2 \leq x \leq 2)$.

(10) $y = \sqrt{4-x^2}$, $(-2 \leq x \leq 2)$.

(11) $y = -\sqrt{4-x^2}$, $(-2 \leq x \leq 2)$.

(12) $y = e^{x^2}$, $(-2 \leq x \leq 2)$.

(13) $y = 2x + \frac{18}{x}$, $(-10 \leq x \leq 10)$.

(14) $y = x^2 + |3-x|$, $(0 \leq x \leq 4)$.

(15) Show that the general cubic function $y = ax^3 + bx^2 + cx + d$ (where we assume $a \neq 0$) has exactly one inflection point.

(16) Show that the most general quadratic function $y = ax^2 + bx + c$ (where we assume $a \neq 0$) has no inflection points and exactly one critical point. Can you describe the concavity for this situation?

Additional exercises for Chapter XI

(1) Prove that an nth degree polynomial $f(x) = a_0 + a_1x + \cdots + a_nx^n, (a_n \neq 0)$ has at most $(n-2)$ inflexion points. (**Hint:** look at problem 15 in the exercises for §11.3).

(2) For the general cubic $f(x) = ax^3 + bx^2 + cx + d \ (a \neq 0)$ find conditions on a, b, c, d to assure that f is always increasing or decreasing on $-\infty < x < \infty$

(3) Find all critical points of $f(x) = x\tan(x)$ for $-\pi/2 < x < \pi/2$ and classify them as inflexion points or points of nondifferentiability. Check by graphing on your CAS.

(4) Determine the concavity, inflexion points, and maxima/minima for $f(x) = \sin(x^2), 0 \leq x \leq 2\pi$. Graph on your CAS to check your answer.

(5) Graph $f(x) = x^{1/3}(x+4)^{4/3}$ with your CAS and determine the critical points from the graph. Verify them by taking the derivative.

(6) Let

$$f(x) = \begin{cases} -x+1, & \text{if } x \leq 1 \\ -(x-\frac{3}{2})^2 + \frac{1}{4}, & \text{otherwise} \end{cases}$$

Find all critical points on the interval $0 \leq x \leq 2$. Determine their nature and check by graphing on your CAS.

(7) Suppose $f'(x) = \begin{cases} -x+1, & \text{if } x \leq 1 \\ -(x-\frac{3}{2})^2 + \frac{1}{4}, & \text{otherwise.} \end{cases}$

(a) What are the critical values of f?

(b) Where is f increasing, decreasing?

(c) Where is f concave up, concave down?

(d) Find a possible function expression for f and check with your CAS.

(8) Graph $f(x) = \frac{3x^2-4}{\exp(x)}$ with your CAS and find all minima and maxima.

(9) Suppose x_0 is a minima for both f and g. Is it a critical point for $f+g$, $f-g$, fg, $\frac{f}{g}$(assuming $g \neq 0$)? What kind if at all? Explain for each case.

(10) Make a rough sketch of the function f around the point u based on the following information:

(a) $\lim_{x \to u^-} f'(x) = +\infty, f''(x) > 0$, for $x < u$;

$\lim_{x \to u^+} f'(x) = 0, f''(x) < 0$, for $x > u$;

(b) $\lim_{x \to u^-} f'(x) = +\infty, f''(x) > 0$, for $x < u$;

$\lim_{x \to u^+} f'(x) = -\infty, f''(x) > 0$, for $x > u$.

(11) Sketch a smooth curve $y = f(x)$ so that the following properties are satisfied: $f(1) = 0$, and $f'(x) > 0$ for $x < 1$, $f'(x) < 0$ for $x > 1$.

(12) Sketch a smooth curve $y = f(x)$ so that the following properties are satisfied: $f(1) = 2$, and $f''(x) > 0$ for $x < 1$, $f''(x) < 0$ for $x > 1$.

(13) Determine the value of b so that the curve $x^3 + bx^2 + x + 4$ has an inflexion point at $x = 3$.

(14) Determine a, b so that the curve $ax^3 + bx^2 + x + 4$ has an inflexion point at $(2, 2)$.

Chapter XII
Asymptotes

12.1 Generalities on Asymptotes

Two curves are termed **asymptotic** on the interval, $a \leq x \leq b$, if they come arbitrarily close to each other but never intersect on the interval. For example if we assume the curves C_1 and C_2 below continue indefinitely and get closer and closer to each other (without touching), then they are asymptotic.

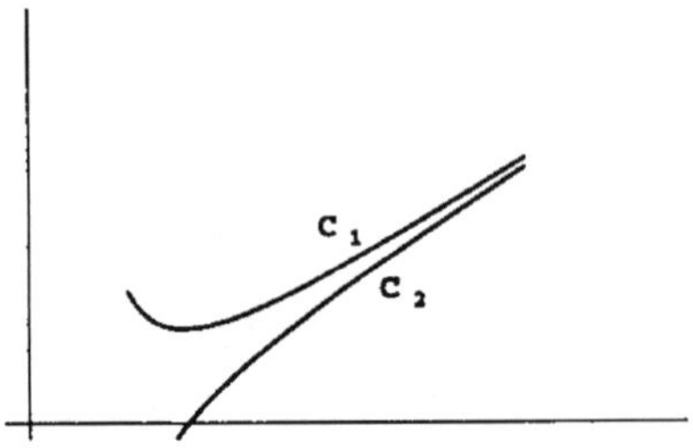

Of particular interest is the case where a curve is asymptotic to a straight line.

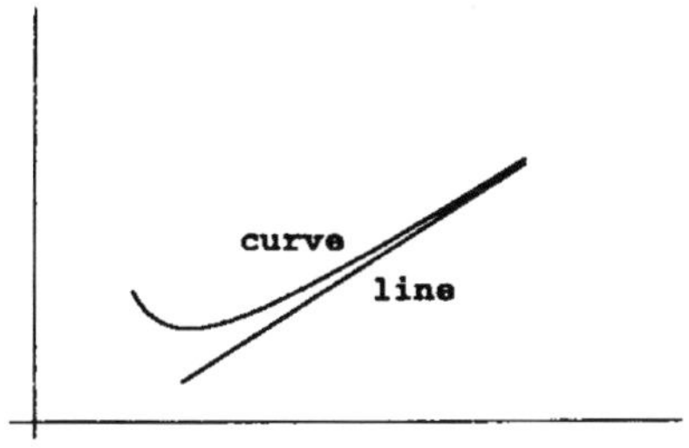

If the line happens to be vertical,

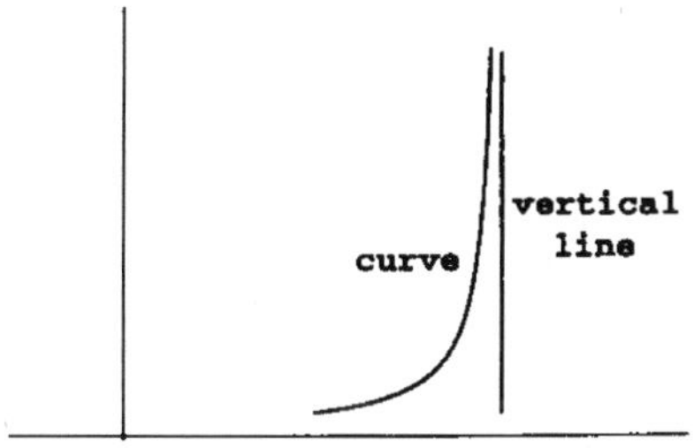

we refer to it as a **vertical asymptote**, while if the line is horizontal,

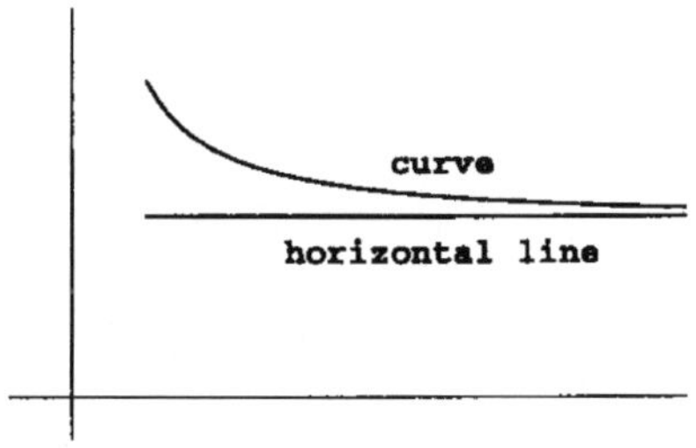

it is called a **horizontal asymptote**.

Remark: The line (horizontal or vertical) *is* the asymptote and is termed an asymptote of the curve.

12.2 Vertical Asymptotes

In chapter 11 we focused our attention on various sketching techniques which are applicable to functions which are well defined and differentiable on an interval $a \leq x \leq b$. In nature, however, functions appear which are not so well behaved. If, for example, we plot the stock market average in the 1920's we observe the infamous crash.

We now broaden our class of functions which we would like to consider by including functions which **blow up** (become infinite in either the positive or negative direction) at some point. A basic example of this behavior is the function

$$f(x) = \frac{1}{x-3}$$

which blows up as $x \to 3$: the graph of $f(x)$ in the region $0 \leq x \leq 6$ is depicted below.

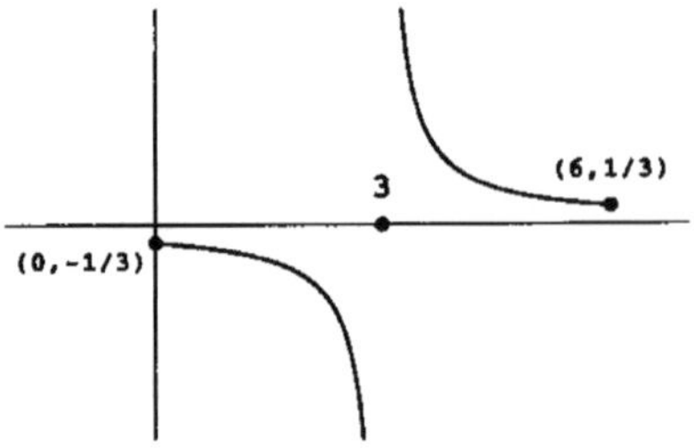

This can be seen by graphing the points $f(0) = -1/3$, $f(1) = -1/2$, $f(2.5) = -2$, $f(2.95) = -20$, etc.

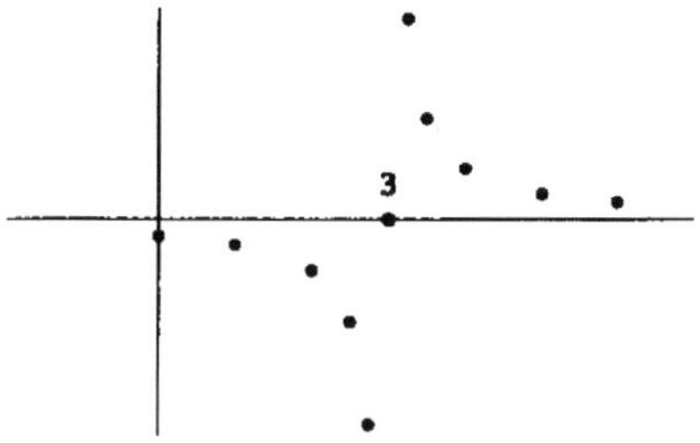

and the points $f(6) = 1/3$, $f(5) = 1/2$, $f(4) = 1$, $f(3.5) = 2$, $f(3.05) = 20$, etc. It is pictorially very helpful to draw in a dotted vertical line at $x = 3$.

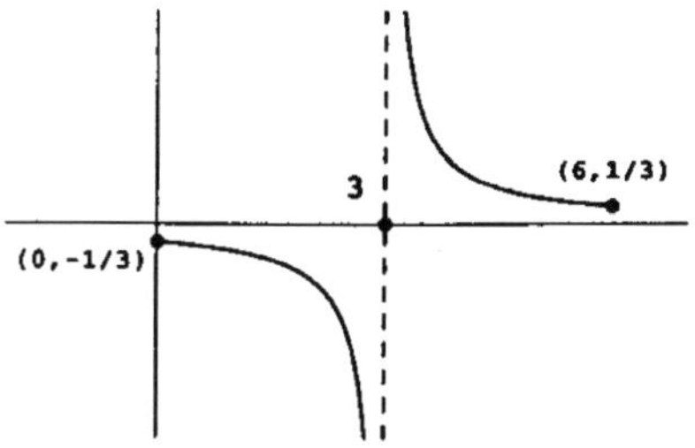

In fact this dotted line will be a vertical asymptote to our curve $y = \frac{1}{x-3}$.

We next consider a somewhat more complex function,

$$f(x) = \frac{2x+1}{(x-2)(x+1)}$$

whose graph

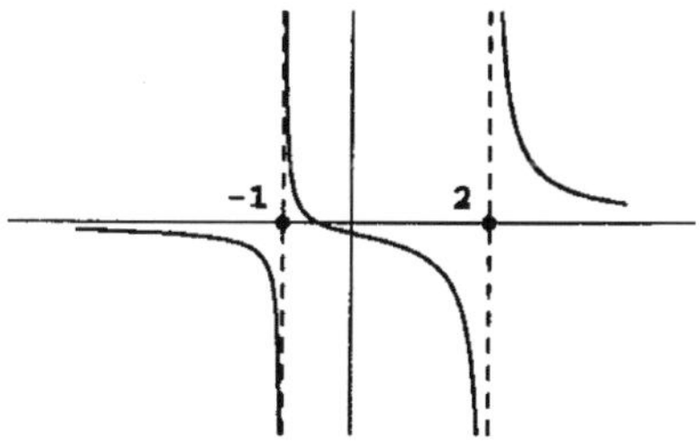

includes the vertical asymptotes at $x = 2, -1$.

We first observe that $f(x)$ blows up at $x = 2, -1$ and no where else. This implies that the vertical asymptotes are located there. Now for $x > 2$, $f(x) > 0$, and hence for $x > 2$ and near 2 the graph must locally look like the following.

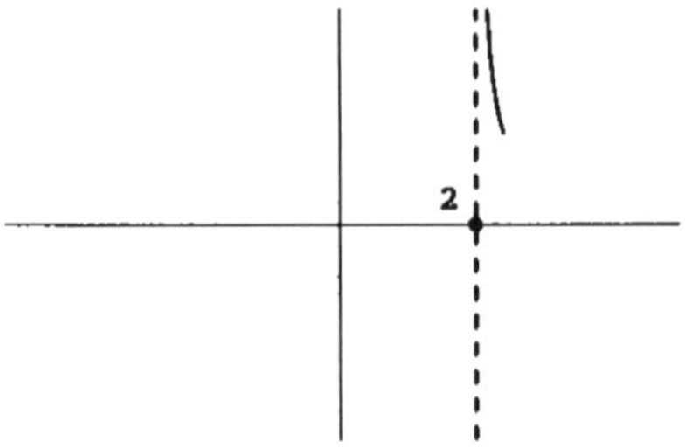

In addition, for $x < 2$ and near 2, $f(x) < 0$ and we obtain another piece of the graph.

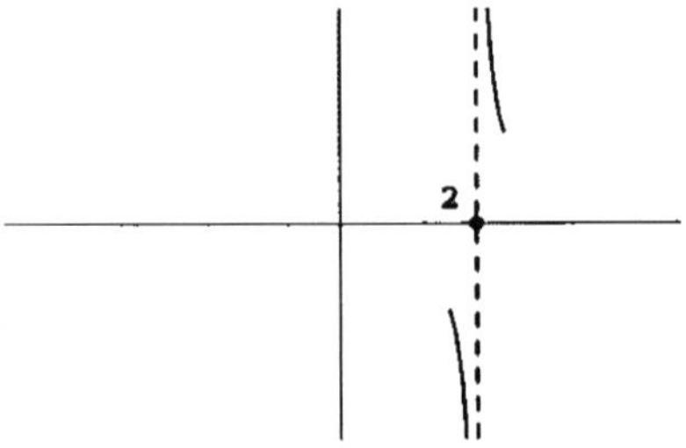

A similar analysis in the neighborhood of $x = -1$ yields,

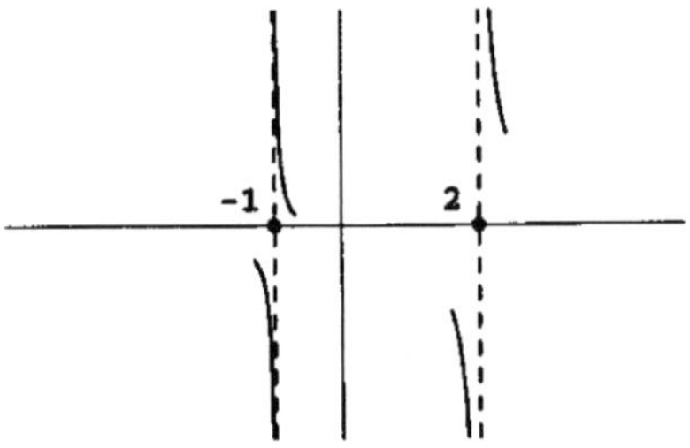

which may now be completed using the methods of §11.1.

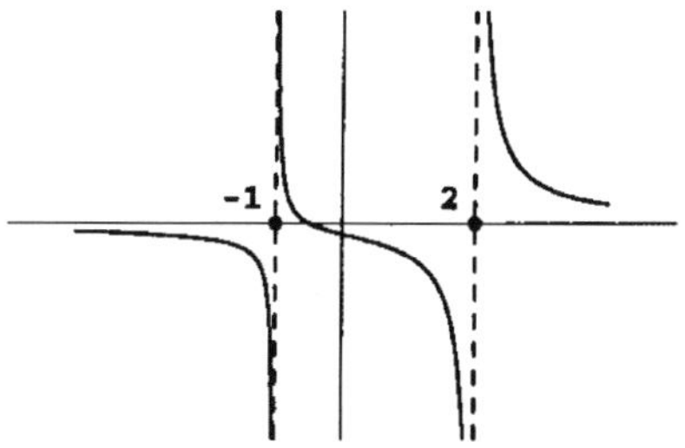

We summarize this procedure.

The vertical asymptote test and sketching method:

Step 1. Given $y = f(x)$ defined on an interval $a \leq x \leq b$ find all points $a \leq a_1, a_2, \ldots, a_n \leq b$ where $f(x)$ blows up.

Step 2. Draw in dotted vertical lines at $x = a_1, a_2, \ldots, a_n$.

Step 3. At each a_i $(i = 1, \ldots, n)$ determine whether $f(x)$ is positive or negative for $x > a_i$ and $x < a_i$ (but x near a_i). Sketch in the corresponding pieces of the curve.

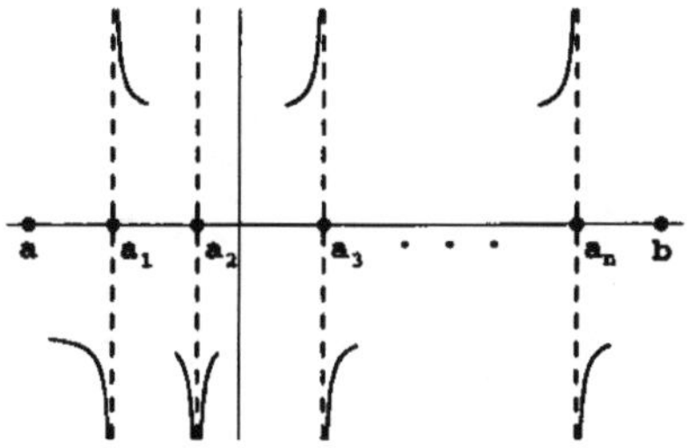

Complete the graph using the four step method detailed in §11.1.

Remarks: After sketching the various possible blow ups of the curve (termed **singularities**) we are left to sketch $f(x)$ where it is well defined and differentiable (and thus the methods from §11.1 apply). To obtain the singularities $x = a_1, a_2, \ldots, a_n$ attempt to express $f(x)$ as a ratio of two functions, i.e., $f(x) = p(x)/q(x)$. In general, the points where $q(x)$ is zero give us the $x = a_1, a_2, \ldots, a_n$.

Exercises for §12.2

For each of the following functions $f(x)$, determine every real number a where $f(a) = \pm\infty$. Check your answers by having the CAS compute $f(a)$. You should get an error message. Plot these functions with your CAS to see if vertical asymptotes appear at $x = a$.

(1) $f(x) = \frac{3x-1}{x^2-3x-4}$.

(2) $f(x) = \frac{4x^3+11}{2x^2-3x-1}$.

(3) $f(x) = \frac{\cos(x)}{(x-11)}$.

(4) $f(x) = e^{\frac{1}{x}}$.

(5) $f(x) = \frac{\sin(x)}{x^2+2x}$.

(6) $f(x) = \frac{x-1}{x^3-x}$.

(7) $f(x) = \frac{x^2+2}{\ln(x)}$, $(0 < x < \infty)$.

(8) $f(x) = \frac{2x+1}{\cos(x)}$.

(9) $f(x) = \frac{x-1}{x^3-1}$.

Sketch the following functions in the indicated interval using the vertical asymptote test and sketching method. Draw in dotted lines at each vertical asymptote. Use your CAS to compute the necessary values of the function and its derivative. Check your work by having the CAS plot the function.

(10) $f(x) = \frac{x^2-3}{x-1}$, $(-10 \le x \le 10)$.

(11) $f(x) = \frac{2x+3}{x^2-3x-4}$, $(-5 \le x \le 5)$.

(12) $f(x) = \frac{x+3}{x^2-1}$, $(-4 \le x \le 4)$.

(13) $f(x) = \frac{4}{2x^2-3x-1}$, $(-2 \le x \le 2)$.

(14) $f(x) = \frac{x^2+1}{x^3-3x^2}$, $(-4 \le x \le 4)$.

(15) $f(x) = \frac{2x+3\sin(x)}{x-2}$, $(-1 \le x \le 4)$.

(16) $f(x) = \tan\left(x - \frac{\pi}{2}\right)$, $(\frac{\pi}{2} \le x \le 2\pi)$.

(17) $f(x) = e^{\frac{1}{x}}$, $(-1 \le x \le 1)$.

(18) $f(x) = \frac{\ln(x)}{x-1}$, $(\frac{1}{2} \le x \le 2)$.

(19) $f(x) = \frac{1}{\sin(x)}$, $(-\frac{\pi}{2} \le x \le \frac{\pi}{2})$.

12.3 Horizontal Asymptotes

In §12.1 we introduced the horizontal asymptote to a curve depicted below.

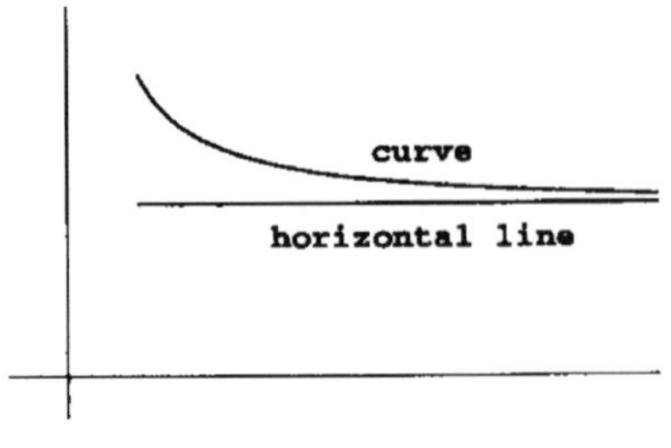

The key point to understanding this phenomena is that a horizontal asymptote cannot occur in a finite interval $a \le x \le b$. As long as we restrict ourselves to such a finite interval there are only two possibilities: either the horizontal line and curve meet,

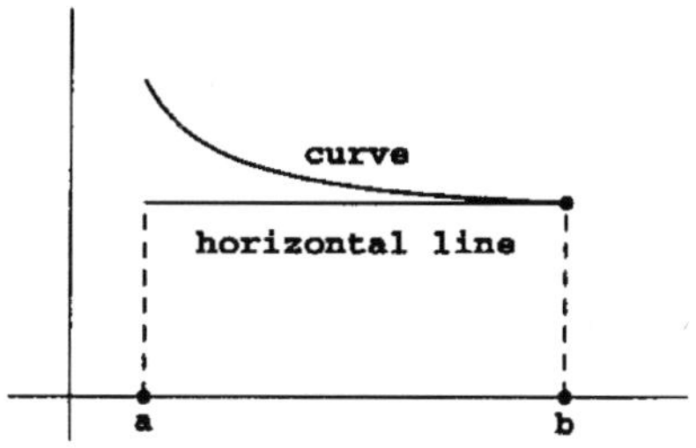

or they do not.

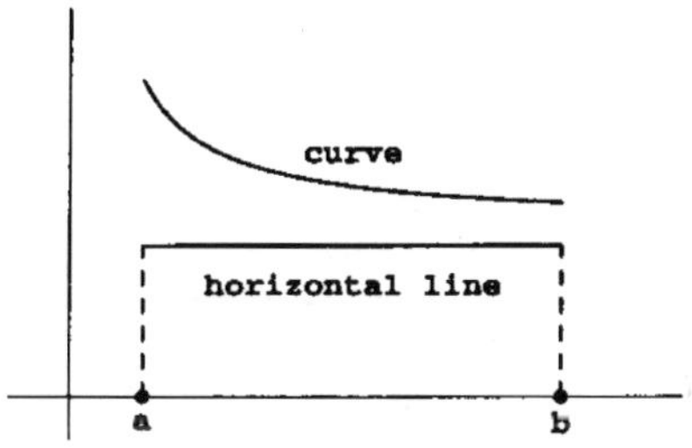

Since by definition a curve and its horizontal asymptote must come arbitrarily close but never intersect we must consider an infinite interval of the type $a \leq x \leq \infty$,

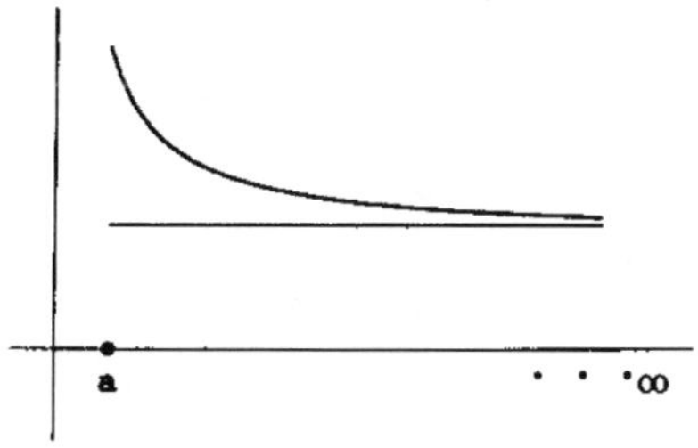

or of the type $-\infty \leq x \leq a$.

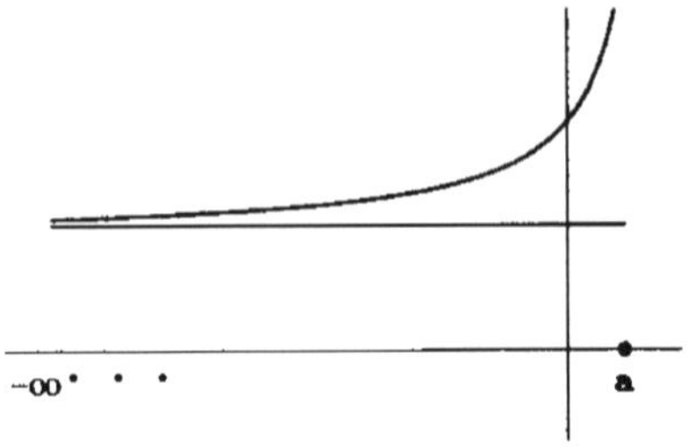

Example [12.1] Consider the function $f(x) = \frac{2x-1}{x+1}$. If we graph this over a finite interval $-3 \leq x \leq 3$ using the techniques of §12.2 and §11.1 we obtain the following graph.

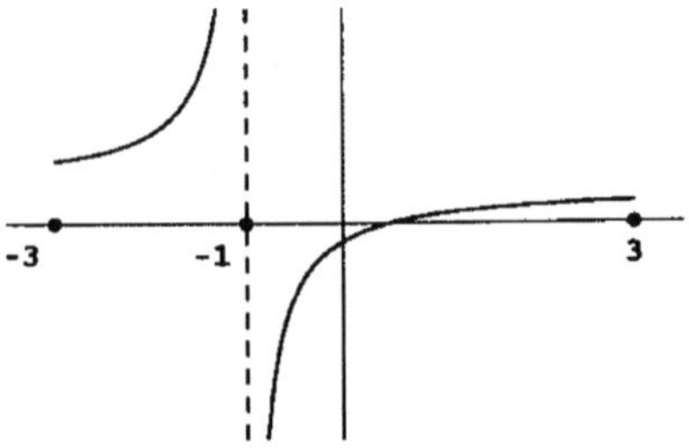

Extending the interval to $-10 \leq x \leq 10$ we obtain,

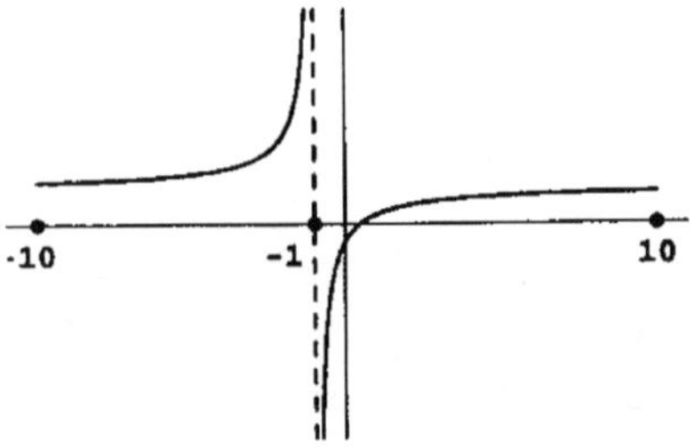

and extending once more to $-100 \leq x \leq 100$ we can visualize the horizontal asymptote at $y = 2$.

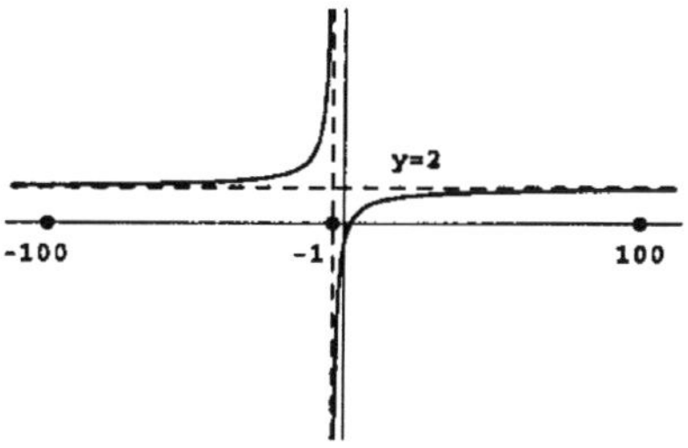

Notice in this example since $f(x) = \frac{2x-1}{x+1}$, we have that $f(3) = \frac{5}{4} \approx 1.25$, and that $f(10) = \frac{19}{11} \approx 1.73$, $f(100) = \frac{199}{101} \approx 1.97$, $f(1000) = \frac{1999}{1001} \approx 1.997$, etc. Clearly $f(x)$ is less than 2 but becomes arbitrarily close to 2 as $x \to \infty$. We compute the limit formally as follows:

$$\lim_{x\to\infty} \frac{2x-1}{x+1} = \lim_{x\to\infty} \frac{2x-1}{x+1} \cdot \overbrace{\frac{1/x}{1/x}}^{\text{The trick: multiply by 1}}$$
$$= \lim_{x\to\infty} \frac{2-1/x}{1+1/x}$$
$$= \frac{2}{1} = 2,$$

since $\frac{1}{x} \to 0$ as $x \to \infty$.

This example motivates the following propositions:

Proposition [12.2] *Let $f(x)$ be a function defined on an infinite interval $a \le x \le \infty$. If $\lim_{x\to\infty} f(x) = c$ and $f(x) \ne c$ for $a \le x \le \infty$, then $f(x)$ has a horizontal asymptote at $y = c$.*

Proposition [12.3] *If $\lim_{x\to-\infty} f(x) = c$, for a function $f(x)$ defined on an interval $-\infty \le x \le a$, and $f(x) \ne c$ for $-\infty \le x \le a$, then $f(x)$ has a horizontal asymptote at $y = c$.*

Proof: We focus on the first proposition in that the latter is analogous. By definition, $\lim_{x\to\infty} f(x) = c$ is equivalent to saying that $|f(x) - c|$ becomes arbitrarily small as the variable x approaches infinity.

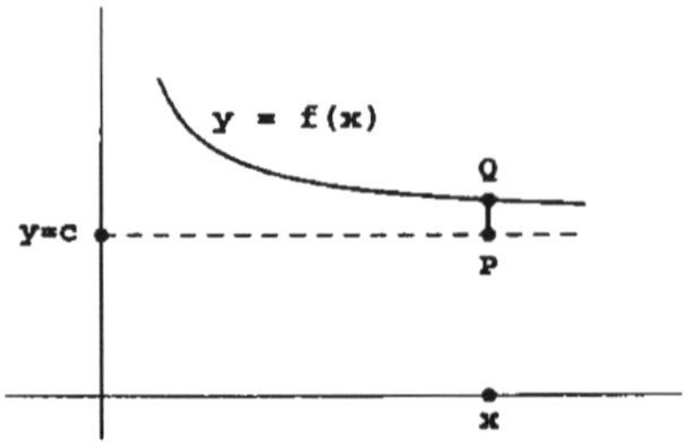

The distance between the curve and the horizontal asymptote (PQ in the diagram above), is in fact $|f(x)-c|$. Since this distance becomes arbitrarily small as $x \to \infty$ the proposition is proved. The condition $f(x) \neq c$ for $a \leq x \leq \infty$ insures that the curve $y = f(x)$ and the line $y = c$ never meet on the interval.

The horizontal asymptote test and sketching method:

Step 1. Given $f(x)$ defined on an interval $a \leq x \leq \infty$, compute $\lim_{x \to \infty} f(x)$.

Step 2. If this limit exists and equals c then the line $y = c$ is a horizontal asymptote provided that the curve $y = f(x)$ does not intersect the line $y = c$ for $x > a$.

Step 3. Draw a horizontal dotted line at $y = c$ and determine if $f(x) > c$ or $f(x) < c$ for $x \to \infty$. Sketch in the corresponding piece of the curve,

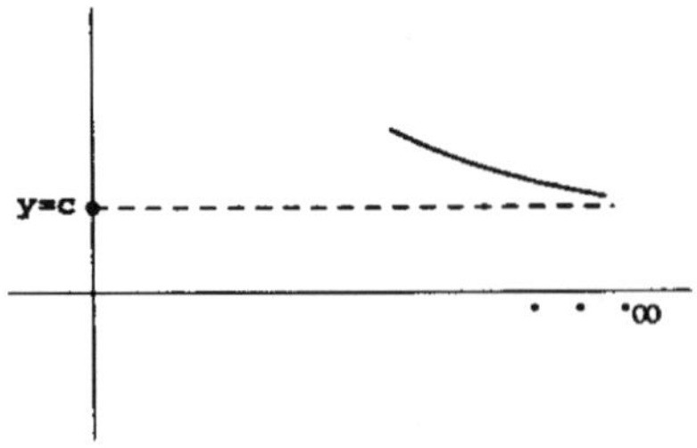

and complete the graph using the vertical asymptote test and sketching methods of §11.1 and 12.2.

Remarks: The same method applies to an interval $-\infty \leq x \leq a$. If we wish to consider the function over the entire real axis $-\infty \leq x \leq \infty$, it is necessary to find the two possible asymptotes by computing the two limits $\lim_{x \to \infty} f(x)$ and $\lim_{x \to -\infty} f(x)$. If neither limit exists then there are no horizontal asymptotes.

Example [12.4] The function $f(x) = x^2$

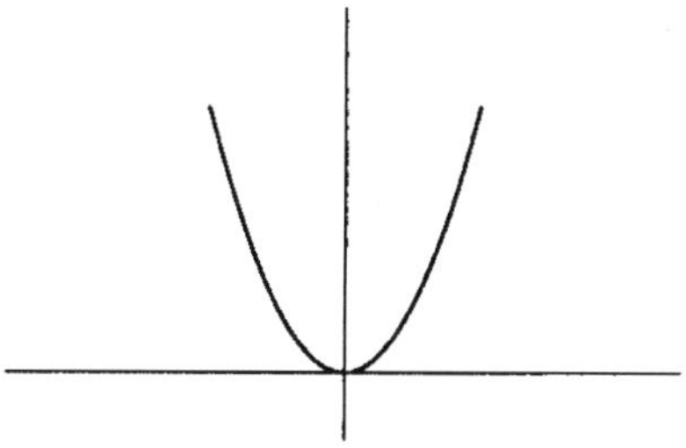

has no horizontal asymptotes since $\lim_{x \to \pm\infty} x^2 = \infty$.

Example [12.5] Find the horizontal asymptotes of the function $f(x) = \frac{x}{\sqrt{x^2+1}}$.

In this case

$$\lim_{x \to \infty} \frac{x}{\sqrt{x^2+1}} = +1,$$

and

$$\lim_{x \to -\infty} \frac{x}{\sqrt{x^2+1}} = -1.$$

Thus there are horizontal asymptotes at $y = 1$ and $y = -1$. Plotting with the CAS we obtain the graph.

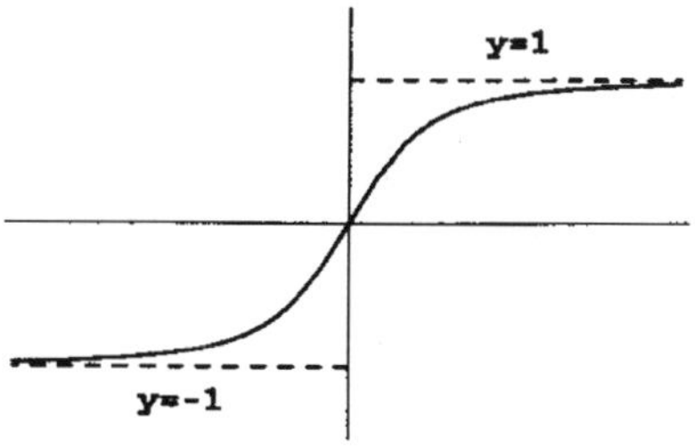

Exercises for §12.3

For each of the following functions $f(x)$, compute the limits

$$\lim_{x\to+\infty} f(x), \qquad \lim_{x\to-\infty} f(x).$$

If at least one of the limits exist the graph of $f(x)$ will have a horizontal asymptote. Check your answers by plotting the graph with your CAS.

(1) $f(x) = \frac{3x^2-2x+11}{x^2+1}$.

(2) $f(x) = \frac{-6x^2+x+9}{2x^2+3}$.

(3) $f(x) = \frac{x^3+2x+9}{2x^3-1}$.

(4) $f(x) = e^x$.

(5) $f(x) = \frac{\ln(x^2)}{x-1}$.

(6) $f(x) = e^{-x^2}$.

(7) $f(x) = e^{\frac{1}{x}}$.

(8) $f(x) = \frac{x^2\cdot\cos(x)}{3x^2-2x+3}$.

Sketch the graphs of the following functions. Include all vertical and horizontal asymptotes. Use your CAS to compute the required values of the function and its derivative. Compare your sketch with the graph the CAS plots.

(9) $f(x) = \frac{2x}{x+4}$.

(10) $f(x) = x + \frac{2}{x}$.

(11) $f(x) = \frac{x-3}{x+3}$.

(12) $f(x) = \frac{x}{2x-5}$.

(13) $f(x) = \frac{x^2+1}{x^2-1}$.

(14) $f(x) = \frac{x^2+x}{x-1}$.

(15) $f(x) = \frac{2x^3}{\sqrt{3x^6+10}}$.

(16) $f(x) = \frac{2x^2-3}{x^2+1}$.

(17) $f(x) = \frac{e^x}{x^2-1}$.

(18) $f(x) = \frac{e^{\frac{1}{x}}}{x-3}$.

(19) $f(x) = \frac{e^{-x^2}}{x^2-1}$.

(20) $f(x) = \sqrt{x^2+2} - \sqrt{x^2+1}$.

(21) Consider the function $y = \frac{\sin(x)}{x}$. Show that the curve intersects the line $y = 0$ infinitely often but $\lim_{x\to\pm\infty} \frac{\sin(x)}{x} = 0$. Plot this curve with your CAS. Is $y = 0$ a horizontal asymptote?

(22) Show that the curve $y = \frac{2x^2+3x+1}{x}$ is asymptotic to the line $y = 2x+3$. Visually check your answer by graphing with the CAS on an appropriate interval.

(23) Show that the curve $y = \frac{x^4+3x^2+x-1}{x^2}$ is asymptotic to the curve $y = x^2+3$. Visually check your answer by graphing with the CAS on an appropriate interval.

Additional exercises for Chapter XII

Sketch the graph of the following functions.Include all vertical and horizontal asymptotes. Use your CAS to computethe required function values and derivatives at critical values. Compare your sketch with the graph the CAS plots,

(1) $y = \frac{x^3 - x^2 + 3}{x^2 - 1}$.

(2) $y = \tan(x)$. (Find all asymptotes for $x > 0$).

(3) $y = \tanh(x)$.

(4) $y = \ln(x^2) - 1$.

(5) $y = 2x^2 - x^{2/3}$.

(6) $y = \frac{\cos^2(x)}{1+\cos(x)}$.

(7) $y = \frac{\sin(x)}{1+\cos(x)}$.

(8) $y = x^{2/3} - x^{2/5}$.

(9) $y = x^{2/3} - (1 - x)^{2/3}$.

(10) $y = \frac{x^2+3}{x^3-x+9}$.

(11) Show that a continuous curve $y = f(x)$ is asymptotic to a line $y = mx + b$ if

$$\lim_{x\to\infty} (f(x) - (mx + b)) = 0$$

or

$$\lim_{x\to-\infty} (f(x) - (mx + b)) = 0.$$

In exercises (12)-(17) determine if the given curve $y = f(x)$ is asymptotic to a line. Find all such asymptotic lines.

(12) $y = \frac{x^3}{x^2-3}$.

(13) $y = 7x - 2 + \frac{4x}{x^2-1}$.

(14) $y = \frac{x}{\sqrt{x^2+1}}$.

(15) $y = \frac{2x^2-6x+4}{x-3}$.

(16) $y = \frac{x^2}{3x+2}$.

(17) $y = x^3 - 2x + 1$.

(18) Find all lines asymptotic to the hyperbola $\left(\frac{x}{a}\right)^2 - \left(\frac{y}{b}\right)^2 = 1$.

Chapter XIII

The Integral as Area

13.1 Intuitive Definition of the Integral as an Area

Question: What is area?

Intuitive as this concept may seem, this is a very difficult question. It is the aim of this chapter to constructively provide a method of computing areas of regions whose boundaries are curved and not simply polygons (see §13.3). In sections 13.1 and 13.2 we will refer to area in an intuitive manner, much the same way the ancient Greeks referred to the area of a circle.

Let $f(x)$ be a continuous function defined on an interval $a \leq x \leq b$. Assume that for all x in the interval, $f(x) > 0$. Then the graph of $y = f(x)$ lies above the x–axis and resembles the figure below.

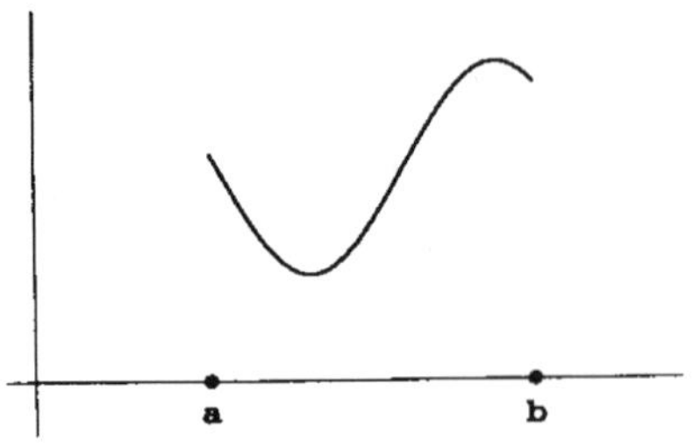

Definition: (Intuitive) *Under the assumption that the graph of the continuous function $f(x)$ lies above the x–axis, then the* ***integral from** a **to** b **of** $f(x)$*, *denoted $\int_a^b f(x)\,dx$, is defined to be the area of the region bounded by the curve $y = f(x)$, the x–axis, and the vertical lines $x = a$ and $x = b$.*

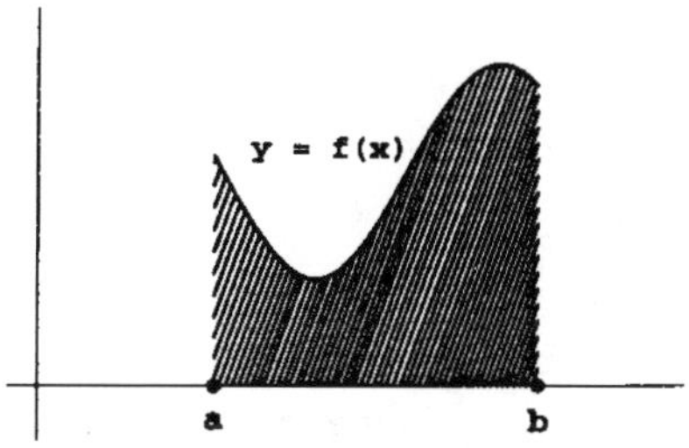

Remark: Pictorially $\int_a^b f(x)\,dx$ is simply the area of the shaded region above.

Example [13.1] What is $\int_0^1 x\,dx$? If we graph the curve $y = x$ on the interval $0 \le x \le 1$ we obtain the line segment depicted below. The integral is, by definition, the area of the shaded triangle (which is $\frac{1}{2}$ the area of the one by one square). We conclude

$$\int_0^1 x\,dx = \frac{1}{2}.$$

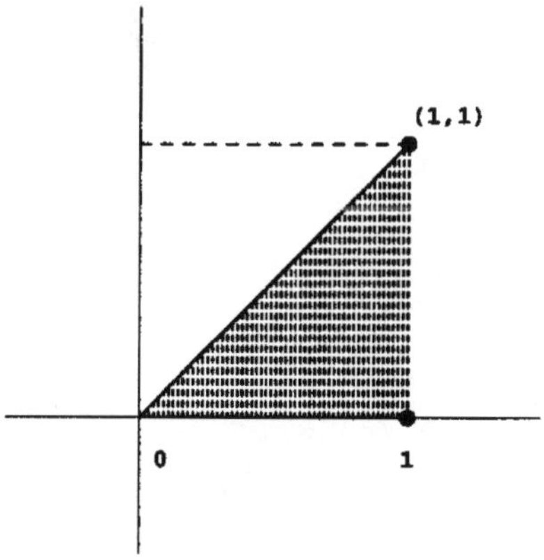

Example [13.2] What is $\int_1^4 2\,dx$? The graph of $y = 2$ on the interval $1 \le x \le 4$ is a horizontal line segment of length 3. The shaded region

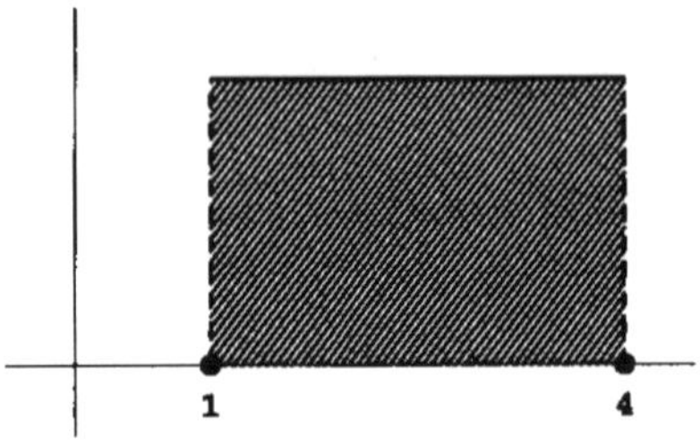

is thus a rectangle of area 6, i.e.,

$$\int_1^4 2\,dx = 6.$$

Question: What is the motivation for the notation $\int_a^b f(x)\,dx$?

Answer: The symbol $\int$ is called the **integral sign.** Historically it derived from an elongated S which was used to represent a sum. The integral may be looked at as a sum in the following manner (which is reminiscent of the classical Greek method for approximating the area of the circle). The area under the curve $y = f(x)$ may be broken up into rectangular type pieces.

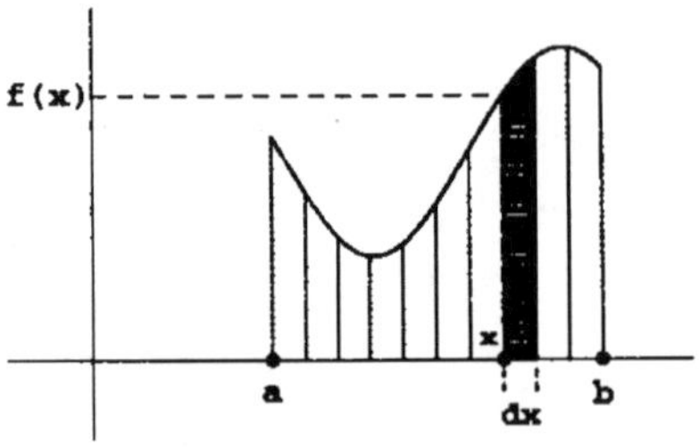

Traditionally the small widths (or increments) depicted above were denoted dx. We can think of dx as being a very small number, say 0.0001. The shaded region above, since it is almost rectangular, has approximate area $f(x){\cdot}dx$. The integral $\int_a^b f(x)\,dx$ may thus be approximated by the sum (from $x = a$ to $x = b$) of the rectangular type pieces whose areas are approximately $f(x) \cdot dx$. Notice that if we magnify the shaded area,

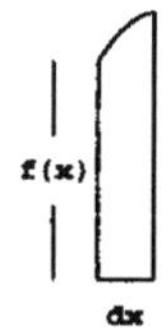

it is, in fact, curved and slanted at the top. By allowing $dx \to 0$, however, the region will look more and more like a rectangle, and the sum of the regions will approach the exact value of the integral. We will formalize this method in section §13.3.

Example [13.3] Approximate the value of $\int_0^1 x\,dx$ by breaking up the triangular region into 10 approximately rectangular pieces.

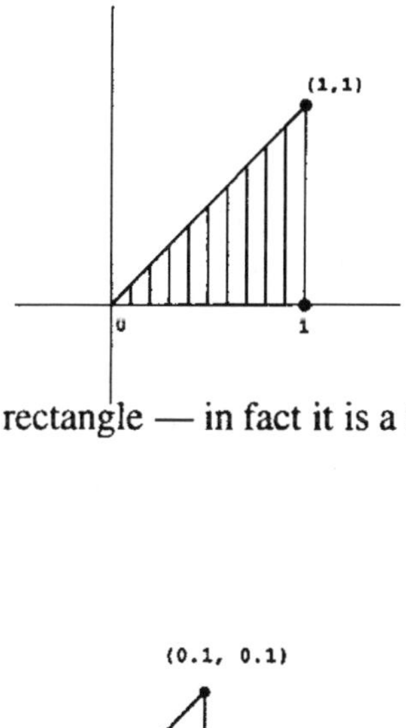

The first piece is not even a rectangle — in fact it is a triangle.

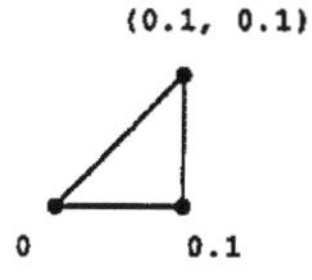

The second and subsequent pieces resemble rectangles.

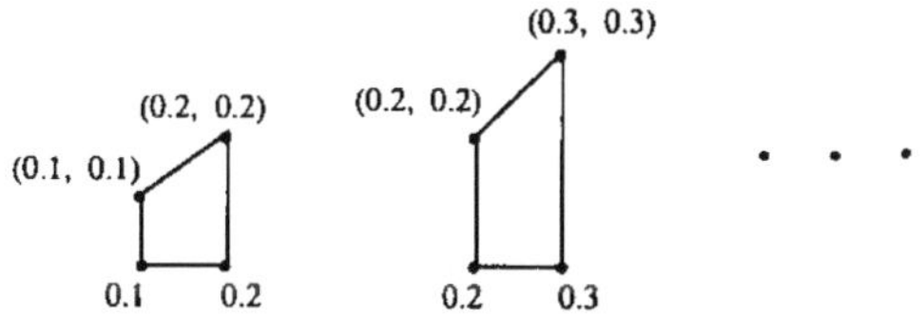

Since we are dividing the region into 10 pieces the value for dx is 0.1. The heights of the pieces increase linearly (since the function is a line) and are given by $0.1, 0.2, 0.3, \ldots, 1$. We can,therefore, approximate our integral:

$$\int_0^1 x\,dx \approx (0.1)(0.1) + (0.1)(0.2) + (0.1)(0.3) + \ldots + (0.1)(1)$$
$$= .55.$$

Note that this approximation is only $\frac{5}{100}$ greater than the exact area.

Exercises for §13.1

Compute the following integrals using the intuitive definition of the integral as area.

(1) $\int_0^5 3\,dx.$

(2) $\int_{-1}^2 5\,dx.$

(3) $\int_0^3 x\,dx.$

(4) $\int_0^3 (2x+1)\,dx.$

(5) $\int_1^3 (2x+1)\,dx.$

(6) $\int_2^4 5(3x-1)\,dx.$

(7) Approximate the integral $\int_0^2 x\,dx$ by breaking up the region into 10 approximately rectangular pieces. Compare your answer with the actual area. How close is the approximation?

(8) Approximate the integral $\int_0^2 (x+1)\,dx$ by breaking up the region into 10 approximately rectangular pieces. Compare your answer with the actual area. How close is the approximation?

13.2 The Integral of an Arbitrary Function

In §13.1 we restricted our attention to functions $y = f(x)$ whose graph (on the interval $a \le x \le b$) lie above the x–axis.

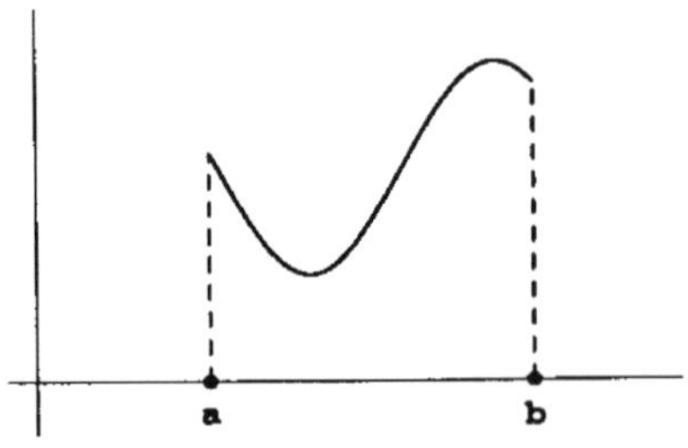

We now abandon this restriction and consider arbitrary continuous functions (whose graph may drop below the x–axis) defined on the interval $a \le x \le b$.

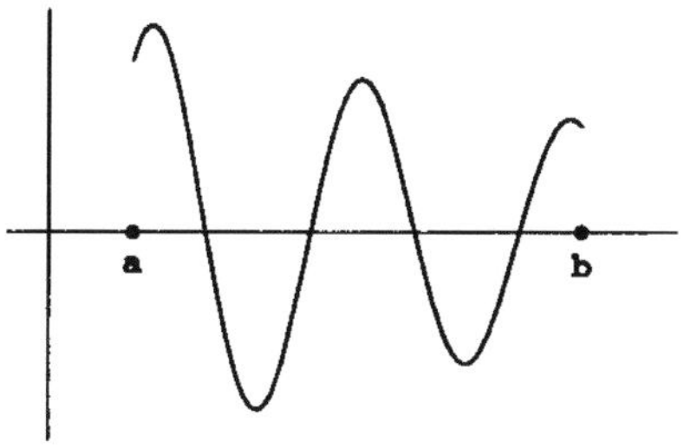

Definition: (Intuitive) *Let $f(x)$ be a continuous function defined on an interval $a \le x \le b$. The symbol $\int_a^b f(x)\,dx$ is defined to be the sum of the areas (taken positively when f lies above the x–axis and negatively when f lies below the x–axis) bounded by the curve $y = f(x)$, the x–axis, and the vertical lines $x = a$ and $x = b$.*

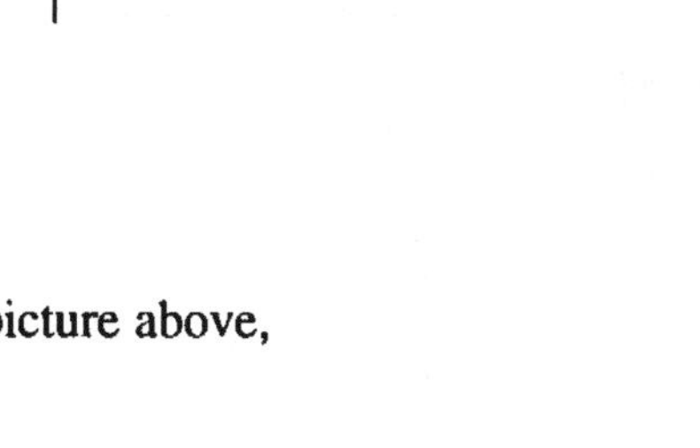

Example [13.4] In the picture above,

$$\int_a^b f(x)\,dx = \text{area}(A) + \text{area}(B) + \text{area}(C) - \text{area}(D) - \text{area}(E).$$

Example [13.5] Compute $\int_0^1 (3x-1)\,dx$. The graph of $y = 3x - 1$ on the interval $0 \le x \le 1$ is given below.

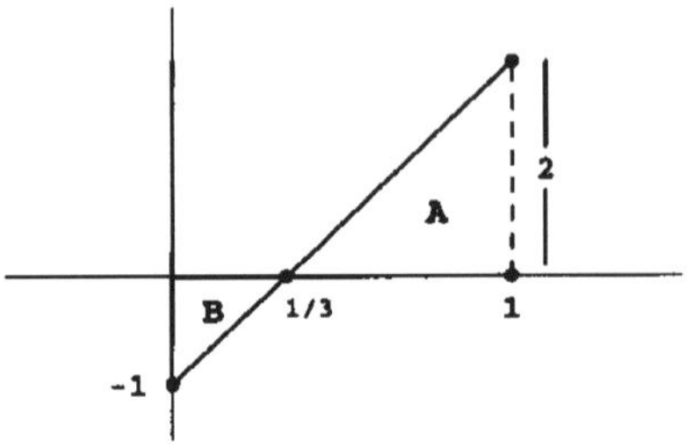

Since area$(A) = \frac{2}{3}$ and area$(B) = \frac{1}{6}$ we have that $\int_0^1 (3x - 1)\,dx = \frac{2}{3} - \frac{1}{6} = \frac{1}{2}$.

Exercises for §13.2

Compute the following integrals using the intuitive definition of the integral as area.

(1) $\int_1^3 (2x - 4)\,dx$.

(2) $\int_{-1}^2 (3x - 5)\,dx$.

(3) $\int_{-3}^3 x\,dx$.

(4) $\int_{-2}^3 (4x + 1)\,dx$.

(5) $\int_{-2}^{-1} (2x + 1)\,dx$.

(6) $\int_{-1}^0 (3x - 1)\,dx$.

13.3 The Integral as a Limit of a Sum

Let $y = f(x)$ be an arbitrary continuous function which is defined on an interval $a \le x \le b$. In order to simplify our exposition we temporarily assume that the graph of f lies above the x–axis.

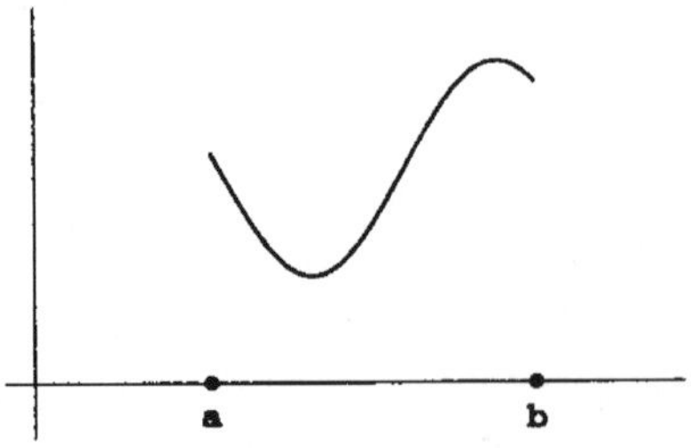

If we break the interval $a \le x \le b$ into N pieces (where N is a large positive integer), then each of the resulting smaller intervals will have length $\frac{b-a}{N}$.

The area under the curve $y = f(x)$ can also be broken up into N rectangular type pieces: for example, if $N = 5$, we get

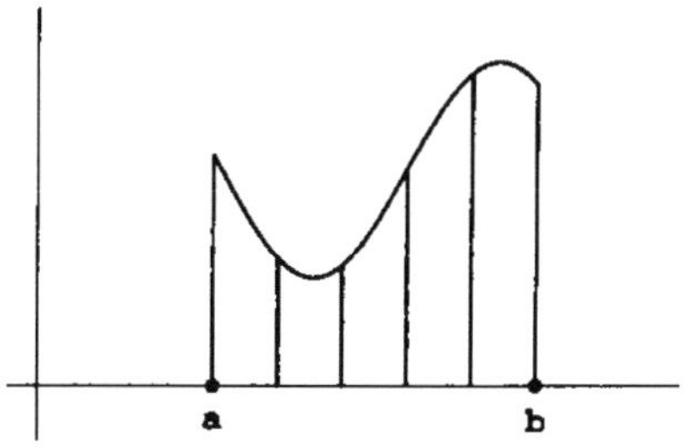

where each rectangular piece has width $\frac{b-a}{5}$. In order to analyze the above diagram in detail we need to label the points on the x–axis. In the case of $N = 5$ we obtain

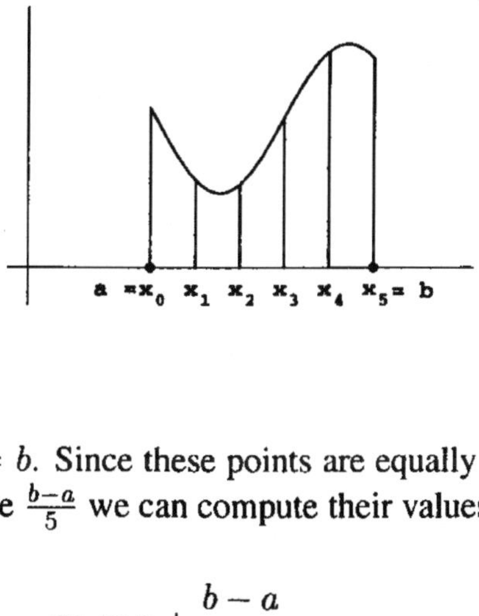

where $x_0 = a$ and $x_5 = b$. Since these points are equally spaced and the distance between any two must be $\frac{b-a}{5}$ we can compute their values explicitly:

$$x_1 = a + \frac{b-a}{5}$$

$$x_2 = a + \frac{2 \cdot (b-a)}{5}$$

$$x_3 = a + \frac{3 \cdot (b-a)}{5}$$

$$x_4 = a + \frac{4 \cdot (b-a)}{5}$$

$$x_5 = a + \frac{5 \cdot (b-a)}{5} = b.$$

More generally, when we break the area into N rectangular type pieces and we label the corresponding points $x_0, x_1, \ldots, x_N$ on the x–axis

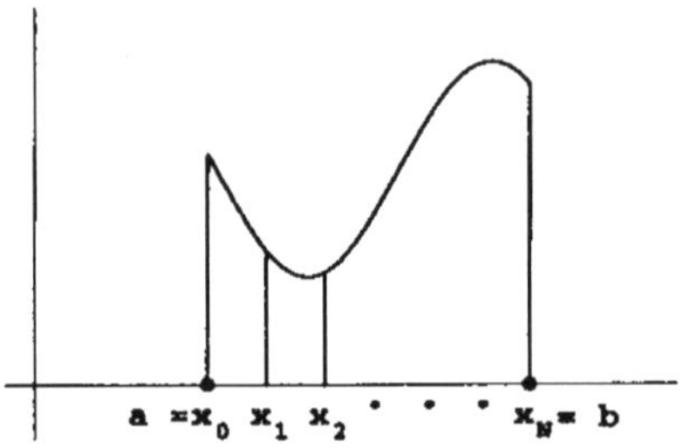

we see that

$$x_0 = a, \quad x_1 = a + \frac{b-a}{N}, \quad x_2 = a + \frac{2 \cdot (b-a)}{N}, \ldots, \quad x_i = a + \frac{i \cdot (b-a)}{N},$$

which holds for any x_i with $0 \le i \le N$.

If we now draw rectangles (as compared to rectangular type pieces) into our graph

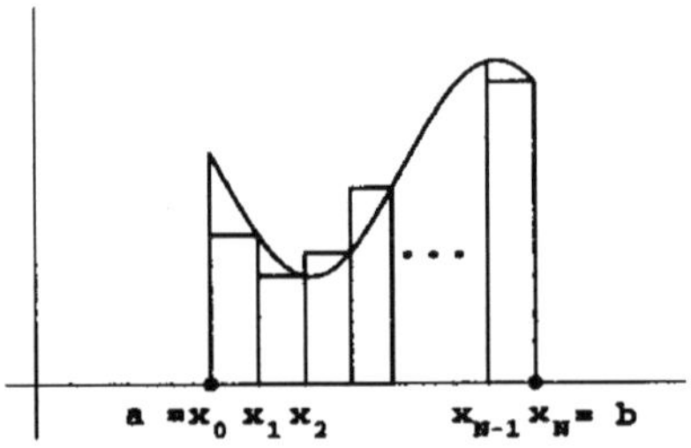

then we will have superimposed N rectangles of width $\frac{b-a}{N}$ where the ith rectangle (for $1 \le i \le N$)

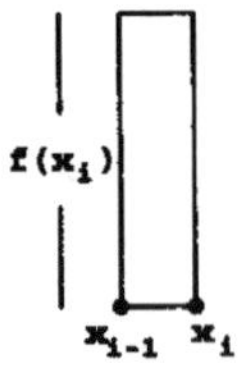

will have area $f(x_i) \cdot \frac{b-a}{N} = f\left(a + \frac{i \cdot (b-a)}{N}\right) \cdot \frac{b-a}{N}$. From this we see that the sum of the areas of these rectangles is given by

$$\Big[f(x_1) + f(x_2) + \cdots + f(x_N)\Big] \cdot \frac{b-a}{N}, \tag{13.1}$$

which may also be expressed as

$$\left[f\left(a+\frac{b-a}{N}\right)+f\left(a+\frac{2\cdot(b-a)}{N}\right)+\cdots+f\left(a+\frac{N\cdot(b-a)}{N}\right)\right]\cdot\frac{b-a}{N}.$$

This sum will be a good approximation of the precise area between the curve and the x–axis when N is very large. The discrepancy between the true area and the area covered by the rectangles is depicted by the shaded area below:

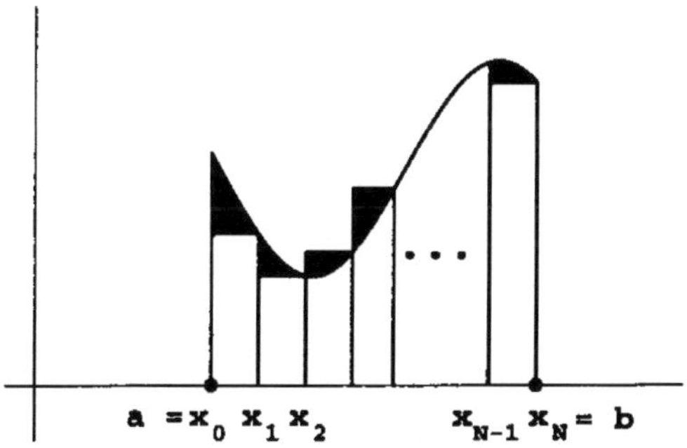

It is intuitively clear that as $N \to \infty$ (i.e., the number of rectangles tends to infinity) the shaded area above becomes minute and the sum (13.1) should approach the true area.

We now formally give:

Definition: *Let $f(x)$ be any function defined on the interval between a and b (where we do not necessarily assume $a < b$). Then we define*

$$\int_a^b f(x)\,dx = \lim_{N\to\infty}\left[f\left(a+\frac{b-a}{N}\right)+f\left(a+\frac{2\cdot(b-a)}{N}\right)+\cdots\right.$$
$$\left.\cdots+f\left(a+\frac{N\cdot(b-a)}{N}\right)\right]\cdot\frac{b-a}{N}.$$

provided the above limit exists. We then say f is integrable on the interval between a and b. When this limit does not exist the integral is not defined.

Remark: This definition is consistent with our intuitive definition of the integral even if the graph of f falls below the x–axis at some points.

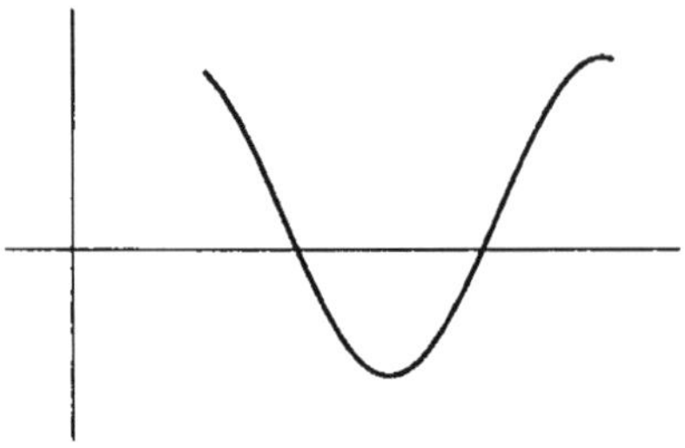

Filling in the region between the curve and the x–axis with N rectangles,

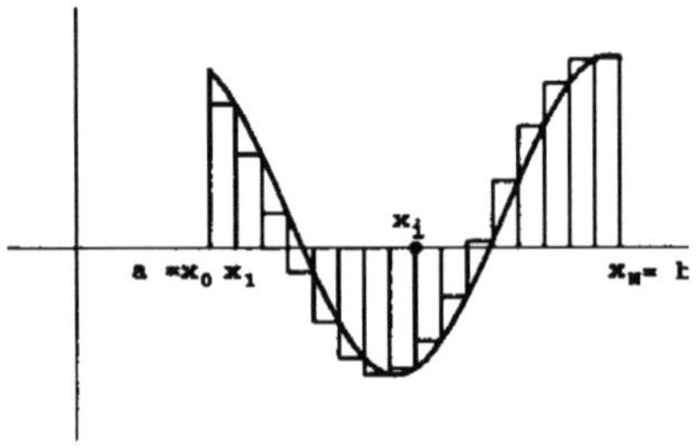

we now see that if the ith rectangle lies below the x–axis

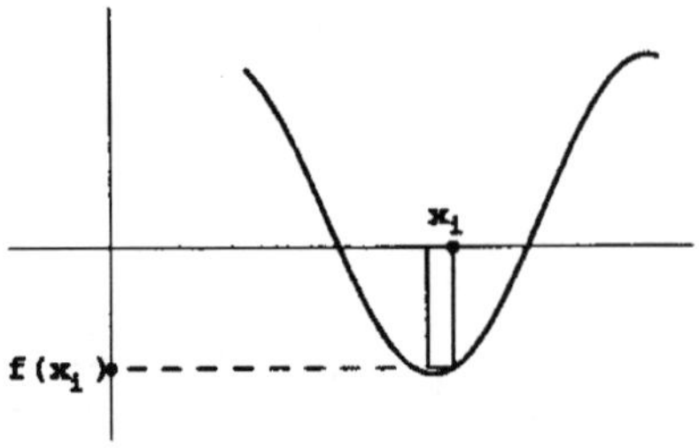

then $f(x_i) < 0$ and $f(x_i) \cdot \frac{b-a}{N}$ is -1 times the area of the ith rectangle. All such terms appearing in the definition combine to give the area (taken negatively) of the shaded region below the x–axis.

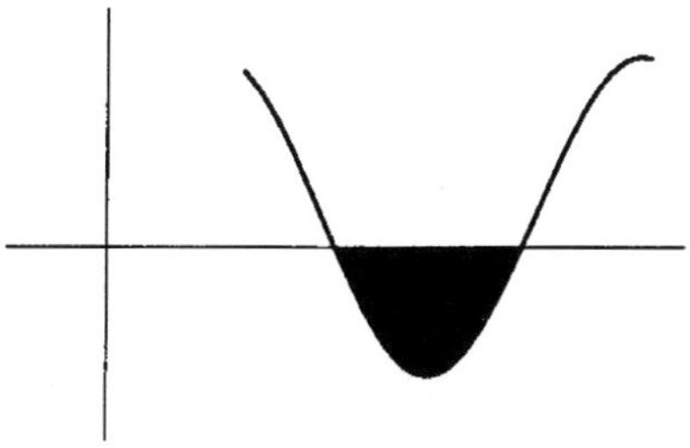

Example [13.6] Compute the integral $\int_0^2 x\,dx$ by employing the classical formula: $1 + 2 + \cdots N = \frac{N(N+1)}{2}$ (which we will prove in the following chapter). In this example $f(x) = x$ and $a = 0, b = 2$. Thus, by definition

$$\begin{aligned}\int_0^2 x\,dx &= \lim_{N\to\infty}\left[\frac{2}{N} + \frac{4}{N} + \frac{6}{N} + \cdots + \frac{2N}{N}\right]\cdot\frac{2}{N}\\ &= \lim_{N\to\infty}\frac{4}{N^2}[1+2+3+\cdots+N]\\ &= \lim_{N\to\infty}\frac{4}{N^2}\frac{N(N+1)}{2}\\ &= \lim_{N\to\infty}\left(2+\frac{2}{N}\right) = 2.\end{aligned}$$

Example [13.7] Compute $\int_1^5 (2x-1)\,dx$.

Here $f(x) = 2x - 1$, and $a = 1, b = 5$. Then

$$f\left(a + \frac{b-a}{N}\right) = f\left(1+\frac{4}{N}\right) = 2\left(1+\frac{4}{N}\right) - 1 = 1 + \frac{8}{N}.$$

Similarly,

$$\begin{aligned}f\left(a + 2\frac{b-a}{N}\right) &= f\left(1+\frac{8}{N}\right) = 2\left(1+\frac{8}{N}\right) - 1 = 1 + \frac{16}{N},\\ f\left(a + 3\frac{b-a}{N}\right) &= f\left(1+\frac{12}{N}\right) = 2\left(1+\frac{12}{N}\right) - 1 = 1 + \frac{24}{N},\\ &\ddots\end{aligned}$$

and we may compute:

$$
\begin{aligned}
\int_1^5 (2x-1)\,dx &= \lim_{N\to\infty}\left[\left(1+\frac{8}{N}\right)+\left(1+\frac{16}{N}\right)+\cdots+\left(1+\frac{8N}{N}\right)\right]\cdot\frac{4}{N}\\
&= \lim_{N\to\infty}\left[N+\frac{8}{N}(1+2+\cdots N)\right]\cdot\frac{4}{N}\\
&= \lim_{N\to\infty}\left[4+\frac{32}{N^2}\cdot\frac{N(N+1)}{2}\right]\\
&= \lim_{N\to\infty}\left(4+16+\frac{16}{N}\right)\\
&= 20.
\end{aligned}
$$

Warning: *Not* every function is integrable. For example, the function $\frac{1}{x-1}$ is not integrable on any interval which contains 1 since the area under the curve becomes infinite!

It is an interesting and difficult problem to determine exactly which functions are integrable. In the following proposition we examine some hypotheses which are sufficient to prove integrability and provide a partial solution to this problem. We require a definition.

Definition: *A function $f(x)$ is said to be* ***nondecreasing*** *(respectively* ***nonincreasing****) on an interval $a \leq x \leq b$ provided that for all d, c with $a \leq d \leq c \leq b$ we have $f(d) \leq f(c)$ (respectively $f(d) \geq f(c)$).*

We now consider an important proposition.

Proposition [13.8] *If a function $f(x)$ is both continuous and nondecreasing (or nonincreasing) on an interval then it is integrable on that interval.*

Remark: In general when we analyze the graph of a continuous function we will observe that it is nondecreasing or nonincreasing on various subintervals. Therefore Proposition [13.8] encompasses a wide range of functions.

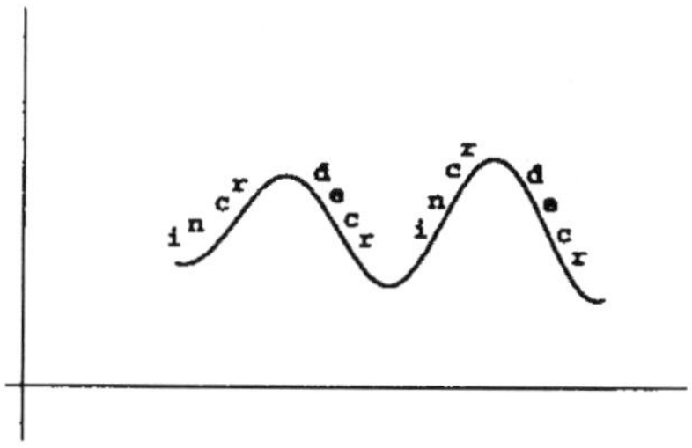

We give a quick sketch of the proof of Proposition [13.8] in the case of nondecreasing functions which lie above the x–axis . A similar proof can be given when the function lies below the x–axis which is left to the reader. For the most general case we divide the region between the curve and the x–axis into pieces which lie either below or above the x–axis and add up the pieces appropriately. The case of nonincreasing functions is entirely analogous and the proof is, therefore, omitted.

Since f is nondecreasing its graph must resemble the figure below,

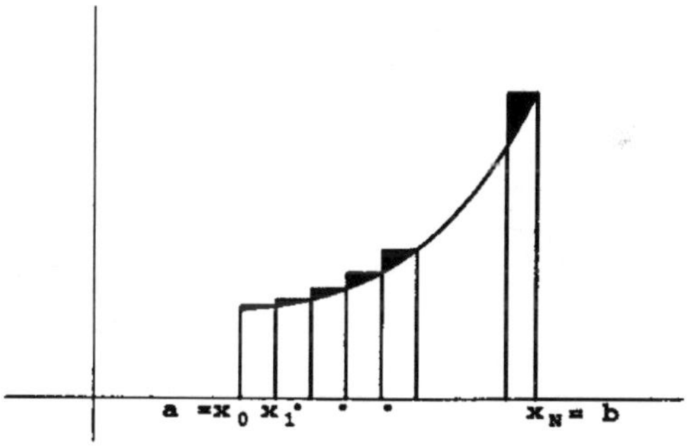

and the shaded regions in the above diagram must all lie above the curve. Note that the maximum *thickness* of the ith shaded region is just $f(x_i) - f(x_{i-1})$.

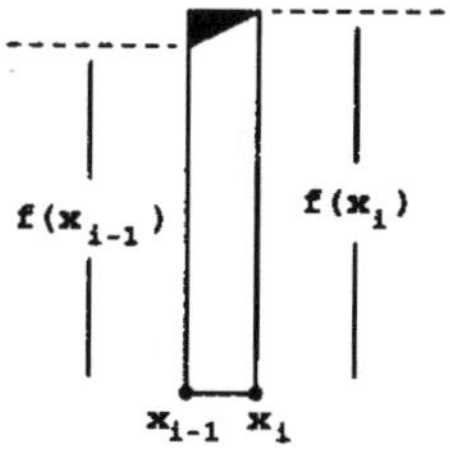

Since $x_i - x_{i-1} = \frac{1}{N}$ becomes arbitrarily small as $N \to \infty$, the continuity of f insures that the maximum thickness of the ith shaded region also becomes arbitrarily small as $N \to \infty$. From this we see that the total area of all the shaded regions must approach zero as $N \to \infty$. For example, when $N = 4, 12$, we see a dramatic decrease in the total area of the shaded regions.

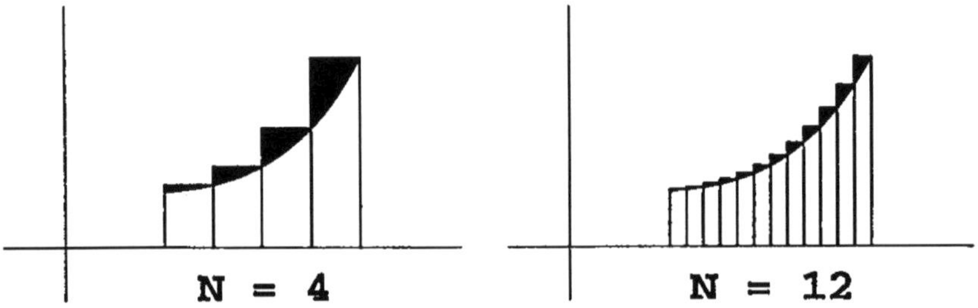

A simple geometric proof that the sum of the areas of the shaded regions tends to zero as $N \to \infty$ can be seen in the figure below. Namely, by sliding all of the small dark areas to the right, the resulting shaded region has area less than the area of a rectangle of base $\frac{b-a}{N}$ and height $f(b) - f(a)$.

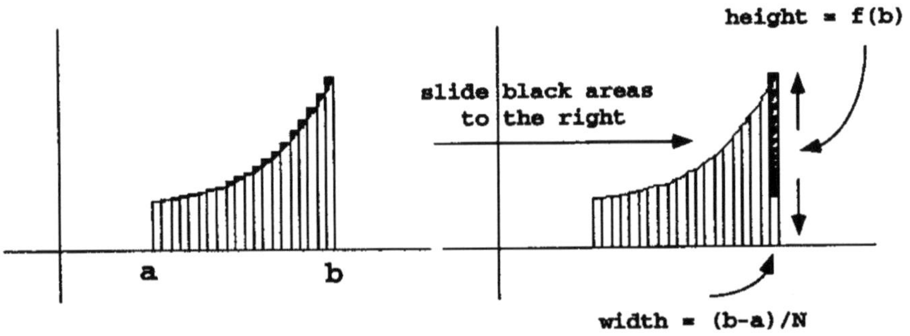

It follows that, as $N \to \infty$, the sum of the areas of the N rectangles

$$\left[f\left(a+\frac{b-a}{N}\right)+f\left(a+\frac{2\cdot(b-a)}{N}\right)+\cdots f\left(a+\frac{N\cdot(b-a)}{N}\right)\right]\cdot\frac{b-a}{N},$$

is decreasing and approaching a definite limit, the area under the curve.

Exercises for §13.3

Compute the following integrals using the definition of the integral as the limit of a sum. You may wish to use the formula $1 + 2 + \cdots + N = \frac{N(N+1)}{2}$.

(1) $\int_0^5 x\,dx$.

(2) $\int_{-1}^2 x\,dx$.

(3) $\int_1^2 (4x-1)\,dx$.

(4) $\int_{-3}^3 x\,dx$.

(5) $\int_{-2}^{-1} (2x+1)\,dx$.

(6) $\int_{-1}^0 (3x-1)\,dx$.

13.4 Properties of Integrals

The integral satisfies several basic properties which follow directly from the definition.

Property 1. *Let $f(x)$ be integrable on the interval between a and b, and let K be a constant. Then*

$$\int_a^b K f(x)\,dx = K \int_a^b f(x)\,dx.$$

Property 2. *Let $f(x)$ and $g(x)$ be integrable functions on the interval between a and b. Then*

$$\int_a^b \Big(f(x) \pm g(x)\Big)\,dx = \int_a^b f(x)\,dx \pm \int_a^b g(x)\,dx.$$

Property 3. *Let $a \le c \le b$. Then*

$$\int_a^b f(x)\,dx = \int_a^c f(x)\,dx + \int_c^b f(x)\,dx.$$

Proof: (by picture)

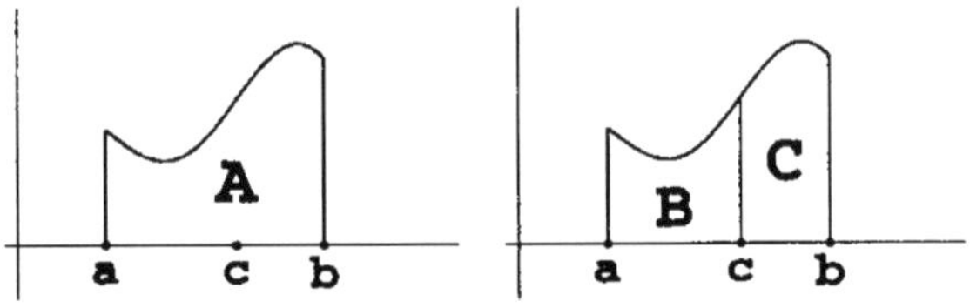

The area under the curve $y = f(x)$ where $a \leq x \leq b$ (denoted A in the above diagram) is simply the area B plus the area C.

Property 4. *Given any real numbers a, b,*

$$\int_a^b f(x)\,dx = -\int_b^a f(x)\,dx.$$

Proof: Beginning with the basic definition we have, since $b - a = -(a - b)$,

$$\sum_{n=1}^{N} f\left(a + \frac{(b-a)\,n}{N}\right) \cdot \frac{b-a}{N} = -\sum_{n=1}^{N} f\left(\frac{aN + (b-a)\,n}{N}\right) \cdot \frac{a-b}{N}.$$

Every number n with $1 \leq n \leq N$ can be written in the form $n = N - m$ where $1 \leq m \leq N$. Plugging this in and simplifying we get

$$\begin{aligned}\sum_{n=1}^{N} f\left(a + \frac{(b-a)\,n}{N}\right) \cdot \frac{b-a}{N} &= -\sum_{n=1}^{N} f\left(\frac{aN + (b-a)(N-m)}{N}\right) \cdot \frac{a-b}{N} \\ &= -\sum_{n=1}^{N} f\left(\frac{bN + (a-b)\,m}{N}\right) \cdot \frac{a-b}{N} \\ &= -\sum_{n=1}^{N} f\left(b + \frac{(a-b)\,m}{N}\right) \cdot \frac{a-b}{N}.\end{aligned}$$

The proof now follows from the definition.

Proposition [13.9] *Let $f(x)$ be an integrable function on the interval between a and b. If there exist numbers m and M such that*

$$m \le f(x) \le M$$

for all x in the interval then

$$m\,(b-a) \;\le\; \int_a^b f(x)\,dx \;\le\; M\,(b-a).$$

Remark: The numbers, m and M, referred to in Proposition [13.2] are termed (respectively) **lower** and **upper** bound for the function $f(x)$.

Proof: (by a picture)

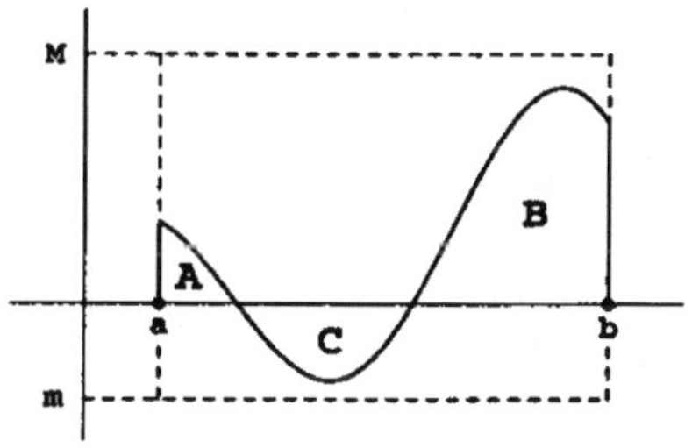

Observe that from the above diagram

$$\begin{aligned}\int_a^b f(x)\,dx &= \text{Area}(A) + \text{Area}(B) - \text{Area}(C)\\ &\le \text{Area}(A) + \text{Area}(B)\\ &\le M\cdot(b-a),\end{aligned}$$

and

$$\begin{aligned}\int_a^b f(x)\,dx &\ge -\text{Area}(C)\\ &\ge m\cdot(b-a)\end{aligned}$$

since in this picture m is negative. The above picture represents a special situation, but typifies the most general case.

Additional exercises for Chapter XIII

(1) Let $f(x) = \begin{cases} 1 - x, & \text{if } -1 \leq x \leq 1, \\ 0, & \text{otherwise.} \end{cases}$

Use your CAS to graph f. Can you find an upper and lower bound for f? Are they finite? What do you conclude about the integrability of f? Can you obtain the area under f? What can you say about the differentiability of f at $x = -1$ and $x = 1$?

(2) Show that if $f(x) \leq g(x)$ on some interval $a \leq x \leq b$ then $\int_a^b f(x)dx \leq \int_a^b g(x)dx$. **Hint:** Partition $a \leq x \leq b$ into n equal parts $a = x_0 < a_1 < x_1 < \cdots < x_n = b$, where $x_i - x_{i-1} = h = \frac{b-a}{n}$ and use the method of Example 13.7.

(3) Let $f(x)$ be a fixed function defined on an interval $a \leq x \leq b$. Suppose the points $a = x_0 < a_1 < x_1 < \cdots < x_n = b$ are given. Define

$$s(x) = f(x_{i-1}); \quad (\text{for } x_{i-1} \leq x < x_i).$$

The function $s(x)$ is called a **step function.** On your CAS, graph $s(x)$ and obtain $\int_a^b s(x)dx$ for the following step functions:

(a)

1	2	3	4	5	6	$\leftarrow x$
0	1	2	1	1	0	$\leftarrow s(x)$

(b)

0	1/2	1	3/2	2	5/2	$\leftarrow x$
−1	−2	−1	2	1	−2	$\leftarrow s(x)$

(c)

−1	0	1	2	3	4	$\leftarrow x$
1/4	0	1/2	1/2	1	1/4	$\leftarrow s(x)$

In the following problems we use the notation $\sum_{i=1}^{n} f(i) = f(1) + f(2) + \cdots + f(n)$.

(4) Using the formula $\sum_{i=1}^{n} i^2 = \frac{n^3}{3} + \frac{n^2}{2} + \frac{n}{6}$ compute $\int_0^b x^2 dx$ and compare your result to the one given by your CAS.

(5) By using the formula for the first n cubes $\sum_{i=1}^{n} i^3 = \frac{1}{4}n^2(n+1)^2$, evaluate the integral $\int_0^b x^3 dx$.

(6) Generalize the previous two problems to $\int_0^b x^m dx$ for $m \geq 1$ (**Hint:** Use your CAS for $m = 1, 2, \ldots$ Do you see a pattern? Prove your pattern formula by induction).

(7) Approximate $\int_0^1 \exp(\frac{-x^2}{2})dx$. Take $n = 10, 20, 30$. **Hint:** Evaluate $\sum_{i=0}^{n} f(a+\frac{b-a}{n}i)\frac{b-a}{n}$, for $a = 0, b = 1$. Compare your result with the one given by your CAS. How does the approximation change as n grows?

(8) Same as (7) with $f(x) = \sin(x^2)$.

(9) Same as (7) with $f(x) = x/\ln(x)$.

(10) Same as (7) with $f(x) = x^2 \exp(-x^2)$.

(11) Approximate $\int_0^1 \frac{dx}{(1+\frac{x^2}{3})^2}$. Use $n = 10, 20$. (**Hint:** Observe that the integrand is an even function).

(12) It is known that $\frac{\sqrt{2\pi}}{2} = \int_0^\infty \exp(\frac{-x^2}{2})dx$. Can you verify this result by approximating the integral between 0 and N using $n = 10, 40$. Verify your result with your CAS by choosing increasing values of N to integrate.

(13) Approximate $\int_0^2 \sqrt{4-x^2}\, dx$ using $n = 10, 30$. Notice that this will approximate the area of a quarter of a circle. Can you deduce an approximation of π?

Chapter XIV

Sums, Induction, and Computation of Integrals

14.1 Sums

In Chapter 13 we defined the integral as a limit of a sequence of sums. Since we will be working with these sums in detail it is convenient to introduce a shorthand notation for summation (so that formulae don't take up a whole page).

Let f be a function which is defined on the integers. We want to introduce a notation so that we can easily write down long sums such as

$$f(1) + f(2) + f(3) + f(4) + f(6) + f(7) + f(8).$$

Traditionally, the Greek letter sigma, Σ, stands for sum and the above expression is written

$$\sum_{n=1}^{8} f(n),$$

where n is the **index of summation**. Similarly, the sum

$$f(4) + f(5) + f(6) + f(7) + f(8) + f(9),$$

can be compactly written in the form $\sum_{n=4}^{9} f(n)$. The general definition is:

Definition: *Let f be a function which is defined on the integers, and let $M \leq N$ be any two integers. We define*

$$\sum_{n=M}^{N} f(n) = f(M) + f(M+1) + \cdots + f(N).$$

Example [14.1] Let $f(x) = x^3$, and let $M = 1$, $N = 3$. Then

$$\sum_{n=1}^{3} f(n) = \sum_{n=1}^{3} n^3 = 1 + 8 + 27 = 36.$$

Example [14.2] Let $f(x) = x^2 - 2$, and let $M = 3$, $N = 5$. Then

$$\sum_{n=3}^{5} f(n) = \sum_{n=3}^{5} (n^2 - 2) = (9-2) + (16-2) + (25-2) = 44.$$

Example [14.3] Let $f(x) = \cos\left(\frac{\pi x}{4}\right)$, and let $M = 1$, $N = 4$. Then

$$\begin{aligned} \sum_{n=1}^{4} f(n) &= \sum_{n=1}^{4} \cos\left(\frac{\pi n}{4}\right) \\ &= \cos\left(\frac{\pi}{4}\right) + \cos\left(\frac{2\pi}{4}\right) + \cos\left(\frac{3\pi}{4}\right) + \cos(\pi) \\ &= \frac{\sqrt{2}}{2} + 0 - \frac{\sqrt{2}}{2} - 1 \\ &= -1. \end{aligned}$$

The beautiful formula

$$\sum_{n=1}^{N} n = 1 + 2 + 3 + \cdots + N = \frac{N(N+1)}{2}, \tag{14.1}$$

was discovered by K. F. Gauss at the age of 9. It can be easily proved by rewriting the terms of the sum backwards and summing with the original terms:

$$\begin{array}{cccccc} 1 & +2 & +3 & + & \dots & +N \\ + & & & & \dots & \\ N & +(N-1) & +(N-2) & + & \dots & +1 \\ \hline \end{array}$$

$$\underbrace{(N+1)+ \quad (N+1)+ \quad (N+1)+ \quad + \quad \dots \quad +(N+1)}_{N \text{ times}}$$

For example, when $N = 100$, $1 + 2 + 3 + \cdots + 100 = \frac{100 \cdot 101}{2} = 5050$ since

$$[1 + 2 + 3 + \cdots + 100] + [100 + 99 + 98 + \cdots + 1] = \underbrace{[101 + 101 + \cdots + 101]}_{100 \text{ times}}.$$

Exercises for §14.1

Use the SUM function on your CAS to compute the following sums.

(1) $\sum_{n=1}^{10} n^3$. (4) $\sum_{n=1}^{10} \sin\left(\frac{\pi n}{3}\right)$. (7) $\sum_{n=1}^{5} 2^n$.

(2) $\sum_{n=10}^{20} n^2$. (5) $\sum_{n=1}^{10} \cos\left(\frac{2\pi n}{7}\right)$. (8) $\sum_{n=1}^{5} 2^{-n}$.

(3) $\sum_{n=1}^{8} (2n^3 + 3n - 1)$. (6) $\sum_{n=1}^{10} \frac{1}{n^3}$. (9) $\sum_{n=1}^{7} (3 + 5 \cdot 2^{-n})$.

(10) Evaluate the sum $\sum_{n=1}^{N} \frac{1}{n^2}$ for $N = 10,\ 100,\ 1000$. Observe that this sum approaches $\frac{\pi^2}{6}$.

(11) Evaluate the sum $\sum_{n=1}^{N} \frac{1}{n}$ for $N = 10,\ 100,\ 1000$. Is $\ln(N)$ a good approximation for this sum?

14.2 Induction

There is another way to prove Gauss' formula (14.1). This is the method of induction which we now illustrate by presenting a second proof of (14.1).

We can easily verify that (14.1) holds for small values of N. For example

$$\begin{aligned} 1 &= \frac{1 \cdot (1+1)}{2} \\ 1 + 2 &= \frac{2 \cdot 3}{2} \\ 1 + 2 + 3 &= \frac{3 \cdot 4}{2}. \end{aligned} \tag{14.2}$$

Suppose we can demonstrate the following statement is true for any positive integer N:

$$\sum_{n=1}^{N} n = \frac{N(N+1)}{2} \quad \text{implies} \quad \sum_{n=1}^{N+1} n = \frac{(N+1)(N+2)}{2} \tag{14.3}$$

then we claim that Gauss' formula is proved. To see the reasoning here first picture an (infinite) ladder. If we know that we can (1) climb onto the first rung, and (2) once we are on the Nth rung we can climb onto the $(N+1)^{\text{st}}$ rung, then we can in fact climb to any level of the ladder. We now prove Gauss' formula by verifying (14.3) (since we have already checked the initial cases in (14.2)). Assuming now that

$$\sum_{n=1}^{N} n = 1+2+3+\cdots+N = \frac{N(N+1)}{2}$$

holds for some positive integer N, it follows that

$$\begin{aligned}
\sum_{n=1}^{N+1} n &= (1+2+\cdots+N) + (N+1) \\
&= \frac{N(N+1)}{2} + N + 1 \\
&= \frac{N^2+N+2(N+1)}{2} \\
&= \frac{N^2+3N+2}{2} \\
&= \frac{(N+1)(N+2)}{2},
\end{aligned}$$

which was to be established.

Problem: Find a formula for $\sum_{n=1}^{N} n^2$.

Solution: We take an experimental approach. Motivated by Gauss' formula we can hypothesize a formula of the form

$$\sum_{n=1}^{N} n^2 = \alpha N^3 + \beta N^2 + \gamma N + \delta,$$

for some constants α, β, γ, and δ. Plugging in $N = 1, 2, 3, \ldots$ yields

$$\begin{aligned} 1 &= \alpha + \beta + \gamma + \delta \\ 1 + 4 &= 8\alpha + 4\beta + 2\gamma + \delta \\ 1 + 4 + 9 &= 27\alpha + 9\beta + 3\gamma + \delta \\ &\vdots \end{aligned}$$

We now need to find the correct values of α, β, γ, and δ, using the CAS. Just as Edison passed electricity through hundreds of metals before discovering that tungsten glowed, we must also work experimentally and try to find the correct values for α, β, γ, and δ. They turn out to be

$$\alpha = \frac{1}{3}, \quad \beta = \frac{1}{2}, \quad \gamma = \frac{1}{6}, \quad \delta = 0,$$

and we prove, by induction, that for $N \geq 1$

$$\sum_{n=1}^{N} n^2 = \frac{N^3}{3} + \frac{N^2}{2} + \frac{N}{6}.$$

The initial cases are easily checked:

$$\begin{aligned} 1 &= \frac{1}{3} + \frac{1}{2} + \frac{1}{6} \\ 1 + 4 &= \frac{2^3}{3} + \frac{2^2}{2} + \frac{2}{6} \\ 1 + 4 + 9 &= \frac{3^3}{3} + \frac{3^2}{2} + \frac{3}{6}. \end{aligned}$$

Now assuming $\sum_{n=1}^{N} n^2 = \frac{N^3}{3} + \frac{N^2}{2} + \frac{N}{6}$ holds for some $N > 1$ we must show

$$\sum_{n=1}^{N+1} n^2 = \frac{(N+1)^3}{3} + \frac{(N+1)^2}{2} + \frac{(N+1)}{6}.$$

This follows easily (with a little help from the CAS):

$$\begin{aligned} \sum_{n=1}^{N+1} n^2 &= \left(\sum_{n=1}^{N} n^2 \right) + (N+1)^2 \\ &= \frac{N^3}{3} + \frac{N^2}{2} + \frac{N}{6} + N^2 + 2N + 1 \\ &= \frac{(N+1)^3}{3} + \frac{(N+1)^2}{2} + \frac{(N+1)}{6}. \end{aligned}$$

Remark: It is important to remember that if we did not find the correct values of α, β, γ, and δ, then the induction would not work.

To recapitulate, we have now found the formulae

$$\sum_{n=1}^{N} 1 = N$$

$$\sum_{n=1}^{N} n = \frac{N(N+1)}{2} \tag{14.4}$$

$$\sum_{n=1}^{N} n^2 = \frac{N^3}{3} + \frac{N^2}{2} + \frac{N}{6}.$$

Combining these formulae algebraically with arbitrary constants a, b, and c we obtain

$$\sum_{n=1}^{N}(an^2 + bn + c) = a \cdot \left(\frac{N^3}{3} + \frac{N^2}{2} + \frac{N}{6}\right) + b \cdot \left(\frac{N(N+1)}{2}\right) + c \cdot N.$$

Exercises for §14.2

(1) Find a formula for $\sum_{n=1}^{N} \frac{1}{n(n+1)}$ by collecting experimental data. Prove your formula by induction.

(2) Find a formula for $\sum_{n=1}^{N} \frac{1}{2^n}$ by collecting experimental data. Prove your formula by induction.

(3) Find a formula for $\sum_{n=1}^{N} \frac{1}{3^n}$ by collecting experimental data. Prove your formula by induction.

(4) Prove by induction that $\sum_{n=1}^{N} n^3 = (\frac{N(N+1)}{2})^2$.

(5) Prove by induction that $\sum_{n=1}^{N} n^4 = \frac{N^5}{5} + \frac{N^4}{2} + \frac{N^3}{3} - \frac{N}{30}$.

(6) Prove by induction that for any $a \neq 1$, $\sum_{n=1}^{N} a^n = \frac{a^{N+1}-1}{a-1}$.

14.3 Computation of Integrals

Using the summation notation we have just developed we can rewrite the definition of the integral in the following form:

$$\int_a^b f(x)\,dx = \lim_{N\to\infty}\left[\sum_{n=1}^{N} f\left(a + \frac{n(b-a)}{N}\right)\right] \cdot \frac{b-a}{N}.$$

Formulae of the type displayed in (14.3) can often be used to evaluate integrals. We give three examples.

Example [14.4] Evaluate $\int_0^2 x^2\,dx$.

Here $a = 0$, $b = 2$, and $f(x) = x^2$. Hence

$$\begin{aligned}\int_0^2 x^2\,dx &= \lim_{N\to\infty}\left[\sum_{n=1}^{N}\left(\frac{2n}{N}\right)^2\right]\cdot\frac{2}{N}\\ &= \lim_{N\to\infty}\frac{8}{N^3}\sum_{n=1}^{N} n^2\\ &= \lim_{N\to\infty}\frac{8}{N^3}\left(\frac{N^3}{3}+\frac{N^2}{2}+\frac{N}{6}\right)\\ &= \lim_{N\to\infty}\left(\frac{8}{3}+\frac{4}{N}+\frac{4}{3N^2}\right) = \frac{8}{3}.\end{aligned}$$

Example [14.5] Evaluate $\int_0^B x^2\,dx$.

Here $a = 0$, $b = B$, and $f(x) = x^2$. Computing much as in Example [14.4] we have

$$\begin{aligned}\int_0^B x^2\,dx &= \lim_{N\to\infty}\left[\sum_{n=1}^{N}\left(\frac{nB}{N}\right)^2\right]\cdot\frac{B}{N}\\ &= \lim_{N\to\infty}\frac{B^3}{N^3}\sum_{n=1}^{N} n^2\\ &= \lim_{N\to\infty}\frac{B^3}{N^3}\left(\frac{N^3}{3}+\frac{N^2}{2}+\frac{N}{6}\right)\\ &= \frac{B^3}{3}.\end{aligned}$$

Example [14.6] Evaluate $\int_{-1}^{3}(x^2-3x)\,dx$.

In this last example, $a=-1$, $b=3$, and $f(x)=x^2-3x$. Applying our method as before:

$$
\begin{aligned}
\int_{-1}^{3}(x^2-3x)\,dx &= \lim_{N\to\infty}\sum_{n=1}^{N}\left(\left(-1+\frac{4n}{N}\right)^2-3\left(-1+\frac{4n}{N}\right)\right)\cdot\frac{4}{N}\\
&= \lim_{N\to\infty}\frac{4}{N}\sum_{n=1}^{N}\left(4-\frac{20n}{N}+\frac{16n^2}{N^2}\right)\\
&= \lim_{N\to\infty}\left(\frac{16}{N}\sum_{n=1}^{N}1-\frac{80}{N^2}\sum_{n=1}^{N}n+\frac{64}{N^3}\sum_{n=1}^{N}n^2\right)\\
&= \lim_{N\to\infty}\left(16-\frac{80}{N^2}\frac{N(N+1)}{2}+\frac{64}{N^3}\left(\frac{N^3}{3}+\frac{N^2}{2}+\frac{N}{6}\right)\right)\\
&= \lim_{N\to\infty}\left(16-40-\frac{40}{N}+\frac{64}{3}+\frac{32}{N}+\frac{64}{6N^2}\right)\\
&= 16-40+\frac{64}{3}=\frac{-8}{3}.
\end{aligned}
$$

Exercises for §14.3

Evaluate the following integrals using the definition of the integral as a limit of a sum. In exercises (4), (5), and (6), A, B are arbitrary real numbers.

(1) $\int_1^4 x^2\,dx$.

(2) $\int_{-1}^{3}(x^2-x)\,dx$.

(3) $\int_0^3 x^3\,dx$.

(4) $\int_0^B x^3\,dx$.

(5) $\int_A^B x^2\,dx$.

(6) $\int_A^B x^3\,dx$.

(7) Using the formula $\sum_{n=1}^{N} n^4=\frac{N^5}{5}+\frac{N^4}{2}+\frac{N^3}{3}-\frac{N}{30}$, evaluate the integral $\int_A^B x^4\,dx$ for any values A, B.

(8) Using the formula $\sum_{n=1}^{N} n^k=\frac{N^{k+1}}{k+1}+\frac{N^k}{2}+p(N)$ where $p(N)$ is a polynomial whose degree is less than k, show that $\int_0^B x^k\,dx=\frac{B^{k+1}}{k+1}$ for any B.

14.4 Approximate Computation of Integrals

It may not always be possible to find the exact value of an integral. For example, if you attempt to evaluate

$$\int_0^1 2^{-x^2}\, dx$$

on your CAS it will be unable to provide an exact answer. Nevertheless, if you specify 5 digit accuracy, for example, you will get the answer .81002. How was this result obtained? Many methods of numerical integration have been devised, and the method your particular CAS used is not always public information. To demonstrate an example of such a method we will present the well known **Trapezoid Rule** which has the distinguished features of being simple and fast A trapezoid is a 4–sided polygon

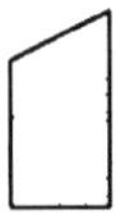

which can be broken up into a rectangle and a triangle as in the figure below.

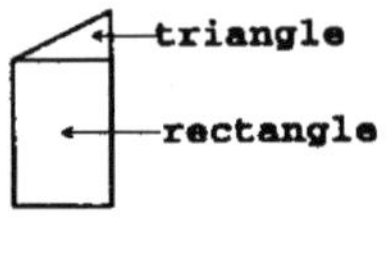

If the sides of the trapezoid have lengths a, b, c, d,

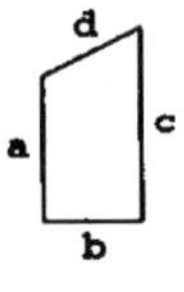

then the area is $\frac{1}{2}(a+c)\cdot b$ (since the area of the rectangular part is $a \cdot b$ and the area of the triangular part is $\frac{1}{2}(c-a)\cdot b$).

The basic idea of the trapezoid rule is to approximate the area under a curve $y = f(x)$

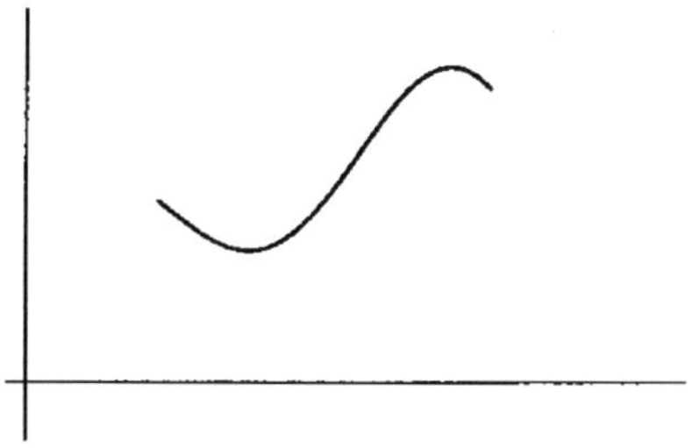

by N trapezoids of equal width $\frac{1}{N}$,

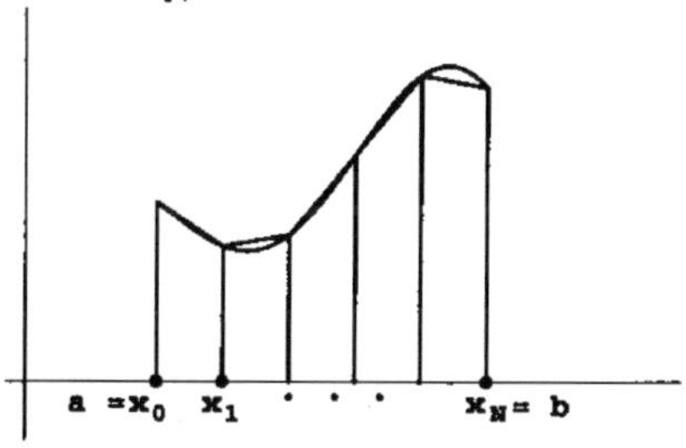

instead of using N rectangles

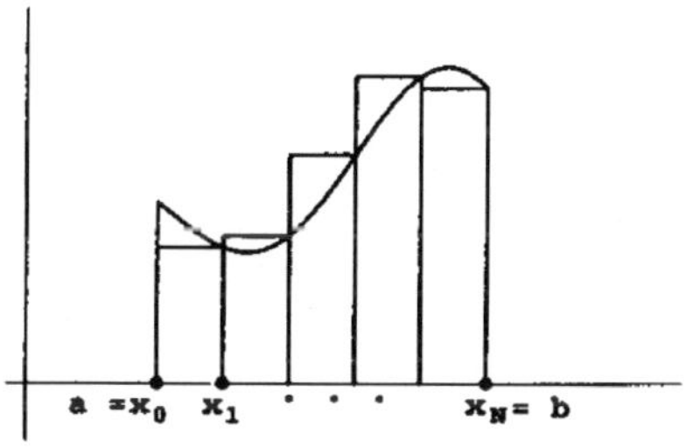

as we did earlier. The area of the nth trapezoid

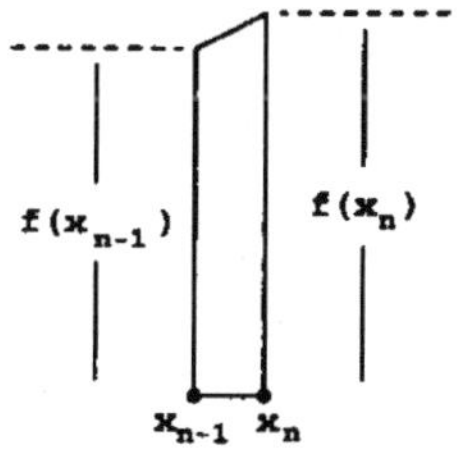

is given by the formula

$$\frac{1}{2}\left[f(x_{n-1}) + f(x_n)\right] \cdot \frac{b-a}{N}.$$

Summing the areas of the trapezoids gives us the approximation

$$\int_a^b f(x)\,dx \approx \sum_{n=1}^{N} \frac{1}{2}\left[f(x_{n-1}) + f(x_n)\right] \cdot \frac{b-a}{N},$$

which will be fairly accurate when N is large. Note the use of the symbol $\approx$ to denote **approximately equal.**

Example [14.7] Approximate $\int_0^1 2^{-x^2}\,dx$ using $N = 4$ trapezoids.

Here $f(x) = 2^{-x^2}$, and $a = 0$, $b = 1$, $x_0 = 0$, $x_1 = \frac{1}{4}$, $x_2 = \frac{2}{4}$, $x_3 = \frac{3}{4}$, and $x_4 = 1$. The trapezoid approximation gives us

$$\begin{aligned}\int_0^1 2^{-x^2}\,dx &\approx \left[\frac{1}{2}\left(2^{-0} + 2^{-\frac{1}{16}}\right) + \frac{1}{2}\left(2^{-\frac{1}{16}} + 2^{-\frac{1}{4}}\right)\right.\\ &\qquad \left. + \frac{1}{2}\left(2^{-\frac{1}{4}} + 2^{-\frac{9}{16}}\right) + \frac{1}{2}\left(2^{-\frac{9}{16}} + 2^{-1}\right)\right] \cdot \frac{1}{4}\\ &= \frac{1}{8}\left[1 + 2\cdot 2^{-\frac{1}{16}} + 2\cdot 2^{-\frac{1}{4}} + 2\cdot 2^{-\frac{9}{16}} + \frac{1}{2}\right]\\ &\approx .80641,\end{aligned}$$

which comes fairly close to .81002 (which is the correct answer to five decimal places).

Example [14.8] Approximate $\int_1^3 \frac{1}{x}\,dx$ using 5 trapezoids.

In this final example $f(x) = \frac{1}{x}$, $a = 1$, $b = 3$, $x_0 = 1$, $x_1 = 7/5$, $x_2 = 9/5$, $x_3 = 11/5$, $x_4 = 13/5$, and $x_5 = 3$.

$$\begin{aligned}\int_1^3 \frac{1}{x}\,dx &\approx \left[\frac{1}{2}\left(1 + \frac{5}{7}\right) + \frac{1}{2}\left(\frac{5}{7} + \frac{5}{9}\right) + \frac{1}{2}\left(\frac{5}{9} + \frac{5}{11}\right)\right.\\ &\qquad \left. + \frac{1}{2}\left(\frac{5}{11} + \frac{5}{13}\right) + \frac{1}{2}\left(\frac{5}{13} + \frac{1}{3}\right)\right] \cdot \frac{2}{5}\\ &\approx 1.1103.\end{aligned}$$

We conclude this section with **Simpson's Rule**. Thomas Simpson (1710–1761), who was trained to be a weaver, found a method that usually approximates integrals even better than the trapezoid rule. The idea is to break the region between the curve $y = f(x)$ and the x–axis into $2N$ rectangular type pieces,

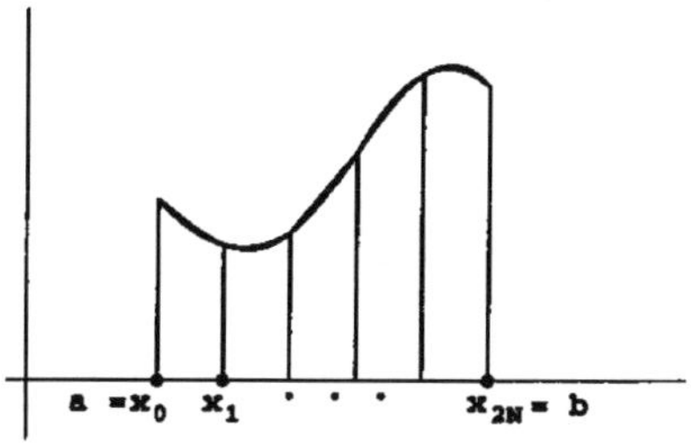

where the ith piece is of the form,

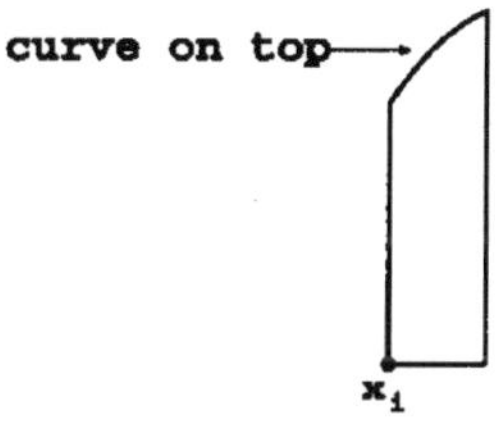

and the curve on top is given by a parabola instead of a straight line (which appears in the trapezoid rule). This very often results in a tighter fit, and hence the sum of all these rectangular type regions gives a better approximation to the integral. The method is now known as Simpson's rule and the formula for the approximation takes the form

$$\int_a^b f(x)\,dx \approx \sum_{n=1}^{N} \frac{1}{3}\left[f(x_{2n-2}) + 4f(x_{2n-1}) + f(x_{2n})\right] \cdot \frac{b-a}{2N}.$$

Example [14.9] Approximate $\int_1^3 \frac{1}{x}\,dx$ using Simpson's rule with $2N = 6$.

Here $f(x) = \frac{1}{x}$, $a = 1$, $b = 3$, $x_0 = 1$, $x_1 = \frac{4}{3}$, $x_2 = \frac{5}{3}$, $x_3 = 2$, $x_4 = \frac{7}{3}$,

$x_5 = \frac{8}{3}$, and $x_6 = 3$. Simpson's rule gives us the following approximation:

$$\int_1^3 \frac{1}{x}\,dx \approx \frac{1}{3}\left[1 + 4\cdot\frac{3}{4} + \frac{3}{5}\right]\cdot\frac{2}{6} + \frac{1}{3}\left[\frac{3}{5} + 4\cdot\frac{1}{2} + \frac{3}{7}\right]\cdot\frac{2}{6}$$

$$+\frac{1}{3}\left[\frac{3}{7} + 4\cdot\frac{3}{8} + \frac{1}{3}\right]$$

$$\approx 1.0989\ldots$$

The correct answer to 5 decimal places is in fact $1.0986\ldots$, so that in this case Simpson's method is accurate to within an error which is less than 4 ten thousandths! Compare this to what we obtained with the trapezoid rule in Example [14.8].

Exercises for §14.4

In the following exercises, approximate the value of the integral using the trapezoid rule and Simpson's rule with $N = 4, 10, 50$. Compare the answers the two methods yield.

(1) $\int_{-1}^{3}(x^2 + 2x)\,dx$.

(2) $\int_{-2}^{1}(3x^2 + 4)\,dx$.

(3) $\int_0^2 \sqrt{2+x}\,dx$.

(4) $\int_0^3 \frac{1}{2+x^2}\,dx$.

(5) $\int_0^2 2^{-x}\,dx$.

(6) $\int_{-1}^{1} e^{-2x^2}\,dx$.

(7) $\int_0^1 \frac{4}{1+x^2}\,dx$.

(8) $\int_{-2}^{2} \frac{dx}{\sqrt{3+x^4}}$.

(9) $\int_{-2}^{2} \frac{e^{-x^2}}{\sqrt{1+x^2}}\,dx$.

(10) $\int_0^2 x\cos(\pi x)\,dx$.

(11) $\int_0^3 \frac{\sin(x)}{x}\,dx$.

(12) $\int_{-2}^{3} \frac{\sin(x)}{\sqrt{1+x^4}}\,dx$.

(13) Let $y = ax^2 + bx + c$ be a parabola. Use Simpson's rule with $N = 3$ to show that Simpson's rule gives the exact value of the integral $\int_0^1(ax^2 + bx + c)dx$.

Additional exercises for Chapter XIV

(1) Calculate $\int_{-1}^{3}(x^2 + 2x)dx$ with $N = 4, 10$ and use the left endpoints for the heights of rectangles. Recalculate with right endpoints as the heights of the approximating rectangles. Take the average of the two and compare with your answers from the trapezoidal method. What can you conclude?

(2) Use the fact that the area of the circle of radius R is πR^2 to estimate π. That is, suppose $x^2 + y^2 = 9$, so in the first quadrant $y = \sqrt{9 - x^2}$. Now apply Simpson's rule with $N = 4, 10, 50$ to estimate $9\frac{\pi}{4} = \int_0^3 \sqrt{9 - x^2}dx$

(3) Use the relationship

$$\frac{\pi}{4} = \int_0^1 \frac{dx}{1 + x^2}$$

and Simpson's rule with $N = 10$ to estimate π.

(4) Let $f(x) = x^2$ for $0 \leq x \leq 2$.

(a) Using rectangles and left endpoints, determine a lower bound for $\int_0^2 x^2 dx$ with $N = 4$.

(b) With $N = 4$ rectangles and right endpoints, determine an upper bound for $\int_0^2 x^2 dx$.

(c) Show that the difference between the two estimates is 1.

(5) Suppose $f(x)$ is an increasing function on the interval $a \leq x \leq b$. Instead of trapezoids, we may approximate $\int_a^b f(x)dx$ with N rectangles. The height of each rectangle is either the left or the right endpoint. If we use only the left endpoint, we get a lower bound and if we use the right endpoint we have an upper bound. Show that the difference between the upper and lower bound is $(f(b) - f(a))\frac{b-a}{N}$. This expression is called the error from the left hand rule.

(6) For $f(x) = \exp(x^2)$ on the interval $1 \leq x \leq 3$, find N so that the error from the left hand rule (defined in (5)) is less than $\frac{1}{50}$. Check with your CAS.

(7) Let L be any number such that $|f''(x)| \leq L$ and M any number so that $|f^{(4)}(x)| \leq M$ for all $a \leq x \leq b$. It is known that the absolute value of the error resulting from Simpson's rule with N subintervals is at most $M\frac{(b-a)^5}{180N^4}$, and the absolute value of the error from the trapezoidal rule is at most $L\frac{(b-a)^3}{12N^2}$.

(a) Using $N = 10$ compute the absolute value of the error for $\int_1^3 \exp(x^2)dx$.

(b) How large should N be to ensure that the absolute value of the error in the approximation of $\int_1^3 \exp(x^2)dx$ by Simpson's rule is at most .00002.

(8) Yet another estimate of π. How large should N be to ensure that the absolute value of the error of approximation of $\int_0^1 \frac{dx}{1+x^2} = \pi/4$ is within 10^{-3}?

(9) Let $f(x) = \begin{cases} -x + 1, & \text{if } x \leq 1 \\ -(x - \frac{3}{2})^2 + \frac{1}{4}, & \text{otherwise} \end{cases}$

(a) Use Simpson's rule with $N = 10, 40$ to estimate $\int_0^5 f(x)dx$.

(b) Though $f(x)$ is not differentiable, can you find a bound on the absolute error? (**Hint:** use your CAS).

Prove the following formulas using induction.

(10) $$\sum_{n=1}^{N}(2n-1) = N^2.$$

(11) $$\sum_{n=1}^{N} n^5 = \frac{1}{12}N^2(N+1)^2(2N^2+2N-1).$$

(12) $$\sum_{n=1}^{N}(2n-1)^2 = \frac{1}{3}N(4N^2-1).$$

(13) $$\sum_{n=1}^{N} n(n+1)^2 = \frac{1}{12}N(N+1)(N+2)(3N+5),$$

(14) $$\sum_{n=1}^{N} n!\cdot n = (N+1)!-1.$$

(15) $$\sum_{n=1}^{N}\frac{1}{(n^2-1)} = \frac{3}{4} - \frac{2N+1}{2N(N+1)}.$$

Chapter XV

The Integral as an Antiderivative

15.1 The Fundamental Theorem of Calculus

We have shown in example [14.5] that

$$\int_0^X x^2\,dx = \frac{X^3}{3},$$

and, further, in Exercises (4), (7), and (8) of §14.3 we saw that

$$\int_0^X x^3\,dx = \frac{X^4}{4}, \quad \int_0^X x^4\,dx = \frac{X^5}{5}, \quad \int_0^X x^k\,dx = \frac{X^{k+1}}{k+1},$$

where $k = 0, 1, 2, \ldots$. In all the above examples the derivative of the integral turns out to be the function we are integrating. This suggests a connection between integration and differentiation. Recall that integration is the computation of area and differentiation is the computation of the slope of the tangent line to a curve. That these two seemingly diverse computational processes should be related is remarkable! It is entirely inobvious that integration is the inverse of differentiation (the anti–derivative). This is made precise in the Fundamental Theorem of Calculus which we now state and prove.

Theorem [15.1] (Fundamental Theorem of Calculus) *Let $f(x)$ be a function which is continuous and integrable on an interval between a and b. Then there exists a differentiable function $F(x)$ with the property that on this interval*

$$F'(x) = f(x),$$

and

$$\int_a^b f(x)\,dx = F(b) - F(a).$$

Remark: The function $F(x)$ referred to in the Fundamental Theorem is called an **antiderivative**. It is natural to ask if the antiderivative is unique. In fact if there are two functions F_1 and F_2 such that $F_1'(x) = F_2'(x) = f(x)$ then necessarily

$$\frac{d}{dx}(F_1(x) - F_2(x)) = 0,$$

and thus by proposition [7.3]

$$F_1(x) - F_2(x) = c,$$

for some constant c. Notice that regardless of which antiderivative F we have the value of $F(b) - F(a)$ will always be the same. Given antiderivatives F_1 and F_2, since $F_1(x) - F_2(x) = c$ for some constant c,

$$F_1(b) - F_1(a) = (F_2(b) + c) - (F_2(a) + c) = F_2(b) - F_2(a).$$

The Fundamental Theorem allows us to compute many integrals rapidly and easily without recourse to sums and limits.

Example [15.2] Compute $\int_{-1}^3 x^4\,dx$.

Here $f(x) = x^4$ and one possible choice for $F(x)$ is $\frac{x^5}{5}$.

$$\begin{aligned}\int_{-1}^3 x^4\,dx &= F(3) - F(-1)\\ &= \frac{3^5}{5} - \frac{(-1)^5}{5} = \frac{244}{5}.\end{aligned}$$

Example [15.3] Compute $\int_0^1 \left(x^3 - 2x + 1\right)\,dx$.

Here $f(x) = x^3 - 2x + 1$ and we can choose the antiderivative to be $F(x) = \frac{x^4}{4} - x^2 + x$. We obtain

$$\int_0^1 [x^3 - 2x + 1]\,dx = F(1) - F(0) = \left(\frac{1}{4} - 1 + 1\right) - 0 = \frac{1}{4}.$$

Example [15.4] Compute $\int_{-\frac{\pi}{2}}^{3\pi} \cos(3x)\,dx$.

In this last example $f(x) = \cos(3x)$ and we choose $F(x) = \frac{1}{3}\sin(3x)$. This gives us

$$\begin{aligned}\int_{-\frac{\pi}{2}}^{3\pi} \cos(3x)\,dx &= F(3\pi) - F\left(\frac{-\pi}{2}\right) \\ &= \frac{1}{3}\sin(9\pi) - \frac{1}{3}\sin\left(\frac{-3\pi}{2}\right) \\ &= -\frac{1}{3}.\end{aligned}$$

Notation: It has now become standard practice to write out computation of integrals (using antiderivatives) in the following manner: If we are evaluating $\int_a^b f(x)\,dx$ and $F(x)$ is an antiderivative of $f(x)$, i.e., $F'(x) = f(x)$, then we write

$$\int_a^b f(x)\,dx = F(x)\Big]_a^b = F(b) - F(a).$$

For example,

$$\int_1^3 x\,dx = \frac{x^2}{2}\Big]_1^3 = \frac{9}{2} - \frac{1}{2} = 4,$$

and

$$\int_0^{10} (x^3 + 1)\,dx = \left(\frac{x^4}{4} + x\right)\Big]_0^{10} = \frac{10^4}{4} + 10 - 0 = 2510.$$

Example [15.5] Evaluate

$$\frac{d}{dX}\int_1^X (x^2 + \sin(x))e^{-x^2}\,dx.$$

If $F(x)$ is a function such that $\frac{d}{dx}F(x) = (x^2 + \sin(x))e^{-x^2}$ then since

$$\int_1^X (x^2 + \sin(x))e^{-x^2}\,dx = F(X) - F(1),$$

we have

$$\frac{d}{dX}\int_1^X (x^2+\sin(x))e^{-x^2}\,dx = (X^2+\sin(X))e^{-X^2}.$$

Proof of the Fundamental Theorem

We present a proof for functions whose graph lies above the x–axis. This restriction can easily be dropped and we leave the general case to the reader.

For any X in the interval $a \leq X \leq b$, define a function $F(X)$ by the rule

$$F(X) = \int_a^X f(x)\,dx.$$

Note that here X is a variable which is entirely independent of the integration variable x. Then $F(X)$ represents the area under the curve $y = f(x)$ for $a \leq x \leq X$.

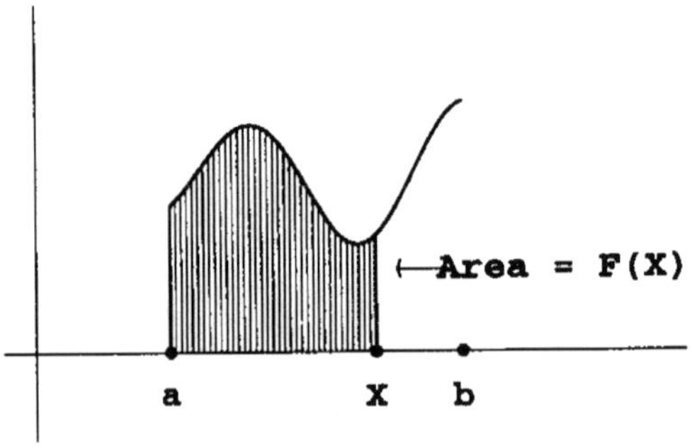

We see that for h small, $F(X+h) - F(X)$ is the area of the rectangular strip of width h:

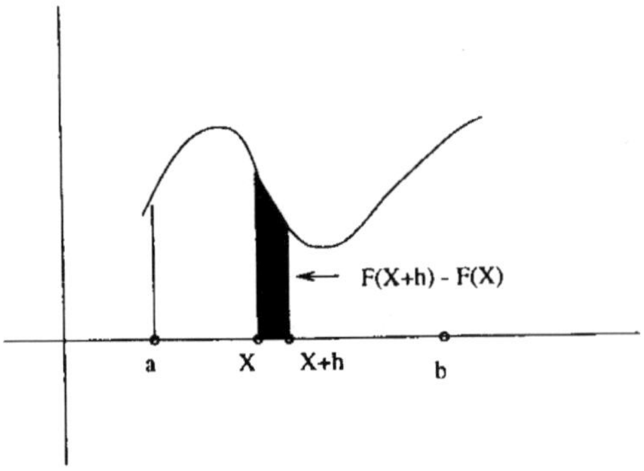

Hence, since f is continuous, for h sufficiently small

$$F(X+h) - F(X) \;\approx\; h \cdot f(X).$$

It follows that

$$\lim_{h \to 0} \frac{F(X+h) - F(X)}{h} \;=\; f(X).$$

We have thus shown that $F'(x) = f(x)$, and

$$\int_a^X f(x)\,dx \;=\; F(X). \qquad (15.1)$$

When $X = a$, notice that $F(a) = \int_a^a f(x)\,dx \;=\; 0$, and when we choose $X = b$ we see that

$$\int_a^b f(x)\,dx \;=\; F(b) = F(b) - F(a),$$

since (as we just saw) $F(a) = 0$.

Exercises for §15.1

Compute the following definite integrals.

(1) $\int_1^3 13\,dx$.

(2) $\int_1^5 \sqrt{x}\,dx$.

(3) $\int_1^7 \frac{1}{x^4}\,dx$.

(4) $\int_2^{10} (\frac{1}{x^4} - \frac{1}{x^3})\,dx$.

(5) $\int_0^3 (1+x)^3\,dx$.

(6) $\int_{\frac{1}{2}}^4 (1+x)^4\,dx$.

(7) $\int_1^5 \frac{7}{\sqrt{x}}\,dx$.

(8) $\int_{-4}^{-2} \frac{7}{x^5}\,dx$.

(9) $\int_0^1 x\sqrt[3]{x}\,dx$.

(10) $\int_0^3 x(\sqrt{x} + x^3)\,dx$.

(11) $\int_1^2 (x + \frac{1}{x^2})\,dx$.

(12) $\int_1^2 (x + \frac{1}{x^2})^2\,dx$.

(13) $\int_1^{10} \frac{x-1}{\sqrt{x}}\,dx$

(14) $\int_0^1 x^{1.7}\,dx$.

(15) $\int_2^4 \frac{x^2-1}{x-1}\,dx$.

(16) $\int_0^1 (x^7 + x + 3) \cdot (x^2 - 3)\,dx$.

(17) $\int_0^2 (x^7 + x + 3) \cdot (\sqrt{x} - 3)\,dx$.

(18) $\int_0^{\frac{\pi}{3}} \sin(x)\,dx$.

(19) $\int_0^{\frac{\pi}{3}} \cos(4x)\,dx$.

(20) $\int_0^{\frac{\pi}{2}} (\cos(x) + \sin(x))\,dx$.

(21) $\int_{\frac{\pi}{4}}^{\pi} (2\sin(x) + \sec^2(x))\,dx$.

(22) $\int_{\frac{2\pi}{3}}^{\pi} \sec(x)\tan(x)\,dx$

(23) Suppose that we know that for a certain function $f(t)$ and some constant c

$$\cos(x) + 1 \;=\; \int_c^x f(t)\,dt.$$

Find $f(t)$ and c.

(24) Suppose that we know that for a certain function $f(t)$ and some constant c

$$x^6 + 8 = \int_c^{x^2} f(t)\,dt.$$

Find $f(t)$ and c.

In the following exercises find the derivative of the following functions using the Fundamental Theorem of Calculus. Note that we have used t for the integration variable.

(25) $h(x) = \int_1^x \sqrt{t^6 - 1}\,dt.$

(26) $h(x) = \int_1^x \sin^2(t)\,dt.$

(27) $h(x) = \int_x^1 \cos(t)\,dt.$

(28) $h(x) = \int_1^{x^2} \frac{t^2}{t^7 - \sin(t)}\,dt.$

(29) $h(x) = \int_{\cos(x)}^{x^4} \sin(t^2)\,dt.$

(30) $h(x) = \int_{2x}^{10x} (3 - \sqrt[3]{t})^2\,dt.$

(31) Verify that if f is a continuous function and $g_1(x)$ and $g_2(x)$ are differentiable functions then the derivative of the function

$$h(x) = \int_{g_1(x)}^{g_2(x)} f(t)\,dt$$

is

$$f(g_2(x))g_2'(x) - f(g_1(x))g_1'(x).$$

15.2 The Indefinite Integral

We now define the symbol

$$\int f(x)\,dx$$

to be the **general antiderivative of** f (recall that an antiderivative is a function $F(x)$ satisfying $F'(x) = f(x)$) where we assume the antiderivative is defined on some interval (which we may or may not specify). If $F(x)$ denotes any particular antiderivative of $f(x)$ then, if C is any constant, $F(x) + C$ is also an antiderivative and the most general antiderivative will take this form. Thus

$$\int f(x)\,dx = F(x) + C,$$

and $\int f(x)\,dx$ is termed the **indefinite integral**. When limits a and b are put on the integral sign we then obtain

$$\int_a^b f(x)\,dx = F(b) - F(a),$$

which is termed the **definite integral.**

Example [15.6] Some simple examples of indefinite integrals are

$$\text{(a)} \int \cos(x)\,dx = \sin(x) + C,$$

$$\text{(b)} \int \frac{1}{x}\,dx = \ln(x) + C$$

$$\text{(c)} \int x^n\,dx = \frac{x^{n+1}}{n+1} \quad (\text{if } n \neq -1).$$

Exercises for §15.2

Compute the following indefinite integrals. Check your answer by differentiating the result with the CAS.

(1) $\int \frac{x^2-x-1}{x}\,dx.$

(2) $\int (3-\sqrt[3]{x})^2\,dx.$

(3) $\int (\sin(x) + 2\sec^2(x))\,dx.$

(4) $\int (\sqrt{x} + \frac{1}{\sqrt[3]{x}})^2\,dx.$

(5) $\int (1-x)(1-x^2)(1-x^3)\,dx.$

(6) $\int \sec(x)\tan(x)\,dx.$

(7) $\int (\frac{1}{x} + \frac{1}{x^2} + \frac{1}{x^3})\,dx.$

(8) $\int \sqrt[3]{x^{-2}}\,dx.$

(9) $\int \sqrt[4]{x^5}\,dx.$

(10) $\int \sqrt{x}(x^3 - \frac{1}{\sqrt[3]{x}})\,dx.$

(11) $\int \sqrt{x}(x^3 - \frac{1}{\sqrt[3]{x}})^2\,dx.$

(12) $\int (1-x)^4\,dx.$

(13) $\int e^{-2x}\,dx.$

(14) $\int \cos(5x)\,dx.$

(15) $\int \frac{2}{x}\,dx.$

(16) $\int \sinh(x)\,dx.$

(17) $\int (e^{3x} + x^2)\,dx.$

(18) $\int 2x\cos(x^2)\,dx.$

(19) $\int 3x^2\sin(x^3)\,dx.$

(20) $\int 4x^3 e^{x^4}\,dx.$

15.3 Integration by the Method of Substitution

Let $F(x)$ and $u(x)$ be differentiable functions. Recall that by the chain rule

$$\frac{d}{dx}F\big(u(x)\big) = F'\big(u(x)\big)\cdot u'(x).$$

It immediately follows that

$$\begin{aligned} \int F'\Big(u(x)\Big) \cdot u'(x)\,dx &= F\Big(u(x)\Big) + C \\ \int_a^b F'\Big(u(x)\Big) \cdot u'(x)\,dx &= F\Big(u(b)\Big) - F\Big(u(a)\Big) \end{aligned} \tag{15.2}$$

Example [15.7] Evaluate $\int (x^3 + 5x - 1)^{100} \cdot (3x^2 + 5)\,dx$.

We see here that the given integral is of the form

$$\int u(x)^{100} \cdot u'(x)\,dx,$$

with $u(x) = x^3 + 5x - 1$. If we let $F'(x) = x^{100}$ (and hence $F(x) = x^{101}/101$) the integral takes the form (15.2). We see that

$$\begin{aligned} \int u(x)^{100} \cdot u'(x)\,dx &= F\Big(u(x)\Big) + C \\ &= \frac{u(x)^{101}}{101} + C \\ &= \frac{(x^3 + 5x - 1)^{101}}{101} + C. \end{aligned}$$

Example [15.8] Evaluate $\int \cos(x^9 - e^x + 11) \cdot (9x^8 - e^x)\,dx$.

Upon scanning this integral we quickly see that $9x^8 - e^x$ is the derivative of $u(x) = x^9 - e^x + 11$, and the integral takes the form

$$\int \cos\Big(u(x)\Big) \cdot u'(x)\,dx.$$

Now setting $F(x) = \sin(x)$ (and hence $F'(x) = \cos(x)$) we obtain an integral of the form (15.2) and thus

$$\int \cos(x^9 - e^x + 11) \cdot (9x^8 - e^x)\,dx = \sin(x^9 - e^x + 11) + C.$$

The previous two examples illustrate the essence of the method of substitution. In order to facilitate quick computation we use the following (very clever) notational trick. When we rewrite the notation

$$\frac{du}{dx} = u'(x)$$

as

$$du = u'(x)\,dx,$$

then we may rewrite equation (15.2) as

$$\int F'\Big(\underbrace{u(x)}_{u}\Big)\cdot\underbrace{u'(x)\,dx}_{du} = \int F'(u)\,du \tag{15.3}$$

$$= F(u) + C.$$

With this formula in mind we return to the indefinite integral from Example [15.8],

$$\int \cos(x^9 - e^x + 11)\cdot(9x^8 - e^x)\,dx.$$

If we set $u = x^9 - e^x + 11$ and $du = (9x^8 - e^x)\,dx$ then

$$\begin{aligned}\int \cos(\underbrace{x^9 - e^x + 11}_{u})\cdot\underbrace{(9x^8 - e^x)\,dx}_{du} &= \int \cos(u)\,du\\ &= \sin(u) + C\\ &= \sin(x^9 - e^x + 11) + C.\end{aligned}$$

Question: How do you program a CAS to be able to integrate a function with the method of substitution?

Answer: On scanning an integral such as

$$\int e^{3x^4+2x^3+3}\cdot(12x^3 + 6x^2)\,dx,$$

the CAS must try to make a substitution. There are basically two visible possibilities:

(1) $u = 3x^4 + 2x^3 + 3$, and $du = (12x^3 + 6x^2)\,dx$, or

(2) $u = 12x^3 + 6x^2$, and $du = (36x^2 + 12x)\,dx$.

Clearly the second choice will lead us astray (it's rather like trying to open a lock with the wrong key). The first choice transforms the integral into a manageable form

$$\begin{aligned}\int e^u\,du &= e^u + C\\ &= e^{3x^4+2x^3+3} + C.\end{aligned}$$

Remark: Actually programming such a procedure on a CAS is highly nontrivial, but it can be done with a method called pattern matching.

Example [15.9] Evaluate $\int_0^1 \sin\left(\pi(x^8 - 2x)\right) \cdot (4x^7 - 1)\ dx$.

Here we set $u = \pi(x^8 - 2x)$, and $du = \pi(8x^7 - 2)\ dx = 2\pi \cdot (4x^7 - 1)\ dx$. When $x = 0$, $u = 0$, and when $x = 1$, $u = -\pi$. The integral transforms to

$$\int_0^1 \sin(\underbrace{\pi(x^8 - 2x)}_{u}) \cdot \underbrace{(4x^7 - 1)\ dx}_{du} = \int_0^{-\pi} \sin(u)\ \frac{du}{2\pi}$$

$$= \frac{-\cos(u)}{2\pi}\Bigg]_0^{-\pi}$$

$$= \frac{-\cos(-\pi) - (-\cos(0))}{2\pi}$$

$$= \frac{1}{\pi}.$$

Example [15.10] Evaluate the integral

$$\int_1^3 \frac{e^t}{(1+e^t)^2}\ dt.$$

Here we let $u(t) = 1 + e^t$, and $du = e^t\ dt$. When $t = 1$, $u = 1 + e$, and when $t = 3$, $u = 1 + e^3$. Thus

$$\int_1^3 \frac{e^t}{(1+e^t)^2}\ dt = \int_{1+e}^{1+e^3} \frac{du}{u^2}$$

$$= -u^{-1}\Bigg]_{1+e}^{1+e^3}$$

$$= \frac{-1}{1+e^3} + \frac{1}{1+e}.$$

Remark: In this particular instance the other visible substitution will also work. If we let $u = e^t$ then the integral is transformed into

$$\int_e^{e^3} \frac{du}{(1+u)^2},$$

and the substitution $v = 1 + u$, $dv = du$ will complete the problem.

Warning: The method of substitution is not the appropriate tool for every occasion and will not always work. For example, if we consider the integral

$$\int (x^5 + 3x + 2)(x^2 + 5x)\, dx,$$

no helpful substitution meets the eye. Nevertheless, this can be integrated directly (if not so cleverly) by simply multiplying

$$(x^5 + 3x + 2)(x^2 + 5x) = x^7 + 5x^6 + 3x^3 + 17x^2 + 10x,$$

and then integrating piece be piece,

$$\int [x^7 + 5x^6 + 3x^3 + 17x^2 + 10x]\, dx = \frac{x^8}{8} + \frac{5x^7}{7} + \frac{3x^4}{4} + \frac{17x^3}{3} + 5x^2 + C.$$

Exercises for §15.3

Compute the following definite and indefinite integrals. Check your answer by differentiating the anti–derivative you obtain with the CAS.

(1) $\int (3x+4)^7\, dx$.

(2) $\int_0^{\frac{\pi}{4}} \cos(2x)\, dx$.

(3) $\int_{-1}^{2} \frac{2x+4}{x^2+4x+18}\, dx$.

(4) $\int_0^{\pi} x \sin(2x^2 - 3)\, dx$.

(5) $\int x^2 \sin(x^3)\, dx$.

(6) $\int \frac{x}{\sqrt{x^2-1}}\, dx$.

(7) $\int x e^{2x^2}\, dx$.

(8) $\int e^{x+e^x}\, dx$.

(9) $\int_0^2 e^x(1-e^x)\, dx$.

(10) $\int_0^{\frac{\pi}{4}} e^{\sin x} \cos(x)\, dx$.

(11) $\int x\sqrt{x-1}\, dx$.

(12) $\int_{-2}^{-3} \frac{e^{2x}+1}{e^{2x}}\, dx$.

(13) $\int \frac{x^3-1}{x^3-x^2}\, dx$.

(14) $\int (1+\sqrt{x})^2\, dx$.

(15) $\int \frac{e^{\sqrt{x}}}{\sqrt{x}}\, dx$.

(16) $\int_3^4 \frac{x}{\sqrt{x-2}}\, dx$.

(17) $\int \frac{\sin(x)}{(1+\cos(x))^7}\, dx$.

(18) $\int (\sin(x))^3\, dx$.

(19) $\int \sqrt{1+\sqrt{x}}\, dx$.

(20) $\int x\sqrt{x^2+a^2}\, dx$.

(21) $\int_0^1 \frac{x}{\sqrt[3]{x^2+2}}\, dx$.

(22) $\int \frac{\sin(x)}{(\cos(x))^2}\, dx$.

(23) $\int x\sqrt{x-4}\, dx$.

(24) $\int_{-\frac{\pi}{2}}^{\frac{\pi}{2}} \frac{\cos(x)}{(\sin(x))^2}\, dx$.

(25) $\int_0^{\frac{\pi}{4}} \tan(x)\, dx$.

(26) $\int (x^3+x-1)^2 x^{-\frac{1}{2}}\, dx$.

(27) $\int_1^3 \frac{(1+\sqrt{x})^6}{\sqrt{x}}\, dx$.

(28) $\int_0^2 x\sqrt{x^2+1}\, dx$.

(29) $\int_0^1 \frac{e^{2x}}{e^{2x}+2}\, dx$

(30) $\int e^x \sin(e^x)\, dx$.

(31) $\int e^x \cos(e^x)\cos(\sin(e^x))\,dx.$

(32) $\int \sec^2(5x)\,dx.$

(33) $\int (x^2+\frac{1}{3})\cdot\cos(1+x^3+x)\,dx.$

(34) $\int_0^1 \sinh(2x)\,dx.$

(35) $\int \cosh(x)\sinh^2(x)\,dx.$

15.4 Integration by Parts

Let $u(x)$ and $v(x)$ be differentiable functions on an interval $a \le x \le b$. The product rule tells us that

$$u'(x)\,v(x) \;+\; u(x)\,v'(x) \;=\; \frac{d}{dx}\Big(u(x)\cdot v(x)\Big).$$

Integrating both sides we obtain

$$\int u'(x)\,v(x)\,dx \;+\; \int u(x)\,v'(x)\,dx \;=\; u(x)\cdot v(x) \;+\; C$$

$$\int_a^b u'(x)\,v(x)\,dx \;+\; \int_a^b u(x)\,v'(x)\,dx \;=\; u(b)\,v(b) \;-\; u(a)\,v(a).$$

The above formulae are usually written in the compact form

$$\int u\cdot v' = u\cdot v \;-\; \int u'\cdot v, \tag{15.4}$$

which has the advantage of being simple to remember. This is called **integration by parts**.

Example [15.11] Evaluate $\int x\,e^{-2x}\,dx$.

Here we let $u = x$ and $v' = e^{-2x}$. Then $u' = 1$ and $v = \frac{-e^{-2x}}{2}$. It follows that

$$\int \underbrace{x}_{u}\,\underbrace{e^{-2x}}_{v'}\,dx \;=\; \underbrace{x}_{u}\,\underbrace{\frac{-e^{-2x}}{2}}_{v} \;-\; \int \underbrace{1}_{u'}\cdot\underbrace{\frac{-e^{-2x}}{2}}_{v}\,dx$$

$$= -\frac{x}{2}e^{-2x} \;-\; \frac{e^{-2x}}{4} \;+\; C.$$

Remark: You should check your answer by showing that

$$\frac{d}{dx}\Big(-\frac{x}{2}e^{-2x} \;-\; \frac{e^{-2x}}{4} \;+\; C\Big) = x\,e^{-2x}.$$

This problem is very instructive. Whenever you are integrating a product of two functions an attempt at integration by parts is worthwhile (often the first attempt is not successful but a second one is). As we see in the next example, even if the method seems to fail, that is to say you are left with another inobvious integral to compute, integration by parts may be tried on this second integral to complete the computation.

Example [15.12] Evaluate $\int x^2 \sin(2x)\, dx$.

If we let $u = x^2$ and $v' = \sin(2x)$ then $u' = 2x$, and $v = -\cos(2x)/2$. Integration by parts then gives us

$$\int \underbrace{x^2}_{u} \underbrace{\sin(2x)}_{v'}\, dx = \underbrace{x^2}_{u} \cdot \underbrace{(-\cos(2x)/2)}_{v} - \int \underbrace{(2x)}_{u'} \cdot \underbrace{(-\cos(2x)/2)}_{v}\, dx$$

$$= \frac{-x^2 \cos(2x)}{2} + \int x \cos(2x)\, dx.$$

At this point it doesn't look like we have accomplished very much, but look again. Our original integral, which had an x^2 in it, has now been transformed into a similar integral with a lower power of x in it (namely x to the first power). This suggests that proceeding in the same way a second time may do the trick;

$$\int \underbrace{x}_{u} \underbrace{\cos(2x)}_{v'}\, dx = \underbrace{x}_{u} \cdot \underbrace{(\sin(2x)/2)}_{v} - \int \underbrace{(1)}_{u'} \cdot \underbrace{(\sin(2x)/2)}_{v}\, dx$$

$$= \frac{x \sin(2x)}{2} + \frac{\cos(2x)}{4} + C,$$

and thus

$$\int x^2 \sin(2x)\, dx = \frac{-x^2 \cos(2x)}{2} + \frac{x \sin(2x)}{2} + \frac{\cos(2x)}{4} + C.$$

Example [15.13] Evaluate $\int x \ln(x)\, dx$.

In this case were we to choose $u = x$ and $v' = \ln(x)$ we would truly not get anywhere (see Example [15.15]). If, however, we try $u = \ln(x)$ and $v' = x$, then

$u' = 1/x$, $v = x^2/2$, and

$$\int \underbrace{x}_{v'} \cdot \underbrace{\ln(x)}_{u}\, dx = \underbrace{\ln(x)}_{u} \cdot \underbrace{\frac{x^2}{2}}_{v} - \int \underbrace{\frac{1}{x}}_{u'} \cdot \underbrace{\frac{x^2}{2}}_{v}\, dx$$

$$= \ln(x) \cdot \frac{x^2}{2} - \int \frac{x}{2}\, dx$$

$$= \frac{x^2 \ln(x)}{2} - \frac{x^2}{4} + C$$

Example [15.14] Evaluate $\int \ln(x)\, dx$.

At first glance this integral does not appear to take the form of a product. But there is an attractive trick which will allow us to use integration by parts: let $v' = 1$ and $u = \ln(x)$. Then $u' = 1/x$, $v = x$, and

$$\int \underbrace{\ln(x)}_{u} \cdot \underbrace{1}_{v'}\, dx = \underbrace{\ln(x)}_{u} \cdot \underbrace{x}_{v} - \int \underbrace{\frac{1}{x}}_{u'} \cdot \underbrace{x}_{v}\, dx$$

$$= x \ln(x) - x + C.$$

Integration by parts can, of course, be used to compute definite integrals as seen in the following example.

Example [15.15] Evaluate $\int_1^4 x^2(\ln(x) + 2)\, dx$.

Using experience as a guide, we let $u = \ln(x) + 2$ and $v' = x^2$, hence $u' = 1/x$

and $v = x^3/3$, and we integrate by parts:

$$\int_1^4 \underbrace{x^2}_{v'} \underbrace{(\ln(x)+2)}_{u} \, dx = \underbrace{(\ln(x)+2)}_{u} \underbrace{x^2}_{v} \Big]_1^4 - \int_1^4 \underbrace{\frac{1}{x}}_{u'} \cdot \underbrace{\frac{x^3}{3}}_{v} \, dx$$

$$= (\ln(4)+2)\cdot 16 - 2 - \int_1^4 \frac{x^2}{3}\, dx$$

$$= 16\ln(4) + 30 - \frac{4^3}{9} + \frac{1}{9}.$$

Our final example is an ingenious demonstration of integration by parts.

Example [15.16] Evaluate $\int (\cos(x))\, e^{-x}\, dx$.

If we let $u = \cos(x)$ and $v' = e^{-x}$ we transform the integral into an integral of similar structure except that the cos is replaced by a sin. This suggests that integrating by parts a second time may yield an integral with yet again the same structure but involving a cos.

First integration by parts

$$\int \underbrace{(\cos(x))}_{u} \underbrace{e^{-x}}_{v'} \, dx = \underbrace{(\cos(x))}_{u} \cdot \underbrace{(-e^{-x})}_{v} - \int \underbrace{(-\sin(x))}_{u'} \cdot \underbrace{(-e^{-x})}_{v} \, dx.$$

Second integration by parts

$$\int \underbrace{(\sin(x))}_{u} \underbrace{e^{-x}}_{v'} \, dx = \underbrace{(\sin(x))}_{u} \cdot \underbrace{(-e^{-x})}_{v} - \int \underbrace{(\cos(x))}_{u'} \cdot \underbrace{(-e^{-x})}_{v} \, dx$$

$$= -(\sin(x))\, e^{-x} + \int (\cos(x)) \cdot e^{-x}\, dx$$

By combining the two previous formulae we obtain the remarkable identity

$$\int (\cos(x))\, e^{-x}\, dx = (-\cos(x) + \sin(x))\, e^{-x} - \int (\cos(x))\, e^{-x}\, dx,$$

which may be rewritten as

$$2\int (\cos(x))\, e^{-x}\, dx \;=\; (-\cos(x) \;+\; \sin(x))\, e^{-x} \;+\; C,$$

so that

$$\int (\cos(x))\, e^{-x}\, dx \;=\; \frac{1}{2}(-\cos(x) \;+\; \sin(x))\, e^{-x} \;+\; C.$$

Exercises for §15.4

Compute the following indefinite integrals by integrating by parts. Check your answer by differentiating the result with the CAS. Some problems may require a preliminary substitution before integrating by parts.

(1) $\int x\cos(x)\, dx.$

(2) $\int (2x+1)e^{-5x}\, dx.$

(3) $\int x^2 e^{3x}\, dx.$

(4) $\int x\ln(2x^3)\, dx.$

(5) $\int (\ln(x))^2\, dx.$

(6) $\int \cos(\sqrt{x})\, dx.$

(7) $\int e^{\sqrt{x}}\, dx.$

(8) $\int x\sinh(x)\, dx.$

(9) $\int (\sin(x))^2\, dx$

(10) $\int x^2\ln(x)\, dx.$

(11) $\int x\cosh(5x)\, dx.$

(12) $\int \cos(4x)\sin(2x)\, dx.$

(13) $\int (5x-3)\ln(3x)\, dx.$

(14) $\int x^9 e^{x^5}\, dx.$

(15) $\int x^{21}\cos(2x^{11})\, dx.$

(16) Show that

$$\int x^n e^x\, dx \;=\; x^n e^x \;-\; n\int x^{n-1} e^x\, dx$$

for all $n \neq 0$.

(17) Show that

$$\int_0^{\pi/2} (\sin(x))^n\, dx \;=\; \frac{n-1}{n}\int_0^{\pi/2} (\sin(x))^{n-2}\, dx$$

for all $n \geq 1$. For n an odd integer, $n > 2$ deduce Wallis' formula

$$\int_0^{\pi/2} (\sin(x))^n\, dx \;=\; \frac{2\cdot 4\cdot 6\cdots(n-1)}{3\cdot 5\cdot 7\cdots n}.$$

15.5 Basics on Differential Equations

Let $y = f(x), y' = f'(x), y'' = f''(x), \ldots$ be the successive derivatives of an unknown function $y = f(x)$. We view $y, y', y'', \ldots$ as unknown variables.

Definition: *A* ***differential equation*** *is an equation involving the derivatives $y, y', y'', \ldots,$ and x.*

Example [15.17] The equations

$$y' - 2x = 0, \quad y'' + y' - 3x^2 - 6x = 0, \quad y'' - (4x^2 + x)y = 0,$$

are differential equations.

Definition: *A* ***solution*** *of a differential equation is a function $f(x)$ (which is differentiable many times) with the property that the differential equation becomes an identity when the substitutions $y = f(x)$, $y' = f'(x)$, $y'' = f''(x)$, etc. are made.*

Example [15.18] Solve the differential equation $y' - 2x = 0$.

To solve this differential equation we require a function $y = f(x)$ such that $f'(x) = 2x$. By integrating both sides of this identity with respect to x we obtain

$$\begin{aligned} \int f'(x)\, dx &= \int 2x\, dx \\ f(x) &= x^2 + C, \end{aligned}$$

for an arbitrary constant C. To check that $y = x^2 + C$ is the solution we simply replace y by $f(x) = x^2 + C$ in the differential equation

$$\underbrace{(x^2 + C)'}_{y'} - 2x = 2x - 2x = 0,$$

and we do, in fact, obtain an identity.

Example [15.19] Verify that the function $y = e^{x^2}$ is a solution to the differential equation $y'' - (4x^2 + 2)y = 0$.

Replacing y by e^{x^2} in the differential equation yields the desired identity

$$\begin{aligned} y'' - (4x^2+2)y &= \underbrace{(4x^2+2)e^{x^2}}_{y''} - (4x^2+2)\underbrace{e^{x^2}}_{y} \\ &= 0. \end{aligned}$$

We have already seen that even when the differential equation seems to take a simple form (as it did in Example [15.18] when $y' - 2x = 0$), there are infinitely many solutions. In fact, $y = x^2 + C$ will be a solution for any constant C. To determine the value of C an additional piece of information must be given. In this example $y(0) = C$, thus if the value of the function at zero is given we can obtain a unique solution to the differential equation.

Definition: *A set of* ***initial conditions*** *for a differential equation is a set of known values of the derivatives of the solution:*

$$y(0), y'(0), y''(0), \ldots$$

It can be shown in general, that if a differential equation has a solution, then, in fact, it has infinitely many solutions unless some initial conditions are given.

Example [15.20] Solve the differential equation

$$y' - 3(x^2+1) = 0$$

given the initial condition $y(0) = 2$.

We are searching for a function $y = f(x)$ such that $f(0) = 2$ and

$$f'(x) = 3x^2 + 3.$$

By integrating both sides of this equation we see that

$$f(x) = x^3 + 3x + C.$$

We can identify the constant C by evaluating the function f at $x = 0$. Thus we see that $f(0) = 2 = C$ and we conclude that the solution to the differential equation is $y = x^3 + 3x + 2$.

There is no known method for solving an arbitrary differential equation and the problem of finding an algorithm to solve a given differential equation is immensely difficult. There are, however, a vast array of techniques which have been developed to solve specific types of differential equations. As an illustration of a basic method in this subject we will consider the method of **separation of variables.**

Consider a differential equation of the form

$$y' - p(x)q(y) = 0,$$

where $p(x)$ is any function involving only x and $q(y)$ is any function involving only y. Suppose $y = f(x)$ is a solution to this differential equation. It immediately follows that $f'(x) = p(x)q(f(x))$, or

$$\frac{f'(x)}{q(f(x))} = p(x).$$

Integrating both sides of this equation yields

$$\int \frac{f'(x)}{q(f(x))}\,dx = \int p(x)\,dx. \qquad (15.5)$$

Focusing on the integral on the left hand side for a moment, there is a natural substitution at hand: when we let $y = f(x)$, $dy = f'(x)dx$, and (15.5) becomes

$$\int \frac{dy}{q(y)} = \int p(x)\,dx. \qquad (15.6)$$

If we can compute the integrals in (15.6) we will have solved our differential equation. We summarize this discussion in the following algorithm.

Separation of Variables Algorithm

Step 1. Rewrite the differential equation $\frac{dy}{dx} - p(x)q(y) = 0$ in the form $\frac{dy}{q(y)} = p(x)dx$.

Step 2. Integrate both sides of the equation: $\int \frac{dy}{q(y)} = \int p(x)\,dx$.

Step 3. Solve for y.

Example [15.21] Solve the equation $y' - 6xy = 0$.

Step 1. $\frac{dy}{y} = 6xdx$.

Step 2. $\int \frac{dy}{y} = \ln(y) = \int 6x\,dx = 3x^2 + C.$

Step 3. $y = e^{3x^2+C}.$

Example [15.22] Solve the differential equation $y' - (3x^2 + 3)y^2 = 0$ with the initial condition $y(0) = 2$.

Step 1. $\frac{dy}{y^2} = (3x^2 + 3)\,dx.$

Step 2. $\int \frac{dy}{y^2} = -\frac{1}{y} = \int (3x^2 + 3)\,dx = x^3 + 3x + C.$

Step 3. $y = \frac{-1}{x^3+3x+C}.$

The initial condition $y(0) = 2$ implies $C = -\frac{1}{2}$.

Exercises for §15.5

Solve the following differential equations. Give the general solution (with an arbitrary constant C). and determine C if an initial condition is given. Check your answer with the CAS by plugging it back into the differential equation.

(1) $y' = y^2 + 2y + 1.$

(2) $y' = 5y + 2, \ (y(0) = 1).$

(3) $\frac{dy}{dx} = 1 + 2x + x^4.$

(4) $\frac{dy}{dx} - \frac{x^3}{2} = 1, \ (y(0) = 1).$

(5) $\frac{dy}{dx} = \frac{-2x}{y}.$

(6) $\frac{dy}{dx} = \sqrt{\frac{x}{y+2}}, \ (y(0) = 2).$

(7) $\frac{dy}{dx} = x^2y(x^3 + 3)^{\frac{1}{4}}.$

(8) $\frac{dy}{dx} - \frac{2x\cos(3x)}{y^2}, \ (y(0) = 1).$

(9) $\frac{dy}{dx} + (x^3 + 1)\frac{\sqrt{2y^2+1}}{y} = 0.$

(10) $\frac{dy}{dx} - 3x^2y = 0, \ (y(0) = 3).$

(11) $\frac{dy}{dx} = \frac{e^x}{y+3}.$

(12) $\frac{dy}{dx} = \frac{\ln(x)}{y^2}, \ (y(1) = 8).$

15.6 Exponential Growth and Decay

Let $f(t)$ denote a function of time t. The statement:

$$f(t) \text{ is changing at a rate proportional to } f(t) \tag{15.7}$$

indicates that there exists a constant K (termed the constant of proportionality) such that

$$f'(t) = Kf(t).$$

Functions that satisfy (15.7) appear frequently in science, engineering, and economics. If we write $y = f(t)$, then (15.7) leads to the differential equation $y' = Ky$ whose solution is

$$y = Ce^{Kt}.$$

Notice that the constant C can be obtained by setting $t = 0$ since $y(0) = C$. It is important to understand that whenever the statement (15.7) is encountered it may be expressed as a differential equation.

Example [15.23] (Exponential Growth) A census taken in a certain country showed that the population was 3 million people. A year later the population was found to be 3.3 million people. Assuming the population grows at a rate proportional to its size, find the population 5 years after the original census.

Let $p(t)$ denote the population t years after the census. Then $p(0) = 3, p(1) = 3.3$ measured in millions of people. The assumption that *the population grows at a rate proportional to its size* leads to the differential equation

$$\frac{dp}{dt} = Kp$$

for some constant K. Solving this differential equation, we obtain

$$\begin{aligned} \int \frac{dp}{p} &= \int K dt \\ \ln(p(t)) &= Kt + C \\ p(t) &= C_0 e^{Kt}, \end{aligned}$$

where $C_0 = e^C$. The condition $p(0) = 3$ implies $C_0 = 3$, and hence $p(t) = 3e^{Kt}$. To find K we use the hypothesis $p(1) = 3.3$:

$$3.3 = 3e^K$$

$$e^K = \frac{3.3}{3} = 1.1$$

$$K = \ln(1.1).$$

We conclude $p(t) = 3e^{\ln(1.1)t} = 3 \cdot (1.1)^t$, and $p(5) \approx 4.8315$ million people.

Example [15.24] (Exponential Decay) An Alice–in–Wonderland rabbit is originally 3 feet high and shrinks to a 2 foot height in 10 seconds. Assuming the rabbit is shrinking at a rate proportional to its height, how long will it take it to become 6 inches tall?

Let $h(t)$ denote the height of the rabbit after t seconds. Then $h(0) = 3, h(10) = 2$. In addition we are told that $\frac{dh}{dt} = Kt$ for some constant K. Solving this differential equation as before yields $h(t) = c_0 e^{Kt}$, where $C_0 = h(0) = 3$, and thus $h(t) = 3e^{Kt}$. In order to find K we use the hypothesis $h(10) = 3e^{10K} = 2$, which yields

$$e^{10K} = \frac{2}{3}$$

$$10K = \ln\left(\frac{2}{3}\right)$$

$$K = \frac{1}{10}\ln\left(\frac{2}{3}\right).$$

We conclude that $h(t) = 3 \cdot \left(\frac{2}{3}\right)^{t/10}$. To see how long it will take the rabbit to shrink to 6 inches (i.e., $\frac{1}{2}$ a foot) in height, we must solve the equation

$$\frac{1}{2} = 3\left(\frac{2}{3}\right)^{t/10}.$$

Dividing both sides of the equation by 3 and taking the natural logarithm yields $\ln\left(\frac{1}{6}\right) = \frac{t}{10}\ln\left(\frac{2}{3}\right)$, i.e., $t = 10 \cdot \frac{\ln\left(\frac{1}{6}\right)}{\ln\left(\frac{2}{3}\right)} \approx 44.19$ seconds.

A generalization of (15.7) is the statement

$$f(t) \textit{ is changing at a rate proportional to } f(t) - b \tag{15.8}$$

where b is a constant. This leads to the differential equation

$$f'(t) = K(f(t) - b),$$

where K is the constant of proportionality. Setting $y = f(t)$, our new differential equation takes the form

$$y' = Ky - Kb,$$

whose solution is given by

$$y = Ce^{Kt} + b,$$

where C is a constant (which can be determined provided initial conditions are specified).

Example [15.25] It is 70° Fahrenheit in a café. A waiter pours a cup of (almost) boiling hot tea (i.e., at 210°). Assume the rate at which the tea cools is proportional to the difference between the temperature of the tea and temperature of the surrounding air. Assume further that after 1 minute the tea cools down to 140°. Determine the temperature of the tea at any given time.

Let $T(t)$ denote the temperature of the tea at the time t. Then by hypothesis,

$$T'(t) = K(T(t) - 70)$$

for some K. Thus $T(t) = Ce^{Kt} + 70$. Since $T(0) = 210$ we see that $C = 140$. Furthermore, since $T(1) = 140$, we have that

$$140 = 140e^{K} + 70,$$

i.e., $K = \ln\left(\frac{1}{2}\right)$. We conclude that $T(t) = 140 \cdot 2^{-t} + 70$.

Exercises for §15.6

(1) A census showed that the population in a certain country was 10 million people. The population doubled to 20 million after 15 years. Assuming the population grows at a rate proportional to its size, how long will it take to double again to 40 million? How long will it take the country to have 1 billion people?

(2) The amount of bacteria in a culture increases at a rate proportional to the amount of bacteria present. Assume the culture starts with $10,000$ bacteria and increases to $30,000$ after 2 hours. What will the population be after 12 hours?

(3) The **half life** of radioactive carbon is 5568 years. This is the amount of time it takes for a given amount of the substance to decay to one–half of its original size. It is also known that the rate of decay is proportional to the amount present at any given time.

Some radioactive carbon was found at an archeological site and decayed to 10% of its original size. Determine the age of the archeological site assuming the sample of radioactive carbon was put into the site at its formation.

(4) A Texas oil company has designed an oil pump with the property that it continuously pumps oil out of a well at a rate proportional to the amount of oil left in the well. Initially, the well contains 3 million gallons of oil and after 5 years there are 2.5 million gallons of oil.

(a) How long will it take to remove 2 million gallons of oil from the well?

(b) When there are only $\frac{1}{2}$ million gallons of oil remaining in the well, at what rate will the oil be pumped out?

(5) Newton's law of cooling states that the rate at which a body cools is proportional to the difference between the temperature $T(t)$ of the body and the ambient temperature T_a of the surrounding medium.

(a) Write down and solve the differential equation for $T(t)$.

(b) If it is 70°F in a restaurant and the waiter pours a boiling cup of tea (at 212°F). The tea cools down to 160°F in 1 minute. How long will it take the tea to reach 120°F?

(6) A software manufacturer is trying to find a mathematical model to determine how long it takes the average person to master its new software. They have have found that the rate of learning is proportional to $100 - L(t)$ where $L(t)$ denotes the percentage ($0 - 100\%$) of the program learned in time t. That is to say learning is at first rapid and then tapers off as $L(t)$ approaches 100%.

(a) Write down and solve the differential equation for $L(t)$.

(b) Assume that initially the average person knows nothing about this software and can master 25% of it in 3 hours. How long will it take to master 90%?

(c) If the new software is purchased by a previous user of the company's other software products, it was found that 30% of the new software is already known. Assuming it takes 2 hours to learn 50%, how long will it take to learn 90%?

(d) Explain why this mathematical model breaks down for very large values of t.

Additional exercises for Chapter XV

---

(1) Let $f(x) = \frac{1}{2} - \exp(-x^2)$ and $F(x) = \int_0^x f(t)dt$

(a) Where is F increasing, decreasing?

(b) Where is F concave up, concave down?

(c) Where are the critical points of F? Check with your CAS.

(2) Show that $F(x) = \begin{cases} x, & \text{if } x > 0 \\ -x, & \text{if } x < 0 \end{cases}$ and $G(x) = \begin{cases} x+2, & \text{if } x > 0 \\ -x+4, & \text{if } x < 0 \end{cases}$ are both antiderivatives of $f(x) = \begin{cases} 1, & \text{if } x > 0 \\ -1, & \text{if } x < 0 \end{cases}$,

But $G(X) \neq F(x)$. Explain.

(3) Solve the initial value problem $y' - x\exp(y) = 2\exp(y)$, $y(0) = 1$.

(4) Find a curve in the xy-plane that passes through the point $(1, 1)$ and whose normal at a point (x, y) has slope $-\frac{2y}{3x^2}$.

(5) Prove that

$$\int x^n \, dx = \frac{1}{n+1} x^{n+1}.$$

(Hint: use induction.)

Suppose $F(x) = \int_0^x f(t)\, dt$, where $f(t)$ is pictured below.

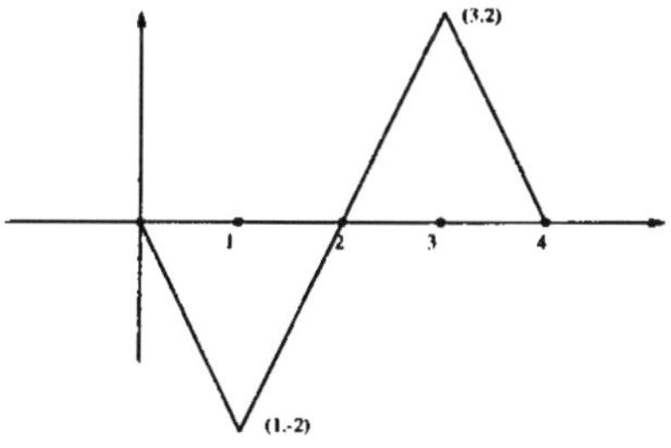

Answer the following questions

(6) Evaluate $F(1)$, $F(2)$, $F(3)$ and $F(4)$.

(7) Use the Fundamental Theorem to find the critical points of F.

(8) Where is F increasing?, Decreasing?

(9) Can you write an algebraic expression for F.

Quite often it is very difficult if not impossible to solve a differential equation. However, we may be able to get a very good picture of possible solutions by analyzing the differential equation itself. For example consider the equation $y' = y(2-y)$, $y(0) = 3/2$. Then $y'(0) = y(0)(2-y(0)) = 3/4 > 0$, so $y(x)$ is increasing for x near 0. Since $y' = y(2-y)$, we see that $y' > 0$ for $y < 2$, so $y(x)$ is increasing for $3/2 \leq y < 2$. Can y be larger than 2? This is not possible since $y' = 0$ when $y = 2$, while $y' > 0$ for $y < 2$ (y near 2), and $y' < 0$ for $y > 2$; and by the first derivative test $y = 2$ would have to be a maximum.

By taking the derivative, we get $y'' = y(2-y)(2-2y)$. Thus $y'' < 0$ for $3/2 \leq y < 2$ and the graph is concave down. We can now sketch the solution:

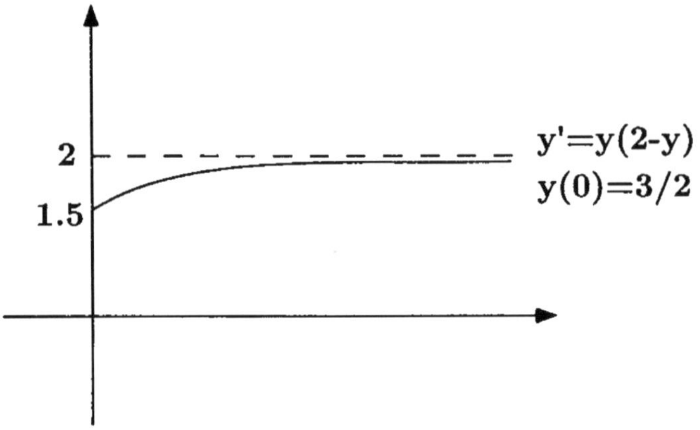

For the following differential equations sketch possible solutions corresponding to each of the two given initial conditions.

(10) $y' = (y-2)(y-4), \ y(0) = 1, \ y(0) = 6.$

(11) $y' = -y(1-y), \ y(0) = -2, \ y(0) = 1/4.$

(12) $y' = y(y-2)(y-5)(y+3), \ y(0) = 1, \ y(0) = 3.$

Chapter XVI

Basic Applications of the Integral

16.1 The Average Value of a Function

Given a collection of numbers (for example test scores) such as

$$100, 87, 92, 53, 67, 66, 42, 92, 85, 73, 76,$$

we may easily compute (using the CAS) the average value

$$\frac{1}{11}(100 + 87 + 92 + 53 + 67 + 66 + 42 + 92 + 85 + 73 + 76) \approx 75.73.$$

Despite its simplicity, the average value is a crucial quantity from the point of view of statistics. The natural generalization of the average of a collection is the concept of the average value of a continuous function $f(x)$ over an interval $a \leq x \leq b$. One way to try to compute this average is to take N equally spaced points

$$a < x_1 < x_2 < \ldots x_N \;=\; b,$$

and compute the average value

$$\frac{1}{N}\sum_{n=1}^{N} f(x_n).$$

By hypothesis the points x_i are given by

$$x_1 = a + \frac{b-a}{N}, \quad x_2 = a + 2\frac{b-a}{N}, \ldots, \; x_N = a + N \cdot \frac{b-a}{N} = b,$$

and the average is given by

$$\frac{1}{N}\sum_{n=1}^{N} f(x_n) = \frac{1}{N}\sum_{n=1}^{N} f\left(a + n\frac{b-a}{N}\right). \tag{16.1}$$

When we allow $N \to \infty$ the above sum (16.1) yields a better and better estimate for the average value of the function $f(x)$. When we compare (16.1) with the definition of the integral as a limit of a sum (see §13.3) we are led to defining the average value of $f(x)$ (for $a \leq x \leq b$) to be

$$\frac{1}{b-a}\int_a^b f(x)\, dx.$$

Example [16.1] What is the average value of $y = x^2$ for $2 \leq x \leq 5$?

The average value is given by the integral

$$\frac{1}{5-2}\int_2^5 x^2\, dx = \frac{1}{3}\frac{x^3}{3}\Big]_2^5 = \frac{117}{9} = 13.$$

Example [16.2] What is the average value of $y = \cos(\pi x)$ for $0 \leq x \leq \frac{1}{3}$?

Here the average value is

$$\frac{1}{\frac{1}{3}-0}\int_0^{\frac{1}{3}} \cos(\pi x)\, dx = \frac{3}{\pi}\sin(\pi x)\Big]_0^{\frac{1}{3}} = \frac{3\sqrt{3}}{2\pi}.$$

Example [16.3] Compute the average value of e^x for $1 \leq x \leq 21$?

In this case the average value is (with the help of our CAS)

$$\frac{1}{21-1}\int_1^{21} e^x\, dx = \frac{1}{20}(e^{21} - e) \approx 65940786.59.$$

Exercises for §16.1

Evaluate the average value of the following continuous functions $f(x)$ over the indicated intervals.

(1) $f(x) = x^3, 10 \leq x \leq 15.$

(2) $f(x) = \sqrt{x}, 100 \leq x \leq 121.$

(3) $f(x) = \cos(x), -\frac{\pi}{3} \leq x \leq \frac{\pi}{3}.$

(4) $f(x) = e^x, 2 \leq x \leq 7.$

(5) $f(x) = xe^{-x^2}, 1 \leq x \leq 3.$

(6) $f(x) = x\sin^2(\pi x), 0 \leq x \leq 2.$

(7) $f(x) = \frac{1}{x}, e \leq x \leq e^2.$

(8) $f(x) = 10^x, 1 \leq x \leq 5.$

(9) $f(x) = \sin(x), 0 \leq x \leq \pi.$

(10) $f(x) = x^{\frac{1}{4}}, 100 \leq x \leq 200.$

16.2 Computing Area

Let $f(x)$ and $g(x)$ be integrable functions on an interval $a \leq x \leq b$. There are two possible scenarios. Either they don't intersect (cross one another)

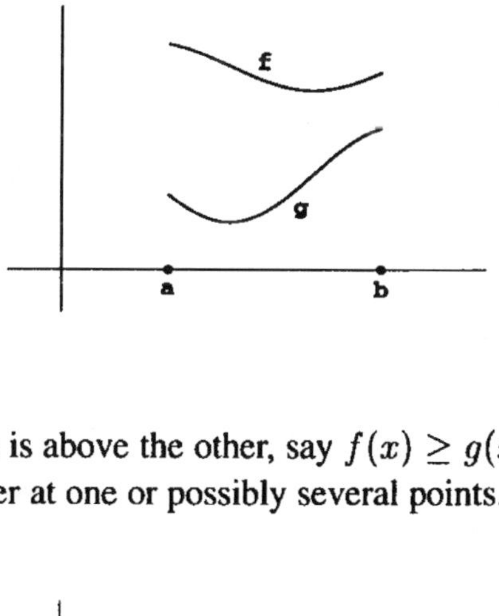

in which case one graph is above the other, say $f(x) \geq g(x)$ for all $a \leq x \leq b$, or they intersect one another at one or possibly several points.

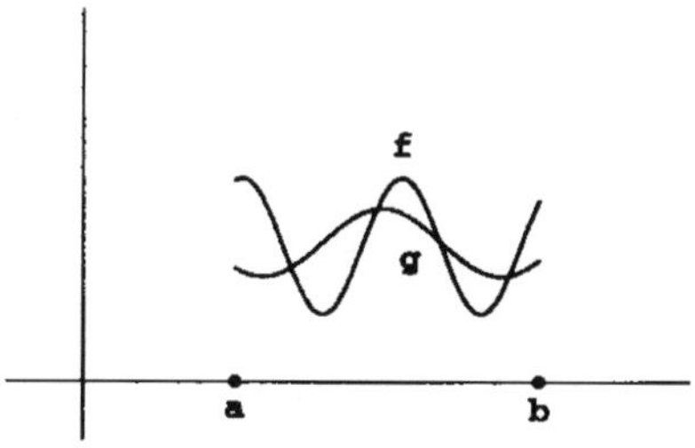

In the first case, $\int_a^b f(x)\ dx$ is the area below $y = f(x)$ and $\int_a^b g(x)\ dx$ is the area below $y = g(x)$.

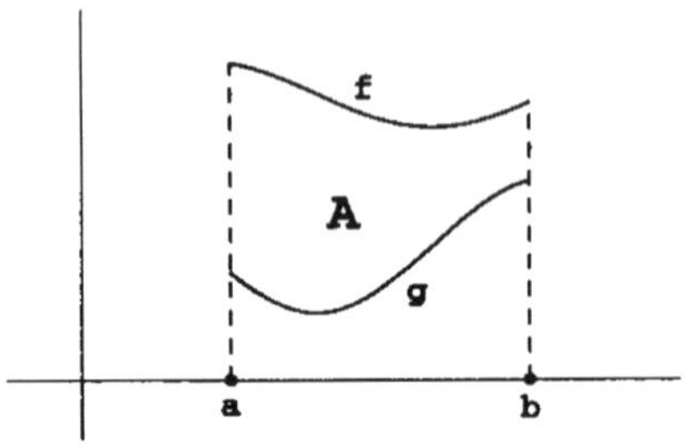

Hence the area of the region A between the graphs of f and g is given by subtracting:

$$\text{Area}(A) = \int_a^b (f(x) - g(x))\ dx.$$

Remark: This formula works with a slight modification even if one or both curves fall below the x–axis. For example, in the situation

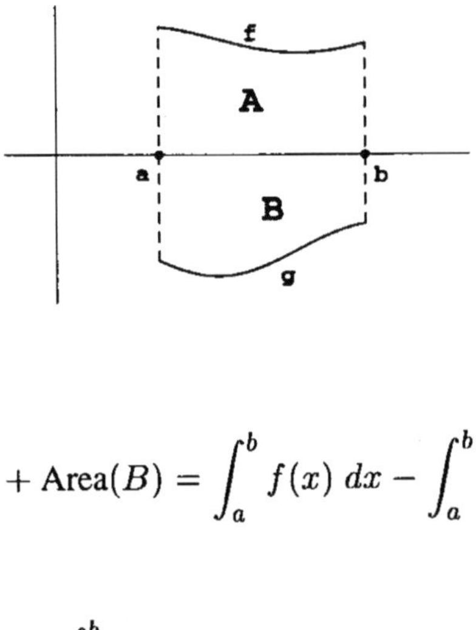

$$\text{Area}(A) + \text{Area}(B) = \int_a^b f(x)\ dx - \int_a^b g(x)\ dx,$$

since

$$\int_a^b g(x)\ dx = -\text{Area}(B).$$

If we insert absolute values signs around the integral we obtain a formula that is valid for any two *nonintersecting* functions (regardless of which function lies above the other or whether they both lie below the x–axis),

$$\left| \int_a^b (f(x) - g(x))\ dx \right|. \tag{16.2}$$

Example [16.4] Find the area of the region between the graphs $y = x^4$ and $y = 2x + 1$ when $2 \leq x \leq 3$.

First notice that $x^4 > 2x + 1$ whenever $2 \leq x \leq 3$ as seen in the figure below.

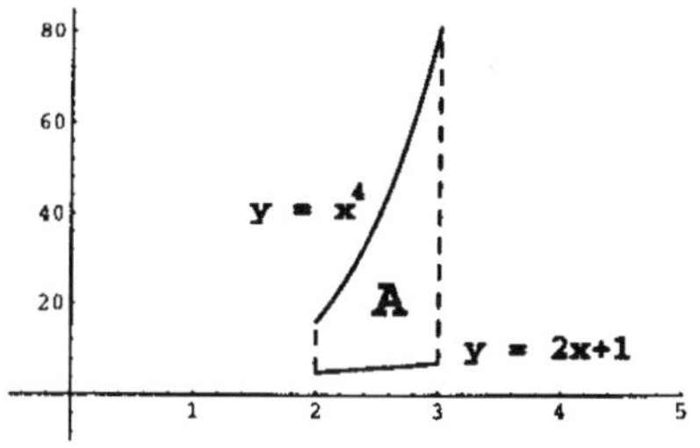

Then Area(A) can be computed using our formula:

$$\begin{aligned}\text{Area}(A) &= \int_2^3 \left(x^4 - (2x+1)\right)\,dx \\ &= \left(\frac{x^5}{5} - x^2 - x\right)\Big]_2^3 \\ &= \frac{181}{5}.\end{aligned}$$

We now move to our second scenario, the case when the curves in question intersect each other (at possibly several points).

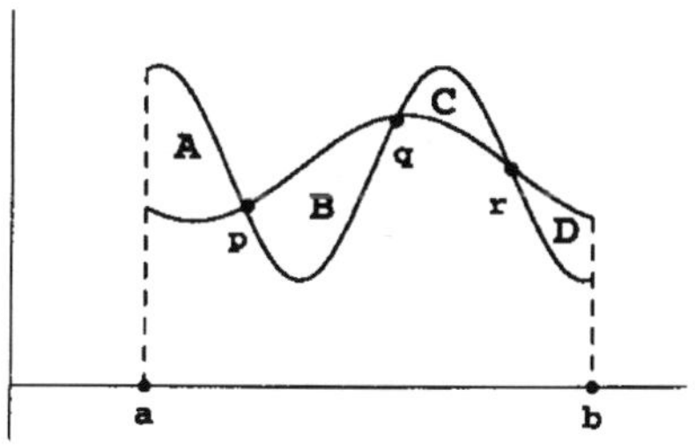

In this situation the area between the curves is simply given by the sum of the smaller areas depicted in the above diagram, i.e.,

$$\text{Area}(A) + \text{Area}(B) + \text{Area}(C) + \text{Area}(D).$$

In order to compute this area efficiently we begin by first finding the points where the curves intersect, in this case p, q, and r, and then compute the various areas involved using the formula we derived for curves which do not intersect at all. This amounts to computing

$$\begin{aligned} \text{Area}(A) &= \left| \int_a^p (f(x) - g(x))\, dx \right| \\ \text{Area}(B) &= \left| \int_p^q (f(x) - g(x))\, dx \right| \\ \text{Area}(C) &= \left| \int_q^r (f(x) - g(x))\, dx \right| \\ \text{Area}(D) &= \left| \int_r^b (f(x) - g(x))\, dx \right|. \end{aligned} \tag{16.3}$$

Note that we have avoided the problem of assessing which curve lies above the other by putting absolute value signs around our integrals.

Example [16.5] Find the area between the curves $y = \sin(x)$ and $y = \cos(x)$ in the region $0 \leq x \leq 2\pi$.

If we simultaneously plot the two curves on our CAS then we will obtain the graph

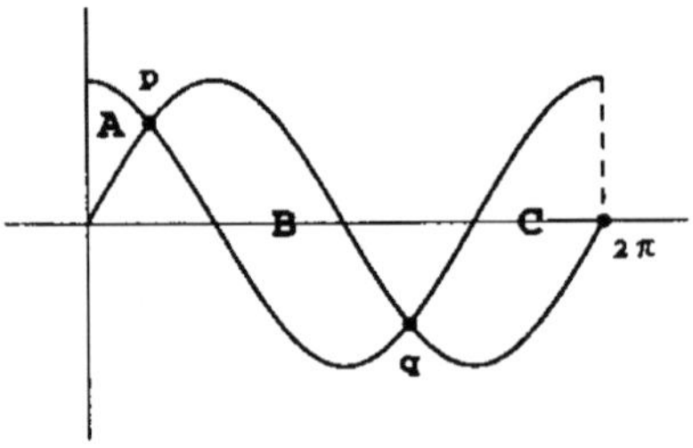

The curves intersect at two points p and q and we find three separate areas A, B, and C to compute. The points p and q correspond to the solutions of the equation

$$\sin(x) = \cos(x) \qquad\qquad (0 \leq x \leq 2\pi).$$

Hence $p = \left(\frac{\pi}{4}, \frac{\sqrt{2}}{2}\right)$, and $q = \left(\frac{5\pi}{4}, -\frac{\sqrt{2}}{2}\right)$ and we have

$$\begin{aligned}
\text{Area}(A) &= \left| \int_0^{\frac{\pi}{4}} (\sin(x) - \cos(x))\, dx \right| \\
&= \left| -\cos(x) - \sin(x) \Big]_0^{\frac{\pi}{4}} \right| \\
&= \sqrt{2} - 1 \\
\text{Area}(B) &= \left| \int_{\frac{\pi}{4}}^{\frac{5\pi}{4}} (\sin(x) - \cos(x))\, dx \right| \\
&= 2\sqrt{2} \\
\text{Area}(C) &= \left| \int_{\frac{5\pi}{4}}^{2\pi} (\sin(x) - \cos(x))\, dx \right| \\
&= \sqrt{2} + 1.
\end{aligned}$$

The area in question is then obtained by taking the sum:

$$\text{Area}(A) + \text{Area}(B) + \text{Area}(C) = 4\sqrt{2}.$$

Example [16.6] The curves $y = x^2 - 3$ and $y = x - 1$

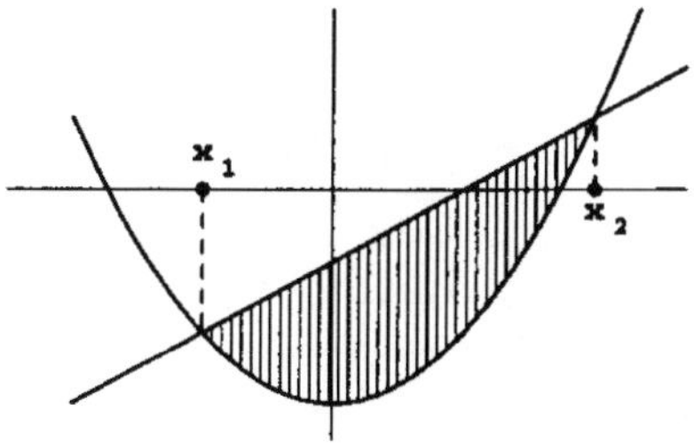

cross each other at exactly two points which we denote x_1 and x_2. Find the area of the region between the graphs for $x_1 \leq x \leq x_2$.

First we need to find x_1 and x_2. To do so we solve the equation $x^2 - 3 = x - 1$, i.e.,

$$x^2 - x - 2 = 0,$$

which tells us $x_1 = -1$ and $x_2 = 2$. The desired area between the graphs is thus given by the integral

$$\begin{aligned}\int_{-1}^{2} [(x-1) - (x^2 - 3)]\, dx &= \left(\frac{x^2}{2} - x - \frac{x^3}{3} + 3x\right)\Big]_{-1}^{2} \\ &= \frac{9}{2}.\end{aligned}$$

Alternately, if we didn't know that the line $y = x - 1$ is on top we could have computed

$$\left| \int_{-1}^{2} [(x^2 - 3) - (x - 1)]\, dx \right|,$$

since the presence of the absolute values ensures that we will obtain the same answer.

Cavalieri's Principle: Let $y = f(x)$, and $y = g(x)$ be integrable functions on the interval $a \leq x \leq b$. By definition

$$|f(x) - g(x)| = \begin{cases} f(x) - g(x), & \text{if } f(x) \geq g(x) \\ g(x) - f(x), & \text{if } g(x) \geq f(x) \end{cases}$$

and we see that, regardless of where the curves intersect (or how many times they do), the area between the two curves is given by the formula

$$\text{Area} = \int_a^b |f(x) - g(x)|\, dx.$$

Note the crucial difference between this formula and (16.2). Here the absolute values are *inside* the integral (which does seriously effect the function we are integrating), while in (16.2) the absolute values are outside and the formula is only valid if the functions do not intersect. This principle is easily verified and the reader is invited to do so.

Exercises for §16.2

In the following exercises plot the functions $f(x)$ and $g(x)$ on the indicated interval with your CAS. Compute the area of the region between the graphs of f and g.

(1) $f(x) = x^2 + 3, g(x) = 1 - x, (0 \leq x \leq 1).$

(2) $f(x) = x^2 + 3, g(x) = 1 - x, (0 \leq x \leq 2).$

(3) $f(x) = x^3 + 2x + 1, g(x) = x^2 + 2x + 1, (-1 \leq x \leq 2).$

(4) $f(x) = x^3 - x, g(x) = 2 - 2x^2, (-1 \leq x \leq 2).$

(5) $f(x) = x^4, g(x) = 6x^2 - 4, (-3 \leq x \leq 3).$

(6) $f(x) = e^{2x}, g(x) = \frac{1}{x}, (2 \leq x \leq 10).$

(7) $f(x) = \frac{3}{4}x\sqrt{x^2+1}, g(x) = 3 - x, (0 \leq x \leq 2).$

(8) $f(x) = \cos(x), g(x) = \sin(x), (0 \leq x \leq 2\pi).$

(9) $f(x) = \cos(\pi x), g(x) = 1 - 2x, (-1 \leq x \leq 2).$

(10) $f(x) = \tan(\pi x), g(x) = x^2, (-\frac{1}{4} \leq x \leq \frac{1}{4}).$

(11) Give a geometric argument to explain why

$$\int_0^{\frac{\pi}{2}} \cos(x)\,dx = \int_0^{\frac{\pi}{2}} \sin(x)\,dx.$$

(12) Find the area of the football shaped region

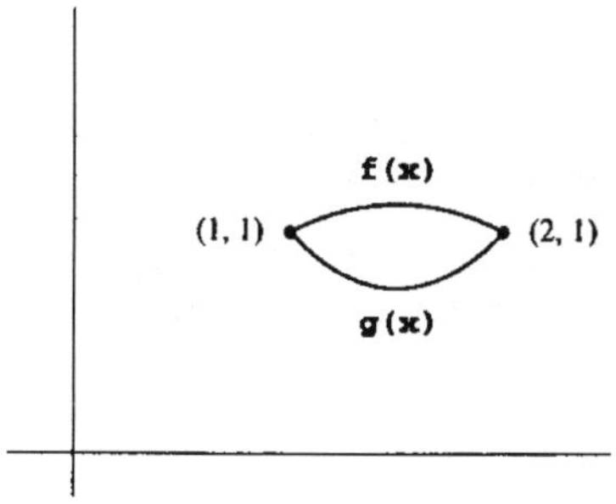

where $f(x) = ax^2 + bx$ and $g(x) = x^2 + cx + d$ for certain constants a, b, c, d. **Hint:** Compute the constants a, b, c, d first.

(13) Using your CAS, plot the three curves $y = 1, y = (1+4x)^{-\frac{1}{2}}$, and $y = -\frac{2}{3}x + \frac{5}{3}$. You will discover that there are exactly 3 points of intersection forming a triangular region in the first quadrant. Find the area of this region.

(14) The functions $f(x) = x^2$ and $g(x) = \sqrt{x}$ are inverses when $x \geq 0$. Their graphs intersect when $x = 0$ and $x = 1$.

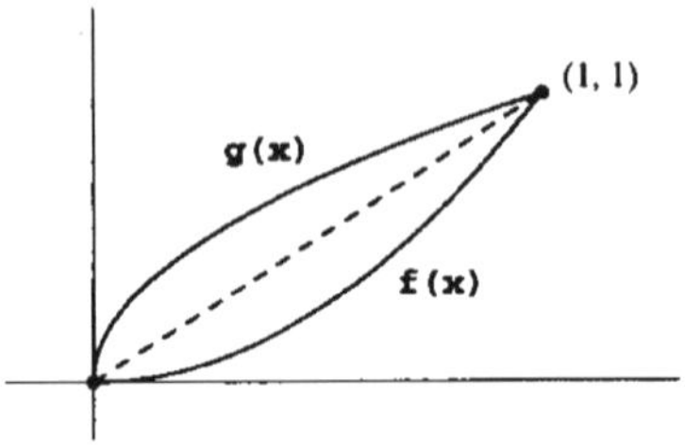

Show that the dotted lines between $(0,0)$ and $(1,1)$ bisect the area of the region between the curves.

16.3 Computing Arc Length

Let $y = f(x)$ be a function with a continuous derivative for $a \le x \le b$.

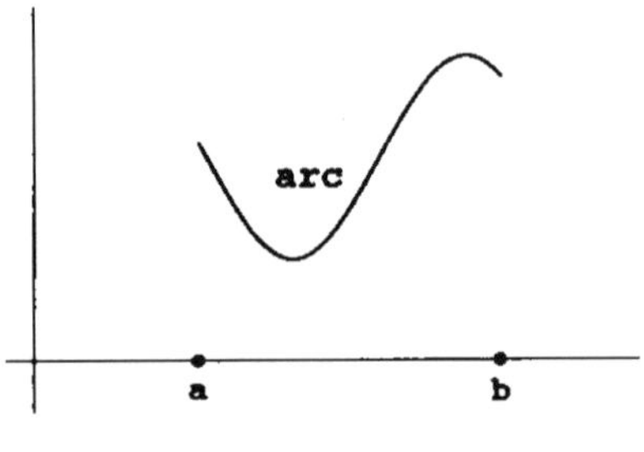

If we imagine this portion of the graph of $y = f(x)$ as a piece of string then, by stretching it out, we can measure its length. This length is termed the **arc length**. There is a beautiful formula for the arc length which is given by the following integral:

$$\int_a^b \sqrt{1 + (f'(x))^2}\, dx. \tag{16.4}$$

Example [16.7] Compute the circumference of a circle whose radius is R.

Consider a circle of radius R. The points on the circle (x, y) are the solutions of the equation $x^2 + y^2 = R^2$. The function $f(x) = \sqrt{R^2 - x^2}$ is well defined for $-R \le x \le R$ and its graph is the semicircle of radius R,

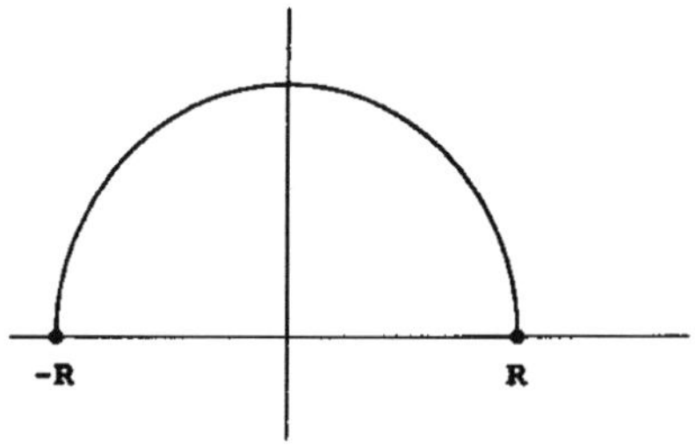

Our formula dictates that the arc length is given by

$$\int_{-R}^{R} \sqrt{1+f'(x)^2}\,dx = \int_{-R}^{R} \sqrt{1+\left(\frac{1}{2}(R^2-x^2)^{-\frac{1}{2}}\cdot(-2x)\right)^2}\,dx$$

$$= \int_{-R}^{R} \sqrt{1+\frac{x^2}{R^2-x^2}}\,dx$$

$$= \int_{-R}^{R} \sqrt{\frac{R^2}{R^2-x^2}}\,dx.$$

This integral can be computed with the substitution $x = R\sin(u), dx = R\cos(u)\,du$ and $-R \le x \le R$ becomes $-\frac{\pi}{2} \le u \le \frac{\pi}{2}$. The arc length of the semicircle is then computed as follows:

$$\int_{-R}^{R} \sqrt{\frac{R^2}{R^2-x^2}}\,du = \int_{-\frac{\pi}{2}}^{\frac{\pi}{2}} \sqrt{\frac{R^2}{R^2-R^2\sin^2(u)}}\cdot R\cos(u)\;du$$

$$= R\int_{-\frac{\pi}{2}}^{\frac{\pi}{2}} \sqrt{\frac{1}{1-(\sin(u))^2}}\cdot\cos(u)\;du$$

$$= R\int_{-\frac{\pi}{2}}^{\frac{\pi}{2}} du$$

$$= \pi R.$$

Remark: Technically speaking, the above integral is **improper**. See §17.5 for further details.

Example [16.8] Find the arc length of the curve $y = 2x^{3/2} - 5$ for $0 \le x \le 1$.

Here $f(x) = 2x^{3/2} - 5$ and thus $f'(x) = 3x^{\frac{1}{2}}$, $f'(x)^2 = 9x$. The arc length is thus given by

$$\int_0^1 \sqrt{1 + 9x}\, dx,$$

and using the substitution $u = 1 + 9x$, $du = 9\,dx$, we compute the integral:

$$\begin{aligned} \int_0^1 \sqrt{1+9x}\, dx &= \int_1^{10} \sqrt{u}\, \frac{du}{9}, \\ &= \frac{2}{27}\, u^{3/2}\Big]_1^{10} \\ &= \frac{2}{27}\left(10^{3/2} - 1\right). \end{aligned}$$

Remark: In the previous two examples we were very fortunate that the integrals could be evaluated exactly. This rarely happens in practice and one must usually utilize a numerical integration technique, such as Simpson's rule.

Derivation of the Arc Length Formula

The derivation of the arc length formula has as its root the Pythagorean theorem. Let $y = f(x)$ be a function with a continuous derivative on an interval $a \le x \le b$.

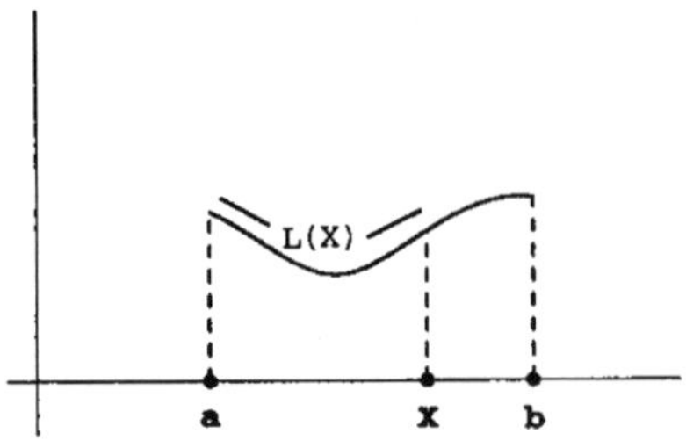

Consider now the arc length function associated to the curve:

$$\text{L}(X) = \text{length of the curve } \ y = f(x) \ \text{ for } \ a \le x \le X.$$

(This function very much resembles the area function defined in the course of the proof of the Fundamental Theorem of Calculus.)

Consider now a small piece of the curve.

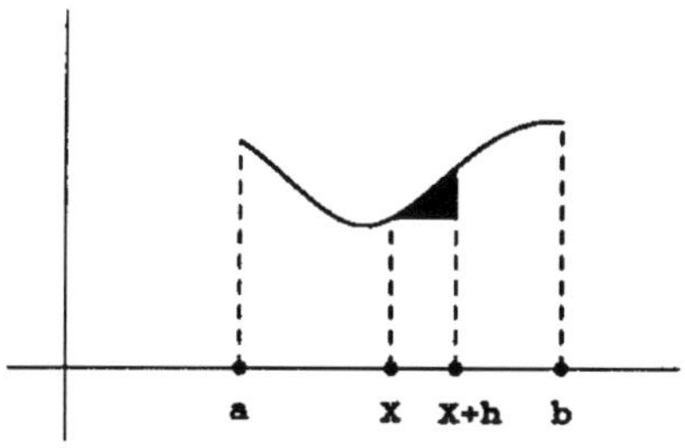

When we magnify the shaded region in the above diagram, we see that it is approximately a right triangle,

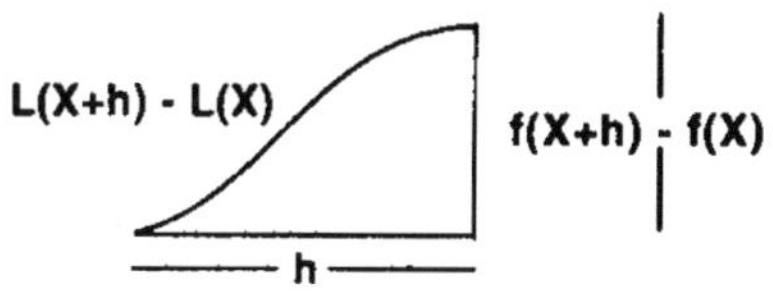

so that (by the Pythagorean theorem)

$$h^2 + (f(X+h) - f(X))^2 \approx (\mathrm{L}(X+h) - \mathrm{L}(X))^2.$$

Dividing by h^2 and taking square roots yields

$$\sqrt{1 + \left(\frac{f(X+h) - f(X))}{h}\right)^2} \approx \frac{(\mathrm{L}(X+h) - \mathrm{L}(X))}{h}.$$

Next, we take the limit as $h \to 0$ (as we have done at opportune moments) to obtain the formula

$$\sqrt{1 + f'(X)^2} = \mathrm{L}'(X).$$

Integrating both sides (over the interval $a \leq x \leq X$) and recalling that, by definition, $\mathrm{L}(a) = 0$

$$\int_a^X \sqrt{1 + f'(x)^2}\, dx = \mathrm{L}(X).$$

Exercises for §16.3

Graph each of the following functions $f(x)$ on the indicated interval with your CAS. Compute the arclength.

(1) $y^2 = 9x^3, (1 \leq x \leq 2).$

(2) $y = \cosh(x), (0 \leq x \leq 3).$

(3) $y = \frac{1}{3}x^3 + \frac{1}{4x}, (2 \leq x \leq 3).$

(4) $y = \frac{x^4}{4} + \frac{1}{8x^2}, (1 \leq x \leq 3).$

Graph each of the following functions $f(x)$ on the indicated interval with your CAS. Express the arclength as an integral and evaluate this integral to 3 decimal places with your CAS.

(5) $y^2 = \ln(x), (1 \leq x \leq e).$

(6) $y = x^2, (0 \leq x \leq 2).$

(7) $y = \sin(x), (1 \leq x \leq \frac{\pi}{2}).$

(8) $y = \ln(\cos(x)), (0 \leq x \leq \frac{\pi}{4}).$

(9) $y = e^x, (0 \leq x \leq 10).$

(10) $y^2 = \sin(x)e^{\frac{1}{x}}, (\frac{1}{10} \leq x \leq 1).$

(11) Given $p \neq 1$ compute the length of the curve $y = \frac{x^{p+1}}{p+1} + \frac{x^{1-p}}{4(p-1)}$ for $1 \leq x \leq 2$. Leave your answer in terms of p.

16.4 Volume as Summation of Cross–Sectional Area

Consider a solid object in 3–dimensional space, for example the pyramid or sphere.

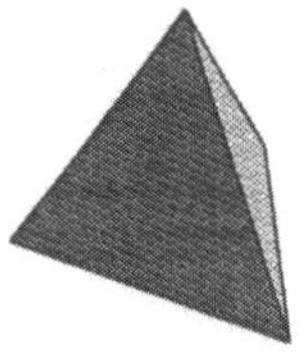

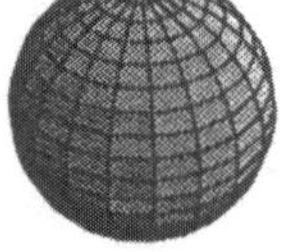

Question: How can we compute the volumes of these and other such objects?

There are various methods for doing such computations. The first method we focus on is very natural and is a procedure which sums cross–sectional area. We first require a fixed frame of reference for 3–dimensional space given by a 3–dimensional coordinate system

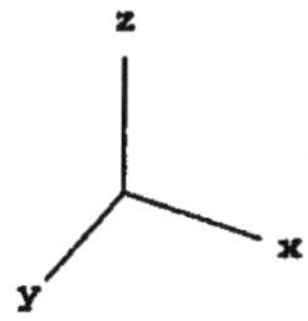

which consists of 3 mutually perpendicular lines, the x, y, and z axes. Every point P in 3–dimensional space is specified by a triple (x_0, y_0, z_0) of coordinates.

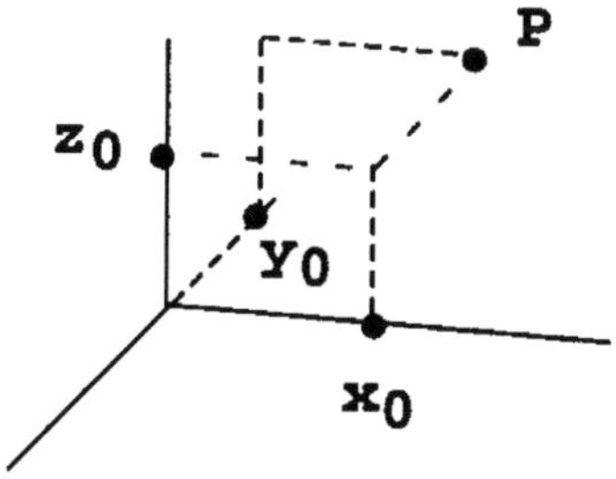

We shall need to introduce the plane at x_0:

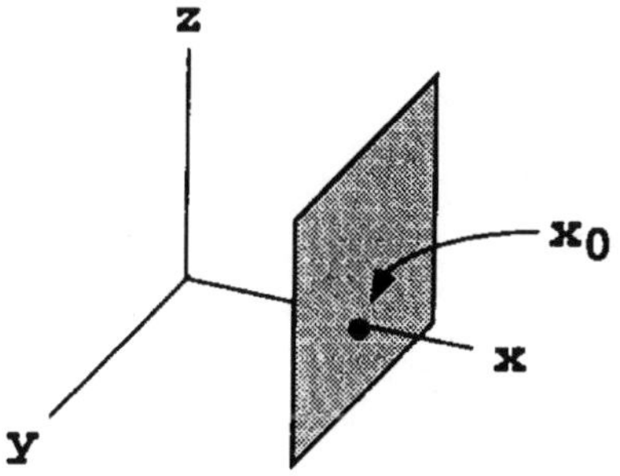

which can be imagined as a sheet of glass perpendicular to the x–axis which passes through the point x_0 on the x–axis.

Now, consider a 3–dimensional solid object Ω lying in our space.

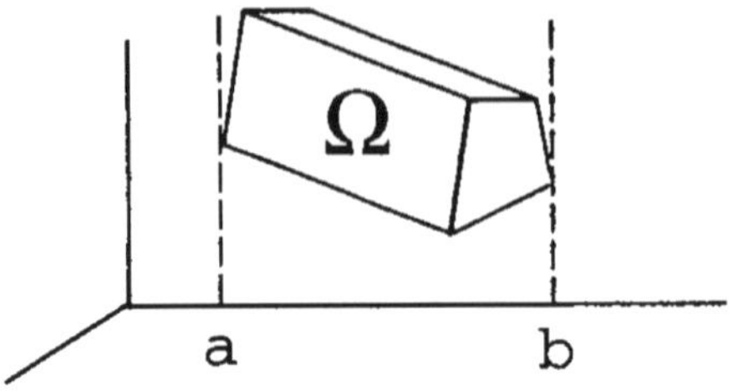

We assume the object Ω lies between planes at a and b. For every point x with $a \leq x \leq b$ we let $A(x)$ (which is shaded in the figure below) denote the area of the cross sectional slice at x, which is simply the intersection of Ω with the plane at x.

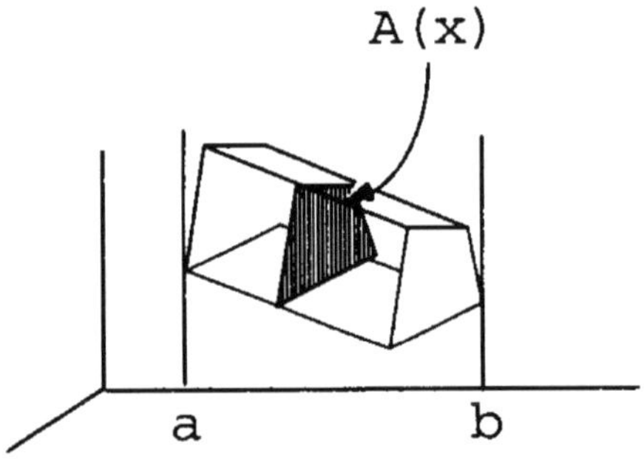

Fix a large integer N and divide the interval $a \leq x \leq b$ into N subintervals of equal length whose endpoints are at

$$x_n = a + n \cdot \frac{(b-a)}{N} \qquad (0 \leq n \leq N).$$

Our object Ω can then be broken up into N box–like pieces where the piece at x_n (seen below) has volume $A(x_n) \cdot \frac{(b-a)}{N}$.

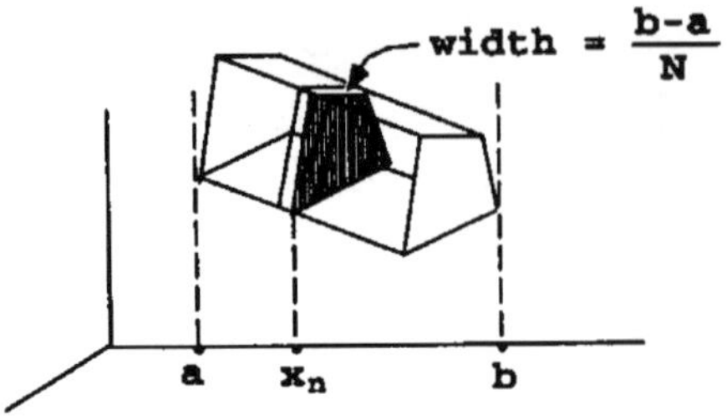

By taking the limit as $N \to \infty$ we obtain

$$\text{Volume}(\Omega) = \lim_{N\to\infty} \sum_{n=1}^{N} A(x_n)\,\frac{(b-a)}{N} = \int_a^b A(x)\,dx.$$

In summary, we have proved:

Proposition [16.1] *Let Ω be a 3–dimensional solid which lies between the planes at a and b on the x–axis where we assume $a < b$. If $A(x)$ is the area of the cross–sectional slice at x, then*

$$\text{Volume}(\Omega) = \int_a^b A(x)\,dx.$$

Example [16.9] What is the volume of a sphere of radius R.

We begin by positioning the sphere so that its center is at the origin.

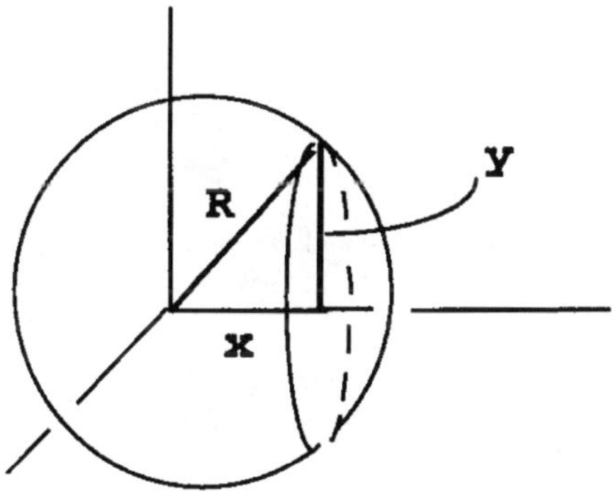

The cross–sectional area at x is πy^2 where $x^2 + y^2 = R^2$. Thus $A(x) = \pi y^2 = \pi(R^2 - x^2)$, and we obtain a volume formula:

$$\begin{aligned}\text{Volume(sphere)} &= \int_{-R}^{R} \pi(R^2 - x^2)\,dx \\ &= \frac{4\pi R^3}{3}.\end{aligned}$$

Exercises for §16.4

Compute the volumes of the following three dimensional objects.

(1) The pyramid of height h whose base is a square of side a.

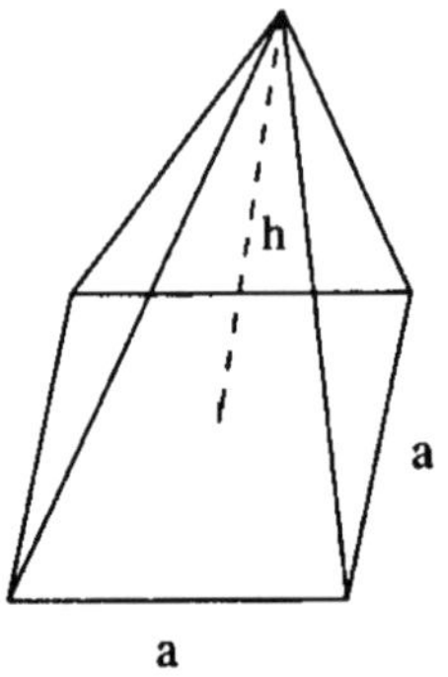

(2) The right circular cone of radius r and height h.

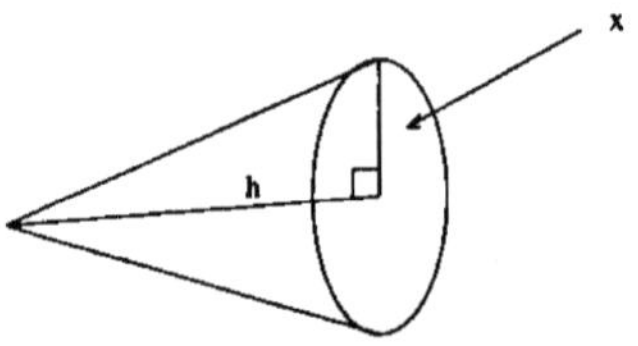

(3) The top portion of a sphere of radius r and height a.

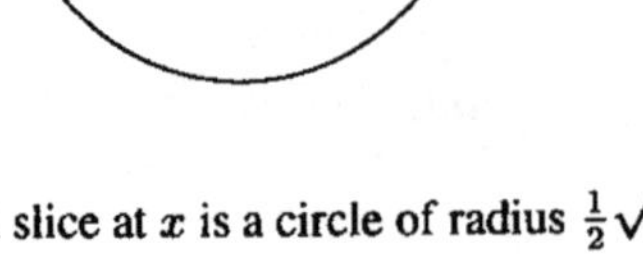

(4) An egg whose cross-sectional slice at x is a circle of radius $\frac{1}{2}\sqrt{1-x^2}$.

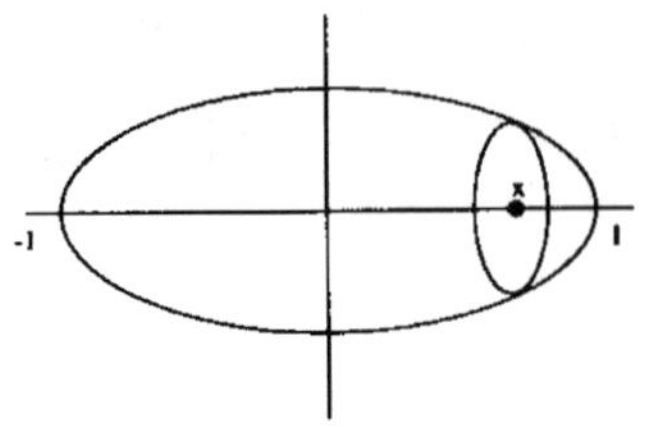

(5) A solid is swept out by moving an equilateral triangle which is always centered and perpendicular to the x–axis. The length of the side of the equilateral triangle $L(x)$ changes according to the formula $L(x) = \frac{1}{4}(1-x)^2$. If the triangle commences at $x = 0$ and moves to $x = 1$, what is the volume of the solid obtained?

(6) A solid is swept out by moving a rectangle which is always centered and perpendicular to the x–axis. The area $A(x)$ of the rectangle varies according to the formula $A(x) = 2x^3 + x^2 - 1$. Find the volume of the solid obtained for $0 \le x \le 2$.

16.5 Volumes of Solids of Revolution

A solid of revolution is obtained by rotating the area below a curve about an axis.

Example [16.10] (The Tuba) Consider the curve $y = x^3$ where $0 \le x \le 2$.

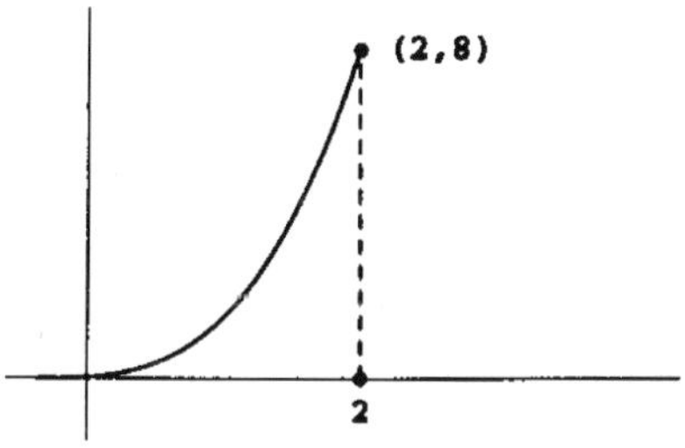

Let's shade in the area between the curve and the x–axis.

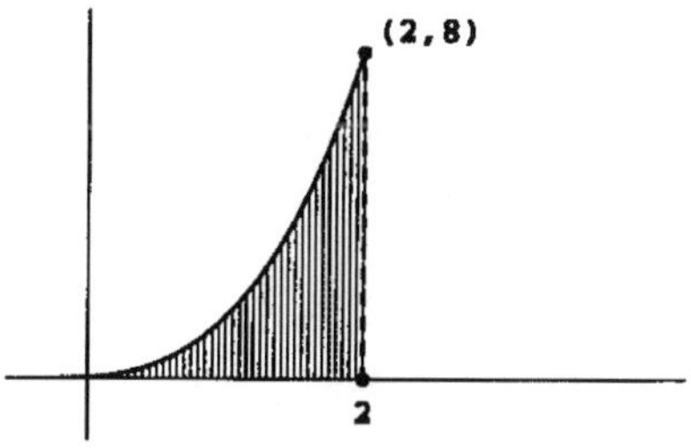

Now imagine that this shaded region is rotating (like an airplane propeller of an early airplane) about the x–axis. We obtain a solid which resembles part of a tuba.

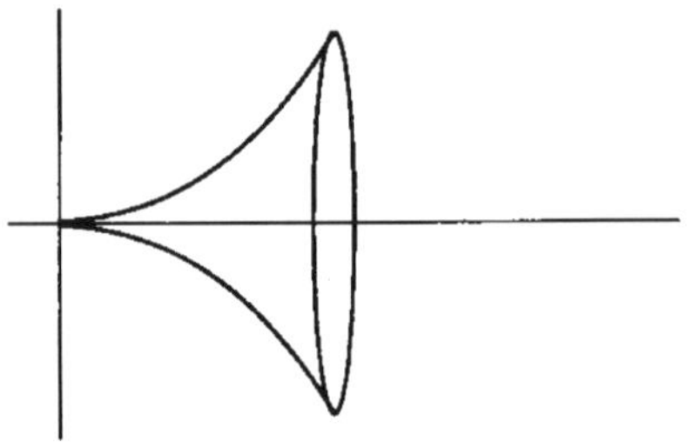

Example [16.11] (The bell) If we restrict the function $y = x^3$ to the domain $\frac{1}{2} \leq x \leq 2$ and then rotate the region between this curve and the x–axis, we obtain

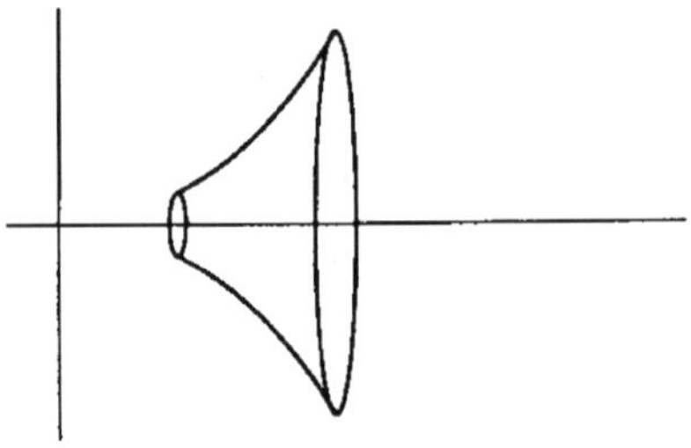

a solid which looks like a bell.

Example [16.12] (The cone) Consider the region between the straight line $y = \frac{3x}{2}$ and the interval $0 \leq x \leq 2$. When we rotate this region about the x–axis

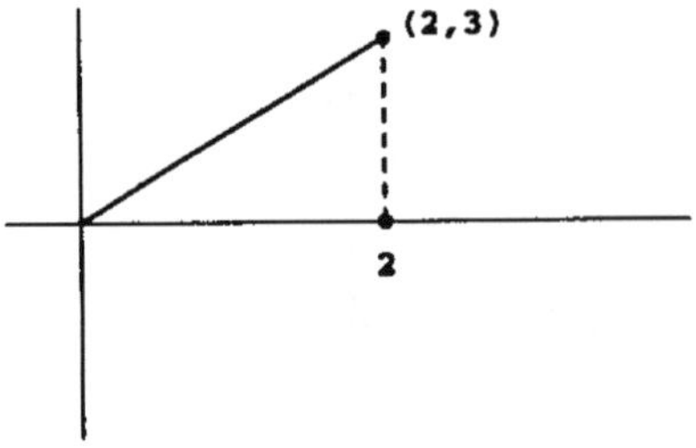

we obtain a cone.

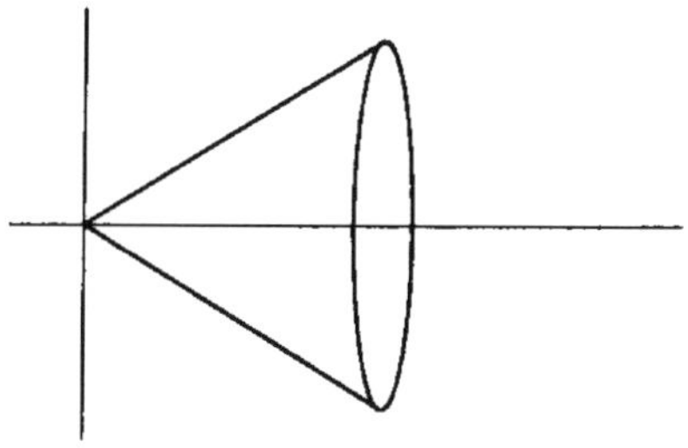

We now give some examples of rotation about the y–axis.

Example [16.13] (The drinking glass) Consider the function $y = 10^x - 1$ on the interval $0 \le x \le 1$. The area between this curve and the x–axis is shaded below.

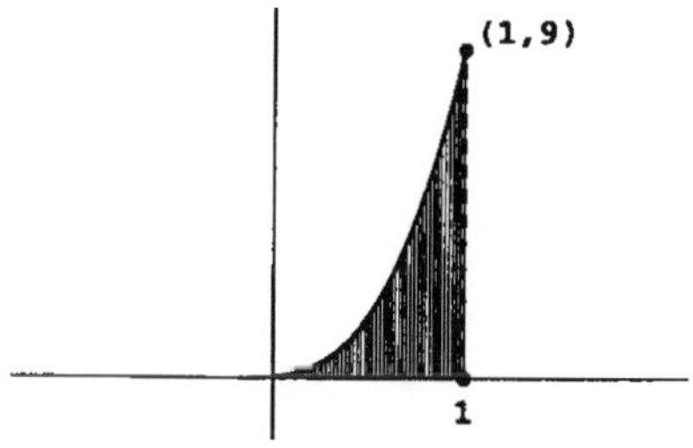

When we revolve this shaded region around the y–axis (like a merry–go–round) we obtain a figure that resembles a drinking glass.

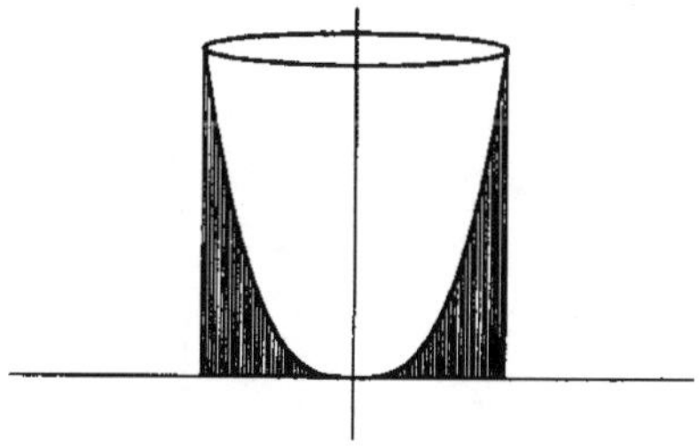

Notice that the interior of the resulting figure is quite empty and is capable of holding liquids.

Example [16.14] (The Salad Bowl) If we choose a function such as $y = \sqrt{x}$ on the interval $0 \le x \le 4$, rotating it about the y–axis will yield a salad bowl.

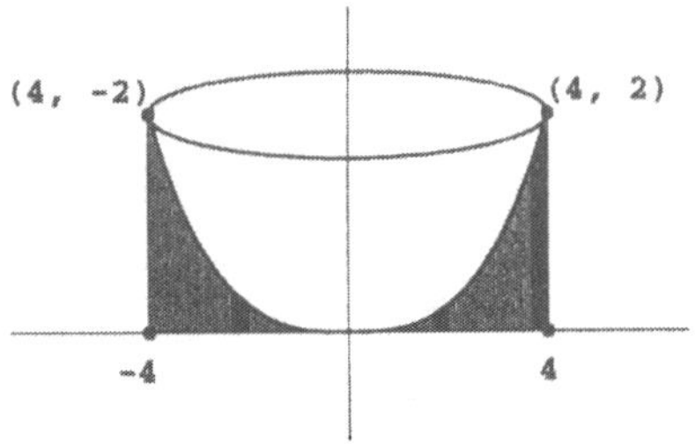

The volumes of each of the previously constructed solids can be computed by evaluating an integral. First we consider solids which are obtained by rotating a function $y = f(x)$ (with $a \le x \le b$) about the x–axis.

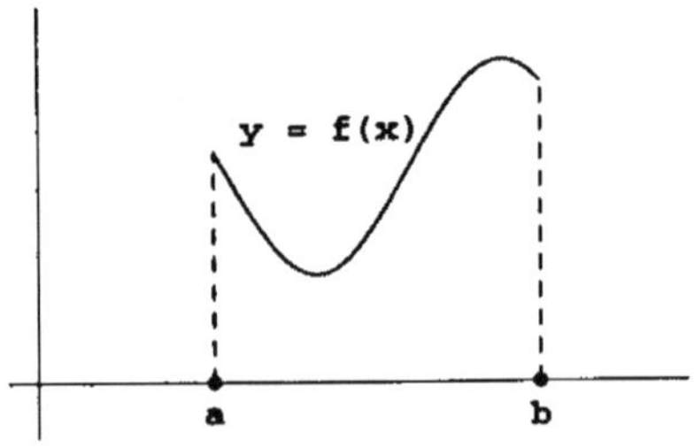

The solid obtained is depicted in the following figure.

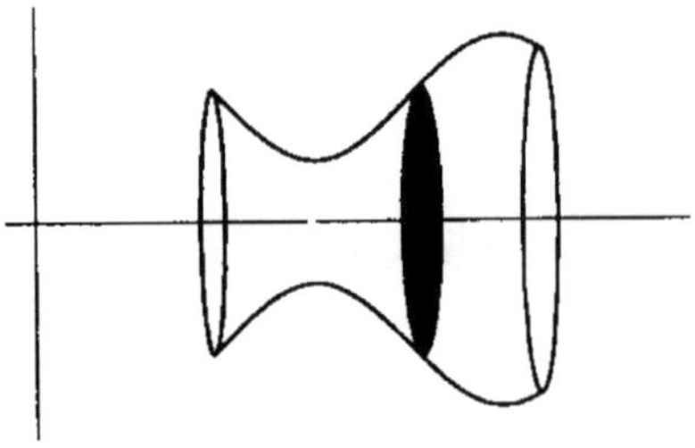

The shaded region above is a cross sectional slice at x on the x–axis.

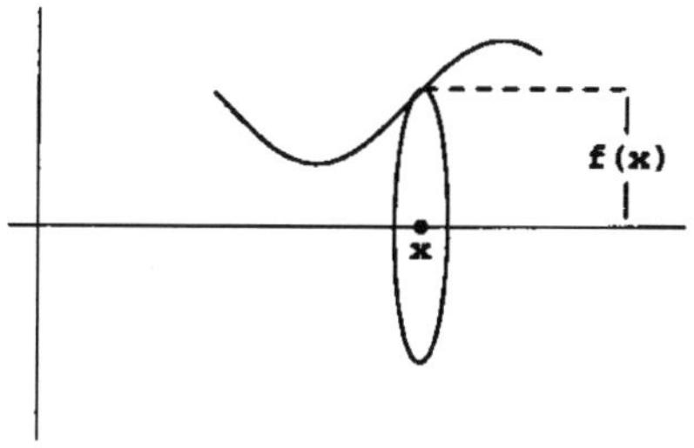

The cross sectional slice is a circle of radius $|f(x)|$ centered at x, and the area of this slice is precisely $\pi f(x)^2$. Fix a large integer N and divide the interval $a \le x \le b$ into N subintervals of equal length whose endpoints are

$$x_n = a + n \cdot \frac{b-a}{N} \qquad (0 \le n \le N).$$

The total volume of the solid can then be broken up into N cross–sectional disk–like pieces of thickness $h = \frac{b-a}{N}$,

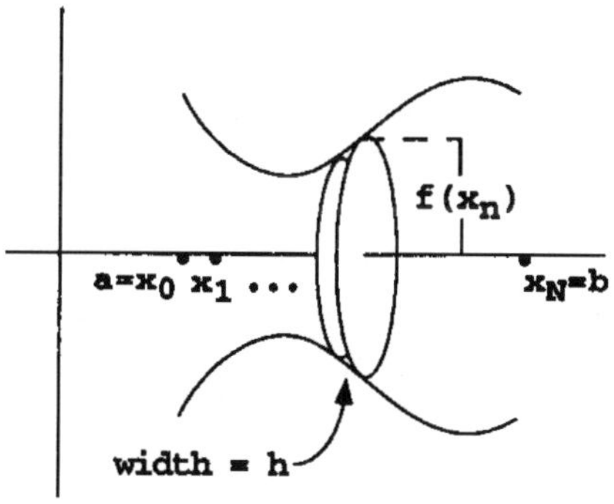

where for $n = 1, 2, \ldots, N$, each 3–dimensional disk–like piece is centered at x_n and has radius approximately $|f(x_n)|$. Summing the volumes of these pieces and letting $N \to \infty$ gives the volume of our solid of revolution,

$$\begin{aligned} \text{Volume} &= \lim_{N\to\infty} \sum_{n=1}^{N} \pi f(x_n)^2 \cdot \frac{b-a}{N} \\ &= \int_a^b \pi f(x)^2 \, dx. \end{aligned}$$

In summary, we have proved

Proposition [16.2] *Let $\mathcal{R}$ denote the region between the curve $y = f(x)$ (for $a \le x \le b$) and the x–axis. The volume of the solid of revolution obtained by revolving $\mathcal{R}$ about the x–axis is given by the formula*

$$\text{Volume} = \int_a^b \pi f(x)^2 \, dx.$$

In our first example, the tuba (see Example [16.10]), the volume is given by

$$\text{Volume(Tuba)} = \int_0^2 \pi x^6 \, dx \; = \; \frac{\pi x^7}{7}\Big]_0^2 = \frac{128\pi}{7}.$$

Similarly, the volume of the bell in Example [16.11] is

$$\text{Volume(Bell)} = \int_{\frac{1}{2}}^2 \pi x^6 \, dx = \frac{128\pi}{7} - \frac{\pi}{7 \cdot 128},$$

while the volume of the cone in Example [16.12] is just

$$\text{Volume(Cone)} = \int_0^2 \pi \frac{9x^2}{4} \, dx = \frac{9\pi}{4} \cdot \frac{x^3}{3}\Big]_0^2 = 6\pi.$$

We now change our perspective and develop a volume formula for curves rotated about the y–axis. In the following discussion we restrict ourselves to functions which lie above the x–axis. Consider the shaded area below the curve $y = f(x)$ for $0 \le a \le x \le b$.

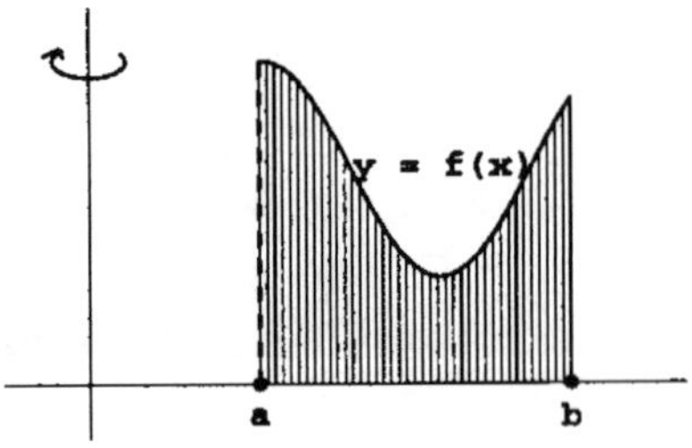

Revolving this region about the y–axis like a merry–go–round results in the Mexican hat–like solid below.

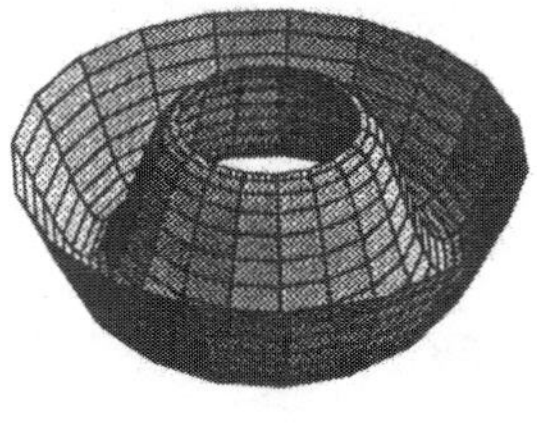

Let's divide our interval (as we did when revolving around the x–axis) into N subintervals of equal length whose endpoints are

$$x_n = a + n \cdot \frac{b-a}{N} \qquad (0 \leq n \leq N).$$

Consider a cylindrical shell at x_n.

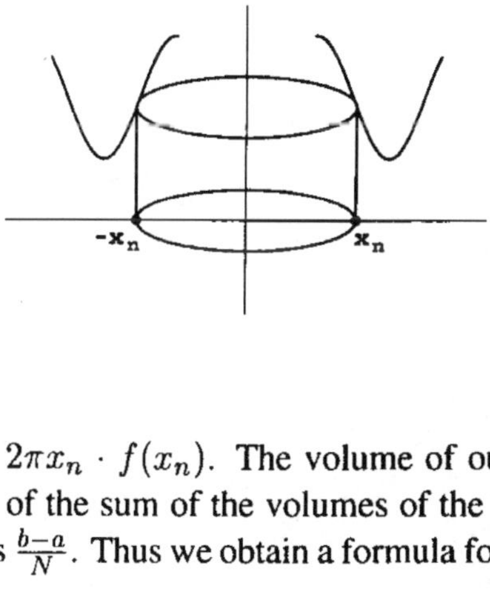

The area of this shell is $2\pi x_n \cdot f(x_n)$. The volume of our solid is then the limit (where we let $N \to \infty$) of the sum of the volumes of the N cylindrical shell–like pieces whose thickness is $\frac{b-a}{N}$. Thus we obtain a formula for the volume of our solid of revolution:

$$\begin{aligned} \text{Volume} &= \lim_{N\to\infty} \sum_{n=1}^{N} 2\pi x_n \cdot f(x_n) \cdot \frac{b-a}{N} \\ &= \int_a^b 2\pi x f(x)\, dx. \end{aligned}$$

Thus we have proved:

Proposition [16.3] *Let $f(x)$ be a continuous integrable function which is non–negative in the region $0 \le a \le x \le b$. Let $\mathcal{R}$ denote the region between the curve $y = f(x)$ (for $0 \le a \le x \le b$) and the x–axis. The volume of the solid of revolution obtained by revolving $\mathcal{R}$ about the y–axis is given by the formula*

$$\text{Volume} = \int_a^b 2\pi x f(x)\, dx.$$

The volume of the drinking glass (Example [16.13]) is computed by

$$\text{Volume(Glass)} = \int_0^1 2\pi x \cdot (10^x - 1)\, dx,$$

while the volume of the salad bowl is just

$$\text{Volume(Bowl)} = \int_0^4 2\pi x \cdot \sqrt{x}\, dx.$$

Exercises for §16.5

Plot each of the following curves (on the indicated interval) with your CAS. By shading in the region between the curve and the indicated axis try to imagine the solid of revolution obtained by revolving the region about that axis. Use your CAS to render a 3–dimensional graphics image of the object. Finally, compute the volume of the solid.

(1) $y = x^4$, $(0 \le x \le 3)$, rotate about the x–axis.

(2) $y = 3x + 1$, $(1 \le x \le 2)$, rotate about the x–axis.

(3) $y = x^3$, $(0 \le x \le 2)$, rotate about the y–axis.

(4) $y = |1 - x|$, $(-\frac{1}{2} \le x \le \frac{1}{2})$, rotate about the x–axis.

(5) $y = |1 - x|$, $(-\frac{1}{2} \le x \le \frac{1}{2})$, rotate about the y–axis.

(6) $y = x^{\frac{2}{3}}$, $(0 \le x \le 10)$, rotate about the x–axis.

(7) $y = \sin(x)$, $(\frac{\pi}{6} \le x \le \frac{\pi}{2})$, rotate about the y–axis.

(8) $y = \sqrt{1 - x^2}$, $(-1 \le x \le 1)$, rotate about the x–axis.

(9) $y = \sqrt{1 - x^2}$, $(0 \le x \le 1)$, rotate about the y–axis.

(10) $y = e^x$, $(1 \le x \le 10)$, rotate about the x–axis.

(11) Find the volume of a torus (donut) of diameter D whose cross sectional slice is a circle of radius R.

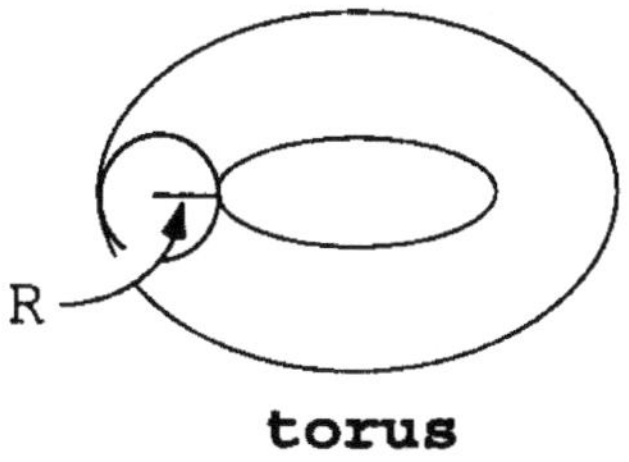

(12) While we live in a 3–dimensional world it is, nevertheless, possible to assume the existence of 4 mutually perpendicular lines which characterizes 4–dimensional space. Find a formula for the volume of the four dimensional sphere of radius R. **Hint:** Graph the semicircle $y = \sqrt{R^2 - x^2}$, $(-R \le x \le R)$. At each point (x, y) on the semicircle, rotate the circle of radius y (which is perpendicular to the x–axis and pictured below) about the fourth dimensional axis. This gives a 3–dimensional sphere protruding into the fourth dimension. Sum (integrate) the volumes of theses spheres for $-R \le x \le R$.

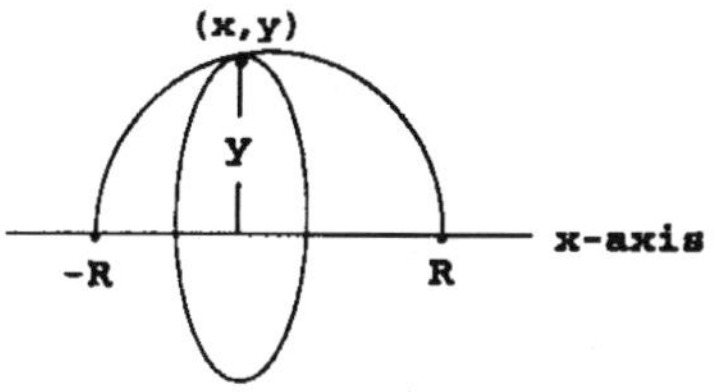

Chapter XVII

Further Topics on Integration

17.1 Integral Representation of the log Function

Recalling that $\frac{d}{dx}\ln(x) = \frac{1}{x}$, the Fundamental Theorem of Calculus dictates that

$$\int \frac{dx}{x} = \ln(x) + C.$$

If in addition we use that fact that $\ln(1) = 0$, we are left with the more precise formulation

$$\ln(X) = \int_1^X \frac{dx}{x}. \tag{17.1}$$

We now show directly that the integral representation (17.1) satisfies the properties of the logarithm function:

Property 1. $\ln(XY) = \ln(X) + \ln(Y)$

Property 2. $\ln(X/Y) = \ln(X) - \ln(Y)$

Property 3. $\ln(X^r) = r\ln(X)$.

To verify Property 1 we begin by noting:

$$\ln(XY) = \int_1^{XY} \frac{dx}{x} = \int_1^{X} \frac{dx}{x} + \int_X^{XY} \frac{dx}{x}$$

$$= \ln(X) + \int_X^{XY} \frac{dx}{x}.$$

To evaluate this last integral we use the substitution

$$u = \frac{x}{X}.$$

Under this substitution the limits of integration change as follows: $x = X$ becomes $u = 1$, and $x = XY$ becomes $u = Y$, and $\frac{dx}{x}$ becomes

$$\frac{dx}{x} = \frac{du}{u}.$$

Thus

$$\int_X^{XY} \frac{dx}{x} = \int_1^{Y} \frac{du}{u} = \ln(Y).$$

Property 2 can be verified in much the same manner. To verify Property 3,

$$\ln(X^r) = r\ln(X),$$

we use the substitution $u = x^{1/r}$. With this substitution $x = 1$ transforms to $u = 1^{1/r} = 1$, and $x = X^r$ becomes $u = (X^r)^{1/r} = X$. Further $du = \frac{1}{r}x^{\frac{1}{r}-1}\,dx$, and hence

$$\frac{dx}{x} = \frac{rx^{1-\frac{1}{r}}\,du}{x} = r\frac{du}{x^{1/r}} = r\frac{du}{u}.$$

We conclude that

$$\int_1^{X^r} \frac{dx}{x} = \int_1^{X} r\frac{du}{u} = r\ln(X).$$

Equation (17.1) can be used to give an alternative definition for the (somewhat elusive) number e. Recall that, by definition, $\ln(e) = 1$.

Definition: *The number* e *is the unique real number greater than* 1 *with the property that*

$$\int_1^{e} \frac{dx}{x} = 1.$$

Pictorially, the number e corresponds to the marked point on the x–axis such that the shaded area (see figure below) under that curve $y = 1/x$ has area equal to precisely 1.

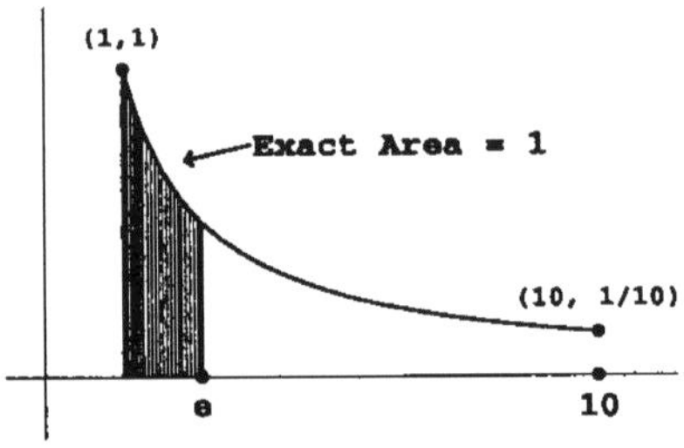

It is possible to *define* the function $\ln(x)$ directly via its integral representation

$$\ln(x) = \int_1^x \frac{dt}{t},$$

and then to define e^x as the inverse of $\ln(x)$. The various properties of the ln function (and hence e^x) that we demonstrated earlier allow us to prove one of the most important rules of differentiation,

$$\frac{d}{dx}x^r = rx^{r-1}.$$

The most direct way to attempt to verify this rule would be to compute the limit

$$\lim_{x\to\infty} \frac{(x+h)^r - x^r}{h}.$$

Unfortunately the complex definition of x^r when r is an arbitrary real number makes working with this limit very impractical unless r is an integer or a rational number. To handle the general case we use the identity $x^r = e^{r\ln(x)}$ and then differentiate with the chain rule as follows:

$$\begin{aligned}\frac{d}{dx}x^r &= \frac{d}{dx}e^{r\ln(x)} \\ &= r\cdot\frac{1}{x}e^{r\ln(x)} \\ &= r\frac{1}{x}x^r \\ &= rx^{r-1}.\end{aligned}$$

Exercises for §17.1

Evaluate the following integrals.

(1) $\int \frac{dx}{x-r}$.

(2) $\int \frac{x^3}{x^4+2}\,dx$.

(3) $\int \frac{\cos(x)}{\sin(x)}\,dx$.

(4) $\int \frac{\ln(x)}{x}dx$.

(5) $\int \frac{x}{1+x^2}\,dx$.

(6) $\int \frac{(\sec(x))^2}{\tan(x)}dx$.

(7) $\int \ln(\sqrt{x})\,dx$.

(8) $\int \sin(\ln(x))dx$.

(9) $\int \frac{x-1}{x^2-1}\,dx$.

(10) $\int \frac{e^x}{1+e^x}dx$

(11) $\int \frac{\sin(x)+\cos(x)}{\sin(x)-\cos(x)}\,dx$.

(12) $\int \frac{\cos(x)\sin(x)}{1-(\cos(x))^2}dx$.

(13) Using the integral representation of $\ln(x)$ show that $\ln(x) \leq (x-1)$ for $x \geq 1$.

(14) Show that $\ln(x) < 0$ for $x < 1$ directly from the integral representation of $\ln(x)$.

(15) Demonstrate that $2 < e < 3$ by showing that $\int_1^3 \frac{dt}{t} > 1$, and $\int_1^2 \frac{dt}{t} < 1$. Explain.

(16) Compute $\ln(2)$ to 3 decimal places by evaluating $\int_1^2 \frac{dt}{t}$ with Simpson's rule.

17.2 Integral Representation of Inverse Trigonometric Functions

In Example (8.13) we saw that the function $y = \arcsin(x)$ has the domain $-1 \leq x \leq 1$ and range $-\frac{\pi}{2} \leq x \leq \frac{\pi}{2}$, and its derivative is given by

$$\frac{d}{dx}\arcsin(x) = \frac{1}{\sqrt{1-x^2}}.$$

This leads to the integral representation

$$\arcsin(x) = \int \frac{1}{\sqrt{1-x^2}}\,dx + C.$$

The function $y = \cos(x)$ is one–to–one on the interval $0 \leq x \leq \pi$,

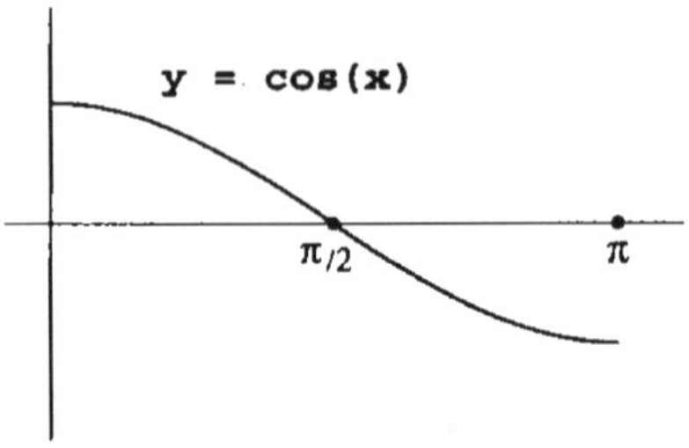

and satisfies the identities

$$\begin{aligned} \arccos(\cos(x)) &= x & 0 \le x \le \pi \\ \cos(\arccos(x)) &= x & -1 \le x \le 1. \end{aligned}$$

In Exercise (5) of §8.4 it was shown that $\frac{d}{dx}\arccos(x) = -\frac{1}{\sqrt{1-x^2}}$, which leads to the integral representation

$$\arccos(x) = -\int \frac{1}{\sqrt{1-x^2}}\,dx + C.$$

The function $y = \tan(x)$ is one–to–one on the interval $-\pi/2 \le x \le \pi/2$ and has vertical asymptotes at $x = \pm\pi/2$.

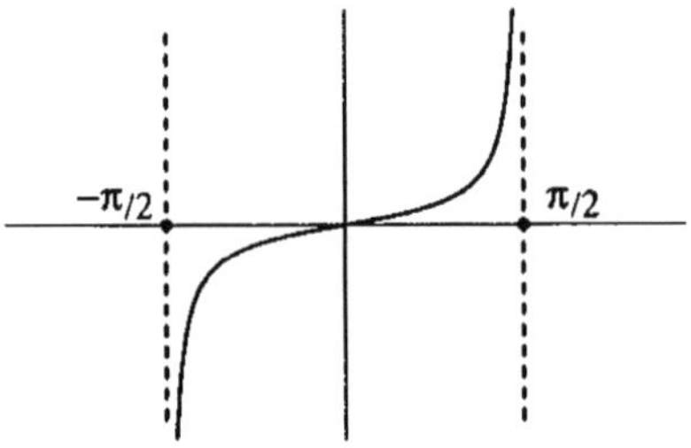

Thus by definition, the inverse function of $\tan(x)$ (denoted $\arctan(x)$) satisfies:

$$\begin{aligned} \arctan(\tan(x)) &= x, & -\pi/2 \le x \le \pi/2 \\ \tan(\arctan(x)) &= x, & -\infty \le x \le \infty, \end{aligned}$$

and

$$\lim_{x\to\infty} \arctan(x) = \frac{\pi}{2}, \qquad \lim_{x\to-\infty} \arctan(x) = -\frac{\pi}{2}.$$

Proposition [17.1] *The following identities hold*

$$\frac{d}{dx}\arctan(x) = \frac{1}{1+x^2}$$

$$\arctan(x) = \int \frac{1}{1+x^2}\,dx + C.$$

Proof: Beginning again with the identity $\tan(\arctan(x)) = x$ we differentiate both sides to obtain the identity

$$\begin{aligned}\frac{d}{dx}\tan(\arctan(x)) &= \frac{1}{(\cos(\arctan(x)))^2}\cdot\frac{d}{dx}\arctan(x)\\ &= 1,\end{aligned}$$

and hence

$$\frac{d}{dx}\arctan(x) = (\cos(w))^2,$$

where $w = \arctan(x)$.

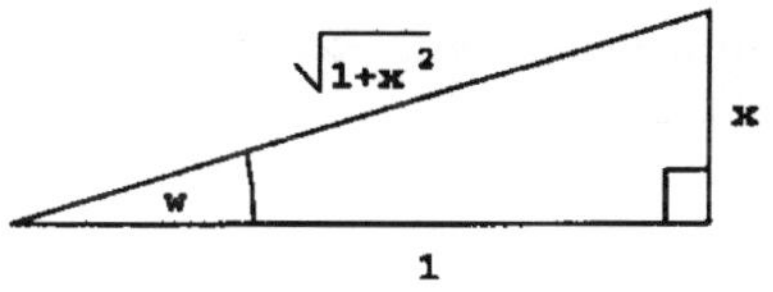

By considering the right triangle above where $\tan(w) = x$ we deduce

$$\cos(w) = \frac{1}{\sqrt{1+x^2}},$$

and thus

$$\frac{d}{dx}\arctan(x) = (\cos(w))^2 = \frac{1}{1+x^2}.$$

Exercises for §17.2

Evaluate the following integrals. Use the CAS to check your work.

(1) $\int_0^1 \frac{dx}{x^2+1}$.

(2) $\int_0^2 \frac{dx}{4+x^2}$.

(3) $\int_{-\frac{1}{2}}^0 \frac{dx}{\sqrt{1-x^2}}$.

(4) $\int_0^{\frac{3}{4}} \frac{dx}{9+x^2}$.

(5) $\int \frac{\sin(x)}{1+(\cos(x))^2}\,dx$.

(6) $\int \frac{e^{\frac{x}{2}}}{1+e^x}\,dx$.

(7) $\int \frac{x^4}{\sqrt{1-x^{10}}}\,dx$.

(8) $\int_0^1 \frac{4\sqrt{x}}{1+x^3}\,dx$.

(9) $\int \frac{x+1}{x^2+1}\,dx$.

(10) $\int_0^{\frac{\sqrt{2}}{2}} \frac{x-1}{x^2+1}\,dx$.

17.3 Integrating Rational Functions

A **polynomial of degree** n (where $n = 1, 2, 3, \ldots$) is a function $P_n(x)$ of the form

$$P_n(x) = a_0 + a_1 x + \cdots a_n x^n$$

where $a_0, a_1, \ldots, a_n$ are fixed real numbers. For example

$$x^7 - 3x^4 + 16x^3 - \pi x + 3.72,$$

is a polynomial of degree 7, and any constant c is a polynomial of degree 0.

Definition: *A* ***rational function*** *is the ratio of two polynomials,*

$$\frac{P(x)}{Q(x)},$$

such that $Q(x)$ *is not identically zero.*

Example [17.1] The function

$$\frac{x^3 - 3}{14x^5 - 2x^3 + 1},$$

is a rational function since it is the ratio of $x^3 - 3$ by $14x^5 - 2x^3 + 1$.

Remark: The term rational function is derived from the term rational number which is a ratio of two integers. The rational number $\frac{19}{6}$ can be rewritten (after long division) in the form $\frac{19}{6} = 3 + \frac{1}{6}$. In a completely analogous manner, the rational function $(x^5 + 2x^3 + x - 1)/(x^2 + 2)$ can be rewritten in the form

$$\frac{x^5 + 2x^3 + x - 1}{x^2 + 2} = x^3 + \frac{x - 1}{x^2 + 2}.$$

More generally, we know that every rational number $\frac{a}{b}$ can be expressed uniquely as

$$\frac{a}{b} = q + \frac{r}{b},$$

where q and r are integers, and r, termed **the remainder** is in the interval $0 \leq r < b$. The analogous statement for rational functions is that every rational function $\frac{a(x)}{b(x)}$ can be put in the form

$$\frac{a(x)}{b(x)} = q(x) + \frac{r(x)}{b(x)}, \tag{17.2}$$

where $q(x)$ and $r(x)$ are polynomials, and $r(x)$, again termed **the remainder**, is a polynomial whose degree is strictly smaller than the degree of $b(x)$.

Example [17.3] We have

$$\frac{x^3+1}{x^2+2} = x + \frac{-2x+1}{x^2+2}.$$

Problem: Let $a(x)$ and $b(x)$ be polynomials. How do we integrate the rational function $a(x)/b(x)$, i.e., what is $\int a(x)/b(x)\,dx$?

We begin an attack on this problem by taking equation (17.2) and then attempt to integrate both sides,

$$\int \frac{a(x)}{b(x)}\,dx = \int q(x)\,dx + \int \frac{r(x)}{b(x)}\,dx.$$

Since any polynomial (in particular $q(x)$) can be easily integrated, we have reduced our problem to the following somewhat simpler problem.

Problem: Let $r(x)$ and $b(x)$ be polynomials where the degree of $r(x)$ is smaller than the degree of $b(x)$. How do we integrate

$$\int \frac{r(x)}{b(x)}\,dx?$$

Continuing with our analogy, consider the fact that integers can always be **factored** (for example $30 = 2 \cdot 3 \cdot 5$) into a product of **prime numbers** (numbers which cannot be factored further). A similar phenomena occurs when we look at polynomials. For example

$$x^3 - 4x = x\,(x-2)\,(x+2),$$

and each of the factors, $x, x-2$, and $x+2$ is a **linear function** (a polynomial of degree one) which cannot be further factored. We shall now give a complete solution to our second problem in the case $b(x)$ can be factored onto k distinct linear factors

$$b(x) = b(x_1+t_1)\cdot(x_2+t_2)\cdots(x_k+t_k), \tag{17.3}$$

where $b \neq 0$. Factorization with repeated or quadratic factors are treated in the later exercises.

The factorization (17.3) implies that there exist real numbers $A_1, A_2, \ldots, A_k$ such that

$$\frac{r(x)}{b(x)} = \frac{A_1}{x+t_1} + \frac{A_2}{x+t_2} + \cdots + \frac{A_k}{x+t_k}. \tag{17.4}$$

Although this may not be immediately evident it will be clarified in the next example. The identity (17.4) allows us to immediately integrate our rational function:

$$\int \frac{r(x)}{b(x)}\,dx = \int \frac{A_1}{x+t_1}\,dx + \int \frac{A_2}{x+t_2}\,dx + \cdots + \int \frac{A_k}{x+t_k}\,dx.$$

Example [17.4] Evaluate the integral

$$\int \frac{x-1}{x^3-4x}\,dx.$$

To compute this integral we must first hunt for numbers A_1, A_2, and A_3 so that

$$\frac{x-1}{x^3-4x} = \frac{x-1}{x\,(x-2)\,(x+2)} = \frac{A_1}{x} + \frac{A_2}{x-2} + \frac{A_3}{x+2}.$$

This equation can be simplified by cross-multiplying by $x\,(x-2)\,(x+2)$ to obtain

$$x-1 = A_1\,(x-2)\,(x+2) + A_2\,x\,(x+2) + A_3\,x\,(x-2).$$

We can then compute A_1, A_2, and A_3 by plugging in convenient values for x. By setting $x=0$, we are left with

$$-1 = -4\,A_1 \qquad \Longrightarrow \qquad A_1 = \frac{1}{4},$$

if $x=2$,

$$1 = 8\,A_2 \qquad \Longrightarrow \qquad A_2 = \frac{1}{8},$$

and if $x = -2$,

$$-3 = 8\,A_3 \qquad \Longrightarrow \qquad A_3 = \frac{-3}{8}.$$

It follows that

$$\int \frac{x-1}{x^3-4x}\,dx = \frac{1}{4}\int \frac{1}{x}\,dx + \frac{1}{8}\int \frac{1}{x-2}\,dx - \frac{3}{8}\int \frac{1}{x+2}\,dx$$

$$= \frac{1}{4}\ln(x) + \frac{1}{8}\ln(x-2) - \frac{3}{8}\ln(x+2) + C.$$

Example [17.5] Evaluate the integral

$$\int \frac{x^3-5x^2+8x-2}{x^2-5x+6}\,dx.$$

In this case, since x^3-5x^2+8x-2 has degree 3 which is larger than the degree of x^2-5x+6, we must first perform long division:

$$\frac{x^3-5x^2+8x-2}{x^2-5x+6} = x + \frac{2x-2}{x^2-5x+6}.$$

Next, we factor our denominator, $x^2-5x+6 = (x-2)(x-3)$, and plugging this into our integral we are left with

$$\int \frac{x^3-5x^2+8x-2}{x^2-5x+6}\,dx = \int x\,dx + \int \frac{2x-2}{x^2-5x+6}\,dx$$

$$= \frac{x^2}{2} + \int \frac{2x-2}{x^2-5x+6}\,dx.$$

To complete our computation we need to find constants A_1 and A_2 so that

$$\frac{2x-2}{x^2-5x+6} = \frac{A_1}{x-2} + \frac{A_2}{x-3},$$

or equivalently

$$2x-2 = A_1\,(x-3) + A_2\,(x-2).$$

By setting $x = 2$ we deduce that $A_1 = -2$, and similarly, by setting $x = 3$ we have that $A_2 = 4$. Inserting these values into our integral we conclude that

$$\int \frac{2x-2}{x^2-5x+6}\,dx = \int \frac{-2}{x-2}\,dx + \int \frac{4}{x-3}\,dx$$

$$= -2\ln(x-2) + 4\ln(x-3) + C,$$

and finally

$$\int \frac{x^3-5x^2+8x-2}{x^2-5x+6}\,dx = \frac{x^2}{2} - 2\ln(x-2) + 4\ln(x-3) + C.$$

Exercises for §17.3

Use your CAS to perform the necessary long division needed to put the following rational functions in the form (17.2).

(1) $\frac{1}{x^2-9}$.

(2) $\frac{x^3+2}{x^2-4}$.

(3) $\frac{x^7+3}{x-2}$.

(4) $\frac{6x^4-2x^3+11}{x^3+2}$.

(5) $\frac{3x^{19}-2x^5+1}{x^3-2x+1}$.

(6) $\frac{x^{11}-13x^3+2x-3}{x^4+2x+3}$.

Evaluate the following integrals of rational functions. Use your CAS to factor polynomials and perform any necessary long division.

(7) $\int \frac{x+2}{x^2-9x}\,dx$.

(8) $\int \frac{x^4+2}{x^2-2x-3}\,dx$.

(9) $\int \frac{3x+1}{x^3-9x}\,dx$.

(10) $\int \frac{x^5}{x^4-5x^2+4}\,dx$.

(11) $\int \frac{x^3+2x^2+3}{x^4+2x^3-9x^2-18x}\,dx$.

(12) $\int \frac{x^8-2x^4+2}{x^5-6x^3+8x}\,dx$.

(13) Find constants A, B_1, B_2 so that

$$\frac{3x+1}{(x-1)(x+3)^2} = \frac{A}{x-1} + \frac{B_1}{x+3} + \frac{B_2}{(x+3)^2}.$$

Use this result to evaluate $\int \frac{3x+1}{(x-1)(x+3)^2}\,dx$.

(14) Find constants A_1, A_2, B_1, B_2, B_3 so that

$$\frac{x-2}{(x^2-1)(x+2)^3} = \frac{A_1}{x-1} + \frac{A_2}{x+1} + \frac{B_1}{x+2} + \frac{B_2}{(x+2)^2} + \frac{B_3}{(x+2)^3}.$$

Use this result to evaluate $\int \frac{x-2}{(x^2-1)(x+2)^3}\,dx$.

(15) Find constants A_1, A_2, B, C so that

$$\frac{x^2}{16-x^4} = \frac{A_1}{2-x} + \frac{A_2}{2+x} + \frac{Bx+C}{x^2+4}.$$

Use this result to evaluate $\int \frac{x^2}{16-x^4}\,dx$.

(16) Find constants A, B, C, D so that

$$\frac{x^2-1}{(x^2+3)^2} = \frac{Ax+B}{x^2+3} + \frac{Cx+D}{(x^2+3)^2}.$$

Use this result to evaluate the integral $\int \frac{x^2-1}{(x^2+3)^2}\,dx$.

17.4 Further Substitutions

Up to this point we have evaluated indefinite integrals by using substitutions, integration by parts, and partial fractions. Since there is no general algorithm to find the anti–derivative of an arbitrary function (it is a nice challenge to see if you can find a function your CAS cannot handle) we present in this section a collection of useful (and inobvious) classical substitutions.

The first class of substitutions are the **trigonometric substitutions** which we illustrate in the following example.

Example [17.6] Evaluate $\int \frac{dx}{\sqrt{9-x^2}}$.

If we let $x = 3\sin(\theta)$ then $dx = 3\cos(\theta)$, and, since $(\sin(\theta))^2 + (\cos(\theta))^2 = 1$ our integral becomes

$$\begin{aligned}\int \frac{3\cos(\theta)}{\sqrt{9-9(\sin(\theta))^2}}\,d\theta &= \int \frac{1}{\cos(\theta)}\cdot\cos(\theta)\,d\theta \\ &= \theta \\ &= \arcsin\left(\frac{x}{3}\right).\end{aligned}$$

While trigonometric substitutions can be utilized in a variety of ways there are some general patterns to look for in the integrand. The following chart is intended as a guide not as a rule:

$$\begin{aligned}&\text{if } \sqrt{a^2-x^2} \text{ appears try } x = a\sin(\theta)\\ &\text{if } \sqrt{a^2+x^2} \text{ appears, try } x = a\tan(\theta)\\ &\text{if } \sqrt{x^2-a^2} \text{ appears, try } x = a\sec(\theta).\end{aligned}$$

The above substitutions will transform the original integral into an integral which involves trigonometric functions. At this point one can either use the CAS or resort to the wide array of trigonometric identities in order to complete the computation. It is important to realize that the suggested substitutions are themselves based on the basic identity $(\sin(\theta))^2 + (\cos(\theta))^2 = 1$.

Example [17.7] Evaluate $\int \frac{dx}{\sqrt{x^2-a^2}}$.

Following our guide we assert $x = a\sec(\theta)$ and thus $dx = a\sec(\theta)\tan(\theta)d\theta$, and our integral becomes

$$\int \frac{dx}{\sqrt{x^2-a^2}} = \int \sec(\theta)\, d\theta.$$

The latter integral is not entirely trivial and does require attention. Notice that

$$\frac{d}{d\theta}(\sec(\theta) + \tan(\theta)) = (\sec(\theta))^2 + \sec(\theta)\tan(\theta).$$

This identity allows us to transform our second integral as follows:

$$\begin{aligned} \int \sec(\theta)\, d\theta &= \int \sec(\theta) \cdot \overbrace{\left(\frac{\sec(\theta)+\tan(\theta)}{\sec(\theta)+\tan(\theta)}\right)}^{\text{The trick: mult. by 1}} d\theta \\ &= \ln(\sec(\theta)+\tan(\theta)) + C. \end{aligned}$$

In order to obtain an answer in the original variable we consider the triangle

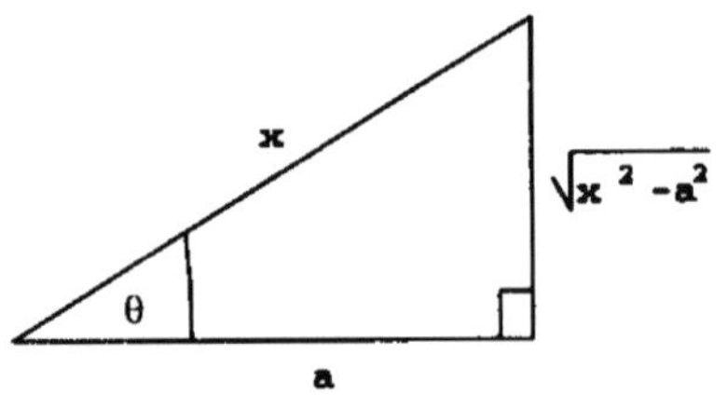

and conclude

$$\int \frac{dx}{\sqrt{x^2-a^2}} = \ln(\sec(\theta)+\tan(\theta)) + C = \ln(x + \sqrt{x^2-a^2}) - \ln(a) + C.$$

The next class of substitutions we shall consider are the so called **hyperbolic substitutions**. Recall that the functions $\cosh(t)$ and $\sinh(t)$ satisfy the identity

$(\cosh(t))^2 - (\sinh(t))^2 = 1$ (see §8.3). In Example [17.7] we could have used the substitution $x = a\cosh(t)$ and $dx = a\sinh(t)$ to reduce the integral to the form

$$\int \frac{dx}{\sqrt{x^2 - a^2}} = \int \frac{a\sinh(t)}{a\sinh(t)}\,dt = t = \operatorname{arccosh}\left(\frac{x}{a}\right).$$

(See exercise (20) of §17.4 for an explicit form of arccosh(t)).

There are occasions where trigonometric substitutions lead to terribly complex integrals (of trigonometric functions) but a hyperbolic substitution will work easily.

Example [17.8] Evaluate

$$\int \sqrt{4 + x^2}\,dx$$

In this case trigonometric substitutions truly make the situation worse. Thus we set $x = 2\sinh(t)$ and thus $dx = 2\cosh(t)dt$ and our integral is readily handled. Since $1 + (\sinh(t))^2 = (\cosh(t))^2$,

$$\begin{aligned}
\int \sqrt{4 + x^2}\,dx &= 4\int \cosh(t)\cdot\cosh(t)\,dt \\
&= 4\int \frac{e^t + e^{-t}}{2}\cdot\frac{e^t + e^{-t}}{2}\,dt \\
&= \int \left(e^{2t} + 2 + e^{-2t}\right)\,dt \\
&= \frac{e^{2t}}{2} + 2t - \frac{e^{-2t}}{2} + C.
\end{aligned}$$

The last substitution we present in this section is the **Weierstrass substitution** (Weierstrass was a prominent 19th century mathematician) which is a means of transforming any rational expression involving trigonometric functions into a rational function involving a single variable t. The substitution is $t = \tan\left(\frac{x}{2}\right)$, $dt = dx/2\left(\cos\frac{x}{2}\right)^2$, which when analyzed with the triangle below,

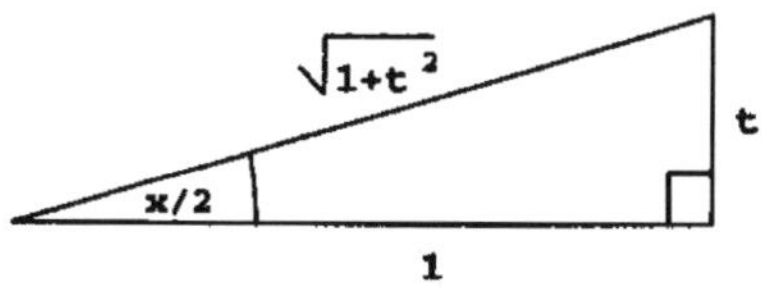

together with the identities $\sin(2\theta) = 2\sin(\theta)\cos(\theta)$ and $\cos(2\theta) = 1-2(\sin(\theta))^2$, yield the identities

$$\sin(x) = \frac{2t}{1+t^2}, \quad \cos(x) = \frac{1-t^2}{1+t^2}, \quad dx = \frac{2dt}{1+t^2}.$$

As we see in the next example, this seemingly complex substitution can be very effective.

Example [17.9] Evaluate

$$\int \frac{1}{1+\sin(x)+\cos(x)}\, dx.$$

Following the Weierstrass method our integral takes the form

$$\int \frac{1}{1+\sin(x)+\cos(x)}\, dx = \int \frac{1}{1+\frac{2t}{1+t^2}+\frac{1-t^2}{1+t^2}} \cdot \frac{2}{1+t^2}\, dt$$

$$= \int \frac{dt}{1+t}$$

$$= \ln(1+t) + C$$

$$= \ln\left(1+\tan\left(\frac{x}{2}\right)\right) + C.$$

Exercises for §17.4

Evaluate the following **trigonometric integrals** with the help of the basic identities

$$(\sin(x))^2 + (\cos(x))^2 = 1$$

$$(\sin(x))^2 = \frac{1}{2}(1-\cos(2x)), \qquad (\cos(x))^2 = \frac{1}{2}(1+\cos(2x)).$$

Check your answer with the CAS.

(1)	$\int(\sin(x))^3\,dx.$	(4)	$\int(\sin(x))^4\,dx.$
(2)	$\int(\sin(x))^3(\cos(x))^2\,dx.$	(5)	$\int(\cos(x))^5\,dx.$
(3)	$\int(\cos(x))^2\,dx.$	(6)	$\int(\sin(x))^2(\cos(x))^4\,dx.$

Evaluate the following integrals using appropriate substitutions, integration by parts, etc. As usual check your work with the CAS.

(7)	$\int \frac{1}{x\sqrt{1-x^2}}\,dx.$	(13)	$\int \frac{e^{3x}}{e^x+1}\,dx.$ **Hint:** Set $x = \ln(u)$.
(8)	$\int \frac{\sqrt{1-x^2}}{x}\,dx.$	(14)	$\int \sqrt{1+\sqrt{x}}\,dx.$
(9)	$\int \frac{1}{x+\sqrt{x}}\,dx.$	(15)	$\int \frac{\sqrt{x}}{\sqrt{1-x}}\,dx.$ **Hint:** Set $x = (\sin(u))^2$.
(10)	$\int \frac{\sqrt{x}}{(1+x^{3/2})}\,dx.$	(16)	$\int x^2\sqrt{x^2+a^2}\,dx.$
(11)	$\int \frac{1}{x^2\sqrt{a^2-x^2}}\,dx.$	(17)	$\int x^3\sqrt{1-x^2}\,dx.$
(12)	$\int \frac{1}{(x^2+1)^2}\,dx.$	(18)	$\int \ln(x+\sqrt{x^2+1})\,dx.$

(19) Demonstrate that if $x = \sinh(t)$ then $t = \ln(x+\sqrt{x^2+1})$. **Hint:** Set $u = e^t$ in the definition of $\sinh(t)$, solve for u, and then deduce that $t = \ln(u)$.

(20) Demonstrate that if $x = \cosh(t)$ then $t = \ln(x+\sqrt{x^2-1})$. **Hint:** Use Exercise (19) as a guide.

(21)	$\int \frac{x^2}{\sqrt{x^2+a^2}}\,dx.$	(29)	$\int \frac{\sqrt{x-x^2}}{x^4}\,dx.$ **Hint:** Try $x = \frac{1}{z}$.
(22)	$\int \sqrt{x^2-1}\,dx.$	(30)	$\int \frac{1}{3-2\cos(x)}\,dx.$
(23)	$\int x\arcsin(x)\,dx.$	(31)	$\int \frac{1}{x\sqrt{3x^2+2x-1}}\,dx.$ **Hint:** Try $x = \frac{1}{z}$.
(24)	$\int \frac{\sqrt{25-x^2}}{x}\,dx.$	(32)	$\int \frac{1}{4\sin(x)}\,dx.$
(25)	$\int x\arccos(x)\,dx.$	(33)	$\int \frac{1}{x^2(4+x^2)}\,dx.$
(26)	$\int \frac{1}{2+\tan(x)}\,dx.$	(34)	$\int \frac{\sqrt{x}}{1+x}\,dx.$
(27)	$\int x^2\ln(x)\,dx.$	(35)	$\int \frac{1}{\sqrt{x}+\sqrt[4]{x}}\,dx.$
(28)	$\int x^2(\ln(x))^2\,dx.$	(36)	$\int \frac{\arctan(x)}{x^2}\,dx.$

17.5 Improper Integrals

Consider a thin infinitely long rectangle:

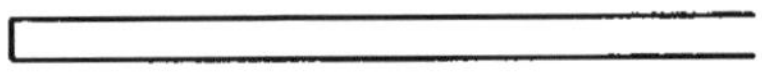

It is intuitively clear that the area of this rectangle is infinite, but is this always the case for any infinite figure? More precisely we pose the following question:

Question: Does there exist a curve C which becomes asymptotic to the x–axis (as $x \to \infty$) such that the area of the region between the curve C and the x–axis is finite?

Answer: Such a situation *is* possible (if not obvious). The essential reason is that it is possible to have an infinite sequence of numbers, for example

$$\frac{1}{2}, \frac{1}{4}, \frac{1}{8}, \ldots,$$

whose sum is finite. In the case of the sequence $\frac{1}{2}, \frac{1}{4}, \frac{1}{8}, \ldots$, the sum is just

$$\frac{1}{2} + \frac{1}{4} + \frac{1}{8} + \ldots = 1,$$

This can be seen by the following limiting procedure.

$$\frac{1}{2}+\frac{1}{4}=\frac{3}{4}$$
$$\frac{1}{2}+\frac{1}{4}+\frac{1}{8}=\frac{7}{8}$$
$$\frac{1}{2}+\frac{1}{4}+\frac{1}{8}+\frac{1}{16}=\frac{15}{16}$$
$$\vdots$$
$$\frac{1}{2}+\frac{1}{4}+\frac{1}{8}+\cdots+\frac{1}{2^N}=\frac{2^N-1}{2^N},$$

and, when we take the limit, we obtain our sum:

$$\lim_{N\to\infty}\left[\frac{1}{2}+\frac{1}{4}+\frac{1}{8}+\cdots+\frac{1}{2^N}\right]=\lim_{N\to\infty}\left(\frac{2^N-1}{2^N}\right)$$
$$=\lim_{N\to\infty}\left(1-\frac{1}{2^N}\right)=1.$$

Now supposing we had a curve C such that the areas of the rectangular type regions $R_1, R_2, R_3, \ldots$

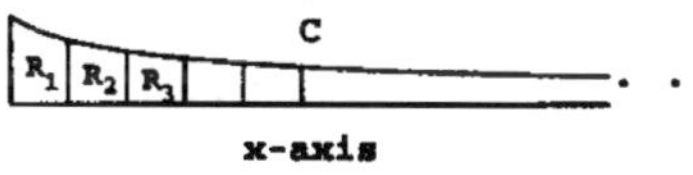

satisfies the condition area$(R_1) = 1/2$, area$(R_2) = 1/4$, area$(R_3) = 1/8, \ldots$ then the area under the curve, which is the sum of all the smaller areas, is simply 1.

Consider now an arbitrary curve $y = f(x)$ which becomes asymptotic to the x–axis as $x \to \infty$.

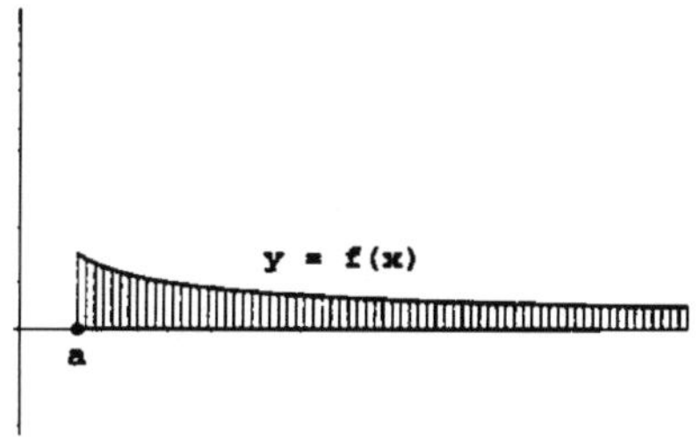

The area of the shaded region (which may be finite or infinite) is given by the integral

$$\int_a^\infty f(x)\,dx.$$

Since we are integrating on the interval $a \le x \le \infty$, the above integral is said to be **improper**. We give meaning to such an integral by defining

$$\int_a^\infty f(x)\,dx = \lim_{B\to\infty}\int_a^B f(x)\,dx.$$

Example [17.10] Evaluate the improper integral

$$\int_1^\infty \frac{1}{x^2}\,dx.$$

We have

$$\begin{aligned}\int_1^\infty \frac{1}{x^2}\,dx &= \lim_{B\to\infty}\int_1^B \frac{1}{x^2}\,dx\\ &= \lim_{B\to\infty} -x^{-1}\Big]_1^B\\ &= \lim_{B\to\infty}\left(-\frac{1}{B}+1\right) = 1.\end{aligned}$$

Question: Does there exist a curve $y = f(x)$ which is asymptotic to the x–axis

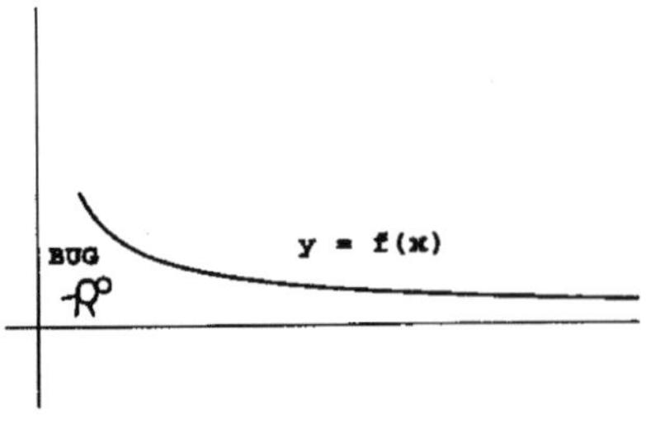

with the following properties:

(1) The area of the region R between the curve $y = f(x)$ and the x–axis is infinite;

(2) The distance between the curve and the x–axis becomes arbitrarily small, i.e., a bug will be unable to crawl all the way through R.

Answer: The curve $y = 1/\sqrt{x}$ will do the trick.

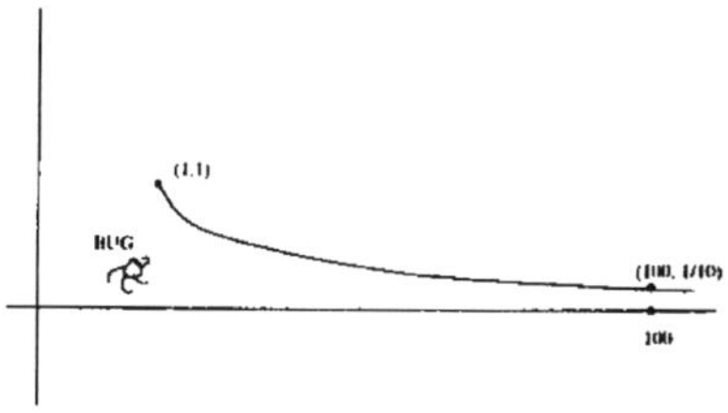

If the bug is about $1/10$ of an inch thick it will get stuck at the point $x \approx 100$. Nevertheless, the area of the region is seen to be infinite by the following computation:

$$\int_0^\infty \frac{1}{\sqrt{x}}\,dx = \lim_{B\to\infty} \int_1^B \frac{1}{\sqrt{x}}\,dx$$

$$= \lim_{B\to\infty} 2x^{\frac{1}{2}}\Big]_1^B$$

$$= \lim_{B\to\infty} (2\sqrt{B} - 2) = \infty.$$

Example [17.11] For which values of p is

$$\int_a^\infty \frac{dx}{x^p}$$

finite.

Isolating the case $p = 1$, the integral is easily seen to be infinite:

$$\int_a^\infty \frac{dx}{x} = \lim_{B\to\infty} \int_a^B \frac{dx}{x} = \lim_{B\to\infty} (\ln(B) - \ln(a)) = \infty.$$

When $p \neq 1$ we have two possibilities:

$$\begin{aligned}
\int_a^\infty \frac{dx}{x^p} &= \lim_{B\to\infty} \int_a^B \frac{dx}{x^p} \\
&= \lim_{B\to\infty} \frac{x^{-p+1}}{-p+1}\Bigg]_a^B \\
&= \lim_{B\to\infty} \left(\frac{B^{1-p}}{1-p} - \frac{a^{1-p}}{1-p}\right) \\
&= \begin{cases} \infty, & \text{if } p < 1 \\ \frac{-a^{1-p}}{1-p}, & \text{if } p > 1. \end{cases}
\end{aligned}$$

Example [17.12] Evaluate $\int_{-\infty}^{3} e^{2x}\,dx$.

In this case the integral is computed as follows:

$$\begin{aligned}
\int_{-\infty}^{3} e^{2x}\,dx &= \lim_{B\to-\infty} \int_B^3 e^{2x}\,dx \\
&= \lim_{B\to-\infty} \left(\frac{e^6}{2} - \frac{e^{2B}}{2}\right) = \frac{e^6}{2}.
\end{aligned}$$

The improper integrals we have discussed up to this point arise from infinitely long horizontal regions associated to a curve C which is asymptotic to the x–axis.

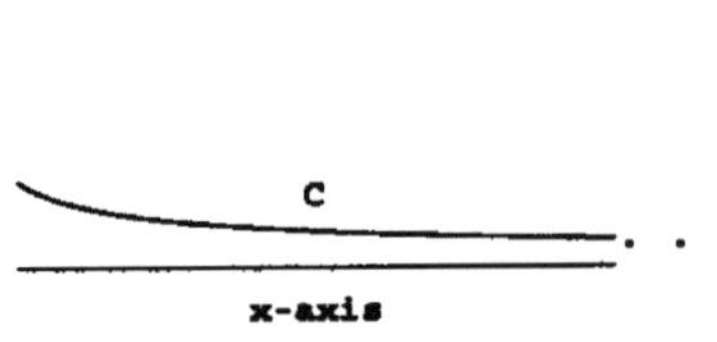

It is also possible to have infinitely long vertical regions coming from a curve $y = f(x)$ with a vertical asymptote (dotted line).

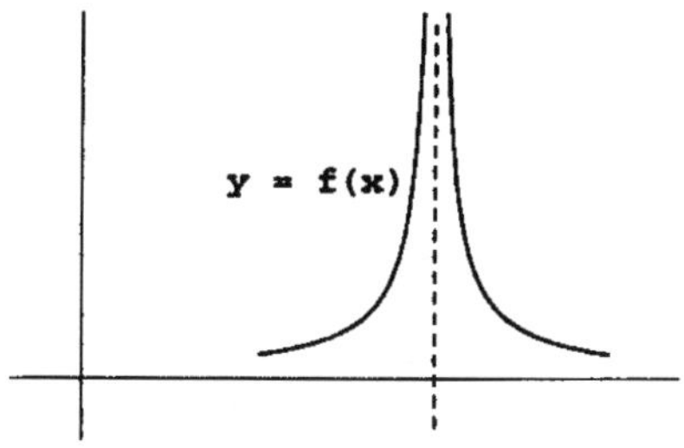

For example, the curve $y = (x - 3/4)^{-2}$ has a vertical asymptote at the point $x = 3/4$.

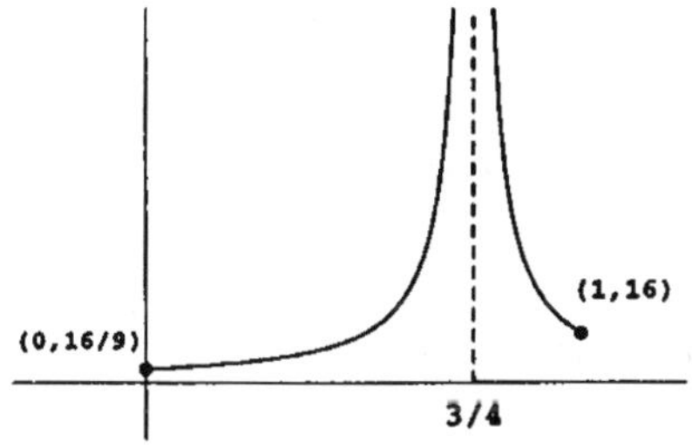

Question: Does the infinitely long vertical region under the curve

$$y = \frac{1}{(x - \frac{3}{4})^2},$$

(for $0 \le x \le 1$) have infinite area?

Suppose, for a moment, that the integral in question is finite. Then by Property 3 in section 13.4, the integral breaks into the sum of two integrals

$$\int_0^1 \frac{dx}{(x - \frac{3}{4})^2} = \int_0^{\frac{3}{4}} \frac{dx}{(x - \frac{3}{4})^2} + \int_{\frac{3}{4}}^1 \frac{dx}{(x - \frac{3}{4})^2}.$$

Focusing on the first of these integrals, we first evaluate for $0 < \beta < 3/4$,

$$\int_0^{\beta} \frac{1}{(x - \frac{3}{4})^2},$$

which represents the shaded area below.

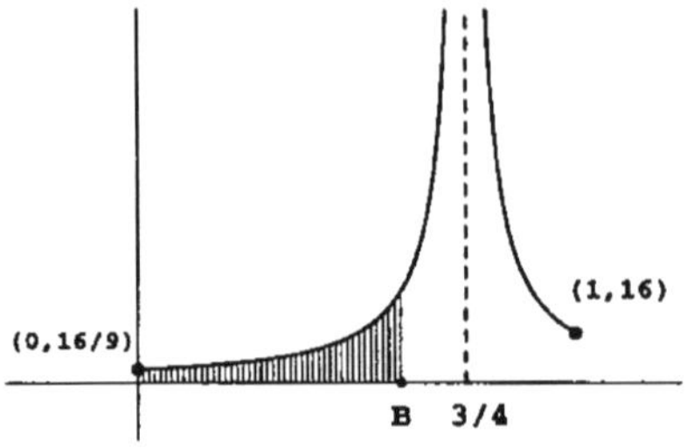

Integrating we have

$$\int_0^\beta \frac{1}{(x-\frac{3}{4})^2} = -(x-\frac{3}{4})^{-1}\Big]_0^\beta$$
$$= \frac{-1}{\beta-\frac{3}{4}} - \frac{4}{3},$$

and when we take the limit as $\beta \to 3/4$ we obtain

$$\lim_{\substack{\beta\to\frac{3}{4}\\ \beta<\frac{3}{4}}} \int_0^\beta \frac{1}{(x-\frac{3}{4})^2} = \lim_{\substack{\beta\to\frac{3}{4}\\ \beta<\frac{3}{4}}} \left(\frac{-1}{\beta-\frac{3}{4}} - \frac{4}{3}\right)$$

$$= \infty.$$

It follows that the original integral (i.e., the area of the infinitely long vertical strip) was indeed infinite.

With these examples in mind we can now give a formal definition of an improper integral.

Definition *An integral* $\int_a^b f(x)\ dx$ *is* ***improper*** *if either* a *or* b *equals* $\pm\infty$, *or the function* $f(x)$ *equals* $\pm\infty$ *at some point* $x = c$ *with* $a \leq c \leq b$.

We have already discussed how to assign a meaning to such an integral if $b = \pm\infty$:

$$\int_a^{\pm\infty} f(x)\ dx = \lim_{\beta\to\pm\infty} \int_a^\beta f(x)\ dx.$$

The case of $\int_{\pm\infty}^{b} f(x)\,dx$ reduces to the above since $\int_a^b f(x)\,dx = -\int_b^a f(x)\,dx$. The case of

$$\int_{-\infty}^{\infty} f(x)\,dx,$$

can be reduced to a sum of integrals seen above by simply breaking the infinite interval $-\infty \le x \le \infty$ at some point c:

$$\int_{-\infty}^{\infty} f(x)\,dx = \int_{-\infty}^{c} f(x)\,dx + \int_{c}^{\infty} f(x)\,dx.$$

We now focus on the case $f(c) = \pm\infty$ for some c in the interval $a < c < b$. Motivated by our analysis of the function $y = \frac{1}{\left(x-\frac{3}{4}\right)^2}$, we note that if $\int_a^b f(x)\,dx$ is finite then it can be expressed as

$$\int_a^b f(x)\,dx = \int_a^c f(x)\,dx + \int_c^b f(x)\,dx.$$

By definition, the integrals appearing on the right hand side must be equal to the following limit (respectively):

$$\lim_{\substack{\beta\to c\\ \beta<c}} \int_a^{\beta} f(x)\,dx, \qquad \lim_{\substack{\beta\to c\\ \beta>c}} \int_{\beta}^{b} f(x)\,dx.$$

We are thus justified by defining

$$\int_a^b f(x)\,dx = \lim_{\substack{\beta\to c\\ \beta<c}} \int_a^{\beta} f(x)\,dx + \lim_{\substack{\beta\to c\\ \beta>c}} \int_{\beta}^{b} f(x)\,dx. \tag{17.5}$$

Remarks: There remain two special cases to consider:

(1) There are several points $a < c_1 < c_2 < \cdots < c_k$ where the function becomes infinite.

(2) The function becomes infinite at an endpoint of the interval.

The integrals in both these cases can be analyzed to obtain formulae which are analogous to (17.5). In the first case, we choose points $a < c_1 < d_1 < c_2 < d_2 < \cdots < d_{k-1} < c_k < b$ such that $f(d_i)$ is finite. Then

$$\int_a^b f(x)\,dx = \int_a^{c_1} f(x)\,dx + \int_{c_1}^{d_1} f(x)\,dx + \int_{d_1}^{c_2} f(x)\,dx + \cdots + \int_{d_{k-1}}^{c_k} f(x)\,dx + \int_{c_k}^{b} f(x)\,dx,$$

and each of the above integrals may be computed by the appropriate limit.

In the final case, i.e., when the function becomes infinite at an endpoint of the interval, say $f(a) = \infty$, the integral may be computed by yet another limit:

$$\int_a^b f(x)\,dx = \lim_{\substack{\beta\to a\\ \beta>a}} \int_\beta^b f(x)\,dx.$$

Example [17.13] Evaluate the integral

$$\int_1^4 \frac{dx}{(x-3)^{1/3}}.$$

When we graph the curve $y = (x-3)^{-1/3}$ we find a vertical asymptote at $x = 3$.

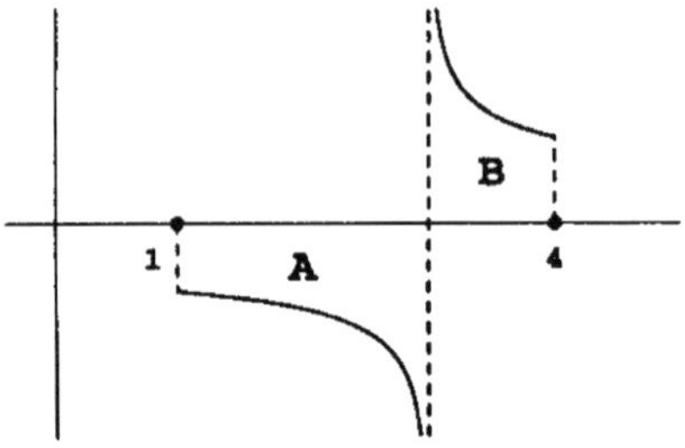

Thus

$$\begin{aligned}
\int_1^4 \frac{dx}{(x-3)^{1/3}} &= -\text{Area}(A) + \text{Area}(B)\\
&= \int_1^3 \frac{dx}{(x-3)^{1/3}} + \int_3^4 \frac{dx}{(x-3)^{1/3}}\\
&= \lim_{\substack{\beta\to 3\\ \beta<3}} \int_1^\beta \frac{dx}{(x-3)^{1/3}} + \lim_{\substack{\beta\to 3\\ \beta>3}} \int_\beta^4 \frac{dx}{(x-3)^{1/3}}\\
&= \lim_{\substack{\beta\to 3\\ \beta<3}} \frac{3}{2}(x-3)^{2/3}\Big]_1^\beta + \lim_{\substack{\beta\to 3\\ \beta>3}} \frac{3}{2}(x-3)^{2/3}\Big]_\beta^4\\
&= -\frac{3}{2}(-2)^{2/3} + \frac{3}{2}\cdot 1\\
&= \frac{3}{2}(1-2^{2/3}).
\end{aligned}$$

Notice that the final answer is negative which reflects the fact that Area$(A) >$ Area(B).

Exercises for §17.5

For each of the following improper integrals use your CAS to graph the area represented, replacing infinity with a large number. Determine whether the integral is finite or infinite. If it is finite compute its value.

(1) $\int_2^\infty \frac{dx}{x^3}$.

(2) $\int_2^\infty \frac{dx}{x^{3/4}}$.

(3) $\int_1^3 \frac{dx}{(x-2)^{1/3}}$.

(4) $\int_2^4 \frac{dx}{(x-3)^3}$.

(5) $\int_1^\infty xe^{-x}\,dx$.

(6) $\int_2^\infty \frac{x}{x+\cos(x)}\,dx$.

(7) $\int_1^\infty \frac{e^{2-x}}{(x-2)^2}\,dx$.

(8) $\int_1^\infty \frac{dx}{x}$.

(9) $\int_0^1 \frac{dx}{x}$.

(10) $\int_2^\infty \frac{dx}{x^2+2x}$.

(11) $\int_1^3 \frac{dx}{\sqrt{x^2+1}}$.

(12) $\int_1^3 \frac{dx}{(x-1)^{\frac{3}{4}}}$.

(13) $\int_1^3 \frac{dx}{x-1}$.

(14) $\int_1^3 \frac{dx}{(x-1)^{\frac{5}{4}}}$.

For each of the following improper integrals use your CAS to graph the area represented by the integral. Determine if the integral is finite or not. Do not try to compute its value. **Hint:** In trying to evaluate $\int_1^3 \frac{e^{x^2}}{(x-1)^2}\,dx$ note that $e < e^{x^2} < e^9$ for $1 \leq x \leq 3$. Hence

$$e\int_1^3 \frac{dx}{(x-1)^2} < \int_1^3 \frac{e^{x^2}}{(x-1)^2}\,dx < e^9\int_1^3 \frac{dx}{(x-1)^2}.$$

Since $\int_1^3 \frac{dx}{(x-1)^2}$ is infinite we can deduce that $\int_1^3 \frac{e^{x^2}}{(x-1)^2}\,dx$ is also infinite.

(15) $\int_{10}^{20} \frac{\cos(3x)}{\sqrt{x-10}}\,dx$.

(16) $\int_1^5 \frac{x^{50}}{(x-1)^{3/2}}\,dx$.

(17) $\int_1^6 \frac{\ln(x)\cdot e^{\frac{1}{x}}}{(x-3)^5}\,dx$.

(18) $\int_{-1}^1 \frac{\arcsin(x)}{\sqrt[3]{x}}\,dx$.

(19) Show that $e^x > x^2$ for $x \geq 1$.

(20) Show that $e^x > x^3$ for $x \geq 6$. **Hint:** Evaluate $\frac{e^5}{5^3}$ on your CAS. Then show that e^x/x^3 is an increasing function for $x \geq 5$ by showing its first derivative is positive.

(21) Show that for any $p > 0$, $e^x > x^p$ for all x which are sufficiently large.

Using exercise (21) determine which of the following improper integrals are finite.

(22) $\int_1^\infty \frac{e^x}{x^{1000}}\,dx$.

(23) $\int_1^\infty x^{1000}e^{-x}\,dx$.

(24) $\int_1^\infty x^{1000}e^{-\sqrt{x}}\,dx$.

(25) $\int_1^\infty \frac{e^{\sqrt[3]{x}}}{x^{1000}}\,dx$.

(26) Define $\Gamma(x) = \int_0^\infty e^{-u}u^{x-1}\,du$. What is $\Gamma(1)$? Show that this integral converges for $x > 0$. Using integration by parts show that $\Gamma(x+1) = x\Gamma(x)$ for all $x \geq 1$. Deduce that $\Gamma(n+1) = n!$ for all positive integers n.

Additional exercises for Chapter XVII

Compute the following integrals and check your answer with your CAS.

(1) $\int \frac{dx}{\sin x - 3\cos x}$.

(2) $\int \frac{dx}{\sin x + \tan x}$.

(3)(a) Find constants a, b such that $x^4 + 1 = (x^2 + ax + 1)(x^2 + bx + 1)$.

(b) Use the result in (a) to show that $\int_0^1 \frac{x}{x^4+1} dx = \pi/8$.

(4) Find constants A_1, B_1, C_1, B_2, C_2, so that $\frac{3x^4+4x^3+16x^2+20x+9}{(x+2)(x^2+3)^2} = \frac{A_1}{x+2} + \frac{B_1x+C_1}{x^2+3} + \frac{B_2x+C_2}{(x^2+3)^2}$. Use this to evaluate $\int \frac{3x^4+4x^3+16x^2+20x+9}{(x+2)(x^2+3)^2} dx$.

(5) Compute $\int_0^1 \arctan(x)dx$ (**Hint:** Set $y = \arctan(x)$ and solve for y). Graph the region on your CAS.

(6) Compute $\int \sqrt{1 + \exp(x)}dx$ (**Hint:** set $u^2 = 1 + \exp(x)$).

(7) Compute $\int_1^3 \frac{\sqrt{9-x^2}}{x} dx$.

(8) A city has a rectangular plot of land of dimensions 800 × 300ft, bordered on one short side by the ocean. The population density as we move away from the ocean is given by $\delta(x) = 500 - \sqrt{x}$. Compute the total population. (**Hint:** in a rectangle of width Δx centered at x_0, and of height 300 , the total population is approximately $\delta(x_0) \cdot 300 \cdot \Delta x$.

(9) With your CAS, graph the region bounded by the curves $y = \ln(x)$ and $y = -\frac{1}{e}x + 2$ and the x-axis. Find the area of the region. (**Hint:** integrate in the y variable).

(10)(a) Evaluate $\int_1^3 \ln(x)dx$ first by finding the antiderivative.

(b) Evaluate again the same integral by use of substitution $x = \exp(y)$.

Use your CAS to perform the necessary long division to put the following rational fractions in the form 17.2.

(11) $\frac{x^4+x^2-2}{x^3+3x+1}$.

(12) $\frac{3x^4+12x-2}{x^2+1}$.

(13) $\frac{x^{14}-x^{12}+2x^{10}+x^8+2x^6-x^4+1}{3x^3+2x+1}$.

(14) $\frac{x^{16}+6x+2}{x^3-6}$.

Use your CAS to factor polynomials and perform any long division. Then evaluate the following integrals.

(15) $\int \frac{x^5+2x^3+x^2-3x-1}{x^3+3x+1} dx.$

(16) $\int \frac{3x^4+12x-2}{x^2+1} dx.$

(17) $\int \frac{dx}{x^4-3x^2-4}.$

Chapter XVIII

Infinite Series

18.1 Geometric Series

As an introduction to this chapter let us contemplate Zeno's paradox which says: *It is impossible to walk from* 0 *to* 1 *on the number line.*

The argument supporting the claim goes like this: in walking from 0 to 1 one must necessarily pass through the midpoint $1/2$ and then pass through the midpoint between $1/2$ and 1, which is $3/4$, etc. Since there are infinitely many midpoints to go through one can never reach the point 1.

How can we refute this argument (after all it is possible to walk between any two points)? Recall first that (as we observed in §17.5)

$$\frac{1}{2} + \frac{1}{4} + \frac{1}{8} + \frac{1}{16} + \cdots = 1.$$

Assuming the distance between 0 and 1 on the number line is one inch and we are traveling at the rate of 1 inch per second, then it will take $1/2$ a second to reach the point $1/2$. Similarly, going from $1/2$ to $3/4$ takes $1/4$ of a second, and continuing in this manner the total traveling time in going from midpoint to midpoint to midpoint, etc., is the sum

$$\frac{1}{2} + \frac{1}{4} + \frac{1}{8} + \frac{1}{16} + \cdots, \qquad (18.1)$$

which we know to be 1 second.

This seemingly simple problem has in fact led us to the following truly profound revelation:

Revelation: *It is possible to add infinitely many numbers and obtain a finite sum.*

The infinite sum (18.1) is a simple example of a **geometric series**. Another example of such a series is

$$\frac{1}{3} + \frac{1}{9} + \frac{1}{27} + \frac{1}{81} + \cdots \qquad (18.2)$$

Question: Is the sum (18.2) finite, and if it is what is its value?

Answer: We claim that, using our CAS, we can show

$$\frac{1}{3} + \frac{1}{9} + \frac{1}{27} + \frac{1}{81} + \cdots = \frac{1}{2}.$$

To begin with we compute the first few sums,

$$\frac{1}{3} + \frac{1}{9} = \frac{4}{9},$$

$$\frac{1}{3} + \frac{1}{9} + \frac{1}{27} = \frac{13}{27},$$

$$\frac{1}{3} + \frac{1}{9} + \frac{1}{27} + \frac{1}{81} = \frac{40}{81}.$$

Is there a pattern here? A sharp eye (and a little experience working with sums) tells us that

$$\frac{1}{3} + \frac{1}{9} + \frac{1}{27} + \frac{1}{81} + \cdots + \frac{1}{3^N} = \frac{(3^N - 1)/2}{3^N}. \qquad (18.3)$$

Can you prove this? **Hint:** Use induction.

Formula (18.3) allows us to sum the infinite series:

$$\begin{aligned}\frac{1}{3} + \frac{1}{9} + \frac{1}{27} + \cdots &= \lim_{N\to\infty}\left(\frac{1}{3} + \frac{1}{9} + \frac{1}{27} + \frac{1}{81} + \cdots + \frac{1}{3^N}\right)\\ &= \lim_{N\to\infty} \frac{(3^N - 1)/2}{3^N}\\ &= \lim_{N\to\infty}\left(\frac{1}{2} - \frac{1}{2\cdot 3^N}\right)\\ &= \frac{1}{2}.\end{aligned}$$

Definition: *Let $-1 < r < 1$ and let a be any fixed number. A* ***geometric series*** *with* ***initial term*** *a and* ***ratio*** *r is the infinite sum*

$$a + ar + ar^2 + ar^3 + ar^4 + \cdots$$

Remarks: In the above definition, r is the constant factor by which terms are multiplied to obtain their successors. Using the summation symbol $\sum$ we can conveniently and compactly denote the geometric series with initial term a and ratio r by

$$\sum_{n=0}^{\infty} a \cdot r^n.$$

Example [18.1] The geometric series with initial term 3 and ratio $-1/7$ is

$$3 - \frac{3}{7} + \frac{3}{49} - \frac{3}{343} + \cdots$$

Remark: If $|r| > 1$ then $|r^n|$ will get bigger and bigger as $n \to \infty$ and thus the geometric series $\sum_{n=0}^{\infty} a \cdot r^n$ can never be finite.

Proposition[18.2] *If $|r| < 1$ then the geometric series is given by*

$$\sum_{n=0}^{\infty} a \cdot r^n = \frac{a}{1-r}.$$

Proof (Formal): It is enough to show that

$$\sum_{n=0}^{\infty} r^n = \frac{1}{1-r},$$

since we may multiply both sides of this equation by a to obtain the general result. Formally

$$\begin{aligned}(1-r)\cdot(1+r+r^2+r^3+\cdots) &= 1\cdot(1+r+r^2+r^3) - r\cdot(1+r+r^2+r^3)\\ &= (1+r+r^2+r^3) - (r+r^2+r^3+\cdots)\\ &= 1.\end{aligned}$$

But then by dividing both sides of the equation

$$(1-r)\cdot(1+r+r^2+r^3+\cdots)=1$$

by $(1-r)$ we obtain the desired result.

How can we make the above proof rigorous? Why doesn't the proof work when $|r|>1$? The answer can be seen in §18.2.

Example [18.3] Express the infinite repeating decimal expansion

$$c = 2.713713713713713713\ldots$$

as a rational number.

By definition we have that

$$c = 2 + \frac{713}{10^3} + \frac{713}{10^6} + \frac{713}{10^9} + \frac{713}{10^{12}} + \cdots$$

Thus $c-2$ is a geometric series with initial term $a = 713/10^3$ and ratio $1/10^3$. Proposition [18.2] tells us that

$$c-2 = \frac{713}{10^3}\cdot\frac{1}{\left(1-\frac{1}{10^3}\right)} = \frac{713}{999},$$

and we conclude that

$$c = 2 + \frac{713}{999} = \frac{2711}{999}.$$

Exercises for §18.1

Sum the following geometric series. Confirm your answers by summing many terms of the series with your CAS.

(1) $\sum_{n=0}^{\infty} 2\cdot 3^{-n}$

(2) $\sum_{n=3}^{\infty} 2\cdot 3^{-2n}$

(3) $\sum_{n=2}^{\infty} 5\cdot 4^{-n}$

(4) $\sum_{n=0}^{\infty} \left(\frac{-3}{4}\right)^n$

(5) $\sum_{n=1}^{\infty} 2\cdot\left(\frac{2}{3}\right)^n$

(6) $\sum_{n=1}^{\infty} (-2)^{-3n}$

(7) $\sum_{n=0}^{\infty} 4\cdot 5^{-2n+2}$

(8) $\sum_{n=1}^{\infty} \left(\frac{-5}{7}\right)^{n-2}$

(9) $\sum_{n=4}^{\infty} \left(\frac{3}{5}\right)^{n-2}$

Express the following infinite repeating decimal expansions as rational numbers (ratios of integers). The bar indicates the repeating part.

(10) $0.\overline{35} = 0.353535\ldots$

(11) $0.\overline{129} = 0.129129129\ldots$

(12) $2.\overline{17} = 2.171717\ldots$

(13) $5.\overline{621} = 5.621621621\ldots$

(14) For which values of x does the series $\sum_{n=0}^{\infty}(\cos(x))^n$ converge? What is the sum if it does converge?

(15) For which values of x does the series $\sum_{n=0}^{\infty}(e^x)^n$ converge? What is the sum if it does converge?

(16) For which values of x does the series $\sum_{n=0}^{\infty}(\ln(x))^n$ converge? What is the sum if it does converge?

(17) A tennis ball is dropped from a height of 15 feet. It is in a total vacuum, so it rebounds to 4/5 of the height from which it falls, and continues to do so indefinitely. Compute the total distance it will travel. What distance will it have traveled after 50 bounces?

18.2 General Infinite Series

let $a_1, a_2, a_3, \ldots$ be an infinite sequence of real numbers. An **infinite series** or simply **series** is just the infinite sum

$$\sum_{n=1}^{\infty} a_n = a_1 + a_2 + a_3 + \cdots$$

For $N \geq 1$ we define the Nth partial sum of the above series by

$$S(N) = \sum_{n=1}^{N} a_n.$$

Definition: *If* $\lim_{N\to\infty} S(N)$ *exists and is equal to* s *then we say that the* ***infinite series converges to*** s*. Otherwise we say the series* ***diverges.***

Remark: We say that an infinite series is convergent if it converges to a finite number, and if that finite number is s we say the sum converges to s.

Problem: Give an example of a convergent infinite series and a divergent series.

The geometric series

$$1 + r + r^2 + r^3 + \cdots = \frac{1}{r-1},$$

converges for $|r| < 1$. The partial sums are simply

$$
\begin{aligned}
S(1) &= 1 = \frac{1-r}{1-r}, \\
S(2) &= 1+r = \frac{1-r^2}{1-r}, \\
S(3) &= 1+r+r^2 = \frac{1-r^3}{1-r}, \\
&\vdots \\
S(N) &= 1+r+r^2+\cdots+r^N = \frac{1-r^{N+1}}{1-r},
\end{aligned}
$$

and thus

$$
\begin{aligned}
\lim_{N\to\infty} S(N) &= \lim_{N\to\infty} \frac{1-r^{N+1}}{1-r} \\
&= \frac{1}{1-r},
\end{aligned}
$$

when $|r| < 1$ since in this case $r^{N+1} \to 0$ as $N \to \infty$. When $|r| > 1$ we see that the above series is divergent. Notice that this argument gives a rigorous solution to the problem posed in §18.1.

Question: What is a very familiar infinite series?

The answer to this question is $\pi = 3.14159\ldots$ When we think about what the decimal expansion really means we see that in fact

$$\pi = 3 + \frac{1}{10} + \frac{4}{100} + \frac{1}{1000} + \frac{5}{10000} + \frac{9}{100000} + \cdots$$

which is clearly an infinite series. In fact every irrational number is an infinite series. It is important to remember that a CAS *cannot* store an infinite sequence of numbers $a_1, a_2, a_3, \ldots$. The CAS can work with irrational numbers and approximate them to any desired degree of accuracy (which amounts to computing a partial sum).

The most pivotal question one can ask about an infinite series is:

Question: How can one determine if an infinite series is convergent or divergent?

Before proceeding to attack this question head on (the answer to this question will encompass the remainder of this chapter) let us convince ourselves that every infinite decimal expansion (such as 398.213715023...) of a real number is a convergent infinite series.

Consider the expansion $a_0.a_1a_2a_3a_4\dots$ where a_0 is any integer and a_n (for $n = 1, 2, 3, \dots$) is an integer between 0 and 9. Then, by definition,

$$a_0.a_1a_2a_3a_4\dots = a_0 + \frac{a_1}{10} + \frac{a_2}{100} + \frac{a_3}{1000} + \frac{a_4}{10000} + \cdots$$

For $N \geq 1$, the Nth partial sum $S(N)$ for this series is

$$\begin{aligned} S(N) &= a_0 + \frac{a_1}{10} + \frac{a_2}{10^2} + \cdots \frac{a_N}{10^N} \\ &\leq a_0 + \frac{9}{10} + \frac{9}{10^2} + \cdots \frac{9}{10^N} \\ &< a_0 + 1, \end{aligned}$$

since the geometric series $\frac{9}{10} + \frac{9}{10^2} + \frac{9}{10^3} + \cdots$ converges to 1. We conclude that the sequence of partial sums is nondecreasing and each term is strictly less that $a_0 + 1$. In general, any sequence meeting these specifications must have a limit, and thus so does ours.

18.3 The Integral Test

Let $f(x)$ be a positive decreasing function for $x \geq 1$. Then necessarily

$$f(1) > f(2) > f(3) > \cdots$$

and we can state our first convergence test.

The Integral Test: *If $f(x)$ is a positive decreasing function which is integrable for $x \geq 1$ then*

$$\sum_{n=1}^{\infty} f(n)$$

converges if an only if

$$\int_1^{\infty} f(x)\, dx$$

converges.

Proof: Suppose the integral converges. Then in the figure below

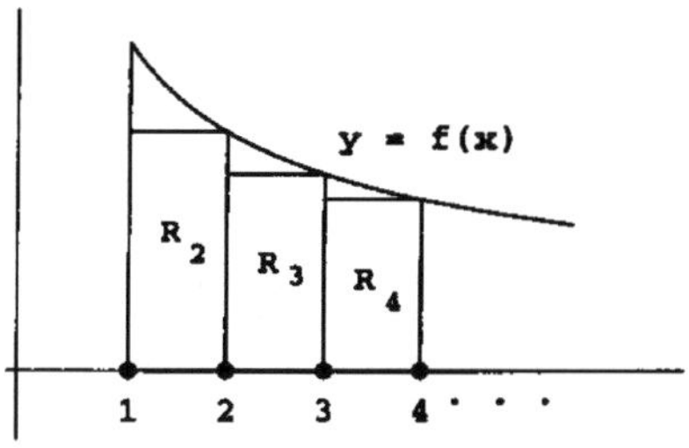

the area of each rectangle $R_2, R_3, R_4, \ldots$ is given by

$$\begin{aligned} \text{Area}(R_2) &= f(2) \cdot (2-1) = f(2) \\ \text{Area}(R_3) &= f(3) \cdot (3-2) = f(3) \\ \text{Area}(R_4) &= f(4) \cdot (4-3) = f(4) \\ &\vdots \end{aligned}$$

Since the sum of all these areas is strictly less than the area between the curve and the x–axis (for $x \geq 1$) we immediately see that

$$\sum_{n=2}^{\infty} f(n) < \int_1^{\infty} f(x)\,dx.$$

Conversely, if $\sum_{n=1}^{\infty} f(n)$ converges then we see in the following figure

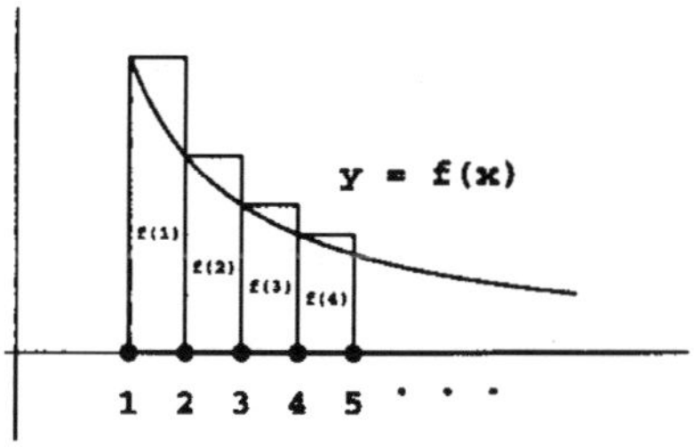

that the area between the curve and the x–axis (for $x \geq 1$) is smaller than the sum of the areas of all the rectangles, and hence

$$\int_1^{\infty} f(x)\,dx < \sum_{n=1}^{\infty} f(n).$$

Example [18.4] For which values of p is the series

$$\sum_{n=1}^{\infty} \frac{1}{n^p}$$

convergent?

By the integral test the series $\sum_{n=1}^{\infty} \frac{1}{n^p}$ converges if and only if $\int_1^{\infty} dx/x^p$ converges. In considering this integral there are two cases:

Case 1: When $p = 1$ the integral becomes

$$\int_1^{\infty} \frac{dx}{x} = \lim_{B\to\infty} \int_1^{B} \frac{dx}{x} = \lim_{B\to\infty} \ln(B) = \infty,$$

and hence $\sum_{n=1}^{\infty} \frac{1}{n}$ diverges.

Case 2: When $p \neq 1$ the integral becomes

$$\begin{aligned}\int_1^{\infty} \frac{dx}{x^p} &= \lim_{B\to\infty} \int_1^{B} \frac{dx}{x^p} \\ &= \lim_{B\to\infty} \frac{x^{1-p}}{1-p}\bigg]_1^B \\ &= \lim_{B\to\infty} \left(\frac{B^{1-p}}{1-p} - \frac{1}{1-p}\right).\end{aligned}$$

Thus $\sum_{n=1}^{\infty} \frac{1}{n^p}$ converges if $p > 1$ and diverges when $p \leq 1$.

Example [18.5] Does the series

$$\sum_{n=1}^{\infty} \frac{n}{n^2+4}$$

converge?

We must check whether the integral

$$\int_1^{\infty} \frac{x}{x^2+4}\, dx$$

converges or not. This integral can be evaluated by using the substitution

$u = x^2 + 4$, $du = 2x\,dx$ as follows:

$$\begin{aligned}
\int_1^\infty \frac{x}{x^2+4}\,dx &= \lim_{B\to\infty} \int_1^B \frac{x}{x^2+4}\,dx \\
&= \lim_{B\to\infty} \int_5^{B^2+4} \frac{1}{u}\frac{du}{2} \\
&= \lim_{B\to\infty} \left(\frac{1}{2} \ln(u) \Big]_5^{B^2+4} \right) \\
&= \lim_{B\to\infty} \frac{1}{2} \Big(\ln(B^2+4) - \ln(5) \Big) \\
&= \infty.
\end{aligned}$$

Hence, the sum $\sum_{n=1}^{\infty} \frac{n}{n^2+4}$ diverges.

Exercises for §18.3

Use the integral test to determine which of the following infinite series converge. Confirm your answers by summing many terms of the series with your CAS.

(1) $\sum_{n=1}^{\infty} \frac{1}{n^{2/3}}$

(2) $\sum_{n=1}^{\infty} \frac{15}{n^{1.01}}$

(3) $\sum_{n=1}^{\infty} \frac{2}{\sqrt{n}}$

(4) $\sum_{n=1}^{\infty} \frac{2}{5n+1}$

(5) $\sum_{n=1}^{\infty} \frac{n}{n^3+1}$

(6) $\sum_{n=1}^{\infty} \frac{n}{2n^2+3}$

(7) $\sum_{n=1}^{\infty} ne^{-n}$

(8) $\sum_{n=1}^{\infty} \ln(n)$

(9) $\sum_{n=1}^{\infty} \frac{1}{1+n^2}$

(10) $\sum_{n=1}^{\infty} \frac{1}{n^2+3n+2}$

(11) $\sum_{n=1}^{\infty} \frac{2n+3}{n^2+3n+2}$

(12) $\sum_{n=1}^{\infty} (\sin(n))^2$

(13) Let $S(N) = \sum_{n=1}^{N} \frac{1}{n}$. Show that $\ln(N+1) < S(N) < 1 + \ln(N)$.

(14) Show that $\sum_{n=1}^{10^{10}} \frac{1}{n} < 25$.

18.4 Other Tests for Convergence

One of the simplest and most powerful ways to test for convergence is to compare one series with another.

The Comparison Test: *If $\sum_{n=1}^{\infty} a_n$ and $\sum_{n=1}^{\infty} b_n$ are infinite series such that $0 < a_n \le b_n$ for all $n = 1, 2, 3, \ldots$ then*

$$\sum_{n=1}^{\infty} a_n \le \sum_{n=1}^{\infty} b_n. \tag{18.4}$$

and we deduce (from inequality (18.4)) that:

(i) if the series $\sum_{n=1}^{\infty} b_n$ converges to some number S then the series $\sum_{n=1}^{\infty} a_n$ converges to some number $\le S$, and

(ii) if $\sum_{n=1}^{\infty} a_n$ diverges then, necessarily, so does $\sum_{n=1}^{\infty} b_n$.

Example [18.6] Show that the series

$$\sum_{n=1}^{\infty} \frac{1}{n^2+n+3}$$

converges.

Since

$$\frac{1}{n^2+n+3} < \frac{1}{n^2}$$

for $n \ge 1$, the comparison test tells us that

$$\sum_{n=1}^{\infty} \frac{1}{n^2+n+3} < \sum_{n=1}^{\infty} \frac{1}{n^2},$$

and we know the latter sum converges by the integral test.

Example [18.7] Show that

$$\sum_{n=1}^{\infty} \frac{1}{n^2+n+3} < 2.$$

The above discussion tells that we need only show $\sum_{n=1}^{\infty} \frac{1}{n^2} \le 2$. By the integral test we know that

$$\sum_{n=2}^{\infty} \frac{1}{n^2} \le \int_1^{\infty} \frac{1}{x^2} = -\frac{1}{x}\Big]_1^{\infty} = 1,$$

and hence

$$\sum_{n=1}^{\infty} \frac{1}{n^2} = 1 + \sum_{n=2}^{\infty} \frac{1}{n^2} \leq 2.$$

Example [18.8] Does

$$\sum_{n=1}^{\infty} \frac{n}{n^4 + 3}$$

converge or diverge?

In this case we can compare our series with the series

$$\sum_{n=1}^{\infty} \frac{n}{n^4} = \sum_{n=1}^{\infty} \frac{1}{n^3}$$

which we know converges. In fact

$$\frac{n}{n^4 + 3} < \frac{n}{n^4},$$

since $n^4 + 3 > n^4$.

One of the most useful convergence tests known is the **ratio test** which we now state.

The Ratio Test: *Let* $a_n > 0$ *for* $n = 1, 2, 3, \ldots$ *If*

$$\lim_{n \to \infty} \frac{a_{n+1}}{a_n} = L$$

exists, then

$$\sum_{n=1}^{\infty} a_n \ \textit{converges if} \ L < 1,$$

and

$$\sum_{n=1}^{\infty} a_n \ \textit{diverges if} \ L > 1.$$

If $L = 1$ *the test provides no information.*

Example [18.9] Does the series

$$\sum_{n=1}^{\infty} \frac{n^7}{2^n}$$

converge?

Here $a_n = \frac{n^7}{2^n}$, $a_{n+1} = \frac{(n+1)^7}{2^{n+1}}$, and the quotient is given by

$$\begin{aligned}\frac{a_{n+1}}{a_n} &= \frac{(n+1)^7/2^{n+1}}{n^7/2^n} \\ &= \frac{(n+1)^7}{n^7} \cdot \frac{2^n}{2^{n+1}} \\ &= \left(\frac{n+1}{n}\right)^7 \cdot \frac{1}{2} \\ &= \left(1+\frac{1}{n}\right)^7 \cdot \frac{1}{2}.\end{aligned}$$

Consequently

$$\begin{aligned}\lim_{n\to\infty} \frac{a_{n+1}}{a_n} &= \lim_{n\to\infty} \left(1+\frac{1}{n}\right)^7 \cdot \frac{1}{2} \\ &= \frac{1}{2},\end{aligned}$$

and since the limit is less than 1 the ratio test tells us that the series converges.

Example [18.10] Use the ratio test to determine whether or not the sum

$$\sum_{n=1}^{\infty} \frac{n!}{n^n}$$

converges.

Despite the apparent complexity of this problem observe that

$$a_n = \frac{n!}{n^n}, \quad a_{n+1} = \frac{(n+1)!}{(n+1)^{(n+1)}},$$

and hence

$$\begin{aligned}
\frac{a_{n+1}}{a_n} &= \frac{(n+1)!/(n+1)^{(n+1)}}{n!/n^n} \\
&= \frac{(n+1)!}{n!} \cdot \frac{n^n}{(n+1)^{(n+1)}} \\
&= (n+1) \cdot \frac{n^n}{(n+1)^n \cdot (n+1)} \\
&= \left(\frac{n}{n+1}\right)^n.
\end{aligned}$$

But

$$\frac{n}{n+1} = \frac{1}{1+\frac{1}{n}},$$

and thus we conclude

$$\frac{a_{n+1}}{a_n} = \frac{1}{\left(1+\frac{1}{n}\right)^n}.$$

It follows that (see Proposition [8.7])

$$\begin{aligned}
\lim_{n\to\infty} \frac{a_{n+1}}{a_n} &= \lim_{n\to\infty} \frac{1}{\left(1+\frac{1}{n}\right)^n} \\
&= \frac{1}{e},
\end{aligned}$$

and since $\frac{1}{e} < 1$, the series converges.

Question: Can we use the ratio test to determine whether or not

$$\sum_{n=1}^{\infty} \frac{1}{n^2}$$

converges?

Answer: In this example $a_n = 1/n^2$, $a_{n+1} = 1/(n+1)^2$, and

$$\lim_{n\to\infty} \frac{a_{n+1}}{a_n} = \lim_{n\to\infty} \left(\frac{n+1}{n}\right)^2 = \lim_{n\to\infty} \left(1+\frac{1}{n}\right)^2 = 1.$$

Thus the ratio test does not provide any useful information in this case and we must look for some other method to analyze this series (as we saw the integral test is appropriate).

Proof of the Ratio Test:

We present the proof when

$$\lim_{n\to\infty} \frac{a_{n+1}}{a_n} = L \tag{18.5}$$

and $L < 1$. The case of $L > 1$ is similar and left to our reader. We begin by choosing a number ℓ such that $L < \ell < 1$. Equation (18.5) tells us that for large n the quotient a_{n+1}/a_n will be very close to L. We can, therefore, choose a number N so large that for all $n \geq N$

$$\frac{a_{n+1}}{a_n} < \ell,$$

which can be rewritten as

$$a_{n+1} < \ell a_n. \tag{18.6}$$

Since equation (18.6) holds for all $n \geq N$, it holds for $n = N$, and thus

$$a_{N+1} < \ell a_N.$$

Similarly, when we let $n = N + 1$ (18.6) yields

$$a_{N+2} < \ell a_{N+1},$$

when we let $n = N + 2$ we get

$$a_{N+3} < \ell a_{N+2},$$

and so on.

If we combine the first two inequalities $a_{N+1} < \ell a_N$ and $a_{N+2} < \ell a_{N+1}$, we obtain

$$a_{N+2} < \ell^2 a_N.$$

Combining this inequality with $a_{N+3} < \ell a_{N+2}$ we obtain

$$a_{N+3} < \ell^3 a_N.$$

Continuing in this manner we obtain

$$\begin{aligned} a_{N+1} &< \ell a_N \\ a_{N+2} &< \ell^2 a_N \\ a_{N+3} &< \ell^3 a_N \\ &\vdots \end{aligned}$$

Summing both sides of this list yields the following crucial inequality:

$$a_N + a_{N+1} + a_{N+2} + \cdots < (1+\ell+\ell^2+\cdots)\cdot a_N.$$

Notice that the right–hand side of this inequality contains the geometric series $1+\ell+\ell^2+\cdots$. Since this geometric series converges only if $\ell < 1$ we can deduce that

$$\sum_{n=1}^{\infty} a_n = \left(\sum_{n=1}^{N-1} a_n\right) + (a_N + a_{N+1} + a_{N+2} + \cdots)$$

$$< \left(\sum_{n=1}^{N-1} a_n\right) + (1+\ell+\ell^2+\cdots)\cdot a_N$$

converges only if $\ell < 1$ to complete our proof.

Exercises for §18.4

Determine which of the following series converge. Confirm your answers by summing many terms of the series with your CAS.

(1) $\sum_{n=1}^{\infty} \frac{2}{n^4+2n}$

(2) $\sum_{n=1}^{\infty} \frac{1}{\sqrt{n}+2}$

(3) $\sum_{n=1}^{\infty} \frac{\left(\cos\left(\frac{n\pi}{8}\right)\right)^2}{n^2+1}$

(4) $\sum_{n=1}^{\infty} \left(\sin\left(\frac{n^2}{3}\right)\right)^2 \cdot \left(\frac{3}{4}\right)^n$

(5) $\sum_{n=1}^{\infty} \frac{\sqrt{n}}{2n-3}$

(6) $\sum_{n=1}^{\infty} \frac{4^n}{n^{100}}$

(7) $\sum_{n=1}^{\infty} \frac{n^{100}}{4^n}$

(8) $\sum_{n=1}^{\infty} \frac{n^3}{n!}$

(9) $\sum_{n=1}^{\infty} \frac{n!}{5^n}$

(10) $\sum_{n=1}^{\infty} \frac{5^n}{n!}$

(11) $\sum_{n=1}^{\infty} \frac{2^n \cdot n^3}{n!}$

(12) $\sum_{n=1}^{\infty} \frac{\left(\cos(\sqrt{n}\cdot\pi)\right)^2 \cdot 3^n}{n!}$

(13) $\sum_{n=1}^{\infty} \frac{2\cdot 4\cdot 6\cdots(2n)}{n!}$

(14) $\sum_{n=1}^{\infty} \frac{n^n}{n!}$

(15) $\sum_{n=1}^{\infty} \frac{n^{100}}{(1.01)^n}$

18.5 Infinite Series with Positive and Negative Terms

We now focus our attention on infinite series $\sum_{n=1}^{\infty} a_n$ where the terms a_n may be either positive or negative. Some simple examples of such series are

$$1 - \frac{1}{4} + \frac{1}{9} - \frac{1}{16} + \frac{1}{25} \cdots \tag{18.7}$$

and

$$1 - \frac{1}{2} + \frac{1}{3} - \frac{1}{4} + \frac{1}{5} \cdots \tag{18.8}$$

Definition: *A series $\sum_{n=1}^{\infty} a_n$ is called **absolutely convergent** if the series $\sum_{n=1}^{\infty} |a_n|$ converges.*

Example [18.11]: Determine if the series (18.7) and (18.8) are absolutely convergent.

Since $1 + \frac{1}{2} + \frac{1}{3} + \frac{1}{4} + \frac{1}{5} \cdots$ diverges (we saw this using the integral test), it follows that $1 - \frac{1}{2} + \frac{1}{3} - \frac{1}{4} + \frac{1}{5} \cdots$ is not absolutely convergent. On the other hand, we know that $1 + \frac{1}{4} + \frac{1}{9} + \frac{1}{16} + \frac{1}{25} \cdots$ does converge and hence $1 - \frac{1}{4} + \frac{1}{9} - \frac{1}{16} + \frac{1}{25} \cdots$ is absolutely convergent.

The motivation for studying absolute convergence can be seen in the following theorem.

Theorem [18.12] *If $\sum_{n=1}^{\infty} |a_n|$ converges then $\sum_{n=1}^{\infty} a_n$ converges, i.e., absolute convergence implies ordinary convergence.*

Proof: At first glance the comparison test appears to be the appropriate tool. The only obstacle here is that the comparison test applies only to series consisting of positive terms. In order to maneuver ourselves into a position where the comparison test does apply we define the following series whose terms are positive. Let

$$\sum_{n=1}^{\infty} a_n^+, \quad \sum_{n=1}^{\infty} a_n^-,$$

be infinite series where

$$a_n^+ = \begin{cases} a_n, & \text{if } a_n \geq 0 \\ 0, & \text{if } a_n < 0, \end{cases}$$

and

$$a_n^- = \begin{cases} 0, & \text{if } a_n \geq 0 \\ -a_n, & \text{if } a_n < 0. \end{cases}$$

By definition $a_n^+, a_n^- \leq |a_n|$, and thus the series $\sum_{n=1}^{\infty} a_n^+, \sum_{n=1}^{\infty} a_n^-$ must both converge by comparison with $\sum_{n=1}^{\infty} |a_n|$. We conclude the proof by observing that

$$a_n = a_n^+ - a_n^-,$$

and hence the series $\sum_{n=1}^{\infty} a_n$ converges to

$$\left(\sum_{n=1}^{\infty} a_n^+\right) - \left(\sum_{n=1}^{\infty} a_n^-\right).$$

Theorem [18.12] allows us to reformulate the ratio test so that it applies to series with both positive and negative terms.

The General Ratio Test: *Let $\sum_{n=1}^{\infty} a_n$ be a general series such that*

$$\lim_{n\to\infty} \left| \frac{a_{n+1}}{a_n} \right| = L$$

exists. If $L < 1$ then the series converges absolutely, if $L > 1$ the series diverges, and if $L = 1$ the test provides no information.

Example [18.13] Does the series

$$\sum_{n=1}^{\infty} \frac{(-1)^{n+1} \cdot n^2}{3^n}$$

converge absolutely?

We compute the required limit:

$$\begin{aligned}
\lim_{n\to\infty}\left|\frac{a_{n+1}}{a_n}\right| &= \lim_{n\to\infty}\frac{|(-1)^{n+2}(n+1)^2/3^{n+1}|}{|(-1)^n n^2/3^n|} \\
&= \lim_{n\to\infty}\frac{\left(1+\frac{1}{n}\right)^2}{3} \\
&= \frac{1}{3}.
\end{aligned}$$

Thus we conclude that the series does converge absolutely.

We now turn our attention to the series:

$$1 - \frac{1}{2} + \frac{1}{3} - \frac{1}{4} + \frac{1}{5}\cdots$$

Does this series converge? To attack this question we begin with some experimentation on our CAS. Let

$$S(N) = \sum_{n=1}^{N}(-1)^{n+1}\cdot\frac{1}{n}$$

denote the sum of the the first N terms of this series. Our CAS tells us that

$$\begin{aligned}
S(1) &= 1 \\
S(2) &= 0.5 \\
S(5) &= 0.78333\ldots \\
S(10) &= 0.64563\ldots \\
S(100) &= 0.68817\ldots \\
S(1000) &= 0.69265\ldots
\end{aligned}$$

This sequence does seem to be converging. Recalling that this series does not converge absolutely, we are motivated to make the following definition.

Definition: *A series $\sum_{n=1}^{\infty} a_n$ is said to be* ***conditionally convergent*** *if $\sum_{n=1}^{\infty} a_n$ converges but $\sum_{n=1}^{\infty} |a_n|$ does not converge.*

Question: Is the series $1 - \frac{1}{2} + \frac{1}{3} - \frac{1}{4} + \frac{1}{5}\cdots$ conditionally convergent?

The answer turns out to concur with our experimentally derived intuition: the series does converge. Since the signs of the terms in the above series alternate (from plus to minus), these series are called **alternating series**. The following remarkable theorem will allow us to analyze a wide range of alternating series which are not absolutely convergent.

Theorem [18.14] *An alternating series* $\sum_{n=1}^{\infty}(-1)^n a_n$ *(where the terms satisfy* $a_n \geq 0$*) converges if*

$$a_1 > a_2 > a_3 > \cdots \to 0,$$

i.e., the terms in the series are decreasing and approach zero in the limit as $n \to \infty$.

Proof: We begin this proof by again considering the partial sum

$$S(N) = \sum_{n=1}^{N}(-1)^{n+1}a_n.$$

We claim that

$$0 < S(2) < S(4) < S(6) < S(8) < \cdots \qquad \textbf{(I)}$$

and

$$S(2m) < a_1 \quad \text{for any } m = 1, 2, 3, \ldots \qquad \textbf{(II)}$$

To verify **I** note that

$$S(2m) = (a_1 - a_2) + (a_3 - a_4) + \cdots + (a_{2m-1} - a_{2m}).$$

Since the sequence of numbers a_n is decreasing to 0 as n gets larger and larger, we see that each of the terms

$$(a_1 - a_2), \ (a_3 - a_4), \ (a_5 - a_6), \ldots$$

are positive, and hence **I** holds. On the other hand, we may rearrange the sum to obtain **II**:

$$\begin{aligned} S(2m) &= a_1 - (a_2 - a_3) - (a_4 - a_3) - \cdots - (a_{2m-1} - a_{2m-2}) - a_{2m} \\ &< a_1. \end{aligned}$$

It follows that the partial sums $S(2m)$ form an increasing sequence each of whose terms are bounded above by a_1 (i.e., are less than a_1). They must, therefore, approach a limit s. We complete the proof by considering the difference

$$S(2m+1) \;-\; S(2m) \;=\; (-1)^{2m+2}a_{2m+1}.$$

Since this difference tends to 0 as $n \to \infty$ we conclude that the sequence of partial sums $S(2m+1)$ must also converge to the limit s.

Example [18.15] Does the series

$$\sum_{n=1}^{\infty}(-1)^n \cdot \frac{1}{\sqrt{n}}$$

converge?

Since

$$1 > \frac{1}{\sqrt{2}} > \frac{1}{\sqrt{3}} > \frac{1}{\sqrt{4}} > \cdots \to 0,$$

the series is conditionally convergent.

Remark: The convergence of the series in Example [18.15] is not at all rapid. This can be seen via a CAS calculation of the sum of the first N terms of the series (denoted $S(N)$ as usual): $S(10) = 0.45073\ldots$, $S(100) = 0.55502\ldots$, $S(1000) = 0.58909\ldots$

Exercises for §18.5

Determine which of the following series converge absolutely or conditionally. Confirm your answers by summing many terms of the series with your CAS. Indicate which convergence test you are using.

(1) $\sum_{n=1}^{\infty} 2\cdot\left(\frac{-5}{8}\right)^n$

(2) $\sum_{n=1}^{\infty} \frac{(-1)^n}{n^{\frac{1}{4}}}$

(3) $\sum_{n=1}^{\infty} \frac{\sqrt{n}}{n+4}$

(4) $\sum_{n=1}^{\infty} \frac{\cos(n\pi)\cdot n^5}{3^n}$

(5) $\sum_{n=1}^{\infty} \frac{(-1)^n\cdot 3^n}{n!}$

(6) $\sum_{n=1}^{\infty} \frac{(-1)^n\cdot e^n}{n^{100}}$

(7) $\sum_{n=1}^{\infty} \frac{15n^{\frac{1}{4}}}{n^2+1}$

(8) $\sum_{n=1}^{\infty} \frac{(-1)^n\cdot n}{n^{\frac{3}{2}}+2}$

(9) $\sum_{n=1}^{\infty} \frac{(-2)^n}{n^{15}}$

(10) $\sum_{n=1}^{\infty} \frac{\cos(n\pi)}{\ln(n)}$

(11) $\sum_{n=1}^{\infty} \frac{\sin\left(\frac{n\pi}{7}\right)}{n^4}$

(12) $\sum_{n=1}^{\infty} \frac{(3n)^n}{n^{3n}}$

(13) $\sum_{n=1}^{\infty} \frac{(-5)^n}{3^{2n}}$

(14) $\sum_{n=1}^{\infty} \frac{1}{2^{\ln(n)}}$

(15) $\sum_{n=1}^{\infty} \frac{(-1)^n}{1+\frac{1}{n}}$

18.6 Power Series

A **power series** is an infinite series of the form

$$\sum_{n=1}^{\infty} a_n x^n = a_0 + a_1 x + a_2 x^2 + a_3 x^3 + \cdots$$

where x is a variable and the numbers a_n are fixed real numbers.

Example [18.16] The infinite sum

$$1 + x + x^2 + \cdots = \frac{1}{1-x}$$

is a power series.

Question: For which values of x does a general power series $\sum_{n=0}^{\infty} a_n x^n$ converge absolutely?

Answer: While there is no one all-encompassing answer to this question the general ratio test will usually provide an answer. If the limit

$$\lim_{n\to\infty} \frac{|a_{n+1}x^{n+1}|}{|a_n x^n|} = \lim_{n\to\infty} \left|\frac{a_{n+1}}{a_n}\right| \cdot |x|,$$

exists then we will have found an interval of x–values where the power series converges.

Example [18.17] For which values of x does the power series

$$\sum_{n=0}^{\infty} (-1)^n \frac{x^n}{3^n}$$

converge?

When we compute the limit appearing in the general ratio test we obtain

$$\lim_{n\to\infty} \left|(-1)^{n+1} \frac{x^{n+1}}{3^{n+1}}\right| \Big/ \left|\frac{(-1)^n x^n}{3^n}\right| = \lim_{n\to\infty} \frac{|x|}{3} = \frac{|x|}{3}.$$

Thus the series converges when $|x| < 3$.

Definition: *If a power series $\sum_{n=0}^{\infty} a_n x^n$ converges absolutely for $|x| < R$ then we define R to be the **radius of convergence** of the power series.*

Warning: There are power series, for example $\sum_{n=0}^{\infty} n!x^n$, which converge only when $x = 0$. In this case we say the radius of convergence is 0.

Example [18.18] Show that the radius of convergence of $\sum_{n=0}^{\infty} n!x^n$ is zero.

Again we use the general ratio test:

$$\lim_{n\to\infty}\left|\frac{(n+1)!x^{n+1}}{n!x^n}\right| = \lim_{n\to\infty}|(n+1)x| = \begin{cases} 0 & \text{if } x = 0 \\ \infty & \text{otherwise.} \end{cases}$$

Hence the series converges only when $x = 0$.

Consider now a general power series

$$f(x) = \sum_{n=0}^{\infty} a_n x^n,$$

with radius of convergence $R > 0$. Then for each x such that $|x| < R$, the infinite series converges to the value $f(x)$ and defines a function $f(x)$. When we differentiate $f(x)$ (term by term) we obtain

$$f(x) = a_0 + a_1 x + a_2 x^2 + a_3 x^3 + \cdots$$

$$f'(x) = a_1 + 2a_2 x + 3a_3 x^2 + \cdots$$

or more compactly

$$f'(x) = \sum_{n=1}^{\infty} n a_n x^{n-1},$$

(It should be noted that we are omitting the detailed justification of taking the derivative of the series). It is natural to ask when $f'(x)$ will converge.

Theorem [18.19] *Consider the power series $f(x) = \sum_{n=1}^{\infty} a_n x^n$ and its derivative $f'(x) = \sum_{n=1}^{\infty} n a_n x^{n-1}$. If $f(x)$ has radius of convergence R then $f'(x)$ also has radius of convergence R.*

Proof: This theorem will follow from the general ratio test together with the small observation

$$\lim_{n\to\infty}\left(\frac{n+1}{n}\right) = \lim_{n\to\infty}\left(1 + \frac{1}{n}\right) = 1.$$

Let R' be the radius of convergence of $f'(x)$. We may compute R' by calculating

$$\begin{aligned}\lim_{n\to\infty}\left|\frac{(n+1)a_{n+1}x^n}{na_nx^{n-1}}\right| &= \lim_{n\to\infty}\left(\frac{n+1}{n}\right)\cdot\left|\frac{a_{n+1}}{a_n}\right|\cdot|x| \\ &= \lim_{n\to\infty}\left|\frac{a_{n+1}}{a_n}x\right|.\end{aligned}$$

But this last limit is the same limit which appears when we apply the general ratio test to the series $f(x)$. Hence $R = R'$.

Example [18.20] Find the power series for the function $\frac{1}{(1-x)^2}$. What is the radius of convergence?

If we first let $f(x) = \frac{1}{1-x}$ then we know

$$f(x) = 1 + x + x^2 + x^3 + \cdots$$

and the radius of convergence is 1. Since $f'(x) = \frac{1}{(1-x)^2}$ we have found a series expansion for our function,

$$\frac{1}{(1-x)^2} = 1 + 2x + 3x^2 + 4x^3 + \cdots$$

Theorem [18.19] dictates that the radius of convergence is again 1.

Finally, if

$$f(x) = \sum_{n=0}^{\infty} a_n x^n$$

is a power series, we may integrate it term by term to obtain (again omitting some details of the proof)

$$\int f(x)\,dx = \left(\sum_{n=0}^{\infty} a_n \frac{x^{n+1}}{n+1}\right) + C,$$

where C is the constant of integration. Alternately, we write

$$\begin{aligned}f(x) &= a_0 + a_1x + a_2x^2 + a_3x^3 + \cdots \\ \int f(x)\,dx &= C + a_0x + \frac{a_1x^2}{2} + \frac{a_3x^3}{3} + \frac{a_4x^4}{4} + \cdots\end{aligned}$$

Notice that the radius of convergence of the new power series is the same as for the original series (this is analogous to Theorem [18.19]).

Example [18.21] Integrate $\frac{1}{x-1}$ to obtain the power series for $\ln(1-x)$.

As before we have

$$\frac{1}{1-x} = 1 + x + x^2 + x^3 + \cdots$$

and integrating gives us

$$\int \frac{1}{1-x}\, dx = C + x + \frac{x^2}{2} + \frac{x^3}{3} + \cdots$$

Since $\int \frac{1}{1-x}\, dx = -\ln(1-x) + C$, we obtain

$$-\ln(1-x) = C + x + \frac{x^2}{2} + \frac{x^3}{3} + \cdots$$

We complete the example by evaluating both sides of the equation at $x = 0$ to see that

$$-\ln(1) = 0 = C,$$

and hence

$$\ln(1-x) = -x - \frac{x^2}{2} - \frac{x^3}{3} - \cdots$$

Exercises for §18.6

Find the radius of convergence of each of the following power series.

(1) $\sum_{n=0}^{\infty} (2x)^n$

(2) $\sum_{n=0}^{\infty} \frac{x^{3n}}{n+1}$

(3) $\sum_{n=0}^{\infty} n^2 x^n$

(4) $\sum_{n=0}^{\infty} \frac{x^n}{n!}$

(5) $\sum_{n=1}^{\infty} \frac{x^n}{n^{1000}}$

(6) $\sum_{n=1}^{\infty} \frac{(3x)^n}{n}$

(7) $\sum_{n=1}^{\infty} \frac{(x-4)^n}{\sqrt{n}}$

(8) $\sum_{n=1}^{\infty} \frac{(2x-1)^n}{n^3}$

(9) $\sum_{n=1}^{\infty} \frac{n! x^n}{2^n}$

Each of the following power series converges for $|x| < 1$. Determine if the series diverges, converges absolutely, or converges conditionally at the points $x = \pm 1$.

(10) $\sum_{n=1}^{\infty} \frac{x^n}{n^2}$

(11) $\sum_{n=1}^{\infty} \frac{x^n}{\sqrt{n}}$

(12) $\sum_{n=1}^{\infty} n x^n$

(13) $\sum_{n=1}^{\infty} \frac{n^2 x^2}{2^n}$

(14) $\sum_{n=1}^{\infty} \frac{x^n}{n^{1.01}}$

(15) $\sum_{n=2}^{\infty} \frac{x^n}{\ln(n)}$

(16) The power series for $\frac{1}{1-x^2}$ is given by $\frac{1}{1-x^2} = \sum_{n=0}^{\infty} x^{2n}$. Compute $\frac{d}{dx}\frac{1}{1-x^2}$ and $\int \frac{x}{1-x^2}\,dx$ and find their power series.

(17) Find the power series for $\frac{1}{1+x^2}$. Compute $\frac{d}{dx}\frac{1}{1+x^2}$ and find its power series. Find the power series for $\arctan(x)$.

(18) Find the power series for $\frac{1}{1-x^4}$. Compute $\frac{d}{dx}\frac{1}{1-x^4}$ and find its power series. Find the power series for $\int \frac{dx}{1-x^4}$.

Additional exercises for Chapter XVIII

---

(1) Show that if $\lim_{N\to\infty}\sum_{n=1}^{N} a_n$ and $\lim_{N\to\infty}\sum_{n=1}^{N} b_n$ both exist, then

$$\lim_{N\to\infty}\sum_{n=1}^{N}(a_n+b_n)=\lim_{N\to\infty}\sum_{n=1}^{N} a_n+\lim_{N\to\infty}\sum_{n=1}^{N} b_n.$$

(2) A series $\sum_{n=1}^{\infty} a_n$ is called Cauchy if for every $\epsilon \geq 0$ there is a number N such that for all n,m,

if $m, n > N$ then $\lim_{m,n\to\infty}\left|\sum_{i=n}^{m} a_i\right| = 0.$

Prove that a Cauchy series is convergent. Is the converse true?

With the help of your CAS, verify whether the following series are Cauchy

(3) $\sum_{n=1}^{\infty}\frac{1}{n^{1+1/n}}.$

(4) $\sum_{n=1}^{\infty}\frac{n^2}{n^4+1}.$

(5) $\sum_{n=1}^{\infty}\frac{n}{n^2+1}.$

(6) $\sum_{n=1}^{\infty}(-1)^n\frac{\ln(n)}{n}.$

(7) $\sum_{n=2}^{\infty}(-1)^n\frac{1}{\ln(n+1)}.$

(8) An infinite product $\prod_{n=1}^{\infty} b_n$ where $b_n \neq 0$ is convergent if the sequence $p_n = \prod_{i=1}^{n}$ converges and $\lim_{n\to\infty} p_n \neq 0.$

(a) Prove that if $\prod_{n=1}^{\infty}(1+a_n) < \infty$ then $\sum_{n=1}^{\infty}\ln(1+a_n) < \infty.$

(b) for $a_n \geq 0$ prove that $\prod_{n=1}^{\infty}(1+a_n) < \infty$ if and only if $\sum_{n=1}^{\infty} a_n < \infty.$

(9) Is the product $\prod_{n=1}^{\infty}(1-\frac{1}{n\sqrt{n}})$ convergent?

Find the sums and the radii of convergence for the following series.

(10) $\cos^2(x)+\cos^4(x)+\cdots+\cos^{2n}(x)+\cdots$

(11) $\frac{1}{2x}+\frac{1}{2^2x^2}+\cdots+\frac{1}{2^nx^n}+\cdots$

(12) $x - \frac{x^2}{2} + \frac{x^3}{3} - \frac{x^4}{4} + \cdots$

(13) $3 + 18x + \cdots + 3^n n x^{n-1} + \cdots$

Find the radius of convergence of the following series.

(14) $\sum_{n=1}^{\infty} n^2 (\cos(x))^n.$

(15) $\sum_{n=1}^{\infty} (-1)^n \frac{x^{2n}}{(2n)!}.$

(16) $\sum_{n=1}^{\infty} (-1)^{n-1} \frac{x^{2n-1}}{(2n-1)!}.$

(17) $\sum_{n=1}^{\infty} (\cos(x))^n n^2.$

(18) $1 + \sum_{n=1}^{\infty} \frac{(\ln(x))^n}{n!}.$

Chapter XIX

Taylor Series

19.1 The Tangent Line Approximation

Let $y = f(x)$ be a differentiable function on an interval $a \leq x \leq b$. The tangent line at $x = a$ has slope $f'(a)$ and the equation of the tangent line is given by

$$y = f'(a)(x - a) + f(a). \tag{19.1}$$

Since the tangent line is very close to the curve for values of x near a, we may use it to approximate our function in the following manner. Let $x = a + h$ where h is small.

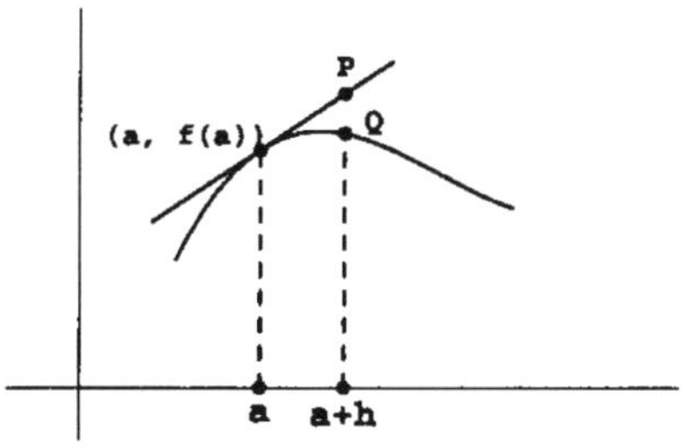

The point P depicted above is on the tangent line and its coordinates are $(a + h, f'(a)h + f(a))$. The point Q is on the curve itself and has coordinates $(a+h, f(a+h))$. By observing that the point P is very close to the point Q when h is small we obtain the approximation

$$f(a+h) \approx f'(a)h + f(a). \tag{19.2}$$

Example [19.1] Use (19.2) to approximate $\sqrt{10}$.

Let $f(x) = \sqrt{x}$ and let $a = 9$ and $h = 1$ in the approximation (19.2). Then $f'(x) = \frac{1}{2}x^{-\frac{1}{2}}$ and $f(9) = 3$, $f'(9) = \frac{1}{6}$. We then see that

$$\sqrt{10} \approx \frac{1}{6} \cdot 1 + 3 = 3.1666\ldots$$

Example [19.2] Approximate $\sqrt[3]{7.5}$ with (19.2).

Let $f(x) = x^{\frac{1}{3}}$. Then $f'(x) = \frac{1}{3}x^{-\frac{2}{3}}$, and by choosing $a = 8$ and $h = -\frac{1}{2}$, $f(8) = 2$, $f'(8) = \frac{1}{12}$, and (19.2) yields

$$\sqrt[3]{7.5} \approx -\frac{1}{12} \cdot \frac{1}{2} + 2 = 1.95833\ldots$$

Example [19.3] Approximate $\cos(33^\circ)$ using the first derivative.

Let $f(x) = \cos(x)$ then $f'(x) = -\sin(x)$. Choosing $a = \frac{\pi}{6}$ and $h = \frac{\pi}{60}$ we see that $f(a) = \frac{\sqrt{3}}{2}$ and $f'(a) = -\sin(\frac{\pi}{6}) = -\frac{1}{2}$. The resulting approximation is

$$\cos(33^\circ) \approx -\frac{1}{2} \cdot \frac{\pi}{60} + \frac{\sqrt{3}}{2}.$$

Exercises for §19.1

Use the tangent line approximation (19.1) to approximate the following functions at the indicated points. Check how accurate your approximations are by computing the actual values with your CAS.

(1)	$\sqrt{15.5}$.	**(5)**	$\tan(34°)$.	**(9)**	$\cos(112°)$.
(2)	$\sqrt[3]{67}$.	**(6)**	$\ln(0.9)$.	**(10)**	$(1.01)^{100}$.
(3)	$\sqrt[5]{1.01}$.	**(7)**	$e^{3.2}$.	**(11)**	$\ln(3)$.
(4)	$\cos(62°)$.	**(8)**	$\sin(87°)$.	**(12)**	$\cosh(\frac{1}{4})$.

19.2 Approximation of Functions by Taylor Polynomials

Let $f(x)$ be a function such as $\sin(x)$, $\ln(x)$, e^x, etc. One of the most striking features of a CAS is its ability to compute $f(x)$ to any desired accuracy. For example, a CAS can compute e^3, $\ln(2)$, or $\sin(\pi/8)$ to N digit accuracy where N is any integer (of course if N is chosen too large, the computation could take years and is infeasible).

Since there are infinitely many functions, and the CAS user can quickly compute almost any function to 100 digit accuracy, it is natural to ask how the CAS accomplishes this. Clearly the CAS designers have utilized a highly efficient approximation algorithm. In this section we will examine a simple algorithm (whose justification we will give later) that will give an approximation of any function $f(x)$ which satisfies the following two properties:

Property 1: *The function $f(x)$ is infinitely differentiable on an interval containing* 0.

Property 2: *The values*

$$f(0), f'(0), f''(0), \dots, f^{(m)}(0),$$

are known and do not grow too rapidly with m as $m \to \infty$.

Remark: The algorithm we present here is based on the idea of replacing $f(x)$ by an infinite series and approximating $f(x)$ by the beginning of this series. The algorithm can only work if the infinite series converges. It is for this reason that the values

$f(0), f'(0), f''(0), \ldots$ cannot be arbitrarily large. A simple growth condition which will assure the success of the algorithm is the assumption that there exists a constant $c \geq 1$ such that $|f^{(m)}(0)| < c^m \cdot m!$ for $m = 1, 2, \ldots$, where we recall the definition $m! = 1 \cdot 2 \cdot \cdots \cdot (m-1) \cdot m$.

Example [19.4] The function $f(x) = e^x$ satisfies properties 1 and 2. In fact e^x can be differentiated as many times as one wants. All of the derivatives of e^x are just e^x, i.e., $f^{(m)}(x) = e^x$ for all $m = 0, 1, 2, \ldots$. In addition, since $f(0) = f'(0) = f''(0) = \cdots = e^0 = 1$, property 2 is quite trivial.

Example [19.5] The polynomial $f(x) = x^3 - 3x^2 + 1$ satisfies properties 1 and 2. We have that $f'(x) = 3x^2 - 6x$, $f''(x) = 6x - 6$, $f'''(x) = 6$, and $f^{(m)} = 0$ for $m > 3$. The values of these derivatives are thus easily computed, $f(0) = 1$, $f'(0) = 0$, $f''(0) = -6$, and $f^{(m)}(0) = 0$ for $m > 3$.

With this large class of functions in hand we proceed to describe the approximation algorithm in three steps.

Step 1. Choose an integer $m > 0$ and compute

$$f(0),\ f'(0),\ \frac{f''(0)}{2!},\ \frac{f^{(3)}(0)}{3!},\ \ldots,\ \frac{f^{(m)}(0)}{m!}.$$

Step 2. Construct the polynomial

$$p_m(x) = f(0) + f'(0)x + \frac{f''(0)}{2!}x^2 + \cdots + \frac{f^{(m)}(0)}{m!}x^m.$$

This is called the **Taylor polynomial of degree** m.

Step 3. For any number x, approximate

$$f(x) \approx p_m(x).$$

Remark: The Taylor polynomial of degree 1 is the tangent line approximation of §19.1.

Example [19.6] We can quickly use our algorithm to approximate the number e. By letting $f(x) = e^x$, we want to approximate $f(1) = e$.

Step 1. Choosing $m = 4$, we immediately see that

$$f(0) = 1, \quad f'(0) = 1, \quad \frac{f''(0)}{2!} = \frac{1}{2}, \quad \frac{f^{(3)}(0)}{3!} = \frac{1}{6}, \quad \frac{f^{(4)}(0)}{4!} = \frac{1}{24}.$$

Step 2. The required Taylor polynomial of degree 4 is

$$p_4(x) = 1 + x + \frac{x^2}{2} + \frac{x^3}{6} + \frac{x^4}{24}.$$

Step 3. Setting $x = 1$ gives us

$$e = f(1) \approx p_4(1) = 1 + 1 + \frac{1}{2} + \frac{1}{6} + \frac{1}{24} \approx 2.708.$$

Notice that we can approximate e^2 using this computation,

$$e^2 = f(2) \approx p_4(2) = 1 + 2 + \frac{4}{2} + \frac{8}{6} + \frac{16}{24} = 7.$$

Example [19.7] To approximate $\ln(1.5)$ we let $f(x) = \ln(1+x)$. Then $f'(x) = \frac{1}{1+x}$, $f''(x) = \frac{-1}{(1+x)^2}$, $f'''(x) = \frac{2}{(1+x)^3}$, and choosing $m = 3$ we can approximate $f(.5) = \ln(1.5)$ with a Taylor polynomial of degree 3.

Step 1. Here $f(0) = \ln(1) = 0, f'(0) = \frac{1}{1+0} = 1$, $f''(0) = \frac{-1}{(1+0)^2} = -1$, and $f'''(0) = \frac{2}{(1+0)^3} = 2$.

Step 2. The Taylor polynomial is given by $p_3(x) = x - \frac{x^2}{2} + \frac{x^3}{3}$.

Step 3. Setting $x = \frac{1}{2}$ we obtain

$$\ln(1.5) \approx p_3\left(\frac{1}{2}\right) = \frac{1}{2} - \frac{\left(\frac{1}{2}\right)^2}{2} + \frac{\left(\frac{1}{2}\right)^3}{3} = \frac{5}{12} = .41666\ldots$$

Example [19.8] As a final example of the algorithm we approximate $\sqrt{2}$. Consider $f(x) = \sqrt{1+x}$. Then $f'(x) = \frac{1}{2}(1+x)^{-\frac{1}{2}}$, $f''(x) = -\frac{1}{4}(1+x)^{-\frac{3}{2}}$, and $f'''(x) = \frac{3}{8}(1+x)^{-\frac{5}{2}}$. Choosing $m = 3$ we can approximate $f(1)$:

Step 1. In this case $f(0) = 1$, $f'(0) = \frac{1}{2}$, $f''(0) = -\frac{1}{4}$, and $f'''(0) = \frac{3}{8}$.

Step 2. The Taylor polynomial of degree 3 is $p_3(x) = 1 + \frac{x}{2} - \frac{x^2}{8} + \frac{x^3}{16}$.

Step 3. Finally we obtain our approximation

$$\sqrt{2} \approx p_3(1) = 1 + \frac{1}{2} - \frac{1}{8} + \frac{1}{16} = 1.4375.$$

Remarks: The choice of m is entirely up to you. The larger the value of m is the more work one has to do, and the result will be a more accurate approximation.

Questions: Why does this algorithm work? Can we predict how accurate the approximation will be as a function of m?

The answer to these questions will comprise the remainder of this chapter and lead to the development of the general theory of Taylor series. The result will be a powerful tool a CAS user can apply to rapidly compute almost any naturally occurring differentiable function.

Exercises for §19.2

Find the Taylor polynomial of degree 4 for each of the following functions.

(1)	e^{3x}.	**(4)**	$\ln(1-x)$.	**(7)**	$\sqrt[3]{1+2x}$.
(2)	$\cos(x)$.	**(5)**	$\tan(x)$.	**(8)**	$\cosh(x)$.
(3)	$\sin(x)$.	**(6)**	$\sqrt{1+x}$.	**(9)**	$\sinh(x)$.

Approximate the following functions at the indicated points by using Taylor polynomials of degree 1, 2, 3, 4. Compare your answers with the values computed by the CAS.

(10)	e.	**(13)**	$\ln(0.9)$.	**(16)**	$\sqrt[3]{9}$.
(11)	$\cos(58°)$.	**(14)**	$\tan(175°)$.	**(17)**	$\ln(3)$.
(12)	$\sin(47°)$.	**(15)**	$\sqrt{1.1}$.	**(18)**	$\cosh(\frac{1}{4})$.

19.3 Maclaurin Series

We are now in a position to justify the algorithm presented in §19.2. The following theorem was first discovered by Maclaurin.

Theorem [19.9] *Let $f(x)$ be represented by a power series*

$$f(x) = \sum_{n=0}^{\infty} a_n x^n,$$

which converges absolutely for $|x| < R$ for some $R > 0$. Then its coefficients are given by the formula

$$a_n = \frac{f^{(n)}(0)}{n!}.$$

Proof: To begin, when we evaluate

$$f(x) = a_0 + a_1 x + a_2 x^2 + \cdots$$

at $x = 0$ we see that

$$f(0) = a_0.$$

Differentiating $f(x)$ once and setting $x = 0$ yields

$$f'(x) = a_1 + 2a_2 x + 3a_3 x^2 + 4a_4 x^3 + \cdots$$

and

$$f'(0) = a_1.$$

Continuing in this manner we see that

$$f''(x) = 2a_2 + 3 \cdot 2a_3 x + 4 \cdot 3a_4 x^2 + \cdots$$

and

$$f''(0) = 2a_2,$$

or

$$a_2 = \frac{f''(0)}{2!}.$$

Clearly this procedure can be repeated indefinitely, and after differentiating $f(x)$ n–times, we obtain

$$f^{(n)}(x) = (n!)a_n + ((n+1) \cdot n \cdots 2a_{n+1})x + ((n+2) \cdots a_{n+2})x^2 + \cdots$$

Setting $x = 0$ then gives the desired result:

$$f^{(n)}(0) = (n!) \cdot a_n.$$

The series

$$\sum_{n=0}^{\infty} \frac{f^{(n)}(0)}{n!} x^n$$

is called the **Maclaurin series** for $f(x)$.

Example [19.10] Find the Maclaurin series for $f(x) = \sin(x)$.

In this example differentiating is very easy: $f'(x) = \cos(x)$, $f''(x) = -\sin(x)$, $f^{(3)}(x) = -\cos(x)$, and $f^{(4)}(x) = \sin(x)$. It follows that

$$f(0) = 0, \quad f'(0) = 1, \quad \frac{f''(0)}{2!} = 0, \quad \frac{f^{(3)}(0)}{3!} = -\frac{1}{3!},$$

$$\frac{f^{(4)}(0)}{4!} = 0, \quad \frac{f^{(5)}(0)}{5!} = \frac{1}{5!}, \quad \frac{f^{(6)}(0)}{6!} = 0, \quad \frac{f^{(7)}(0)}{7!} = -\frac{1}{7!}$$

etc. Thus the Maclaurin series for $\sin(x)$ is

$$\sin(x) = \sum_{n=0}^{\infty} \frac{(-1)^n x^n}{(2n+1)!}.$$

The Maclaurin series we have found has an infinite radius of convergence, i.e., it converges for any real number x.

Example [19.11] Find the Maclaurin series for $\cos(x)$.

Differentiating $f(x) = \cos(x)$ gives us $f'(x) = -\sin(x)$, $f''(x) = -\cos(x)$, $f^{(3)}(x) = \sin(x)$, etc. We thus have

$$f(0) = 1, \; f'(0) = 1, \; \frac{f''(0)}{2!} = \frac{-1}{2!}, \; \frac{f^{(3)}(0)}{3!} = 0, \; \frac{f^{(4)}(0)}{4!} = \frac{1}{4!}, \; \ldots$$

and finally

$$\cos(x) = \sum_{n=0}^{\infty} \frac{(-1)^n}{(2n)!} x^{2n}.$$

Here again the series has infinite radius of convergence.

Example [19.12] Find the Maclaurin series for $\ln(1-x)$.

In this final example

$$f(x) = \ln(1-x), \; f'(x) = \frac{-1}{1-x}, \; f''(x) = \frac{1}{(1-x)^2}, \; f^{(3)}(x) = \frac{-2}{(1-x)^3}, \; \ldots$$

and for $n > 1$,

$$f^{(n)}(x) = \frac{-(n-1)!}{(1-x)^n}.$$

Thus

$$f(0) = 0,\ f'(0) = -1,\ \frac{f''(0)}{2!} = \frac{-1}{2},\ \frac{f^{(3)}(0)}{3!} = \frac{-1}{3},\ \frac{f^{(4)}(0)}{4!} = \frac{-1}{4}, \ldots$$

and we obtain the desired expansion,

$$\ln(1-x) = -\sum_{n=1}^{\infty} \frac{x^n}{n}.$$

Notice that this series converges absolutely for $|x| < 1$, it diverges when $x = 1$, and it converges conditionally for $x = -1$.

Exercises for §19.3

Compute the Maclaurin series for the following functions. Determine the radius of convergence of each series.

(1)	e^{3x}.	**(5)**	$\frac{1}{1+2x}$.	**(9)**	e^{x^2}.
(2)	e^{-x}.	**(6)**	$\frac{1}{1-3x}$.	**(10)**	10^x.
(3)	$\cos(2x)$.	**(7)**	$\cosh(x)$.	**(11)**	$\frac{1}{1-2x^2}$.
(4)	$\ln(e+x)$.	**(8)**	$\sinh(x)$.	**(12)**	$(x-1)^3$.

19.4 Binomial Series

Consider the expansion

$$(x+y)^2 = x^2 + 2xy + y^2,$$

(which is well known and elementary). Using our CAS, we can easily obtain the expansions

$$(x+y)^3 = x^3 + 3x^2y + 3xy^2 + y^3$$

$$(x+y)^4 = x^4 + 4x^3y + 6x^2y^2 + 4xy^3 + y^4$$

$$\vdots$$

ad infinitum. Focus for a moment on the expansion for $(x+y)^3$. If we divide both sides of the expression by x^3 we see that

$$\frac{1}{x^3}(x+y)^3 = \left(1+\frac{y}{x}\right)^3 = 1+3\left(\frac{y}{x}\right)+3\left(\frac{y}{x}\right)^2+\left(\frac{y}{x}\right)^3.$$

This identity elucidates the principle that if we obtain the expansion for

$$(1+x)^3 = 1+3x+3x^2+x^3$$

then, with a little algebraic manipulation, we can recover the expansion for $(x+y)^3$.

An expansion of $(1+x)^r$, where r is any real number, can be obtained by computing the Maclaurin series for the function

$$f(x) = (1+x)^r. \tag{19.3}$$

We begin by computing

$$\begin{aligned}
f(0) &= 1 \\
f'(0) &= r(1+0)^{r-1} = r \\
\frac{f''(0)}{2!} &= \frac{r(r-1)(1+0)^{r-2}}{2!} = \frac{r(r-1)}{2!} \\
\frac{f^{(3)}(0)}{3!} &= \frac{r(r-1)(r-2)(1+0)^{r-3}}{3!} = \frac{r(r-1)(r-2)}{3!} \\
&\vdots
\end{aligned} \tag{19.4}$$

To simplify the above coefficients we introduce the **binomial coefficients,** $\binom{r}{n}$, which are defined by

$$\begin{aligned}
\binom{r}{0} &= 1 \\
\binom{r}{n} &= \frac{r(r-1)\cdots(r-n+1)}{n!} \qquad n = 1,2,3,\ldots
\end{aligned} \tag{19.5}$$

With this done the Maclaurin series is then given by the compact formula

$$(1+x)^r = \sum_{n=0}^{\infty}\binom{r}{n}x^n. \tag{19.6}$$

Remarks: This is called the binomial series. If r is an integer then the binomial coefficient

$$\binom{r}{n} = \frac{r(r-1)\cdots(r-n+1)}{n!}$$

will be zero for $n > r$ and the sum in (19.2) will be finite (as one would expect).

Example [19.13] Find the binomial series for $(1+x)^3$.

Here we do in fact retrieve our original formula:

$$\begin{aligned}(1+x)^3 &= \sum_{n=0}^{3}\binom{3}{n}x^n \\ &= \binom{3}{0}+\binom{3}{1}x+\binom{3}{2}x^2+\binom{3}{3}x^3 \\ &= 1+3x+3x^2+x^3.\end{aligned}$$

Example [19.14] Find the binomial series for $(1+x)^{\frac{1}{2}}$.

Here the expansion is infinite (and inobvious): by our general formula

$$\begin{aligned}(1+x)^{\frac{1}{2}} &= \sum_{n=0}^{\infty}\binom{\frac{1}{2}}{n}x^n \\ &= \binom{\frac{1}{2}}{0}+\binom{\frac{1}{2}}{1}x+\binom{\frac{1}{2}}{2}x^2+\binom{\frac{1}{2}}{3}x^3+\cdots\end{aligned} \qquad (19.7)$$

Since

$$\begin{aligned}\binom{\frac{1}{2}}{0} &= 1,\ \binom{\frac{1}{2}}{1} = \frac{1}{2}, \\ \binom{\frac{1}{2}}{2} &= \frac{\frac{1}{2}(\frac{1}{2}-1)}{2!} = -\frac{1}{8}, \\ \binom{\frac{1}{2}}{3} &= \frac{\frac{1}{2}(\frac{1}{2}-1)(\frac{1}{2}-2)}{3!} = \frac{1}{16}, \\ \binom{\frac{1}{2}}{4} &= \frac{\frac{1}{2}(\frac{1}{2}-1)(\frac{1}{2}-2)(\frac{1}{2}-3)}{4!} = \frac{-5}{128} \\ &\vdots\end{aligned} \qquad (19.8)$$

we conclude that

$$(1+x)^{\frac{1}{2}} = 1+\frac{1}{2}x-\frac{1}{8}x^2+\frac{1}{16}x^3-\frac{5}{128}x^4+\cdots \tag{19.9}$$

Exercises for §19.4

Evaluate the following binomial coefficients. Check your answers by constructing $\binom{r}{n}$ with your CAS.

(1) $\binom{5}{2}$ **(3)** $\binom{\frac{1}{3}}{2}$. **(5)** $\binom{-\frac{1}{2}}{4}$ **(7)** $\binom{0.21}{2}$.

(2) $\binom{2}{8}$. **(4)** $\binom{\frac{1}{2}}{0}$. **(6)** $\binom{-\frac{1}{3}}{3}$. **(8)** $\binom{-3.21}{2}$.

(9) Find the binomial expansions for $(1+x)^{\frac{1}{3}}$, $(1+x)^{-\frac{1}{3}}$, $(1+x)^{\frac{2}{3}}$, and $(1+x)^{-\frac{2}{3}}$.

(10) Find the first 4 terms in the binomial expansion for $(1+x)^{0.23}$.

(11) Estimate $\sqrt[3]{2}$ by using the first 4 terms in the binomial expansion for $(1+x)^{\frac{1}{3}}$. Obtain a better approximation by using the first 50 terms.

(12) Estimate 2^{π} by using the first 4 terms in the binomial expansion for $(1+x)^{\pi}$. Use the approximation $\pi \approx 3.14$. Get a better approximation of 2^{π} by using the first 50 terms in the binomial expansion together with the more accurate approximation $\pi \approx 3.14159$.

19.5 Estimates of Errors in the Taylor Approximation of Functions

In §19.2 we gave an algorithm for approximating a function $f(x)$ (which is differentiable m–times) by a Taylor polynomial of degree m

$$f(x) \approx f(0) + f'(0)x + \frac{f''(0)}{2!}x^2 + \cdots + \frac{f^{(m)}(0)}{m!}x^m. \tag{19.10}$$

How good is this approximation as $m \to \infty$? Remarkably, it is possible to obtain an exact formula for the difference

$$f(x) - \left(f(0) + f'(0)x + \cdot + \frac{f^{(m)}(0)}{m!}x^m\right),$$

in terms of an integral. By estimating this integral we can effectively predict how accurate the approximation (19.3) will be.

Theorem [19.15] *Let $f(x)$ be a function which is differentiable $m+1$–times on an interval $|x| < A$ for some $A > 0$. Then for any x such that $|x| < A$,*

$$f(x) - \left(f(0) + f'(0)x + \cdot + \frac{f^{(m)}(0)}{m!}x^m\right)$$

$$= \int_0^x f^{(m+1)}(t) \cdot \frac{(x-t)^m}{m!}\, dt.$$

Example [19.16] Let $f(x) = e^x$. Then $f(t) = e^t$ and $f^{(m+1)}(t) = e^t$ for all $m = 1, 2, 3, \ldots$, and by Theorem [19.15] we have that

$$e^x - \left(1 + x + \frac{x^2}{2!} + \cdots + \frac{x^m}{m!}\right) = \int_0^x \frac{(x-t)^m}{m!} e^t\, dt.$$

Before proving theorem [19.15] we demonstrate three cases where we can use theorem [19.15] to approximate the error. It should be noted that there is no general method of approximating an integral, the argument in every case must be individualized.

Example [19.17] How accurate is the approximation to e given by

$$e \approx 1 + 1 + \frac{1}{2!} + \frac{1}{3!} + \cdots + \frac{1}{m!}?$$

If we set $x = 1$ in example [19.16] we see that

$$e - \left(1 + 1 + \frac{1}{2!} + \frac{1}{3!} + \cdots + \frac{1}{m!}\right) = \int_0^1 \frac{(1-t)^m}{m!} e^t\, dt.$$

Now, for $0 \le t \le 1$ we have

$$e^t \le e < 3$$

and

$$(1-t)^m \le 1.$$

These simple estimates imply

$$\int_0^1 \frac{(1-t)^m}{m!} e^t\, dt < \int_0^1 3 \cdot \frac{1}{m!}\, dt = \frac{3}{m!},$$

and thus we conclude

$$e - \left(1 + 1 + \frac{1}{2!} + \frac{1}{3!} + \cdots + \frac{1}{m!}\right) < \frac{3}{m!}.$$

Example [19.18] Calculate e to 4 decimal places.

For our approximation of e to be accurate to 4 decimal places it is necessary to be accurate to within an error of 10^{-5}, i.e., we must choose m such that

$$\frac{3}{m!} < 10^{-5}.$$

This can be accomplished by choosing $m = 8$ since $8! = 40320$ and

$$\frac{3}{8!} = \frac{3}{40320} < 10^{-5}.$$

A quick computation with the CAS shows that

$$e \approx 1 + 1 + \frac{1}{2!} + \frac{1}{3!} + \frac{1}{4!} + \frac{1}{5!} + \frac{1}{6!} + \frac{1}{7!} + \frac{1}{8!} \approx 2.718.$$

Example [19.19] Approximate $\sqrt{e}$ to 3 decimal places.

Here we use the formula from Example [19.16] with $x = \frac{1}{2}$ to see that

$$e^{\frac{1}{2}} - \left(1 + \frac{1}{2} + \frac{1}{2! \cdot 2^2} + \frac{1}{3! \cdot 2^3} + \cdots + \frac{1}{m! \cdot 2^m}\right)$$

$$= \int_0^{\frac{1}{2}} \frac{(\frac{1}{2} - t)^m}{m!} e^t \, dt.$$

In this case we again begin with the very simple estimates

$$e^t \le e^{\frac{1}{2}} < 2$$

and

$$(\frac{1}{2} - t)^m < 2^{-m},$$

when $0 \le t \le \frac{1}{2}$, and deduce

$$\int_0^{\frac{1}{2}} \frac{(\frac{1}{2} - t)^m}{m!} e^t \, dt < \frac{2 \cdot 2^{-m}}{m!} \int_0^{\frac{1}{2}} dt$$

$$= \frac{1}{2^m \cdot m!}.$$

The desired accuracy may be obtained by choosing m such that

$$\frac{1}{2^m \cdot m!} < 10^{-4}. \tag{19.11}$$

In this case $m = 6$ will serve our purpose, and we conclude that

$$\sqrt{e} \approx 1.6487\ldots. \tag{19.12}$$

Proof of Theorem [19.15] We prove this theorem by induction on m. The case of $m = 0$ follows immediately from the Fundamental Theorem of Calculus,

$$f(x) - f(0) = \int_0^x f'(t)\,dt. \tag{19.13}$$

We now assume that the theorem is known for $k \leq m$, i.e.,

$$\begin{aligned} f(x) - \left(f(0) + f'(0)x + \cdot + \frac{f^{(k)}(0)}{k!}x^k\right) & \\ = \int_0^x f^{(k+1)}(t) \cdot \frac{(x-t)^k}{k!}\,dt & \end{aligned} \tag{19.14}$$

when $k \leq m$. We must verify the theorem when $k = m + 1$. By integrating the last integral when $k = m$ by parts we obtain

$$\begin{aligned} &\int_0^x f^{(k+1)}(t) \cdot \frac{(x-t)^k}{k!}\,dt \\ &= -\frac{(x-t)^{m+1}}{(m+1)!} f^{(m+1)}(t)\Big]_0^x + \int_0^x f^{(m+2)}(t) \cdot \frac{(x-t)^{(m+1)}}{(m+1)!}\,dt \\ &= \frac{x^{(m+1)}}{(m+1)!} f^{(m+1)}(0) + \int_0^x f^{(m+2)}(t) \cdot \frac{(x-t)^{(m+1)}}{(m+1)!}\,dt. \end{aligned} \tag{19.15}$$

Combining (19.14) and (19.15) yields

$$f(x) - \left(f(0) + f'(0)x + \cdot + \frac{f^{(m)}(0)}{m!}x^m + \frac{f^{(m+1)}(0)}{(m+1)!}x^{(m+1)}\right)$$

$$= \int_0^x f^{(m+2)}(t) \cdot \frac{(x-t)^{(m+1)}}{(m+1)!}\,dt, \tag{19.16}$$

which verifies the theorem in the case $k = m + 1$. Our theorem is thus proved by induction.

Exercises for §19.5

Calculate the following numbers to the desired degree of accuracy. Use your CAS to compute suitable Taylor polynomials of sufficiently large m. Choose m to obtain the desired degree of accuracy.

(1) $\sqrt[3]{e}$ to 3 decimal places.

(2) $\sqrt[3]{e}$ to 10 decimal places.

(3) $\cos\left(\frac{\pi}{7}\right)$ to 3 decimal places.

(4) $\sin\left(\frac{\pi}{7}\right)$ to 10 decimal places.

(5) $\sqrt{5}$ to 10 decimal places.

(6) $\sqrt[3]{2}$ to 20 decimal places.

(7) $\ln(2)$ to 10 decimal places.

(8) $\ln(3)$ to 50 decimal places.

(9) $\cosh(2)$ to 10 decimal places.

(10) $5^{2.93}$ to 20 decimal places.

19.6 The General Taylor Expansion

In §19.2 we examined an algorithm which approximates functions by Taylor polynomials. The method did, however, require that our function $f(x)$ satisfy two properties: (1) $f(x)$ is infinitely differentiable on an interval containing 0, and (2) The values of $f(0)$ and $f^k(0)$ are known for $k = 1, 2, \ldots m$. Unfortunately, there are numerous examples where the second property may fail to hold. For example, suppose we want to approximate $\sqrt{3}$. We could try using the the seemingly nice function

$$f(x) = \sqrt{2+x} \tag{19.17}$$

which is clearly differentiable around 0. When we proceed to evaluate f and its derivative at 0 we obtain

$$f(0) = \sqrt{2}, \ f'(0) = \frac{1}{2\sqrt{2}}, \ f''(0) = -\frac{1}{8\sqrt{2}}, \ldots \tag{19.18}$$

and unless we know the value of $\sqrt{2}$, the approximation of $\sqrt{3}$ cannot proceed. In this case we can of course use an alternate function for the approximation, for example $f(x) = \sqrt{1+x}$.

Very often the values of a function and its derivatives may be known at some point c where $c \neq 0$. For example, the function $f(x) = (x-2)^{\frac{2}{3}}$ is particularly nice at $x = 3$ since $f(3) = 1$, and $f'(3) = \frac{2}{3}$, $f''(3) = -\frac{2}{9}$, etc. It would be useful to generalize our original algorithm to one where we evaluate the function in question and its derivatives at 3 (or in general some convenient point c) to obtain the coefficients of the approximating series.

To begin developing this new algorithm let c be a constant. We define a **power series about** $x = c$ to be the infinite series

$$\sum_{n=0}^{\infty} a_n (x-c)^n, \tag{19.19}$$

where a_n are fixed real numbers for $n = 0, 1, 2, \ldots$ If (19.19) converges absolutely for $|x-c| < R$ and diverges if $|x-c| > R$ then we call R the **radius of convergence** of the power series (19.19). If a function $f(x)$ can be represented as

$$f(x) = \sum_{n=0}^{\infty} a_n (x-c)^n,$$

then necessarily (the argument is analogous to the one given for Theorem [19.9]),

$$a_n = \frac{f^{(n)}(c)}{n!}.$$

The expansion thus takes the form

$$f(x) = \sum_{n=0}^{\infty} \frac{f^{(n)}(c)}{n!}(x-c)^n,$$

and is referred to as the **Taylor expansion of** $f(x)$ **about** $x = c$.

Example [19.20] Find the Taylor expansion of $f(x) = e^{x-2}$ about $x = 2$.

The point $x = 2$ is certainly a good choice: $f(2) = e^{2-2} = 1$, $f'(2) = 1$,

$f''(2) = 1, \ldots$ Hence

$$e^{x-2} = \sum_{n=0}^{\infty} \frac{f^{(n)}(2)}{n!}(x-2)^n$$

$$= \sum_{n=0}^{\infty} \frac{(x-2)^n}{n!}.$$

Example [19.21] Find the Taylor expansion of $\sin(2x-1)$ about $x = \frac{1}{2}$.

With $f(x) = \sin(2x-1)$ we easily see that $f(\frac{1}{2}) = 0$. Further, $f'(x) = 2\cos(2x-1)$, hence $f'(\frac{1}{2}) = 2$, $f''(x) = -4\sin(2x-1)$ yields $f''(\frac{1}{2}) = 0$, and $f^{(3)}(x) = -8\cos(2x-1)$, implies $f^{(3)}(\frac{1}{2}) = -8$. Continuing in this manner we arrive at the desired expansion,

$$\sin(2x-1) = 2\left(x-\frac{1}{2}\right) - \frac{8}{3!}\left(x-\frac{1}{2}\right)^3 + \frac{32}{5!}\left(x-\frac{1}{2}\right)^5 - \cdots$$

Exercises for §19.6

Find the first 4 terms in the Taylor series (about the given point $x = c$) for the following functions. Use your CAS to plot the original function and then its Taylor polynomial of degree 4. Are the graphs close together?

(1)	e^{2x-1} about $x = \frac{1}{2}$.	**(6)**	$\tan(x)$ about $x = \frac{\pi}{4}$.
(2)	$\cos(3x-2)$ about $x = \frac{2}{3}$.	**(7)**	$\ln(x)$ about $x = e$.
(3)	$\sin(x)$ about $x = \frac{2\pi}{3}$.	**(8)**	x^{19} about $x = 1$.
(4)	$\cos(x)$ about $x = 2\pi - \frac{pi}{6}$.	**(9)**	$x^{-\frac{2}{3}}$ about $x = 27$.
(5)	$\sqrt{x}$ about $x = 16$.	**(10)**	$\ln(\sin(x))$ about $x = \frac{\pi}{2}$.

19.7 Complex Taylor Series and Euler's Formula

The equation $x^2 + 1 = 0$ has no solution in the real numbers because there is no real number whose square is -1. Nevertheless, we can *postulate* the existence of the **imaginary number** $i = \sqrt{-1}$. By defining i in this manner we see that $i^2 = -1$, $i^3 = -i$, $i^4 = 1$, $i^5 = -1$, etc. When we adjoin the imaginary number i

to the set of real numbers $\mathbb{R}$ the resulting enlarged number system is denoted

$$\mathbb{C} = \{a + bi \mid a, b \in \mathbb{R}\},$$

and is termed the **complex numbers**.

Complex numbers can be added, subtracted, multiplied, and divided. For example

$$\begin{aligned}(2+3i)\cdot(1-2i) &= (2+3i)\cdot 1 - (2+3i)\cdot 2i \\ &= 2 + 3i - 4i + 6 \\ &= 8 - i.\end{aligned}$$

$$\begin{aligned}\frac{2+3i}{1-2i} &= \frac{2+3i}{1-2i}\cdot\overbrace{\frac{1+2i}{1+2i}}^{\text{The trick}} \\ &= \frac{8-i}{1+4} \\ &= \frac{8}{5} - \frac{i}{5}.\end{aligned}$$

A **complex power series** is a power series of the form

$$\sum_{n=0}^{\infty} a_n x^n,$$

where $a_n \in \mathbb{C}$. Since each $a_n \in \mathbb{C}$ must take the form $a_n = \gamma_n + i\beta_n$ where $\gamma_n, \beta_n \in \mathbb{R}$, it follows that the complex power series can be expressed as the sum

$$\sum_{n=0}^{\infty} a_n x^n = \left(\sum_{n=0}^{\infty} \gamma_n x^n\right) + i\left(\sum_{n=0}^{\infty} \beta_n x^n\right).$$

The theory of functions of a complex variable is a prominent branch of mathematics which is beyond the scope of this course. To give a flavor of this beautiful subject we shall deduce **Euler's formula** which states:

$$e^{i\theta} = \cos(\theta) + i\sin(\theta).$$

The Taylor series for the function e^x is $\sum_{n=0}^{\infty} \frac{x^n}{n!}$. If we formally replace x with $i\theta$ in this Taylor series we will obtain

$$\begin{aligned}
e^{i\theta} &= \sum_{n=0}^{\infty} \frac{(i\theta)^n}{n!} \\
&= 1 + i\theta + \frac{i^2\theta^2}{2!} + \frac{i^3\theta^3}{3!} + \frac{i^4\theta^4}{4!} + \cdots \\
&= 1 + i\theta - \frac{\theta^2}{2!} - i\frac{\theta^3}{3!} + \frac{\theta^4}{4!} + i\frac{\theta^5}{5!} - \frac{\theta^6}{6!} + \cdots \qquad (19.20) \\
&= \left(1 - \frac{\theta^2}{2!} + \frac{\theta^4}{4!} - \cdots\right) + i\left(\theta - \frac{\theta^3}{3!} + \frac{\theta^5}{5!} - \cdots\right) \\
&= \cos(\theta) + i\sin(\theta).
\end{aligned}$$

Example [19.22] $e^{i\pi} = -1$. Without question this is one of most remarkable identities in mathematics. Notice that it relates e, π, and i!

Example [19.23] Use Euler's formula to verify the following trigonometric identities:

$$\begin{aligned}
\cos(2\theta) &= (\cos(\theta))^2 - (\sin(\theta))^2, \\
\sin(2\theta) &= 2\cos(\theta)\sin(\theta).
\end{aligned} \qquad (19.21)$$

Since $e^{2i\theta} = \cos(2\theta) + i\sin(2\theta)$, and by definition

$$e^{2i\theta} = (e^{i\theta})^2 = (\cos(\theta) + i\sin(\theta))^2 = (\cos(\theta))^2 - (\sin(\theta))^2 + 2i\cos(\theta)\sin(\theta), \qquad (19.22)$$

we deduce $\cos(2\theta) = (\cos(\theta))^2 - (\sin(\theta))^2$ and $\sin(2\theta) = 2\cos(\theta)\sin(\theta)$.

Exercises for §19.23

Perform the following arithmetic computations involving complex numbers. Check with your CAS.

(1) $(3 + 2i) + (1 - 5i)$.

(2) $(3 + 2i) \cdot (1 - 5i)$.

(3) $\frac{(3+2i)}{(1-5i)}$.

(4) $(2 + 3i)^2$.

(5) $(2 + 3i)^{-2}$.

(6) $(1 + i)^3$.

(7) Find formulae for $\cos(a+b)$ and $\sin(a+b)$ by applying Euler's formula to the exponential identity $e^{ia} \cdot e^{ib} = e^{i(a+b)}$.

(8) Find formulae for $\cos(3\theta)$ and $\sin(3\theta)$ by applying Euler's formula to $e^{3i\theta} = (e^{i\theta})^3$.

(9) Find formulae for $\cos(N\theta)$ where N is an arbitrary positive integer.

(10) Find formulae for $\cos(\frac{\theta}{N})$ where N is an arbitrary positive integer.

19.8 L'Hospital's Rule

Another striking application of Taylor series is the capacity to evaluate limits such as

$$\lim_{x\to 2} \frac{x-2}{\cos\left(\frac{\pi x}{4}\right)}, \qquad \lim_{x\to 1} \frac{\ln(x)}{x-1}, \qquad \lim_{x\to 0} \frac{e^{2x}-2e^{x}+1}{x^2}. \tag{19.23}$$

These are examples of **indeterminate** limits since they are each of the form

$$\lim_{x\to c} \frac{f(x)}{g(x)}$$

where $f(c) = g(c) = 0$. Marquis de L'Hospital (1661–1704) found a simple rule which enables one to evaluate such limits. Assuming that both $f(x)$ and $g(x)$ have convergent Taylor series expansions in some interval about $x = c$, the rule states that

$$\lim_{x\to c} \frac{f(x)}{g(x)} = \lim_{x\to c} \frac{f'(x)}{g'(x)}.$$

Remark: If $f'(c) = g'(c) = 0$ then $\lim\limits_{x\to c} \frac{f'(x)}{g'(x)}$ is again an indeterminate form and a second application of L'Hospital's rule is called for.

Example [19.24] Evaluate the limits in (19.23) using L'Hospital's rule.

On viewing these computations the utility of L'Hospital's rule becomes clear.

$$\lim_{x\to 2}\frac{x-2}{\cos\left(\frac{\pi x}{4}\right)}=\lim_{x\to 2}\frac{1}{-\frac{\pi}{4}\sin\left(\frac{\pi x}{4}\right)}=-\frac{4}{\pi}$$

$$\lim_{x\to 1}\frac{\ln(x)}{x-1}=\lim_{x\to 1}\frac{\frac{1}{x}}{1}=1$$

$$\lim_{x\to 0}\frac{e^{2x}-2e^{x}+1}{x^2}=\lim_{x\to 0}\frac{2e^{2x}-2e^{x}}{2x}=\lim_{x\to 0}\frac{4e^{2x}-2e^{x}}{2}=1$$

Example [19.25] Evaluate the limit

$$\lim_{x\to 0}\frac{x+\sin(x)}{x-\sin(x)}.$$

In this case we have

$$\lim_{x\to 0}\frac{x+\sin(x)}{x-\sin(x)}=\lim_{x\to 0}\frac{1+\cos(x)}{1-\cos(x)}=\infty.$$

To justify L'Hospital's rule let

$$f(x)=f'(c)(x-c)+\frac{f''(c)}{2!}(x-c)^2+\cdots$$

$$g(x)=g'(c)(x-c)+\frac{g''(c)}{2!}(x-c)^2+\cdots$$

be Taylor series expansions for $f(x)$ and $g(x)$. Then

$$\lim_{x\to c}\frac{f(x)}{g(x)}=\lim_{x\to c}\frac{f'(c)(x-c)+\frac{f''(c)}{2!}(x-c)^2+\cdots}{g'(c)(x-c)+\frac{g''(c)}{2!}(x-c)^2+\cdots}\cdot\overbrace{\frac{\frac{1}{(x-c)}}{\frac{1}{(x-c)}}}^{\text{The trick}}$$

$$=\lim_{x\to c}\frac{f'(c)+\frac{f''(c)}{2!}(x-c)+\cdots}{g'(c)+\frac{g''(c)}{2!}(x-c)+\cdots}$$

$$=\lim_{x\to c}\frac{f'(x)}{g'(x)},$$

unless $f'(c) = g'(c) = 0$. In this case we will have

$$\lim_{x\to c}\frac{f(x)}{g(x)} = \lim_{x\to c}\frac{\frac{f''(c)}{2!}(x-c)^2 + \frac{f'''(c)}{3!}(x-c)^3 + \cdots}{\frac{g''(c)}{2!}(x-c)^2 + \frac{g'''(c)}{3!}(x-c)^3 + \cdots}$$

$$= \lim_{x\to c}\frac{f''(x)}{g''(x)}$$

unless $f''(c) = g''(c) = 0$. It is clear how to proceed in this case.

Exercises for §19.8

Find the following limits using L'Hospital's rule.

(1) $\lim_{x\to 3}\frac{x-3}{x^2-9}$.

(2) $\lim_{x\to 0}\frac{x}{\sin(x)}$.

(3) $\lim_{x\to 3}\frac{x-3}{\cos\left(\frac{\pi x}{6}\right)}$.

(4) $\lim_{x\to 1}\frac{\ln(x)}{x^2-2x+1}$.

(5) $\lim_{x\to 0}\frac{e^x-1-x}{(\sin(x))^2}$.

(6) $\lim_{x\to 0}\frac{\tan(x)-x}{x^3}$.

(7) $\lim_{x\to 0}\frac{\sin(x)-x}{x^3}$.

(8) $\lim_{x\to 0}\frac{\sinh(x)}{\sin(x)}$.

(9) $\lim_{x\to 2}\frac{x^2-2}{\sin(\pi x)}$.

(10) $\lim_{x\to 1}\frac{1}{\ln(x)} - \frac{1}{(x-1)}$.

(11) $\lim_{x\to 0}\frac{a^x-b^x}{x}$.

(12) $\lim_{x\to 0}\frac{x}{\arcsin(4x)}$.

(13) Evaluate

$$\lim_{x\to 0}\frac{1}{x^3}\int_0^x \sin(\pi t^2)\,dt$$

using L'Hospital's rule. You should also do this computation by plugging in the Taylor series expansion for $\sin(\pi t^2)$ and integrating term by term.

(14) Show that L'Hospital's rule also works for limits of the form

$$\lim_{x\to\infty}\frac{f(x)}{g(x)}$$

by setting $x = \frac{1}{y}$ and letting $y \to 0$.

19.9 Solving Differential Equations with Taylor Series

One of the most effective methods for solving a differential equation is to begin by assuming the solution can be expressed as a power series. By making this assumption we can plug the power series into the differential equation and attempt to solve for the coefficients of the power series. We illustrate this method with two examples which will demonstrate the power of this method to the reader.

Example [19.26] Solve the differential equation

$$\frac{dy}{dx} = 2xy.$$

We begin by assuming that a solution $y = f(x)$ is given by a power series, i.e.,

$$y = f(x) = \sum_{n=0}^{\infty} a_n x^n.$$

Then, by definition,

$$\frac{dy}{dx} = a_1 + 2a_2 x + 3a_3 x^2 + \cdots$$

and we are reduced to solving for the coefficients $a_1, a_2, \ldots$. Plugging the power series expansion into the original differential equation yields

$$\begin{aligned} a_1 + 2a_2 x + 3a_3 x^2 + \cdots &= 2x(a_0 + a_1 x + a_2 x^2 + \cdots) \\ &= 2a_0 x + 2a_1 x^2 + 2a_2 x^3 + \cdots \end{aligned}$$

or equivalently,

$$a_1 + (2a_2 - 2a_0)x + (3a_3 - 2a_1)x^2 + (4a_4 - 2a_2)x^3 + \cdots = 0.$$

The only way that this power series can be identically zero is if each coefficient is 0:

$$\begin{aligned} &a_1 = 0 \\ &2a_2 - 2a_0 = 0 \\ &3a_3 - 2a_1 = 0 \\ &4a_4 - 2a_2 = 0 \\ &\vdots \\ &na_n - 2a_{n-2} = 0. \end{aligned}$$

From these equations we see that

$$a_1 = 0$$

$$a_2 = a_0$$

$$a_3 = 0$$

$$a_4 = \frac{a_2}{2} = \frac{a_0}{2}$$

$$a_5 = 0$$

$$a_6 = \frac{a_4}{3} = \frac{a_0}{3 \cdot 2}$$

$$a_7 = 0$$

$$a_8 = \frac{a_6}{4} = \frac{a_0}{4 \cdot 3 \cdot 2}$$

$$\vdots$$

and finally we conclude

$$a_n = \begin{cases} 0, & \text{if } n = 1, 3, 5, 7, \ldots \\ \frac{a_0}{(n/2)!}, & \text{if } n = 0, 2, 3, 6, 8, \ldots \end{cases}$$

The value of a_0 remains undetermined (it can be anything). Since only the even coefficients contribute to the series, when we set $n = 2k$ with $k = 0, 1, 2, \ldots$ the function takes the form

$$f(x) = a_0 \sum_{k=0}^{\infty} \frac{x^{2k}}{k!}.$$

This series happens to be a familiar one. It is in fact the Taylor expansion of the function

$$f(x) = a_0 e^{x^2},$$

and thus $y = a_0 e^{x^2}$ is the most general solution to the equation $\frac{dy}{dx} = 2xy$. The constant a_0 can be determined when an initial condition is given. An example of an initial condition is $y(0) = 10$ in which case $a_0 = 10$.

Example [19.27] Solve the differential equation

$$\frac{d^2y}{dx^2} = y.$$

Again we begin by assuming that

$$y = a_0 + a_1 x + a_2 x^2 + \cdots$$

Differentiating twice yields

$$\begin{aligned} \frac{dy}{dx} &= a_1 + 2a_2 x + 3a_3 x^2 + \cdots \\ \frac{d^2y}{dx^2} &= 2a_2 + 3\cdot 2a_3 x + 4\cdot 3a_4 x^2 + \cdots \end{aligned}$$

and together with the original differential equation we obtain the identity

$$2a_2 + 3\cdot 2a_3 x + 4\cdot 3a_4 x^2 + \cdots = a_0 + a_1 x + a_2 x^2 + \cdots$$

But then

$$(2a_2 - a_0) + (3\cdot 2a_3 - a_1)x + (4\cdot 3a_4 - a_2)x^2 + \cdots = 0,$$

and so

$$\begin{aligned} 2a_2 - a_0 &= 0 \\ 6a_3 - a_1 &= 0 \\ 12a_4 - a_2 &= 0 \\ 30a_6 - a_4 &= 0 \\ &\vdots \end{aligned}$$

We again solve these equations,

$$\begin{aligned} a_2 &= \frac{a_0}{2} + \frac{a_0}{2!} \\ a_3 &= \frac{a_1}{6} = \frac{a_1}{3!} \\ a_4 &= \frac{a_2}{12} = \frac{a_0}{24} = \frac{a_0}{4!} \\ a_5 &= \frac{a_3}{20} = \frac{a_1}{120} = \frac{a_1}{5!} \\ a_6 &= \frac{a_4}{30} = \frac{a_0}{720} = \frac{a_0}{6!} \\ &\vdots \end{aligned}$$

and at this point the pattern emerges:

$$a_n = \begin{cases} \frac{a_0}{n!} & \text{if } n = 1, 2, 4, 6, \ldots \\ \frac{a_1}{n!} & \text{if } n = 1, 3, 5, 7, \ldots \end{cases}$$

Here a_0 and a_1 are undetermined, and the general solution $y = f(x)$ to the differential equation is given by the sum

$$y = a_0 \sum_{k=0}^{\infty} \frac{x^{2k}}{(2k)!} + a_1 \sum_{k=0}^{\infty} \frac{x^{2k+1}}{(2k+1)!}.$$

The constant a_0, a_1 can only be determined if initial conditions, such as $y(0) = 1$ and $y'(0) = 1$, are given. In this case, $y(0) = 1$ implies $a_0 = 1$, and similarly, $y'(0) = 1$ implies $a_1 = 1$ since

$$y = a_0\left(1 + \frac{x^2}{2!} + \frac{x^4}{4!} + \cdots\right) + a_1\left(x + \frac{x^3}{3!} + \frac{x^5}{5!} + \cdots\right)$$

$$y' = a_0\left(x + \frac{x^3}{3!} + \frac{x^5}{5!} + \cdots\right) + a_1\left(1 + \frac{x^2}{2!} + \frac{x^4}{4!} + \cdots\right).$$

Question: Find the function $y = f(x)$ for which $\frac{d^2y}{dx^2} = y$ with the initial condition $y(0) = 1$ and $y'(0) = 1$.

Answer: Having already shown in example [19.27] that

$$y = a_0 \sum_{k=0}^{\infty} \frac{x^{2k}}{(2k)!} + a_1 \sum_{k=0}^{\infty} \frac{x^{2k+1}}{(2k+1)!},$$

the initial conditions $y(0) = 1$ and $y'(0) = 1$ imply that $a_0 = 1 = a_1$. We may, therefore, simplify the sum of the series to obtain

$$y = \sum_{n=0}^{\infty} \frac{x^n}{n!} = e^x.$$

Exercises for §19.9

Solve the following differential equations with the method of Taylor series. Try to identify series solutions and if initial conditions are given use them to determine constants.

(1) $y'' = 1$.

(2) $y''' = x^2, y(0) = 1, y'(0) = 1, y''(0) = 1$.

(3) $y' + 2y = 0.$

(4) $y'' = -y,$
$y(0) = 1, y'(0) = 0.$

(5) $y'' + xy = 0.$

(6) $y'' - xy' + 2y = 0.$

Additional exercises for Chapter XIX

For the following problems use of CAS is highly encouraged.

(1) Write each of the polynomials that follow as polynomials in $(x - 2)$.

(a) $3x^2 - 6x + 4.$

(b) $x^4 - 5x^3 + 9x^2.$

Find the Taylor polynomials of the indicated degree around the indicated point.

(2) $\cos(x); 2n, \pi.$

(3) $\exp(\cos(x)); 3, 0.$

(4) $\frac{1}{1+x}; n, 0.$

(5) $\cos(x); 2n, \pi/2.$

Calculate the following numbers using Taylor polynomials of sufficiently large n, to the indicated degree of accuracy.

(6) $\sin(5)$; error $< 10^{-20}$.

(7) $\arctan(1)$; error $< 10^{-100}$.

(8) Suppose the Taylor polynomials of f and g at the point a are

$$\sum_{i=1}^{n} \frac{f^i(a)}{i!}(x-a)^i$$

and, respectively,

$$\sum_{i=1}^{n} \frac{g^i(a)}{i!}(x-a)^i.$$

Find the coefficients of the Taylor polynomials at a of $f + g$, fg.

(9) Find Taylor polynomials of degree 4 at 0 of $f(x) = \int_{-\infty}^{x} \exp(-t^2/2)\, dt$.

(10) Compute $\exp(-1/2)$ with error $< 10^{-4}$.

(11) Solve the following differential equations using the series method:

(a) $y' = 2t^5 - \frac{3}{t}$,

(b) $y' = \exp(t) + 4t - \frac{5}{t}$.

(7) Suppose a fish population in a pond grows according to the differential equation

$$\frac{dP}{dt} = .001(1000 - P)P.$$

(a) The fish population is at equilibrium if it remains constant with time. Explain what this means in terms of the number of fish in the pond.

(b) We know that $P \geq 0$ and $\frac{dP}{dt}$ depends on P. For what values of P is $\frac{dP}{dt} > 0$? For what values of P is $\frac{dP}{dt} < 0$? Explain what this means in terms of the fish population. Can you see why ecologists would refer to 1000 as the carrying capacity of the pond for fish?

(c) What would be the effect of fishing for the fish population? What would happen if the fish were being caught at the rate of 80 per year? (distributed evenly throughout the year). Set up the differential equation for a rate x per year and find equilibrium level.

(d) Solve the differential equation using the series method.

(8) Solve the differential equation $y' = (y-1)(y-3) + x, y(0) = 0$ using the series method. Let $y = g(x)$ denote the solution. The line tangent to the curve $y = g(x)$ at the point $(3, 5)$ intersects the y-axis at the point $(0, 10)$.

(a) What is $g'(3)$?

(b) Estimate $g(2.95)$ using a linear approximation.

(9) Suppose $f'(x) = \cos(x^2)$ and $f(1) = 0$. Estimate $f(.5)$ using, first, a linear then a quadratic approximation.

(10) The linear approximation $(1+x)^k \simeq 1 + kx$ is quite useful. Find the error we make by using this approximation for

(a) $(1.03)^2$, (b) $(1.001)^{-1}$ (c) $\sqrt{1.0404}$.

(11) Suppose h is a twice differentiable function such that $h(2) = 3$, $h'(2) = -2$ and $-2 \leq h''(x) \leq 1$ for all $0 \leq x \leq 4$. Show that $0 \leq h(3) \leq 2$.

Chapter XX

Vectors in Two and Three Dimensions

20.1 Introduction to Vectors

A **vector** is a mathematical entity which is uniquely determined by two components — magnitude and direction. In the Cartesian plane (i.e., two dimensional space,) we may visualize a vector $\vec{\mathbf{v}}$ to be an arrow:

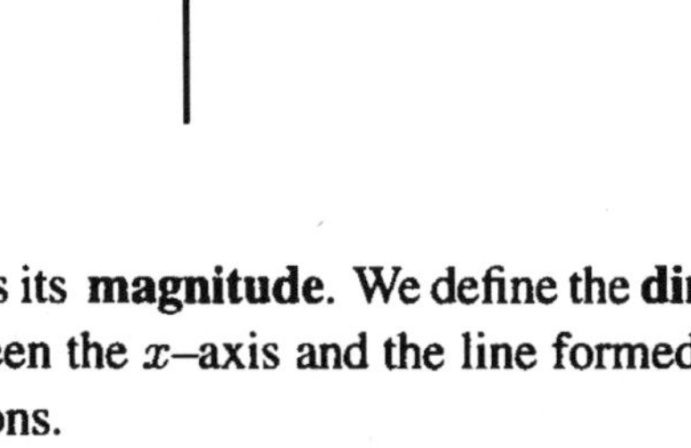

The length of the arrow is its **magnitude**. We define the **direction** to be the measure of angle α formed between the x–axis and the line formed by extending the vector infinitely in both directions.

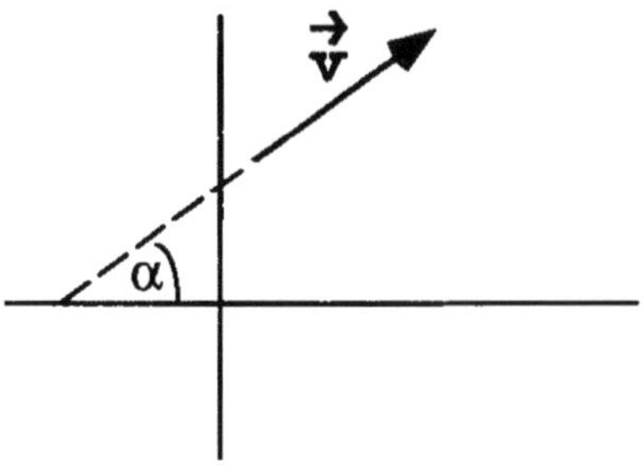

One physical realization of the concept of a vector is that of a force of a given magnitude being applied in a specified direction.

There is one subtle point to grasp here: the vector is *uniquely* determined by its magnitude and direction. It does not depend on anything else. In particular, it does not depend on position in the plane we choose to locate it. In the diagram below we see a given vector in two distinct positions in space. Just as there is a only one Hope diamond (largest blue diamond in the world), whether it is in a museum or in a cave, there is only one vector of a given magnitude and a given direction whether it emanates from the point $(1, 2)$ or the point $(5, 1)$.

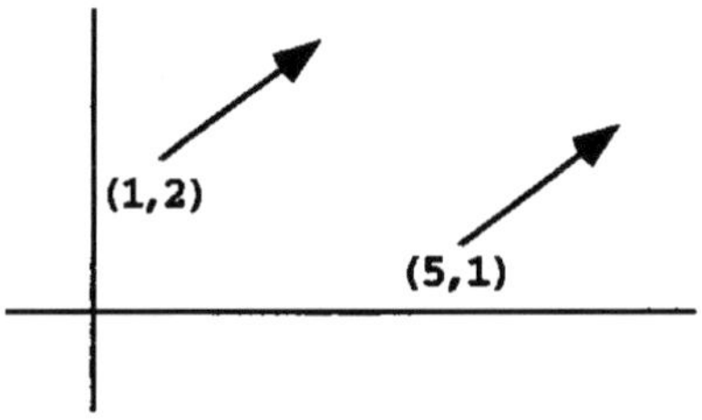

The arrows in the above diagrams are parallel and have the same length — they represent the same vector. The sensory qualities which uniquely define the Hope diamond are independent of its location, and similarly, the mathematical quantities of magnitude and direction uniquely define the vector.

Question: How can we represent vectors mathematically?

Answer: The classical way to represent a vector in 2–dimensional space is by a pair of real numbers (a, b). To signify that this is a vector we will put an arrow over the pair and write

$$\vec{\mathbf{v}} = \overrightarrow{(a, b)}$$

to denote the vector which is represented by an arrow (emanating from any point P) and uniquely determined by the condition that the arrow is the hypotenuse of a right triangle whose sides have length a and b.

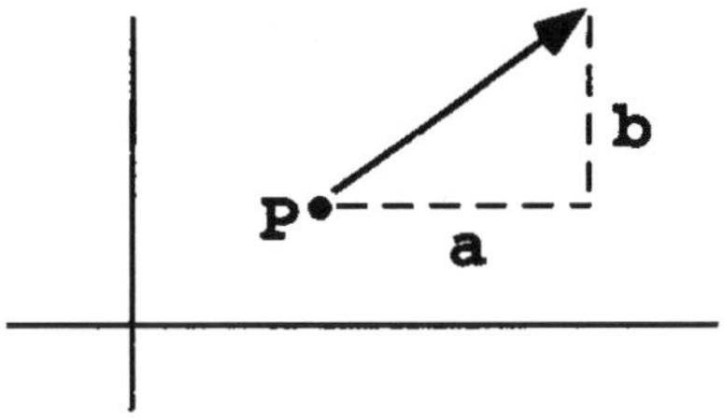

Intuitively we think of the vector $\overrightarrow{(a,b)}$ as the arrow obtained by traveling a steps in the horizontal direction and b steps in the vertical direction. This vector is independent of the point P in the diagram.

Example [20.1] Draw 3 different arrows that represent the $\overrightarrow{(2,3)}$– vector.

The $\overrightarrow{(2,3)}$–vector is uniquely determined by traveling 2 steps in the horizontal direction from some point, and then proceeding 3 steps in the vertical direction. The following 3 arrows all represent the same vector.

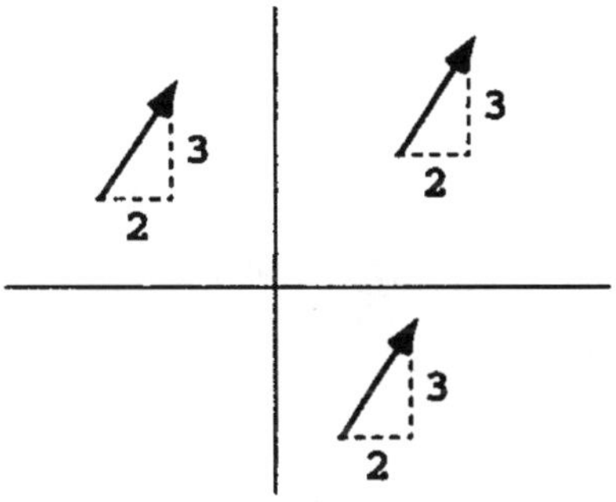

Question: Given a vector $\overrightarrow{(a,b)}$, what is its magnitude and direction?

Answer: The length of $\overrightarrow{(a,b)}$ (which is obtained from the Pythagorean theorem) is simply $\sqrt{a^2+b^2}$. The direction of the vector is simply the angle $\alpha = \arctan\left(\frac{b}{a}\right)$, which is restricted to the range $-\pi \le \alpha \le \pi$.

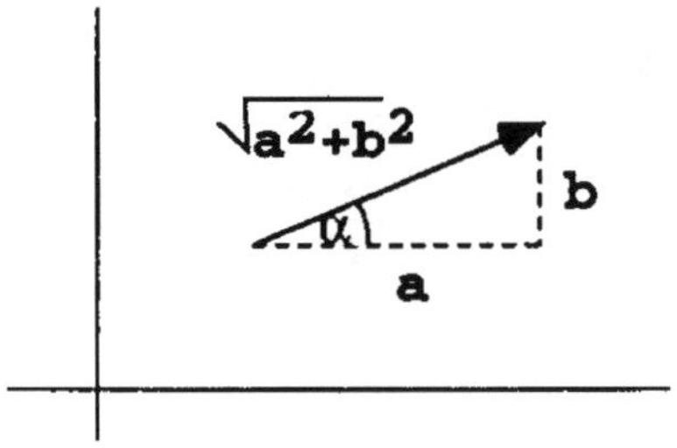

We next consider the three dimensional coordinate system with the axes x, y and z. A triple $\overrightarrow{(a, b, c)}$ of real numbers is a vector in 3–dimensional space. It is represented by an arrow

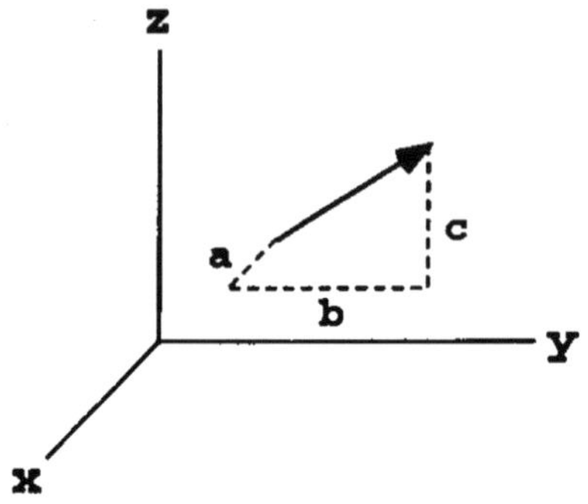

obtained by beginning at some arbitrary point, traveling a steps in the x–direction, b steps in the y–direction, and c steps in the z–direction.

Proposition [20.2] *The 3–dimensional vector* $\overrightarrow{(a, b, c)}$ *has length* $\sqrt{a^2 + b^2 + c^2}$.

Proof: Letting L denote the length of the vector $\overrightarrow{(a, b, c)}$, we refer to the diagram below.

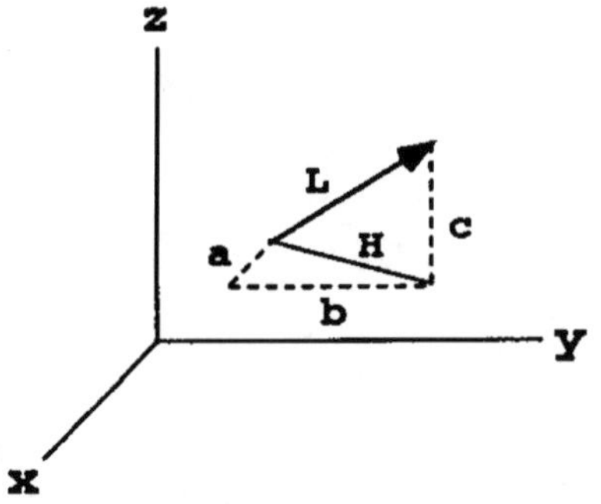

Observing that $H^2 + c^2 = L^2$ and $a^2 + b^2 = H^2$ we see that $L = \sqrt{a^2 + b^2 + c^2}$.

Exercises for §20.1

Find the length of the following vectors:

(1)	$\overrightarrow{(2,5)}$.	(3)	$\overrightarrow{(-1.5,2)}$.	(5)	$\overrightarrow{(\sqrt{2},-\sqrt{3})}$.	(7)	$\overrightarrow{(\pi,1)}$.
(2)	$\overrightarrow{(0,-1,1)}$.	(4)	$\overrightarrow{(\frac{1}{2},-\frac{1}{4},\frac{1}{2})}$.	(6)	$\overrightarrow{(0,\sqrt{2},1)}$.	(8)	$\overrightarrow{(-3,1,0)}$.

(9) Using your CAS construct a function which upon inputting a vector $\overrightarrow{(a,b)}$ outputs the graph of an arrow between the points $(0,0)$ and (a,b).

(10) Using your CAS construct a function which upon inputting a vector $\overrightarrow{(a,b)}$ and a fixed point $P=(p,q)$ outputs the graph of an arrow between the points P and (a,b).

(11) Repeat exercises (9) and (10) in 3–dimensional space.

20.2 The Algebra of Vectors

Just as numbers and functions can be added, subtracted, multiplied, and divided, it is natural to attempt to execute these operations on vectors. With this in mind, we begin with addition.

Vector addition is a very natural process (from the geometric stand point). Given two vectors $\vec{v}$ and $\vec{w}$,

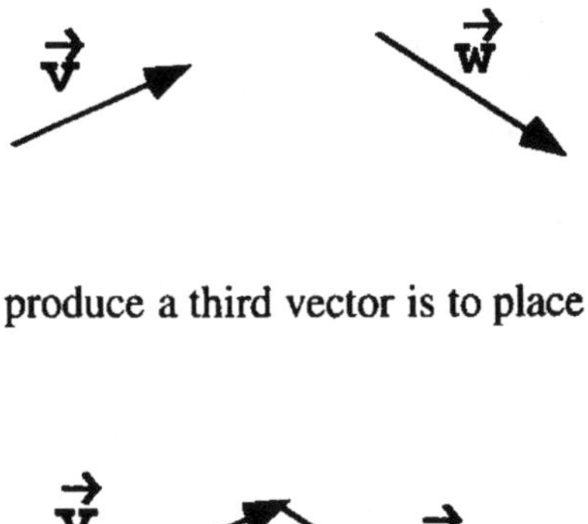

the most obvious way to produce a third vector is to place original vectors head to tail

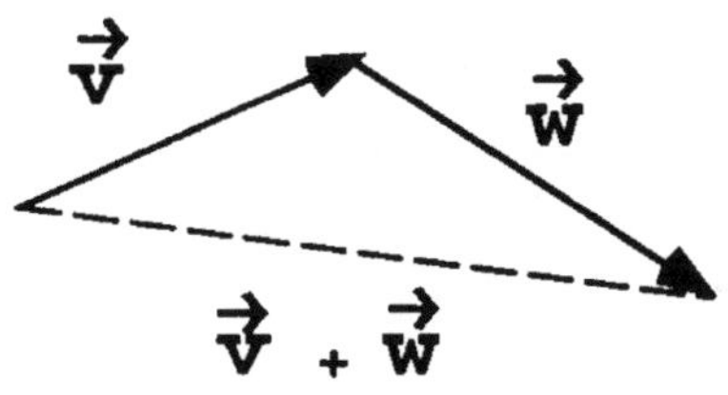

and the resulting dotted vector in the figure above is defined to be $\vec{v}+\vec{w}$.

To represent this process symbolically, we suppose $\vec{\mathbf{v}} = (a, b)$ and $\vec{\mathbf{w}} = (c, d)$, and consider the next diagram. By definition, assuming $\vec{\mathbf{v}}$ originates from some point P, $\vec{\mathbf{v}}$ indicates that we move a steps in the horizontal direction and b steps in the vertical direction arriving at the point Q. Now positioning $\vec{\mathbf{w}}$ so that it originates from Q,

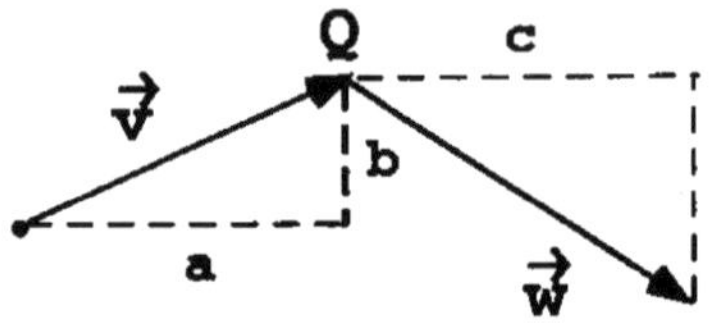

we see that, by definition, $\vec{\mathbf{v}} + \vec{\mathbf{w}}$ is simply the vector $\overrightarrow{(a+c, b+d)}$. Notice that in this particular diagram, $d < 0$.

We summarize the above discussion in the following definition.

Definition: *Let* $\vec{\mathbf{v}} = \overrightarrow{(a, b)}$ *and* $\vec{\mathbf{w}} = \overrightarrow{(c, d)}$ *be any two vectors in 2–dimensional space. We define*

$$\vec{\mathbf{v}} + \vec{\mathbf{w}} = \overrightarrow{(a+c,\ b+d)}.$$

The case of 3–dimensional vectors is similar.

Definition: *Let* $\vec{\mathbf{v}} = \overrightarrow{(a, b, c)}$ *and* $\vec{\mathbf{w}} = \overrightarrow{(d, e, f)}$ *be any two vectors in 3–dimensional space. We define*

$$\vec{\mathbf{v}} + \vec{\mathbf{w}} = \overrightarrow{(a+d,\ b+e,\ c+f)}.$$

Remarks: Notice that addition of vectors is analogous to addition of numbers or functions in that the operation is commutative, i.e., $\vec{\mathbf{v}} + \vec{\mathbf{w}} = \vec{\mathbf{w}} + \vec{\mathbf{v}}$, where the vectors here are 2 or 3–dimensional. In addition the **zero vectors,** $\overrightarrow{(0,0)}$ and $\overrightarrow{(0,0,0)}$ (denoted $\vec{\mathbf{0}}$), have the property that for any vector $\vec{\mathbf{v}}$,

$$\vec{\mathbf{0}} + \vec{\mathbf{v}} = \vec{\mathbf{v}} + \vec{\mathbf{0}} = \vec{\mathbf{v}}.$$

The zero vector has length 0 and unspecified direction.

Example [20.3] Find the sum of the vectors $\overrightarrow{(2, -1, 3)}$ and $\overrightarrow{(1, 4, -5)}$.

The sum of these vectors is $\overrightarrow{(2,-1,3)} + \overrightarrow{(1,4,-5)} = \overrightarrow{(3,3,-2)}$.

Having defined addition of vectors we now proceed to its inverse, subtraction. When we consider subtraction of ordinary numbers we may view $a-b$ as an addition (provided negative numbers have been given meaning), $a - b = a + (-b)$. With this in mind we begin with

Definition: *Let $\vec{\mathbf{v}} = \overrightarrow{(a, b)}$ be any vector in 2–dimensional space. We define $-\vec{\mathbf{v}} = \overrightarrow{(-a, -b)}$. Then $-\vec{\mathbf{v}}$ is simply the vector $\vec{\mathbf{v}}$ with direction reversed.*

Similarly with 3–dimensional vectors,

Definition: *If $\vec{\mathbf{v}} = \overrightarrow{(a, b, c)}$ is any 3–dimensional vector, then we define $-\vec{\mathbf{v}} = \overrightarrow{(-a, -b, -c)}$.*

We can now define subtraction of vectors,

Definition: ***Subtraction of vectors*** *is defined by*

$$\vec{\mathbf{v}} - \vec{\mathbf{w}} = \vec{\mathbf{v}} + (-\vec{\mathbf{w}}).$$

Example [20.4] Let $\vec{\mathbf{v}} = \overrightarrow{(2,1,4)}$ and $\vec{\mathbf{w}} = \overrightarrow{(3,-1,2)}$. Then

$$\begin{aligned} \vec{\mathbf{v}} - \vec{\mathbf{w}} &= \overrightarrow{(2,1,4)} + \overrightarrow{(-3,1,-2)} \\ &= \overrightarrow{(-1,2,2)}. \end{aligned}$$

Question: Given the points $P = (p_1, p_2, p_3)$ and $Q = (q_1, q_2, q_3)$ in 3–dimensional space, what is the vector $\vec{\mathbf{v}}$ which originates at P and terminates at Q?

Answer: While the answer to this question is simply

$$\vec{\mathbf{v}} = \overrightarrow{(q_1 - p_1, q_2 - p_2, q_3 - p_3)},$$

the verification of this fact requires some contemplation.

Consider the vectors $\overrightarrow{OP}$ and $\overrightarrow{OQ}$ which emanate from the origin and end at the points P and Q respectively.

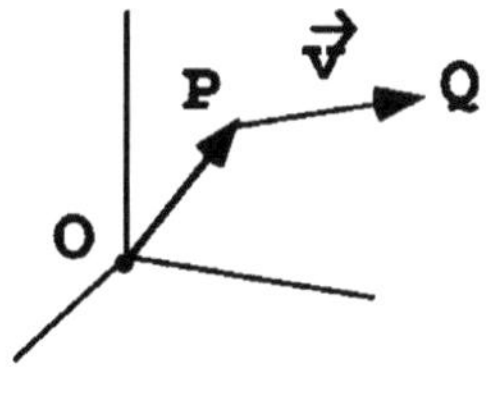

Then clearly $\overrightarrow{OP} + \vec{\mathbf{v}} = \overrightarrow{OQ}$, and hence

$$\begin{aligned}\vec{\mathbf{v}} &= -\overrightarrow{OP} + \overrightarrow{OQ}\\ &= \overrightarrow{(q_1 - p_1, q_2 - p_2, q_3 - p_3)}.\end{aligned}$$

Finally we would like to complete the analogy between numbers (or functions) and vectors by defining multiplication of vectors. This task turns out to be more complex than addition and subtraction. We will in fact defer the definition of multiplying two vectors to §20.4 and §20.5. In the remainder of this section we will define multiplication of a real number and a vector.

Given a vector $\vec{\mathbf{v}}$, we may add it to itself 3 times (which triples its length) to obtain the new vector $3\vec{\mathbf{v}}$. Similarly, we can divide the vector in half.

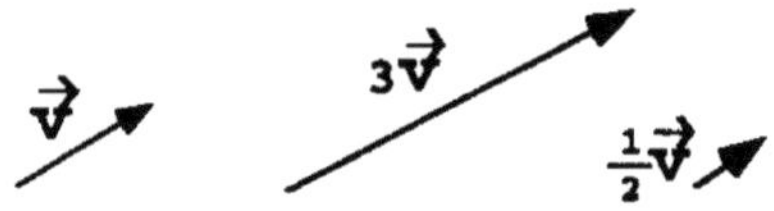

These geometric observations motivate the following definition.

Definition: *Let* $\vec{\mathbf{v}} = \overrightarrow{(a, b)}$ *and* $\vec{\mathbf{w}} = \overrightarrow{(a, b, c)}$ *be vectors in two and three dimensional space, respectively. Given any real number, m, we define*

$$m\vec{\mathbf{v}} = \overrightarrow{(ma, mb)}$$

$$m\vec{\mathbf{w}} = \overrightarrow{(ma, mb, mc)}.$$

Example [20.5] Let $\vec{\mathbf{v}}_1 = \overrightarrow{(1, 2)}$, $\vec{\mathbf{v}}_2 = \overrightarrow{(5, -2)}$, and $\vec{\mathbf{w}} = \overrightarrow{(2, 9, -1)}$. Then

$$4\vec{\mathbf{w}} = \overrightarrow{(8, 36, -4)}, \qquad \frac{1}{3}\vec{\mathbf{w}} = \overrightarrow{\left(\frac{2}{3}, 3, -\frac{1}{3}\right)},$$

and

$$2\vec{\mathbf{v}}_1 - \frac{1}{4}\vec{\mathbf{v}}_2 = \overrightarrow{\left(\frac{3}{4}, \frac{9}{2}\right)}.$$

Exercises for §20.2

Perform the following arithmetic vector operations. Check your answers with your CAS.

(1) $\overrightarrow{\langle 2,-1\rangle} + \overrightarrow{\langle -3,2\rangle}$.

(2) $\overrightarrow{\langle -1,8\rangle} - \overrightarrow{\langle -2,3\rangle}$.

(3) $5 \cdot \overrightarrow{\langle 8,-2\rangle}$.

(4) $\overrightarrow{\langle 3,-2,4\rangle} + \overrightarrow{\langle 6,1,3\rangle}$.

(5) $\overrightarrow{\langle 2,-7,1\rangle} - \overrightarrow{\langle -3,7,3\rangle}$.

(6) $\frac{1}{4} \cdot \overrightarrow{\langle 4,0,-1\rangle}$.

(7) Let $P = (1,-3)$ and $Q = (2,6)$ be points in 2–dimensional space. Find a vector which originates at P and terminates at Q. What is its length. Graph the arrow from P to Q.

(8) Let $P = (-2,3,0)$ and $Q = (1,4,-3)$ be points in 3–dimensional space. Find a vector which originates at P and terminates at Q. What is its length. Graph the arrow from P to Q.

20.3 Basis Vectors in Two and Three Dimensions

In 2–dimensional space, when we define the vectors

$$\vec{\mathbf{i}} = \overrightarrow{\langle 1,0\rangle}, \quad \vec{\mathbf{j}} = \overrightarrow{\langle 0,1\rangle},$$

it is clear that any vector $\vec{\mathbf{v}} = \overrightarrow{\langle a,b\rangle}$ can be expressed as a **linear combination** of $\vec{\mathbf{i}}$ and $\vec{\mathbf{j}}$:

$$\vec{\mathbf{v}} = a\vec{\mathbf{i}} + b\vec{\mathbf{j}}.$$

Note that $\vec{\mathbf{i}}$ and $\vec{\mathbf{j}}$ have magnitude 1. It is easily shown that the expression, $\vec{\mathbf{v}} = a\vec{\mathbf{i}} + b\vec{\mathbf{j}}$, is in fact unique.

In 3–dimensional space we have an analogous definition. Let

$$\vec{\mathbf{i}} = \overrightarrow{\langle 1,0,0\rangle}, \quad \vec{\mathbf{j}} = \overrightarrow{\langle 0,1,0\rangle}, \quad \vec{\mathbf{k}} = \overrightarrow{\langle 0,0,1\rangle}.$$

Then any vector $\vec{\mathbf{v}} = \overrightarrow{\langle a,b,c\rangle}$ in 3–dimensional space can be represented as a linear combination of $\vec{\mathbf{i}}$, $\vec{\mathbf{j}}$, and $\vec{\mathbf{k}}$,

$$\vec{\mathbf{v}} = a\vec{\mathbf{i}} + b\vec{\mathbf{j}} + c\vec{\mathbf{k}}.$$

The vector $\vec{\mathbf{i}}$ and $\vec{\mathbf{j}}$, and $\vec{\mathbf{i}}, \vec{\mathbf{j}}$, and $\vec{\mathbf{k}}$, form a **basis** of 2 and 3–dimensional space respectively, and are called **basis vectors.** A collection of vectors form a basis for a space (such as 3–dimensional space) provided they have the property that any vector in the space can be uniquely represented as a linear combination of vectors in the basis. In general, there is no unique way to choose the basis vectors. For example, in 3–dimensional space, we have chosen $\vec{\mathbf{i}}, \vec{\mathbf{j}}, \vec{\mathbf{k}}$ as basis vectors, but we could have equally well have chosen any other three mutually perpendicular vectors. Regardless of our choice of basis, however, the number of basis vectors remains constant and equals the dimension of the space in question.

20.4 The Dot Product

We now introduce the **dot product** $\vec{\mathbf{v}} \cdot \vec{\mathbf{w}}$ of two vectors.

Definition: *Let* $\vec{\mathbf{v}} = \overrightarrow{(a, b)}$ *and* $\vec{\mathbf{w}} = \overrightarrow{(c, d)}$ *be vectors in 2–dimensional space. The* ***dot product*** *is defined to be*

$$\vec{\mathbf{v}} \cdot \vec{\mathbf{w}} = ac + bd.$$

Definition: *Let* $\vec{\mathbf{v}} = \overrightarrow{(a, b, c)}$ *and* $\vec{\mathbf{w}} = \overrightarrow{(d, e, f)}$ *be vectors in 3–dimensional space. The* ***dot product*** *is defined to be*

$$\vec{\mathbf{v}} \cdot \vec{\mathbf{w}} = ad + be + cf.$$

Warning: The dot product of two vectors *is not* a vector. It is a number (which is often referred to as a **scalar**).

Example [20.6] The following cases demonstrate that the dot product is easily computed:

$$\overrightarrow{(4, -1, 2)} \cdot \overrightarrow{(2, 3, -5)} = 8 - 3 - 10 = -5,$$

$$\overrightarrow{(2, 3)} \cdot \overrightarrow{(-1, 4)} = -2 + 12 = 10,$$

$$\vec{\mathbf{i}} \cdot \vec{\mathbf{k}} = \overrightarrow{(1, 0, 0)} \cdot \overrightarrow{(0, 0, 1)} = 0,$$

$$(2\vec{\mathbf{i}} + \vec{\mathbf{j}}) \cdot \vec{\mathbf{j}} = \overrightarrow{(2, 1)} \cdot \overrightarrow{(0, 1)} = 1.$$

The dot product behaves much the way multiplication of numbers or functions does in that it satisfies the algebraic properties

$$(\vec{\mathbf{v}} + \vec{\mathbf{w}}) \cdot \vec{\mathbf{u}} = \vec{\mathbf{v}} \cdot \vec{\mathbf{u}} + \vec{\mathbf{w}} \cdot \vec{\mathbf{u}}, \tag{20.1}$$

and

$$(\vec{\mathbf{v}} + \vec{\mathbf{w}}) \cdot (\vec{\mathbf{v}} + \vec{\mathbf{w}}) = \vec{\mathbf{v}} \cdot \vec{\mathbf{v}} + 2\vec{\mathbf{v}} \cdot \vec{\mathbf{w}} + \vec{\mathbf{w}} \cdot \vec{\mathbf{w}}. \tag{20.2}$$

Recall that the length of a vector $\vec{\mathbf{v}} = \overrightarrow{(a, b, c)}$, which is often denoted $|\vec{\mathbf{v}}|$, was defined to be $\sqrt{a^2 + b^2 + c^2}$. We can simplify this formula using dot product notation,

$$|\vec{\mathbf{v}}| = \sqrt{\vec{\mathbf{v}} \cdot \vec{\mathbf{v}}}. \tag{20.3}$$

Notice that if we multiply $\vec{\mathbf{v}}$ by a scalar c, then

$$\begin{aligned} |c\vec{\mathbf{v}}| &= \sqrt{c\vec{\mathbf{v}} \cdot c\vec{\mathbf{v}}} \\ &= c\sqrt{\vec{\mathbf{v}} \cdot \vec{\mathbf{v}}} = c|\vec{\mathbf{v}}|. \end{aligned} \tag{20.4}$$

This simple observation, in conjunction with the algebraic properties of the dot product, allow us to prove the following proposition.

Proposition [20.7] *If two vectors $\vec{\mathbf{v}}$ and $\vec{\mathbf{w}}$ are perpendicular (orthogonal), then*

$$\vec{\mathbf{v}} \cdot \vec{\mathbf{w}} = 0.$$

Proof: Notice first that we may assume both vectors are nonzero. Consider the triangle formed with the vectors $\vec{\mathbf{v}}$, $\vec{\mathbf{w}}$, and $\vec{\mathbf{w}} - \vec{\mathbf{v}}$.

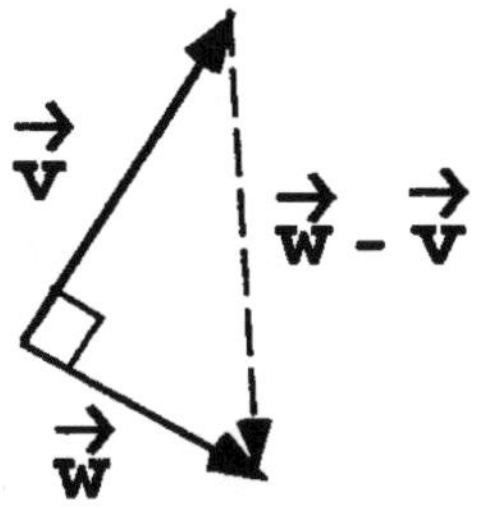

Since $\vec{v}$ is assumed to be perpendicular to $\vec{w}$, we can use the Pythagorean theorem which, together with equation (20.3), gives us the following equality:

$$\begin{aligned} |\vec{v}|^2 + |\vec{w}|^2 &= |\vec{w} - \vec{v}|^2 \\ &= (\vec{w} - \vec{v}) \cdot (\vec{w} - \vec{v}) \\ &= \vec{w} \cdot \vec{w} - 2\vec{w} \cdot \vec{v} + \vec{v} \cdot \vec{v} \\ &= |\vec{w}|^2 - 2\vec{w} \cdot \vec{v} + |\vec{v}|^2. \end{aligned} \tag{20.5}$$

Clearly this equality can only be valid if $\vec{v} \cdot \vec{w} = 0$.

The argument we used to prove Proposition [20.7] can, with a slight modification, give us a stronger (and more important) result.

Theorem [20.8] *Let θ denote the angle between vectors $\vec{v}$ and $\vec{w}$. Then*

$$\vec{v} \cdot \vec{w} = |\vec{v}||\vec{w}| \cos(\theta).$$

Proof: Consider the diagram

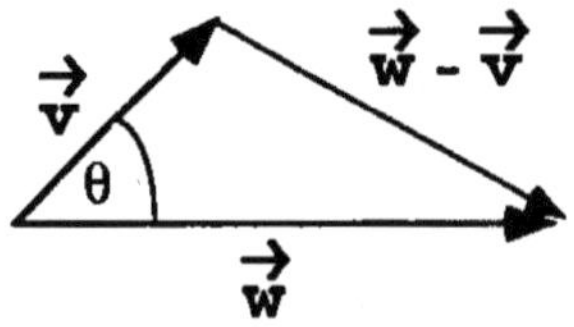

When we combine the Law of Cosines

$$|\vec{w} - \vec{v}|^2 = |\vec{v}|^2 + |\vec{w}|^2 - 2|\vec{v}||\vec{w}| \cos(\theta),$$

with the equality (20.5)

$$|\vec{w} - \vec{v}|^2 = |\vec{w}|^2 - 2\vec{w} \cdot \vec{v} + |\vec{v}|^2,$$

the theorem follows.

Corollary [20.9] *Two vectors, $\vec{v}$ and $\vec{w}$, are perpendicular if and only if*

$$\vec{v} \cdot \vec{w} = 0.$$

Proof: We have already shown in Proposition [20.7] that orthogonality implies the dot product is zero. The converse statement follows immediately from our theorem: if $\vec{\mathbf{v}} \cdot \vec{\mathbf{w}} = 0$, then

$$|\vec{\mathbf{v}}||\vec{\mathbf{w}}| \cos(\theta) = 0,$$

hence $\cos(\theta) = 0$, and the vectors are perpendicular.

Example [20.10] Are the vectors $\overrightarrow{(2,-3,-1)}$ and $\overrightarrow{(4,1,5)}$ perpendicular?

Since $\overrightarrow{(2,-3,-1)} \cdot \overrightarrow{(4,1,5)} = 8-3-5 = 0$ the vectors must be perpendicular.

Example [20.11] Find a vector of length 10 which is perpendicular to the vector $3\vec{\mathbf{i}} - 2\vec{\mathbf{j}}$.

To solve this problem we begin by finding a vector which is perpendicular to $3\vec{\mathbf{i}} - 2\vec{\mathbf{j}}$. The vector $\overrightarrow{(2,3)}$ is perpendicular to the vector $3\vec{\mathbf{i}} - 2\vec{\mathbf{j}} = \overrightarrow{(3,-2)}$ since $\overrightarrow{(2,3)} \cdot \overrightarrow{(3,-2)} = 0$. Unfortunately, the length of $\overrightarrow{(2,3)}$ is

$$|\overrightarrow{(2,3)}| = \sqrt{\overrightarrow{(2,3)} \cdot \overrightarrow{(2,3)}} = \sqrt{13}.$$

To force the vector to have length 10 we multiply $\overrightarrow{(2,3)}$ by $\frac{10}{\sqrt{13}}$, yielding the vector

$$\overrightarrow{\left(\frac{20}{\sqrt{13}}, \frac{30}{\sqrt{13}}\right)}.$$

Clearly this new vector will again be perpendicular to $\overrightarrow{(3,-2)}$ (since it was obtained by stretching another vector which was perpendicular to $\overrightarrow{(3,-2)}$), and the length of the new vector is 10 by equation (20.4).

Example [20.12] Find the angle between the vectors $\overrightarrow{(1,0,1)}$ and $\overrightarrow{(0,1,1)}$.

Applying Theorem [20.8] we have

$$\cos(\theta) = \frac{\overrightarrow{(1,0,1)} \cdot \overrightarrow{(0,1,1)}}{|\overrightarrow{(1,0,1)}| \cdot |\overrightarrow{(0,1,1)}|} = \frac{1}{\sqrt{2} \cdot \sqrt{2}} = \frac{1}{2},$$

and hence $\theta = \frac{\pi}{3}$.

Example [20.13] Find the angle between the vectors $\vec{\mathbf{i}} + \vec{\mathbf{k}}$ and $\vec{\mathbf{k}}$.

In this case we have

$$\begin{aligned}\cos(\theta) &= \frac{(\vec{\mathbf{i}}+\vec{\mathbf{k}})\cdot\vec{\mathbf{k}}}{|\vec{\mathbf{i}}+\vec{\mathbf{k}}|\cdot|\vec{\mathbf{k}}|} \\ &= \frac{\overrightarrow{(1,0,1)}\cdot\overrightarrow{(0,0,1)}}{|\overrightarrow{(1,0,1)}|\cdot|\overrightarrow{(0,0,1)}|} \\ &= \frac{1}{\sqrt{2}},\end{aligned}$$

and hence $\theta = \frac{\pi}{4}$.

The dot product can be used to find the **projection** of a vector $\vec{\mathbf{v}}$ onto another vector $\vec{\mathbf{w}}$.

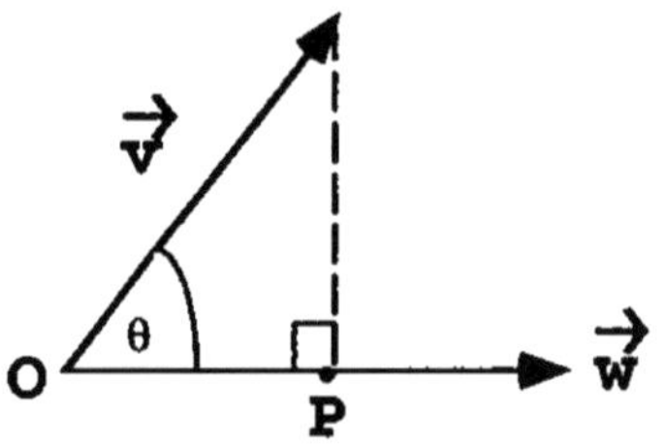

The length of the projection of $\vec{\mathbf{v}}$ onto $\vec{\mathbf{w}}$, labeled $\overrightarrow{OP}$ in the diagram, can be computed via Theorem [20.8]. In fact we know from trigonometry that

$$\cos(\theta) = \frac{|\overrightarrow{OP}|}{|\vec{\mathbf{v}}|},$$

and Theorem [20.8] tells us that

$$\cos(\theta) = \frac{\vec{\mathbf{v}}\cdot\vec{\mathbf{w}}}{|\vec{\mathbf{v}}||\vec{\mathbf{w}}|}.$$

Combining these identities we obtain

$$|\overrightarrow{OP}| = \frac{\vec{\mathbf{v}}\cdot\vec{\mathbf{w}}}{|\vec{\mathbf{w}}|}. \tag{20.6}$$

Example [20.14] Find the projection of the vector $\overrightarrow{(1,1,1)}$ on the vector $\overrightarrow{(2,4,2)}$.

In order to keep a good geometric perspective on the question we begin by drawing the given 3–dimensional vectors.

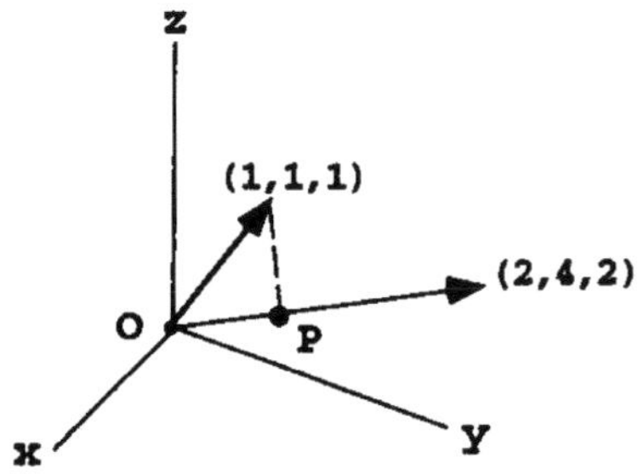

To find the vector $\overrightarrow{OP}$ we begin by computing its length:

$$|\overrightarrow{OP}| = \frac{\overrightarrow{(1,1,1)} \cdot \overrightarrow{(2,4,2)}}{|\overrightarrow{(2,4,2)}|} = \frac{8}{\sqrt{24}} = \frac{24}{\sqrt{6}} = \frac{4\sqrt{6}}{6}.$$

Since the vector $\overrightarrow{OP}$ is parallel to $\overrightarrow{(2,4,2)}$, it is a scalar multiple of $\overrightarrow{(2,4,2)}$. But $\overrightarrow{(2,4,2)}$ has length $2\sqrt{6}$, and thus by multiplying $\overrightarrow{(2,4,2)}$ by $\frac{2}{6}$ we obtain our solution, i.e.,

$$\overrightarrow{OP} = \frac{2}{6}\overrightarrow{(2,4,2)} = \overrightarrow{\left(\frac{2}{3}, \frac{4}{3}, \frac{2}{3}\right)}.$$

Exercises for §20.4

Compute the following dot products.

(1) $\overrightarrow{(2,-4)} \cdot \overrightarrow{(1,3)}$.

(2) $2\vec{i} \cdot \vec{k}$.

(3) $(2\vec{i} - 3\vec{k}) \cdot (2\vec{j} + 4\vec{k})$.

(4) $\overrightarrow{(1,-3,7)} \cdot \overrightarrow{(2,0,-2)}$.

(5) $(\vec{i} + 2\vec{j}) \cdot (\vec{i} + 2\vec{j})$.

(6) $\overrightarrow{(1,-1,3)} \cdot \overrightarrow{(2,-3,4)}$.

Determine if the following pairs of vectors are perpendicular. Confirm your answer visually by graphing the vectors (emanating from some common point).

(7) $\overrightarrow{(-2,9)}$, $\overrightarrow{(1,-\frac{1}{2})}$.

(8) $\overrightarrow{(-2,9)}$, $\overrightarrow{(\frac{3}{2},\frac{1}{3})}$.

(9) $\overrightarrow{(2,4)}$, $\overrightarrow{(2,-1)}$.

(10) $\overrightarrow{(1,3,-2)}$, $\overrightarrow{(2,4,1)}$.

(11) $\overrightarrow{(3,-1,2)}$, $\overrightarrow{(2,4,-1)}$.

(12) $\overrightarrow{(2,1,-4)}$, $\overrightarrow{(1,2,0)}$.

For each of the following vectors find vectors of lengths 1, 5, 10 which are perpendicular to the given vector. Confirm your answer visually by graphing the vectors.

(13) $\overrightarrow{(2,6)}$.

(14) $\overrightarrow{(-3,4)}$.

(15) $\overrightarrow{(1,2,-3)}$.

(16) $\overrightarrow{(-1,2,\frac{1}{2})}$.

Find the angles between the following pairs of vectors.

(17) $\overrightarrow{(3,-1)}$, $\overrightarrow{(2,1)}$.

(18) $\overrightarrow{(1,1,0)}$, $\overrightarrow{(-1,-2,1)}$.

(19) $3\vec{i}+\vec{j}$, $2\vec{i}+4\vec{j}$.

(20) $\overrightarrow{(-6,0,2)}$, $\overrightarrow{(-5,3,-2)}$.

(21) $\overrightarrow{(-1,3,2)}$, $\overrightarrow{(-1,3,2)}$.

(22) $\overrightarrow{(\frac{1}{7},-3)}$, $\overrightarrow{(28,\frac{4}{3})}$.

Find the projection of the vector $\vec{v}$ onto the vector $\vec{w}$ for each of the following pairs of vectors.

(23) $\vec{v}=\overrightarrow{(1,-2)}$, $\vec{w}=\overrightarrow{(2,8)}$.

(24) $\vec{v}=\overrightarrow{(2,-3)}$, $\vec{w}=\overrightarrow{(5,1)}$.

(25) $\vec{v}=\overrightarrow{(1,4,-2)}$, $\vec{w}=\overrightarrow{(2,3,-1)}$.

(26) $\vec{v}=\overrightarrow{(2,1,3)}$, $\vec{w}=\overrightarrow{(1,-3,4)}$.

(27) Find the angle between the vectors $\overrightarrow{(1,4)}$, $\overrightarrow{(2,-3,4)}$ and the y–axis.

(28) Prove that $|\vec{v}\cdot\vec{w}| \le |\vec{v}|\cdot|\vec{w}|$ for any two vectors $\vec{v}$ and $\vec{w}$.

(29) Show that $|\vec{v}+\vec{w}| \le |\vec{v}|+|\vec{w}|$ for any two vectors $\vec{v}$ and $\vec{w}$.

(30) Show that if $\vec{v}$ and $\vec{w}$ are any two non–zero vectors of the same length then $\vec{v}+\vec{w}$ and $\vec{v}-\vec{w}$ are orthogonal.

(31) Let $\vec{u}$ be an arbitrary vector of length 1 (i.e., a unit vector). Show that $\vec{v}-(\vec{v}\cdot\vec{u})\vec{u}$ is always perpendicular to $\vec{u}$ for any vector $\vec{v}$.

20.5 The Cross Product

To complete our discussion of the algebraic structure of vectors we now define the **cross product,** $\vec{v}\times\vec{w}$ of two vectors in 3–dimensional space. The cross product is a method of multiplying two vectors to obtain a third vector. Recall that since $\vec{i}, \vec{j}$, and $\vec{k}$ are basis vectors of 3–dimensional space, it is enough to define the cross product on these vectors. The general definition is obtained by extending these definitions with the laws of algebra.

Definition: *The cross product is defined by the following set of identities.*

$$\vec{i}\times\vec{i}=\vec{j}\times\vec{j}=\vec{k}\times\vec{k}=\vec{0}$$
$$\vec{i}\times\vec{j}=\vec{k},\quad \vec{j}\times\vec{k}=\vec{i},\quad \vec{k}\times\vec{i}=\vec{j}, \tag{20.7}$$
$$\vec{j}\times\vec{i}=-\vec{k},\quad \vec{k}\times\vec{j}=-\vec{i},\quad \vec{i}\times\vec{k}=-\vec{j}.$$

This definition is actually easy to remember. The first three identities simply state that the cross product of a vector with itself is zero. The latter identities can be remembered by positioning the vectors $\vec{i}, \vec{j}$, and $\vec{k}$ around a circle, and then noticing that the cross product is akin to clockwise motion. When we move around the circle in a counterclockwise direction the signs are reversed.

To extend this definition to all vectors we use the identities

$$\begin{aligned} (a\vec{\mathbf{v}} + b\vec{\mathbf{w}}) \times c\vec{\mathbf{u}} &= ac(\vec{\mathbf{v}} \times \vec{\mathbf{u}}) + bc(\vec{\mathbf{w}} \times \vec{\mathbf{u}}) \\ c\vec{\mathbf{u}} \times (a\vec{\mathbf{v}} \times b\vec{\mathbf{w}}) &= ac(\vec{\mathbf{u}} \times \vec{\mathbf{v}}) + bc(\vec{\mathbf{u}} \times \vec{\mathbf{w}}), \end{aligned} \tag{20.8}$$

where $\vec{\mathbf{v}}$, $\vec{\mathbf{w}}$, and $\vec{\mathbf{u}}$ are any vectors, and a, b and c are scalars. This definition of the cross product can be programmed on a CAS (although this is non–trivial) allowing the CAS to calculate the cross product of an arbitrary pair of vectors.

Warning: The cross product is *not* commutative, i.e., in general,

$$\vec{\mathbf{v}} \times \vec{\mathbf{w}} \neq \vec{\mathbf{w}} \times \vec{\mathbf{v}}.$$

Example [20.15] Calculate the cross product $(2\vec{\mathbf{i}} - 3\vec{\mathbf{j}}) \times \vec{\mathbf{k}}$.

We have

$$(2\vec{\mathbf{i}} - 3\vec{\mathbf{j}}) \times \vec{\mathbf{k}} = 2(\vec{\mathbf{i}} \times \vec{\mathbf{k}}) - 3(\vec{\mathbf{j}} \times \vec{\mathbf{k}})$$

$$= -2\vec{\mathbf{j}} - 3\vec{\mathbf{i}}.$$

Example [20.16] Calculate $\overrightarrow{(1,0,1)} \times \overrightarrow{(0,1,1)}$.

To perform this calculation we first express each vector as a sum of basis vectors and then proceed as before.

$$\overrightarrow{(1,0,1)} \times \overrightarrow{(0,1,1)} = (\vec{\mathbf{i}} + \vec{\mathbf{k}}) \times (\vec{\mathbf{j}} + \vec{\mathbf{k}})$$

$$= (\vec{\mathbf{i}} + \vec{\mathbf{k}}) \times \vec{\mathbf{j}} + (\vec{\mathbf{i}} + \vec{\mathbf{k}}) \times \vec{\mathbf{k}}$$

$$= \vec{\mathbf{k}} - \vec{\mathbf{i}} - \vec{\mathbf{j}}$$

$$= \overrightarrow{(-1,-1,1)}.$$

Remarks: It it critical to remember that the dot product and the cross product are entirely different entities. The dot product of two vectors results in a *scalar* while the cross product of two vectors results in a third *vector*.

Having computed a few examples of the cross product it is natural to try to compute the cross product of two arbitrary vectors. Thus consider two vectors

$$\vec{\mathbf{v}} = \overrightarrow{(a_1, b_1, c_1)}$$
$$\vec{\mathbf{w}} = \overrightarrow{(a_2, b_2, c_2)}.$$

By definition the cross product is given by the following (somewhat lengthy) calculation:

$$\begin{aligned}
\vec{\mathbf{v}} \times \vec{\mathbf{w}} &= \overrightarrow{(a_1, b_1, c_1)} \times \overrightarrow{(a_2, b_2, c_2)} \\
&= (a_1\vec{\mathbf{i}} + b_1\vec{\mathbf{j}} + c_1\vec{\mathbf{k}}) \times (a_2\vec{\mathbf{i}} + b_2\vec{\mathbf{j}} + c_2\vec{\mathbf{k}}) \\
&= (a_1\vec{\mathbf{i}} \times a_2\vec{\mathbf{i}} \;+\; b_1\vec{\mathbf{j}} \times a_2\vec{\mathbf{i}} \;+\; c_1\vec{\mathbf{k}} \times a_2\vec{\mathbf{i}}) \\
&\quad + (a_1\vec{\mathbf{i}} \times b_2\vec{\mathbf{j}} \;+\; b_1\vec{\mathbf{j}} \times b_2\vec{\mathbf{j}} \;+\; c_1\vec{\mathbf{k}} \times b_2\vec{\mathbf{j}}) \\
&\quad\quad + (a_1\vec{\mathbf{i}} \times c_2\vec{\mathbf{k}} \;+\; b_1\vec{\mathbf{j}} \times c_2\vec{\mathbf{k}} \;+\; c_1\vec{\mathbf{k}} \times c_2\vec{\mathbf{k}}) \\
&= \Big(0 + b_1a_2(-\vec{\mathbf{k}}) + c_1a_2(\vec{\mathbf{j}})\Big) \qquad (20.9) \\
&\quad + \Big(a_1b_2(\vec{\mathbf{k}}) + 0 + c_1b_2(-\vec{\mathbf{i}})\Big) \\
&\quad\quad + \Big(a_1c_2(-\vec{\mathbf{j}}) + b_1c_2(\vec{\mathbf{i}}) + 0\Big) \\
&= (b_1c_2 - c_1b_2)\,\vec{\mathbf{i}} + (c_1a_2 - a_1c_2)\,\vec{\mathbf{j}} + (a_1b_2 - b_1a_2)\,\vec{\mathbf{k}} \\
&= (b_1c_2 - c_1b_2)\,\vec{\mathbf{i}} - (a_1c_2 - c_1a_2)\,\vec{\mathbf{j}} + (a_1b_2 - b_1a_2)\,\vec{\mathbf{k}}.
\end{aligned}$$

This computation is actually used as a *definition* of the cross product in many texts. In order to make (20.9) easier to remember we now introduce the concept of the determinant. Given a 2×2 array

$$A = \begin{pmatrix} a & b \\ c & d \end{pmatrix}$$

define the **determinant of** A to be

$$\det(A) = \begin{vmatrix} a & b \\ c & d \end{vmatrix} = ad - bc.$$

Notice that differences of this form appear at the end of the calculation (20.9). Now given a 3×3 array

$$B = \begin{pmatrix} a_{11} & a_{12} & a_{13} \\ a_{21} & a_{22} & a_{23} \\ a_{31} & a_{32} & a_{33} \end{pmatrix},$$

we define the **determinant of** B to be

$$\det(B) = \begin{vmatrix} a_{11} & a_{12} & a_{13} \\ a_{21} & a_{22} & a_{23} \\ a_{31} & a_{32} & a_{33} \end{vmatrix} \tag{20.10}$$

$$= a_{11} \cdot \begin{vmatrix} a_{22} & a_{23} \\ a_{32} & a_{33} \end{vmatrix} - a_{12} \cdot \begin{vmatrix} a_{21} & a_{23} \\ a_{31} & a_{33} \end{vmatrix} + a_{13} \cdot \begin{vmatrix} a_{21} & a_{22} \\ a_{31} & a_{32} \end{vmatrix}$$

One way of remembering (20.10) is to focus on a_{11}, remove from the 3×3 array the row and column which include a_{11}, and then take the determinant of the resulting 2×2 array you are left with. Repeating this process with a_{12} and a_{13} and forming the alternating sum gives you the determinant of the 3×3 array.

Combining equations (20.9) with (20.10) gives us the following proposition.

Proposition [20.17] *The cross product of* $\vec{\mathbf{v}} = \overrightarrow{(a_1, b_1, c_1)}$ *and* $\vec{\mathbf{w}} = \overrightarrow{(a_2, b_2, c_2)}$ *is given by the determinant*

$$\vec{\mathbf{v}} \times \vec{\mathbf{w}} = \begin{vmatrix} \vec{\mathbf{i}} & \vec{\mathbf{j}} & \vec{\mathbf{k}} \\ a_1 & b_1 & c_1 \\ a_2 & b_2 & c_2 \end{vmatrix}.$$

Example [20.18] Compute the determinant of $B = \begin{pmatrix} 1 & 2 & 3 \\ 1 & 0 & 2 \\ -1 & 1 & 0 \end{pmatrix}$.

Following the directions given in (20.10) we have

$$\begin{vmatrix} 1 & 2 & 3 \\ 1 & 0 & 2 \\ -1 & 1 & 0 \end{vmatrix} = 1 \cdot \begin{vmatrix} 0 & 2 \\ 1 & 0 \end{vmatrix} - 2 \begin{vmatrix} 1 & 2 \\ -1 & 0 \end{vmatrix} + 3 \begin{vmatrix} 1 & 0 \\ -1 & 1 \end{vmatrix},$$

and, hence, $\det(B) = -2 - 4 + 3 = -3$.

Example [20.19] Calculate $\overrightarrow{(2,-1,4)} \times \overrightarrow{(3,1,0)}$.

Here we apply our proposition to obtain,

$$\overrightarrow{(2,-1,4)} \times \overrightarrow{(3,1,0)} = \begin{vmatrix} \vec{\mathbf{i}} & \vec{\mathbf{j}} & \vec{\mathbf{k}} \\ 2 & -1 & 4 \\ 3 & 1 & 0 \end{vmatrix},$$

and then definition (20.10) tells us that

$$\overrightarrow{(2,-1,4)} \times \overrightarrow{(3,1,0)} = \vec{\mathbf{i}} \cdot \begin{vmatrix} -1 & 4 \\ 1 & 0 \end{vmatrix} - \vec{\mathbf{j}} \cdot \begin{vmatrix} 2 & 4 \\ 3 & 0 \end{vmatrix} + \vec{\mathbf{k}} \cdot \begin{vmatrix} 2 & -1 \\ 3 & 1 \end{vmatrix}.$$

By computing the various 2×2 arrays we obtain our answer,

$$\overrightarrow{(2,-1,4)} \times \overrightarrow{(3,1,0)} = -4\vec{\mathbf{i}} + 12\vec{\mathbf{j}} + 5\vec{\mathbf{k}}.$$

Exercises for §20.5

Compute the cross product of the following pairs of vectors.

(1) $(2\vec{\mathbf{i}} - 3\vec{\mathbf{j}}) \times \vec{\mathbf{i}}$.

(2) $(-2\vec{\mathbf{i}} + \vec{\mathbf{j}} - 4\vec{\mathbf{k}}) \times (2\vec{\mathbf{i}} - \vec{\mathbf{j}})$.

(3) $\overrightarrow{(1,4,-1)} \times \overrightarrow{(2,1,0)}$.

(4) $\overrightarrow{(-3,2,5)} \times \overrightarrow{(0,-1,-3)}$.

(5) $\overrightarrow{(1,-1,1)} \times \overrightarrow{(2,-3,-5)}$.

(6) $\overrightarrow{(2,-3,2)} \times \overrightarrow{(1,2,2)}$.

Compute the following determinants.

(7) $\begin{vmatrix} 1 & 2 \\ 6 & -3 \end{vmatrix}$.

(8) $\begin{vmatrix} 1 & 6 & 2 \\ -1 & 1 & 1 \\ 2 & -3 & 1 \end{vmatrix}$.

(9) $\begin{vmatrix} \vec{\mathbf{i}} & \vec{\mathbf{j}} & \vec{\mathbf{k}} \\ 4 & -1 & 0 \\ 1 & 2 & 1 \end{vmatrix}$.

20.6 Some Basic Properties of the Cross Product

The cross product of 2 vectors $\vec{v}$ and $\vec{w}$ in 3 dimensional space has three distinguished properties which we present in the following form.

Theorem [20.20] *Let $\vec{v}$ and $\vec{w}$ be vectors in three dimensional space.*

1. *The vector $\vec{v} \times \vec{w}$ is perpendicular to both $\vec{v}$ and $\vec{w}$.*

2. *The direction of $\vec{v} \times \vec{w}$ is determined by the right–hand–rule, that is to say if the fingers of the right hand are curled in the direction of the rotation from $\vec{v}$ to $\vec{w}$, then the thumb points in the direction of the vector $\vec{v} \times \vec{w}$.*

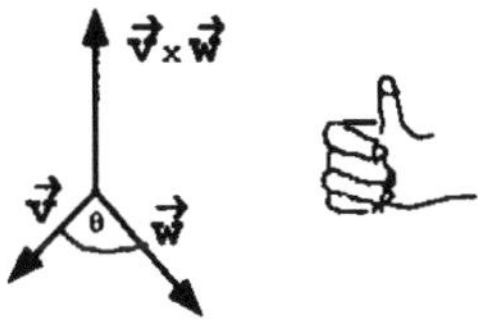

3. *The length of $\vec{v} \times \vec{w}$ is given by the identity*

$$|\vec{v} \times \vec{w}| = |\vec{v}| \cdot |\vec{w}| \cdot \sin(\theta),$$

where θ is the angle between $\vec{v}$ and $\vec{w}$.

Before we prove these properties of the cross product lets examine some of their applications.

Example [20.21] Find a vector of length 3 which is perpendicular to both $\overrightarrow{(1,-1,2)}$ and $\overrightarrow{(1,1,-1)}$.

In order to find the vector with the desired properties we first compute the cross product of the given vectors (yielding a vector which is perpendicular to the given vectors) and then multiply the result by an appropriate scalar (to adjust the size).

The cross product is

$$\overrightarrow{(1,-1,2)} \times \overrightarrow{(1,1,-1)} = \begin{vmatrix} \vec{\mathbf{i}} & \vec{\mathbf{j}} & \vec{\mathbf{k}} \\ 1 & -1 & 2 \\ 1 & 1 & -1 \end{vmatrix}$$

$$= -\vec{\mathbf{i}} + 3\vec{\mathbf{j}} + 2\vec{\mathbf{k}}$$

$$= \overrightarrow{(-1,3,2)},$$

and the length of $\overrightarrow{(-1,3,2)}$ is $\sqrt{1+9+4} = \sqrt{14}$. To force our vector to have length 3 we simply multiply $\overrightarrow{(-1,3,2)}$ by $3/\sqrt{14}$ yielding

$$\overrightarrow{\left(\frac{-3}{\sqrt{14}}, \frac{9}{\sqrt{14}}, \frac{6}{\sqrt{14}}\right)}.$$

Example [20.22] What is the angle $0 \le \theta \le \pi$ between $\overrightarrow{(1,0,1)}$ and $\overrightarrow{(0,1,1)}$?

First we compute the cross product of the vectors,

$$\overrightarrow{(1,0,1)} \times \overrightarrow{(0,1,1)} = -\vec{\mathbf{i}} - \vec{\mathbf{j}} + \vec{\mathbf{k}} = \overrightarrow{(-1,-1,1)}.$$

The third property in our theorem tells us that

$$|\overrightarrow{(-1,-1,1)}| = \sqrt{3}$$

$$= |\overrightarrow{(1,0,1)}| \cdot |\overrightarrow{(0,1,1)}| \sin(\theta)$$

$$= \sqrt{2} \cdot \sqrt{2} \sin(\theta),$$

and hence $\sin(\theta) = \sqrt{3}/2$ and $\theta = \pi/3$.

Proof of Theorem [20.20]

1. To prove $\vec{\mathbf{v}} \times \vec{\mathbf{w}}$ is perpendicular to $\vec{\mathbf{v}}$ we shall demonstrate that the dot product $(\vec{\mathbf{v}} \times \vec{\mathbf{w}}) \cdot \vec{\mathbf{v}} = 0$. In fact this verification is quite easy: assuming $\vec{\mathbf{v}} = \overrightarrow{(v_1, v_2, v_3)}$ and $\vec{\mathbf{w}} = \overrightarrow{(w_1, w_2, w_3)}$ then by (20.9)

$$(\vec{\mathbf{v}} \times \vec{\mathbf{w}}) \cdot \vec{\mathbf{v}} = v_1(v_2w_3 - v_3w_2) - v_2(v_1w_3 - v_3w_1) + v_3(v_1w_2 - v_2w_1).$$

and every term cancels to prove the identity. The proof that $\vec{\mathbf{v}} \times \vec{\mathbf{w}}$ is perpendicular to $\vec{\mathbf{w}}$ follows in just the same way.

2. This is simply a restatement of the definitions $\vec{\mathbf{i}} \times \vec{\mathbf{j}} = \vec{\mathbf{k}}$, etc.

3. To demonstrate this last property we use the identity

$$(\vec{\mathbf{v}} \times \vec{\mathbf{w}}) \cdot (\vec{\mathbf{v}} \times \vec{\mathbf{w}}) = |\vec{\mathbf{v}}|^2 |\vec{\mathbf{w}}|^2 - (\vec{\mathbf{v}} \cdot \vec{\mathbf{w}})^2,$$

which can easily be verified on the CAS (or by hand). Combining this identity with

$$\vec{\mathbf{v}} \cdot \vec{\mathbf{w}} = |\vec{\mathbf{v}}||\vec{\mathbf{w}}| \cos(\theta),$$

yields

$$|\vec{\mathbf{v}} \times \vec{\mathbf{w}}|^2 = |\vec{\mathbf{v}}|^2 |\vec{\mathbf{w}}|^2 (1 - \cos^2(\theta)) = |\vec{\mathbf{v}}|^2 |\vec{\mathbf{w}}|^2 \sin^2(\theta).$$

20.7 Applications of the Cross Product

In this section we discuss three noteworthy applications of the cross product.

The Area of a Parallelogram

Consider a parallelogram spanned by two vectors $\vec{\mathbf{v}}$ and $\vec{\mathbf{w}}$.

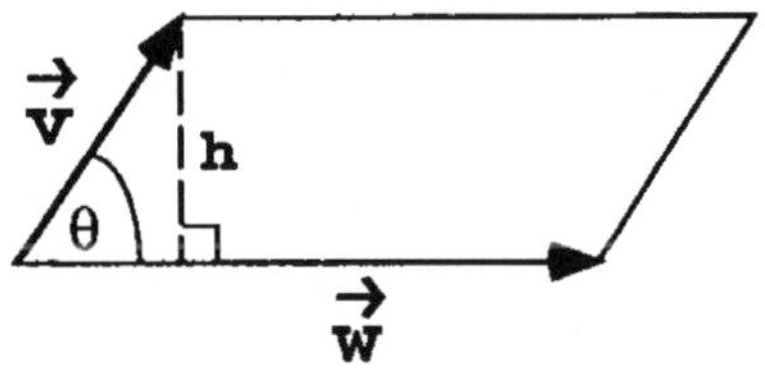

Notice that since $h = |\vec{\mathbf{v}}| \sin(\theta)$ the area of the parallelogram is

$$\begin{aligned} |\vec{\mathbf{w}}| \cdot h &= |\vec{\mathbf{v}}| \cdot |\vec{\mathbf{w}}| \sin(\theta) \\ &= |\vec{\mathbf{v}} \times \vec{\mathbf{w}}|. \end{aligned} \tag{20.11}$$

Example [20.23] Find the area of the triangle whose vertices are $P = (1,1,1)$, $Q = (1,2,3)$, and $R = (3,-1,2)$.

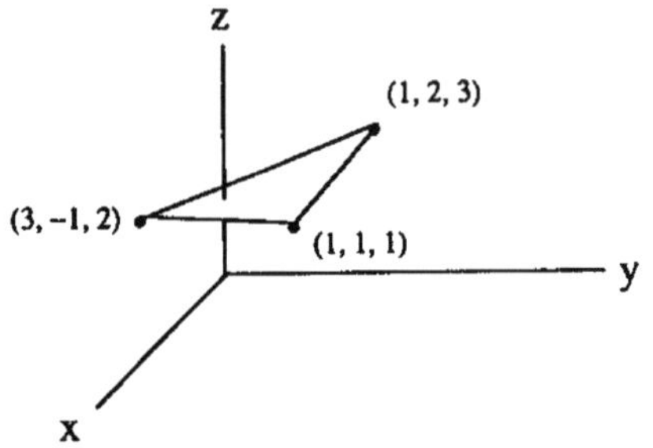

To obtain the area of the triangle notice that the triangle has half the area of the parallelogram spanned by the vectors

$$\overrightarrow{PQ} = \overrightarrow{(1,2,3)} - \overrightarrow{(1,1,1)} = \overrightarrow{(0,1,2)}$$

$$\overrightarrow{PR} = \overrightarrow{(3,-1,2)} - \overrightarrow{(1,1,1)} = \overrightarrow{(2,-2,1)}.$$

Hence

$$\text{Area of the triangle} = \left|\frac{1}{2}(\overrightarrow{PQ} \times \overrightarrow{PR})\right|$$

$$= \frac{1}{2}\begin{vmatrix} \vec{i} & \vec{j} & \vec{k} \\ 0 & 1 & 2 \\ 2 & -2 & 1 \end{vmatrix}$$

$$= \left|\frac{5}{2}\vec{i} + 2\vec{j} - \vec{k}\right|$$

$$= \sqrt{\frac{25}{4} + 4 + 1} = \frac{1}{2}\sqrt{45},$$

Volume of a Parallelepiped

Let $\vec{\mathbf{a}}$, $\vec{\mathbf{b}}$, and $\vec{\mathbf{c}}$ be three vectors in 3–dimensional space.

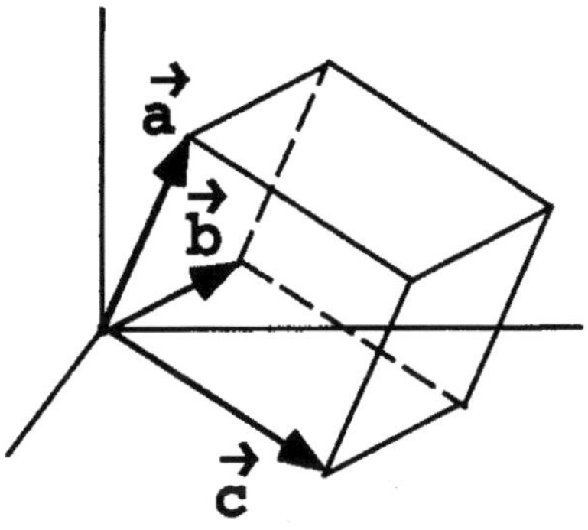

which determine a parallelepiped. The volume of the solid is

$$V = |\vec{\mathbf{a}} \cdot (\vec{\mathbf{b}} \times \vec{\mathbf{c}})| = |d|,$$

where

$$d = \begin{vmatrix} a_1 & a_2 & a_3 \\ b_1 & b_2 & b_3 \\ c_1 & c_2 & c_3 \end{vmatrix}. \tag{20.12}$$

Example [20.24] Find the volume of the parallelepiped determined by the vectors $\vec{\mathbf{i}}, \vec{\mathbf{j}}$, and $2\vec{\mathbf{i}} - 3\vec{\mathbf{j}} + \vec{\mathbf{k}}$.

Following (20.12) we have

$$\begin{aligned} V &= |\vec{\mathbf{i}} \cdot (\vec{\mathbf{j}} \times (2\vec{\mathbf{i}} - 3\vec{\mathbf{j}} + \vec{\mathbf{k}}))| \\ &= |\vec{\mathbf{i}} \cdot (\vec{\mathbf{i}} - 2\vec{\mathbf{k}})| \\ &= 1. \end{aligned}$$

Note that we can also obtain this result by computing the determinant

$$\begin{vmatrix} 1 & 0 & 0 \\ 0 & 1 & 0 \\ 2 & -3 & 1 \end{vmatrix} = 1 \cdot \begin{vmatrix} 1 & 0 \\ -3 & 1 \end{vmatrix} = 1$$

Cramer's Formula

For our final application of the cross product we move from geometry to systems of three linear equations in three unknowns. Thus, consider the system of equations

$$\begin{aligned} a_1x + b_1y + c_1z &= m_1 \\ a_2x + b_2y + c_2z &= m_2 \\ a_3x + b_3y + c_3z &= m_3, \end{aligned} \tag{20.13}$$

where a_i, b_i, c_i and m_i are known constants for $i = 1, 2, 3$.

Using the vectors

$$\vec{\mathbf{a}} = \overrightarrow{(a_1, a_2, a_3)}, \quad \vec{\mathbf{b}} = \overrightarrow{(b_1, b_2, b_3)}, \quad \vec{\mathbf{c}} = \overrightarrow{(c_1, c_2, c_3)}, \quad \vec{\mathbf{m}} = \overrightarrow{(m_1, m_2, m_3)},$$

we first express (20.13) in the form

$$\vec{\mathbf{a}}x + \vec{\mathbf{b}}y + \vec{\mathbf{c}}z = \vec{\mathbf{m}}. \tag{20.14}$$

Recall that $\vec{\mathbf{b}} \times \vec{\mathbf{c}}$ is perpendicular to both $\vec{\mathbf{b}}$ and $\vec{\mathbf{c}}$, which is a consequence of the identity $\vec{\mathbf{b}} \cdot (\vec{\mathbf{b}} \times \vec{\mathbf{c}}) = \vec{\mathbf{c}} \cdot (\vec{\mathbf{b}} \times \vec{\mathbf{c}}) = 0$. It follows from equation (20.14) that

$$\begin{aligned} (\vec{\mathbf{a}}x + \vec{\mathbf{b}}y + \vec{\mathbf{c}}z) \cdot (\vec{\mathbf{b}} \times \vec{\mathbf{c}}) &= x\vec{\mathbf{a}} \cdot (\vec{\mathbf{b}} \times \vec{\mathbf{c}}) \\ &= \vec{\mathbf{m}} \cdot (\vec{\mathbf{b}} \times \vec{\mathbf{c}}). \end{aligned}$$

We can thus *solve* for x in an unexpected and remarkable manner,

$$x = \frac{\vec{\mathbf{m}} \cdot (\vec{\mathbf{b}} \times \vec{\mathbf{c}})}{\vec{\mathbf{a}} \cdot (\vec{\mathbf{b}} \times \vec{\mathbf{c}})}.$$

Similarly

$$y = \frac{\vec{\mathbf{m}} \cdot (\vec{\mathbf{a}} \times \vec{\mathbf{c}})}{\vec{\mathbf{b}} \cdot (\vec{\mathbf{a}} \times \vec{\mathbf{c}})},$$

and

$$z = \frac{\vec{\mathbf{m}} \cdot (\vec{\mathbf{a}} \times \vec{\mathbf{b}})}{\vec{\mathbf{c}} \cdot (\vec{\mathbf{a}} \times \vec{\mathbf{b}})}.$$

Example [20.25] Given the following system of equations, solve for x,

$$\begin{aligned} 2x - y + z &= 1 \\ 3x + 2y - z &= -1 \\ -x + y + 2z &= 0. \end{aligned}$$

Using the method described above we arrive at a solution without effort,

$$x = \frac{\overrightarrow{(1,-1,0)} \cdot \left(\overrightarrow{(-1,2,1)} \times \overrightarrow{(1,-1,2)}\right)}{\overrightarrow{(2,3,-1)} \cdot \left(\overrightarrow{(-1,2,1)} \times \overrightarrow{(1,-1,2)}\right)} = \frac{1}{10}.$$

Exercises for §20.7

(1) Using your CAS graph the triangle with vertices at $(1,-1,2)$, $(3,5,-2)$, $(-2,6,0)$. Find the area of this triangle.

(2) Find the area of the parallelogram spanned by the vectors $(1,2,-2)$ and $(3,-4,1)$. Graph this parallelogram with your CAS.

Compute the volume of the following parallelepipeds determined by the following triples $\vec{a}, \vec{b}, \vec{c}$ of vectors.

(3) $\vec{a} = \overrightarrow{(1,1,1)}$, $\vec{b} = \overrightarrow{(-2,3,2)}$, and $\vec{c} = \overrightarrow{(0,0,1)}$

(4) $\vec{a} = \overrightarrow{(2,-1,3)}$, $\vec{b} = \overrightarrow{(1,-2,-1)}$, and $\vec{c} = \overrightarrow{(1,-1,9)}$.

Solve the following systems of 3 equations in 3 unknowns using Cramer's formula. Check your answers with your CAS.

(5)
$$\begin{aligned} 2x + 4y + z &= 0 \\ x + y + 3z &= 6 \\ 3x - y + z &= 6 \end{aligned}$$

(6)
$$\begin{aligned} 3x - y + z &= 3 \\ 2x + y + 4z &= 2 \\ -2x + 3y - z &= -7. \end{aligned}$$

(7) Show that $|\vec{v} \cdot \vec{w}|^2 + |\vec{v} \times \vec{w}|^2 = |\vec{v}|^2|\vec{w}|^2$ for any two vectors $\vec{v}$ and $\vec{w}$.

(8) Consider an arbitrary triangle with sides of lengths a, b, c and angles α, β, γ as in the diagram below.

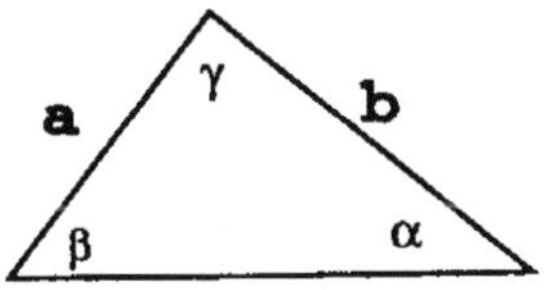

Derive the **law of sines:** $\frac{\sin(\alpha)}{a} = \frac{\sin(\beta)}{b} = \frac{\sin(\gamma)}{c}$. **Hint:** Let $\vec{a}, \vec{b}, \vec{c}$ be vectors associated to the sides of the triangle. The $\vec{u} = \vec{a} + \vec{b} + \vec{c} = 0$, and $\vec{a} \times \vec{u} = \vec{b} \times \vec{u} = \vec{b} \times \vec{u} = 0$.

(9) Show that $\vec{a} \times (\vec{b} \times \vec{c}) = (\vec{a} \cdot \vec{c})\vec{b} - (\vec{a} \cdot \vec{b})\vec{c}$ for any three vectors $\vec{a}, \vec{b}$, and $\vec{c}$.

(10) Using your CAS, verify **Jacobi's identity:**

$$\vec{a} \times (\vec{b} \times \vec{c}) + \vec{b} \times (\vec{c} \times \vec{a}) + \vec{c} \times (\vec{a} \times \vec{b}) = 0.$$

(11) Show that $(2\vec{\mathbf{v}} - \vec{\mathbf{w}}) \cdot (\vec{\mathbf{v}} \times \vec{\mathbf{w}}) = 0$ for any two vectors $\vec{\mathbf{v}}$ and $\vec{\mathbf{w}}$.

Additional exercises for Chapter XX

(1) Let M be the midpoint of side BC of a parallelogram $ABCD$. Prove that AM trisects the diagonal BD.

(2) Let O be the origin in the xy plane and let $\overrightarrow{OA} = \vec{\mathbf{i}} + \vec{\mathbf{j}}, \overrightarrow{OB} = 3\vec{\mathbf{i}} + \vec{\mathbf{k}}, \overrightarrow{OC} = 4\vec{\mathbf{i}} - 3\vec{\mathbf{j}} - 4\vec{\mathbf{k}}$. Show that ABC is a right triangle and find its area.

(3) What does it mean for the cross product of two vectors $\vec{\mathbf{v}}$ and $\vec{\mathbf{w}}$ to be zero.

(4) Find the volume of the parallelepiped determined by the vectors $\vec{\mathbf{i}} + \vec{\mathbf{j}} + \vec{\mathbf{k}}, \vec{\mathbf{i}} - \vec{\mathbf{j}} - 2\vec{\mathbf{k}}, 2\vec{\mathbf{j}} - \vec{\mathbf{k}}$.

(5) Is it true that if $\vec{\mathbf{a}} \times \vec{\mathbf{b}} = 0$ and $\vec{\mathbf{b}} \times \vec{\mathbf{c}} = 0$ then $\vec{\mathbf{a}} \times \vec{\mathbf{c}} = 0$?

(6) Is it true that if $\vec{\mathbf{a}} \cdot \vec{\mathbf{b}} = 0$ and $\vec{\mathbf{b}} \cdot \vec{\mathbf{c}} = 0$ then $\vec{\mathbf{a}} \cdot \vec{\mathbf{c}} = 0$?

(7) Show that the diagonals of a parallelogram bisect each other. Find the areas of the triangles determined by the diagonal and the sides of the parallelogram.

(8) A vector triple product of the vectors $\vec{\mathbf{a}}, \vec{\mathbf{b}}, \vec{\mathbf{c}}$ is defined by $\vec{\mathbf{a}} \times (\vec{\mathbf{b}} \times \vec{\mathbf{c}})$. This vector is perpendicular to $\vec{\mathbf{b}} \times \vec{\mathbf{c}}$.

  (a) What can you say about the vector triple product and the vectors $\vec{\mathbf{b}}$ and $\vec{\mathbf{c}}$?

  (b) Show that there are constants s and t so that $\vec{\mathbf{a}} \times (\vec{\mathbf{b}} \times \vec{\mathbf{c}}) = s\vec{\mathbf{b}} + t\vec{\mathbf{c}}$.

  (c) Express the fact that $\vec{\mathbf{a}} \times (\vec{\mathbf{b}} \times \vec{\mathbf{c}})$ is perpendicular to $\vec{\mathbf{a}}$ in terms of $s, t, \vec{\mathbf{b}}, \vec{\mathbf{c}}$.

  (d) Show that $\vec{\mathbf{a}} \times (\vec{\mathbf{b}} \times \vec{\mathbf{c}}) = (\vec{\mathbf{a}} \cdot \vec{\mathbf{c}})\vec{\mathbf{b}} - (\vec{\mathbf{a}} \cdot \vec{\mathbf{b}})\vec{\mathbf{c}}$.

  (e) Is the vector multiplication associative?

(9) Vector $\overrightarrow{OP}$ has magnitude $2a$ and points to the right in direction $60°$ above the horizontal. What vector combined with it will yield a vertical resultant, $\overrightarrow{OR}$, of magnitude $2\sqrt{3}a$?

(10) If $\vec{\mathbf{u}}$ is of magnitude 3 and points horizontally due east and $\vec{\mathbf{v}}$ is of magnitude 5 and points vertically upwards, describe the vector $\vec{\mathbf{u}} \times \vec{\mathbf{v}}$.

(11) Forces are represented by vectors. What force combined with a force at a point O of $1lb$ pulling to the east will yield a resultant force of $2lb$ pulling in a direction $30°$ north of east ?

(12) Show that $\vec{\mathbf{i}} \cdot (\vec{\mathbf{j}} \times \vec{\mathbf{k}}) = (\vec{\mathbf{i}} \cdot \vec{\mathbf{j}}) \times \vec{\mathbf{k}} = \vec{\mathbf{i}} \times (\vec{\mathbf{j}} \cdot \vec{\mathbf{k}})$.

(13) Show that $\vec{\mathbf{i}} \cdot (\vec{\mathbf{i}} + \vec{\mathbf{j}}) \cdot (\vec{\mathbf{i}} + \vec{\mathbf{j}} + \vec{\mathbf{k}}) = 1$. Interpret the result geometrically in connection with a cube of unit edge and then prove the result geometrically.

(14) Find the angle between $(\vec{\mathbf{i}} + \vec{\mathbf{j}} + \vec{\mathbf{k}})$ and $(\vec{\mathbf{i}} + 2\vec{\mathbf{j}} + 3\vec{\mathbf{k}})$.

(15) Show that $\vec{\mathbf{i}}\cos(\alpha) + \vec{\mathbf{j}}\sin(\alpha)$ is a vector of unit magnitude in the x, y plane making an angle α with $\vec{\mathbf{i}}$. Then by considering the cross product of this vector with a similar one making an angle β with $\vec{\mathbf{i}}$, derive the well-known trigonometric formula

$$\sin(\beta - \alpha) = \sin(\beta)\cos(\alpha) - \cos(\beta)\sin(\alpha).$$

Solve the following systems of three equations with three unknowns using Cramer's formula. Use your CAS to verify the result.

(16)
$$\begin{aligned} 2x + 3y + z &= 14 \\ x + y + z &= 6 \\ 3x + 5y - z &= 19. \end{aligned}$$

(17)
$$\begin{aligned} 3x + 4y - z &= -7 \\ x - y + z &= 4 \\ -2x + 3y - 2z &= -9. \end{aligned}$$

(18)
$$\begin{aligned} 10x - y + 4z &= 8 \\ -2x + y - 4z &= -4 \\ x - 5y - 2z &= 10. \end{aligned}$$

(19)
$$\begin{aligned} 8x + y + 4z &= -9 \\ 24x - y - 3z &= 3 \\ -8x + 3y + 8z &= 9. \end{aligned}$$

(20)
$$\begin{aligned} 3x + 4y + 12z &= 1 \\ x - y + z &= 0 \\ 12x - 2y + 6z &= 3. \end{aligned}$$

Chapter XXI

Two and Three Dimensional Graphics

21.1 Lines in Space

One of the most significant applications of the theory of vectors is to computer aided graphical display. A computer creates an image by placing small dots at various grid points on a screen. The greater the number of grid points the better the resolution of the resulting image.

Let us assume that the CAS screen has three fixed lines (axes) drawn to create an illusion of three dimensional space. Following the right–handed convention, the axes are labeled x, y, z as in the figure below.

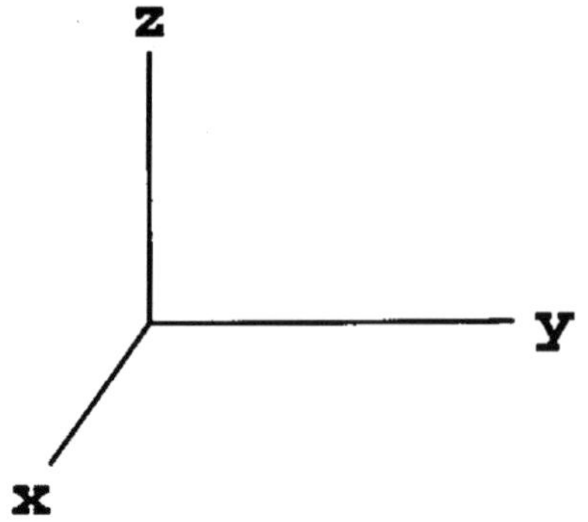

The three axes determine a grid, and given an arbitrary grid point (a, b, c), the CAS is preprogrammed to be able to place a dot there.

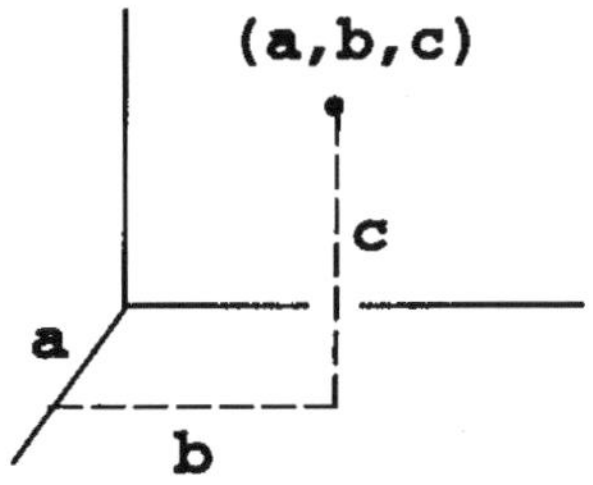

Problem: How do we obtain graphical displays of more complex three–dimensional objects?

Solution: The general approach to attacking this problem involves two steps. First we must describe the object in question by an equation (we have of course done just this for a variety of two dimensional figures). Once the equation is known we direct the CAS to place dots at appropriate grid points (which the equation dictates) to create the desired image.

We begin our exploration of three dimensional figures with the line (much as we began our study of two dimensional figures in Chapter 3).

Example [21.1] What is the equation of the line which joins the distinct points $P = (p_1, p_2, p_3)$ and $Q = (q_1, q_2, q_3)$.

A direct approach to solving this problem uses the theory of vectors which we developed in Chapter 20. Consider the line in question and let (x, y, z) be the coordinates of an arbitrary point on the line.

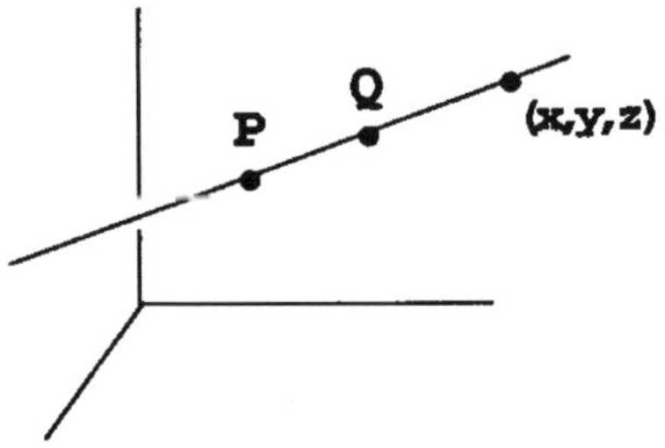

Form the vectors $\overrightarrow{(x, y, z)} - \overrightarrow{(p_1, p_2, p_3)}$, and $\overrightarrow{(q_1, q_2, q_3)} - \overrightarrow{(p_1, p_2, p_3)}$. Clearly these vectors are parallel (they visibly appear on the same line), and thus they can only differ in magnitude, i.e., there exists some real number t such that

$$\overrightarrow{(x - p_1, x - p_2, x - p_3)} = t\,\overrightarrow{(q_1 - p_1, q_2 - p_2, q_3 - p_3)}.$$

Thus we obtain the equations

$$x = p_1 + (q_1 - p_1)\,t, \quad y = p_2 + (q_2 - p_2)\,t, \quad z = p_3 + (q_3 - p_3)\,t. \tag{21.1}$$

By allowing t to run over the set of real numbers the resulting set of points will consist of the entire line which passes through P and Q. The equations given in (21.1) are termed the **parametric equations of the line**.

Example [20.2] Find the equation of the line ℓ which passes through the point $(1, 3, 2)$ and is parallel to the vector $\vec{\mathbf{v}} = \overrightarrow{(1, -2, -2)}$.

Again we consider an arbitrary point (x, y, z) on the line in question and consider the vector $\vec{\mathbf{w}} = \overrightarrow{(x, y, z)} - \overrightarrow{(1, 3, 2)}$.

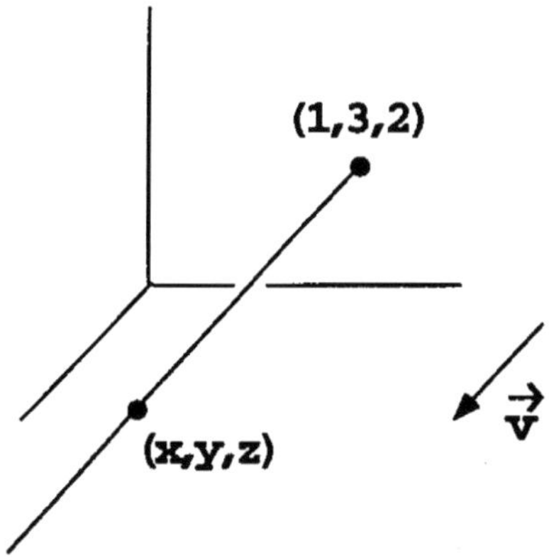

Since $\vec{\mathbf{w}}$ lies on ℓ, and ℓ is parallel to $\vec{\mathbf{v}}$, we see that $\vec{\mathbf{w}}$ is itself parallel to $\vec{\mathbf{v}}$. Thus there is a real number t such that

$$\vec{\mathbf{w}} = \overrightarrow{(x - 1, y - 3, z - 2)} = t\,\vec{\mathbf{v}} = t\,\overrightarrow{(1, -2, -2)},$$

and the parametric equation of the line ℓ is

$$x = 1 + t, \quad y = 3 - 2t, \quad z = 2 - 2t.$$

Notice that the parametric form of the line allows us to obtain points on the line very quickly: setting $t = 0$ we obtain $(1, 3, 2)$, setting $t = 1$ yields $(2, 1, 0)$, $t = 2$ gives $(3, -1, -2)$, and finally if $t = 3$ we get $(4, -3, -4)$.

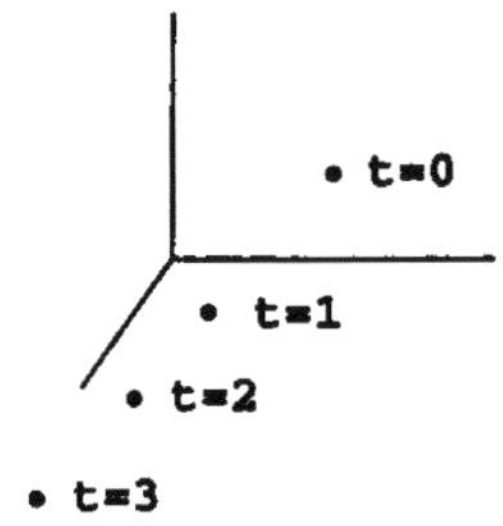

Exercises for §21.1

For each of the following pairs of points find the parametric equation of the line which passes through the points. Confirm your answer by (1) graphing the original points (as dots), (2) picking one other point on the line, and (3) graphing the line segments between the three points.

(1) $(1,3,0)$, $(2,-4,0)$.

(2) $(-1,0,3)$, $(2,0,8)$.

(3) $(1,1,2)$, $(2,-4,0)$.

(4) $(0,-1,3)$, $(4,-1,3)$.

(5) $(-2,9,-3)$, $(4,0,8)$.

(6) $(-2,-1,-3)$, $(1,2,3)$.

(7) Show that the three altitudes (one for each side) of a triangle intersect at a common point.

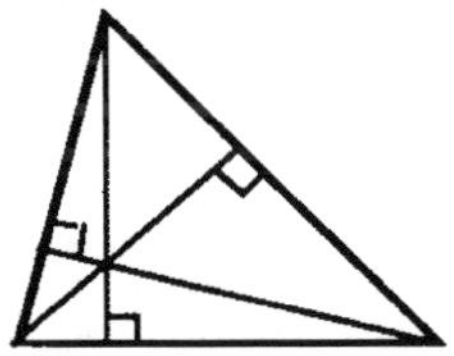

(8) Show that the set of all points (x, y, z) in 3–dimensional space which satisfy the equation

$$\frac{x - x_0}{a} = \frac{y - y_0}{b} = \frac{z - z_0}{c},$$

(where a, b, c, x_0, y_0, z_0 are fixed) lie on a straight line. Express the equation of this line in parametric form. (The above form is termed the **symmetric form** of the line.)

21.2 Planes: Their Equations and Properties

Having successfully derived the equation for a line in 3–dimensional space we move to the next most important object in graphics, the plane. A plane is a flat infinite two dimensional surface which can be visualized by imagining a sheet of tremendously thin glass.

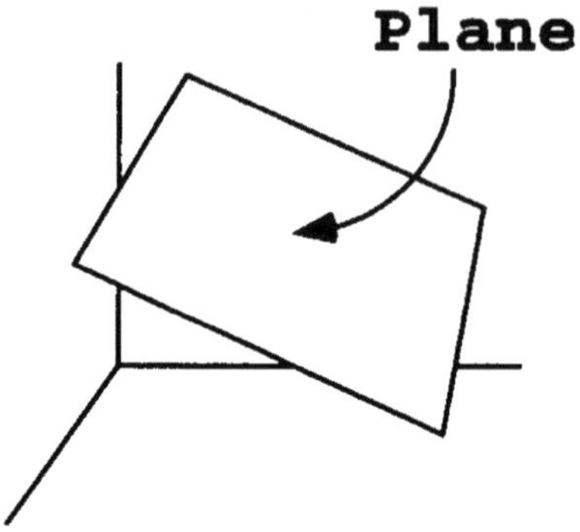

Question: How can a plane be characterized (i.e., defined) mathematically?

Answer: A plane is determined by a point P on the plane and a vector $\vec{n}$ which is perpendicular (normal) to the plane.

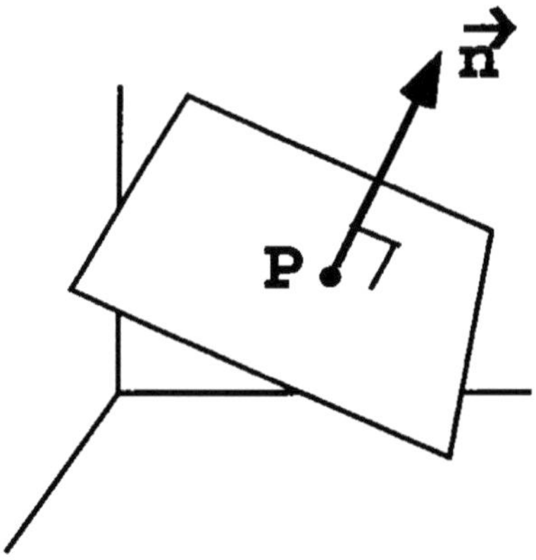

If we imagine the plane anchored at a point P we can rotate the plane in infinitely many directions (three of which are depicted below).

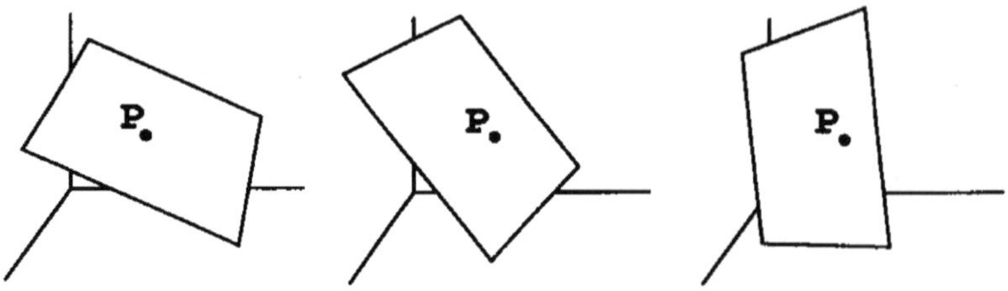

By specifying a normal vector $\vec{n}$ we uniquely determine the position of the plane, and hence the plane itself.

Example [21.3] Find the equation of the plane which passes through the point $P = (p_1, p_2, p_3)$ and is perpendicular to $\vec{n} = \overrightarrow{(n_1, n_2, n_3)}$.

Let (x, y, z) be an arbitrary point on the plane.

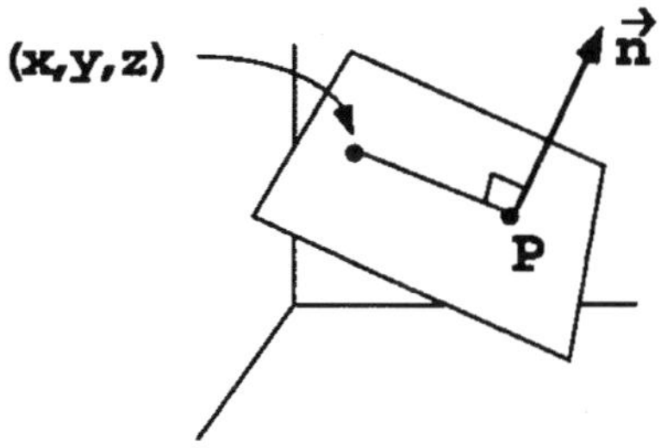

Since the vector $\overrightarrow{(x,y,z)} - \overrightarrow{(p_1,p_2,p_3)} = \overrightarrow{(x-p_1, y-p_2, z-p_3)}$ lies on the plane it must be perpendicular to $\vec{n}$. Recalling Proposition [20.7] we obtain the identity

$$\overrightarrow{(x-p_1, y-p_2, z-p_3)} \cdot \vec{n} = 0,$$

and the plane in question is given by the equation

$$n_1(x-p_1) + n_2(y-p_2) + n_3(z-p_3) = 0. \tag{21.2}$$

Example [21.4] Find the equation of the plane which passes through the point $(1, -2, 4)$ and is perpendicular to the vector $\overrightarrow{(2, -1, 3)}$.

Following equation (21.2) we see that the plane is given by

$$2(x-1) - (y+2) + 3(z-4) = 0,$$

which, when simplified, reduces to the equation

$$2x - y + 3z = 16.$$

Example [21.5] Find three points on the plane $2x - y + 3z = 16$.

There are of course infinitely many points of the plane and they all satisfy the equation $2x - y + 3z = 16$. Our choices are random: if $x = 0, y = 0$, then $z = 16/3$, if $x = 1, z = 5$ then $y = 1$, and if $y = 3, z = 5$, then $x = 2$. We conclude that the points $(0, 0, 16/3), (1, 1, 5)$, and $(2, 3, 5)$ appear on the plane.

Example [21.6] Find the equation of the plane which passes through the points $(3, 0, 0)$, $(0, 2, 0)$, and $(0, 0, 8)$.

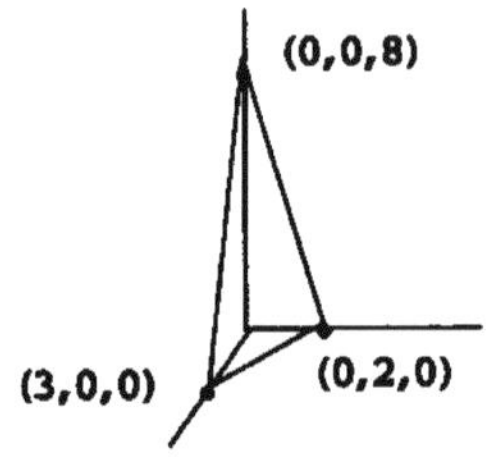

To solve this problem we require a normal vector $\vec{n}$. Recalling that the cross product of two vectors is perpendicular to each of the original vectors (Theorem [20.20]), we consider the vectors which appear naturally on the plane:

$$\overrightarrow{(0,0,8)} - \overrightarrow{(3,0,0)} \text{ and } \overrightarrow{(0,0,8)} - \overrightarrow{(0,2,0)}.$$

When we take the cross product of these vectors we obtain

$$\begin{vmatrix} \vec{i} & \vec{j} & \vec{k} \\ -3 & 0 & 8 \\ 0 & -2 & 8 \end{vmatrix} = 16\vec{i} + 24\vec{j} + 6\vec{k},$$

which is the required normal vector to the plane. Applying (21.2) with $P = (3, 0, 0)$ we obtain the equation of our plane,

$$16x + 24y + 6z = 48.$$

Example [21.7] (Distance from a point to a plane) Let $P = (p_1, p_2, p_3)$ be a point in space and let $ax + by + cz = d$ be a plane. Then $\overrightarrow{(a, b, c)} = \vec{n}$ is a normal vector to the plane. Let $Q = (q_1, q_2, q_3)$ be any point on the plane.

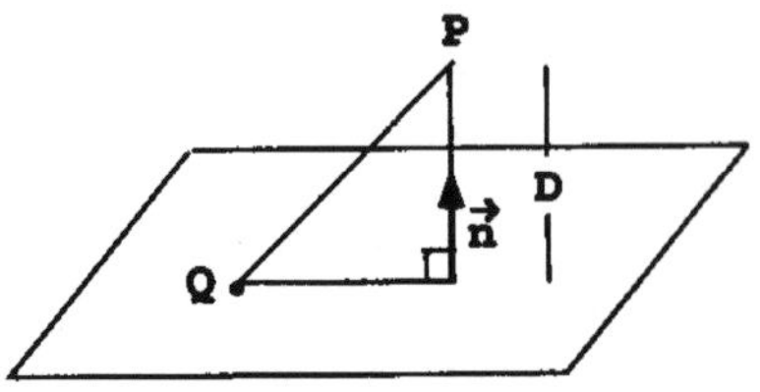

To find the distance, D, from the point P to the plane we must first form a line which passes through P and is perpendicular to the plane. With this done, we then measure the length of the line segment (on this line) from P to the plane. This construction is identical to computing the projection of the vector $\overrightarrow{PQ}$ onto the normal $\vec{\mathbf{n}}$, and then the distance is given by the compact formula

$$D = \frac{\overrightarrow{PQ} \cdot \vec{\mathbf{n}}}{|\vec{\mathbf{n}}|}. \tag{21.3}$$

Example [21.8] Find the distance from the point $(2, -3, 1)$ to the plane $x - 2y - z = 4$.

By (21.2), the normal vector to this plane is $\vec{\mathbf{n}} = \overrightarrow{(1, -2, -1)}$and the length of this vector is simply $|\vec{\mathbf{n}}| = \sqrt{6}$. Choosing a point Q on the plane, for example $Q = (0, 0, -4)$, then the distance from P to the plane is given by (21.3):

$$\begin{aligned} \frac{\overrightarrow{PQ} \cdot \vec{\mathbf{n}}}{|\vec{\mathbf{n}}|} &= \frac{\overrightarrow{(2, -3, 5)} \cdot \overrightarrow{(1, -2, -1)}}{\sqrt{6}} \\ &= \frac{3}{\sqrt{6}}. \end{aligned}$$

Remark: If we fix one of the variables in the equation of a plane,

$$ax + by + cz = d,$$

say $z = z_0$, then the set of points of the form (x, y, z_0) (where $x, y \in \mathbb{R}$) satisfying the equation $ax + by + cz_0 = d$ will form a line in the plane.

Example [21.9] Graph the planes $2x + 3y = 4$ and $x = 1$.

In these example we are graphing all possible triples (x, y, z) which satisfy the equations. Thus even though the equations look like lines in 2–dimensional space, they are in fact planes (i.e., the *context* is critical). When we set $z = 0$ in the first example we are reduced to the equation of a line in the $x\,y$–plane. The planes are depicted below.

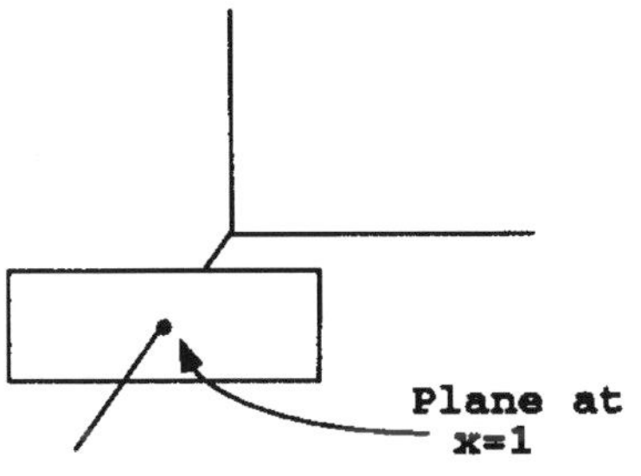

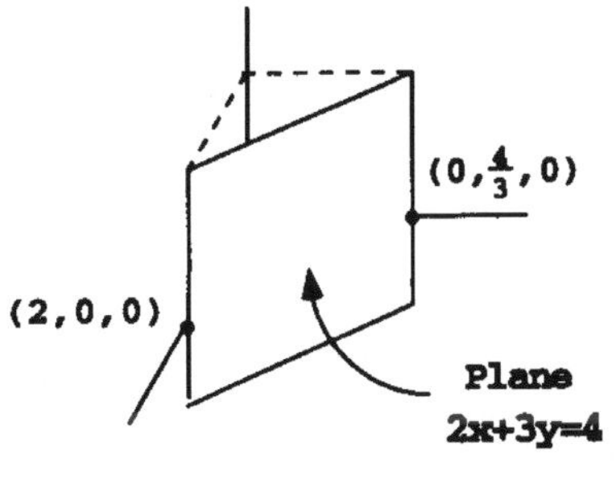

Notice the the plane at $x = 1$ is parallel to the zy–plane and intersects only the x–axis, while the other plane does not intersect the z–axis.

Exercises for §21.2

Find the equation of the plane in each of the following situations.

(1) The plane going through $(2, -3, 1)$ which is perpendicular to $\overrightarrow{(3, -2, 4)}$.

(2) The plane going through $(2, -3, 1)$ which is perpendicular to $\overrightarrow{(-1, 4, 2)}$.

(3) The plane which passes through the points $(-1, 3, 5)$, $(1, -2, 4)$, $(-2, 1, -8)$.

(4) The plane which passes through the points $(2, -2, 3)$, $(1, 1, 8)$, $(0, -1, -3)$.

(5) The plane containing the line $x = 1 + t$, $y = 2 - t$, $z = 3t$ and the point $(1, -1, 5)$.

(6) The plane containing the line $x = 3 - 2t$, $y = -1 + 3t$, $z = 1 + t$ and the point $(-1, 3, 4)$.

(7) The plane which is perpendicular to the plane $3x + y - 2z = 4$ and passes through the points $(0, 0, 1)$ and $(0, -1, 0)$.

Determine which of the following pairs of planes are perpendicular, parallel, and skew (neither parallel or perpendicular).

(8) $2x - 3y + z = 11$, $3x - 3y - 15z = 2$.

(9) $2x - y + z = 1$, $6x - 3y + 3z = 5$.

(10) $x - 3y + 3z = 2$, $5x - y - 2z = 7$.

(11) $7x - y - z = 8$, $2x + 3y + 11z = 2$.

Find the cosine of the angle between the following pairs of planes.

(12) $3x - 7y + 2z = 2$, $-5x + y + 4z = -2$.

(13) $8x + y - z = 1$, $2x - 3y + 2z = 4$.

(14) $x - 5y + 2z = 4$, $x - 3y = 13$.

(15) $z = 1$, $x - 3y + z = 4$.

Find the distance from the point P to the given plane in the following examples.

(16) $P = (1, -1, 5)$, plane: $3x - y + 4z = 3$.

(17) $P = (0, 1, 2)$, plane: $2x + y - z = 5$.

(18) $P = (1, -1, 3)$, plane: $2x + 4y + z = 1$.

(19) $P = (3, -1, 2)$, plane: $x + y + z = 100$.

(20) Show that the distance from the point $P = (p, q, r)$ to the plane $ax + by + cz = d$ is

$$\frac{|ap + bq + cr - d|}{\sqrt{a^2 + b^2 + c^2}}.$$

(21) Find the equation for the line of intersection of the two planes $2x - 4y + 4z = 8$, $8x + 7y + z = 2$.

(22) The planes $ax + by + cz = d_1$ and $ax + by + cz = d_2$ are always parallel. Show that the distance between these planes is

$$\frac{|d_1 - d_2|}{\sqrt{a^2 + b^2 + c^2}}.$$

(23) Let $a_1x + b_1y + c_1z = d_1$ and $a_2x + b_2y + c_2z = d_2$ be any two planes which intersect. Show that (for arbitrary m, n)

$$(ma_1 + na_2)x + (mb_1 + nb_2)y + (mc_1 + nc_2)z = md_1 + nd_2$$

is the most general form of the equation of the plane which passes through the line of intersection of the two original planes.

21.3 Space Curves

The next class of geometric objects we shall analyze are termed **space curves,** i.e., one–dimensional curves which twist and turn through 3–dimensional space.

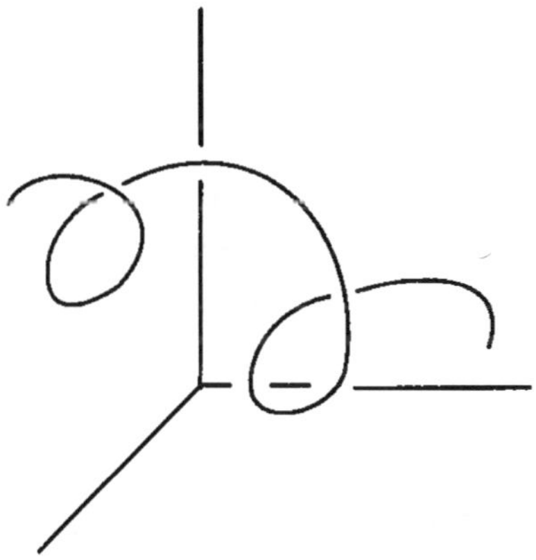

Question: How can we describe space curves mathematically?

Answer: In order to delve into this question we begin with a simple example. In §21.1 we derived the equation of a line in 3–dimensional space and found that every point (x, y, z) on a line must be of the form

$$x = a_1 + b_1\,t, \quad y = a_2 + b_2\,t, \quad z = a_3 + b_3\,t,$$

where a_i, b_i are fixed constants (for $i = 1, 2, 3$), and $t \in \mathbb{R}$. As t varies over the real numbers the points

$$(a_1 + b_1\,t,\; a_2 + b_2\,t,\; a_3 + b_3\,t)$$

form the line. We may thus define the **vector valued function**

$$\overrightarrow{f(t)} = (a_1 + b_1 t)\,\vec{\mathbf{i}} + (a_2 + b_2 t)\,\vec{\mathbf{j}} + (a_3 + b_3 t)\,\vec{\mathbf{k}},$$

which for each real number t yields a vector whose endpoint (when we assume the vector is anchored at the origin) is the t–point on the line.

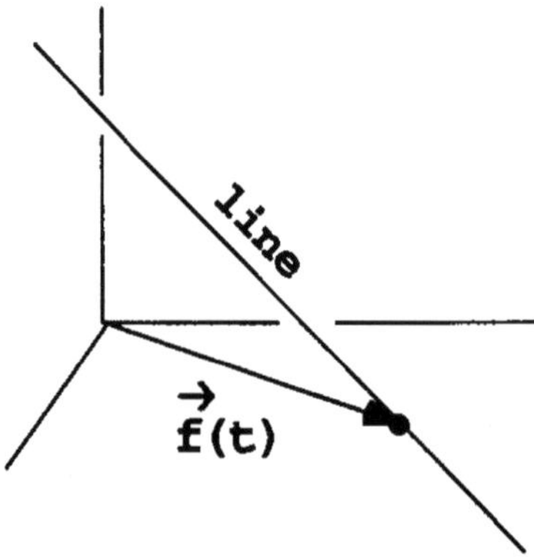

This example motivates the following

Definition: *A vector valued function* $\overrightarrow{f(t)}$ *is a rule which assigns to each real number t a unique vector* $\overrightarrow{f(t)}$*. The graph of the function* $\overrightarrow{f(t)}$ *is defined to be the set of points* $f(t)$ *where t varies over the real numbers (alternately, if we begin each vector in the image of* $\overrightarrow{f(t)}$ *at the origin, then the graph consists of the endpoints of the set of vectors in the image).*

Recalling the general nature of the original definition of a function (which was presented in Chapter 1) we see that vector valued functions are nothing more than a special class of functions. We can in fact construct the most general vector valued function in the following manner. Choose three real–valued functions $x(t), y(t)$, and $z(t)$. The most general vector valued function $\overrightarrow{f(t)}$ must take the form

$$\overrightarrow{f(t)} = x(t)\,\vec{\mathbf{i}} + y(t)\,\vec{\mathbf{j}} + z(t)\,\vec{\mathbf{k}}.$$

Definition: *The graph of a vector valued function*

$$f(\vec{t}) = x(t)\,\vec{\mathbf{i}} + y(t)\,\vec{\mathbf{j}} + z(t)\,\vec{\mathbf{k}}$$

*is termed a **space curve**.* The space curve is termed **continuous** provided that the functions $x(t)$, $y(t)$ and $z(t)$ are themselves continuous.

Example [21.10] Let $x(t) = t^2$, $y(t) = 1 + t$, and $z(t) = 1 - t^2$. The graph of the vector valued function

$$\overrightarrow{f(t)} = x(t)\,\vec{\mathbf{i}} + y(t)\,\vec{\mathbf{j}} + z(t)\,\vec{\mathbf{k}},$$

is sketched below.

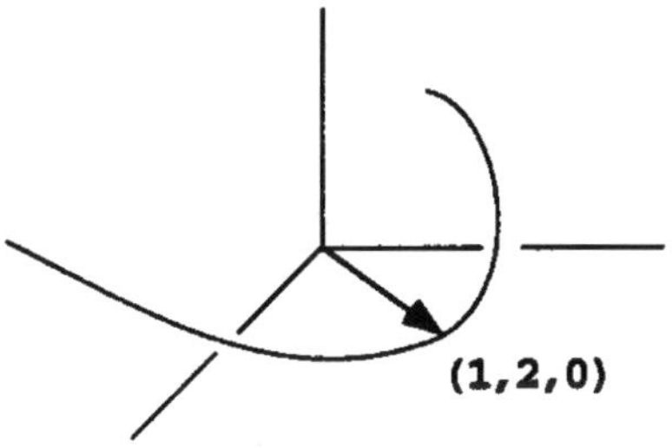

Example [21.11] The graph of the curve $\overrightarrow{f(t)} = 10\,\vec{\mathbf{i}} + \cos(t)\,\vec{\mathbf{j}} + \sin(t)\,\vec{\mathbf{k}}$ is a circle of radius 1 centered at the point $(10, 0, 0)$.

To see that the circle below

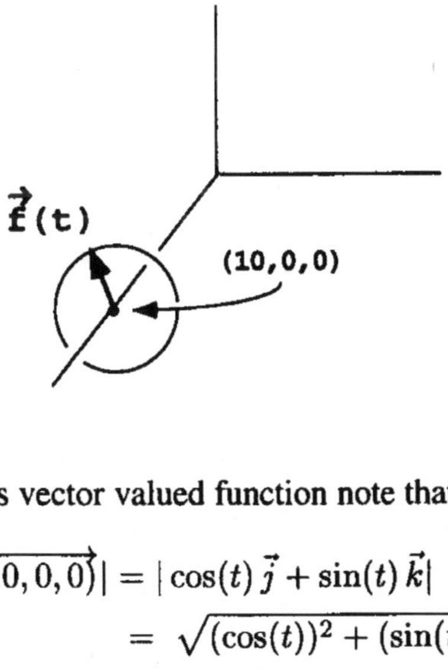

is in fact the graph of this vector valued function note that

$$\begin{aligned} |\overrightarrow{f(t)} - \overrightarrow{(10,0,0)}| &= |\cos(t)\,\vec{j} + \sin(t)\,\vec{k}| \\ &= \sqrt{(\cos(t))^2 + (\sin(t))^2} \;=\; 1. \end{aligned}$$

Thus every point on this curve is at a distance one from the point $(10, 0, 0)$, and, hence, it must be the designated circle.

Example [21.12] Construct a vector valued function whose graph is the circle of radius 3 which is centered at $(0, 8, 0)$ and lies on a plane which is parallel to the $x\,z$–plane.

The function $\overrightarrow{f(t)} = x(t)\,\vec{\mathbf{i}} + y(t)\,\vec{\mathbf{j}} + z(t)\,\vec{\mathbf{k}}$ in this example will be a modified version of the function in Exercise [21.11]. Since the circle is supposed to be parallel to the $x\,z$–plane, the function $y(t)$ must be independent of t. Thus the function we desire is $\overrightarrow{f(t)} = 3\cos(t)\,\vec{\mathbf{i}} + 8\,\vec{\mathbf{j}} + 3\sin(t)\,\vec{\mathbf{k}}$.

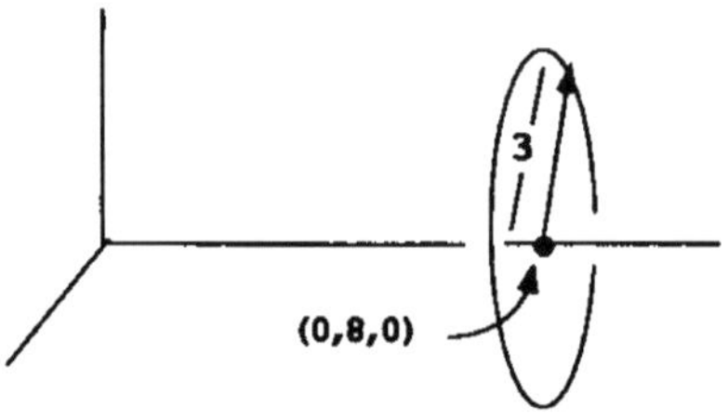

Example [21.13] (Helix) The graph of the function $\overrightarrow{f(t)} = \cos(t)\,\vec{\mathbf{i}} + \sin(t)\,\vec{\mathbf{j}} + t\,\vec{\mathbf{k}}$ is termed the circular helix.

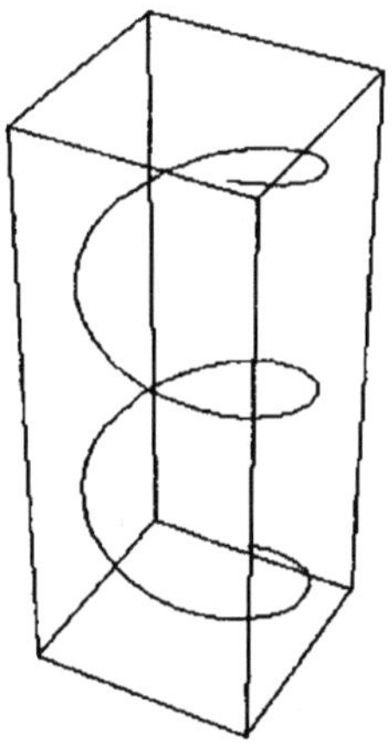

Remark: If $x(t)$ and $y(t)$ are arbitrary real–valued functions of t then the vector valued function $\overrightarrow{f(t)} = \overrightarrow{(x(t), y(t), 0)}$ represents a curve in the $x\,y$–plane.

Example [21.14] Sketch and identify the space curve represented by the function $\overrightarrow{f(t)} = (t^2 - 1)\,\vec{\mathbf{i}} + (t+2)\,\vec{\mathbf{j}}$.

As remarked above, this curve lies in the $x\,y$–plane.

Let's graph a few points:

$$\overrightarrow{f(0)} = \overrightarrow{(-1,2,0)},$$
$$\overrightarrow{f(1)} = \overrightarrow{(0,3,0)},$$
$$\overrightarrow{f(-1)} = \overrightarrow{(0,1,0)},$$
$$\overrightarrow{f(2)} = \overrightarrow{(3,4,0)},$$
$$\overrightarrow{f(-2)} = \overrightarrow{(3,0,0)}.$$

When we graph these points a parabola appears in the $x\,y$–plane.

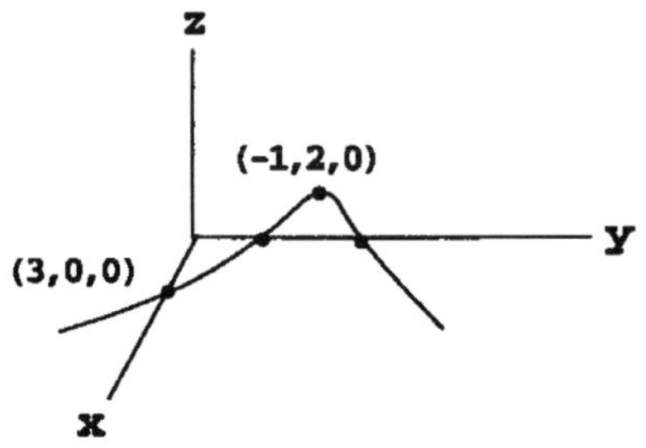

It is natural to ask what the equation of the parabola is in standard form. This is easily attained: since $x = t^2 - 1$ and $y = t + 2$, we have $t = y - 2$, and we conclude that

$$x = (y-2)^2 - 1.$$

Exercises for §21.3

(1) Find the vector valued function whose graph is a circle of radius 4 (centered at $(7,0,0)$) which lies in a plane parallel to the $y\,z$–plane.

(2) Find the vector valued function whose graph is a circle of radius 2 (centered at $(0,-5,0)$) which lies in the plane $y = -5$.

Graph the following vector valued functions.

(3) $\overrightarrow{f(t)} = \overrightarrow{(t, 0, 1 + 2t)}$.

(4) $\overrightarrow{f(t)} = \cos(t)\,\vec{\mathbf{i}} + \sin(t)\,\vec{\mathbf{j}} + e^t\,\vec{\mathbf{k}}$.

(5) $\overrightarrow{f(t)} = \overrightarrow{(1, t^2, 2 - t)}$.

(6) $\overrightarrow{f(t)} = \frac{1}{t}\,\vec{\mathbf{i}} + (t+1)\,\vec{\mathbf{j}} + 10\,\vec{\mathbf{k}}$.

21.4 Polar and Cylindrical Coordinates

In the process of representing a point in two or three dimensional space by a pair or triple of real numbers (respectively) we are specifying what is called a **coordinate system**. For the purposes of graphical display, it is sometimes more convenient to use other coordinate systems. We now introduce two other possible coordinate systems, polar coordinates, and cylindrical coordinates.

Polar Coordinates

Fix a point O (the origin) in a plane and a line (the polar axis) going through O.

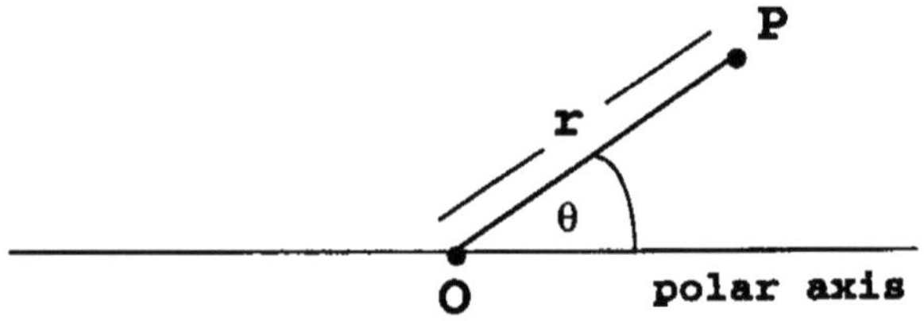

Every point P in the plane can then be specified by the pair (r, θ), where $r \geq 0$ is the distance from O to P, and θ is the angle (measured in radians) between the line OP and the polar axis. If the polar axis is chosen to be the x–axis in the $x\,y$–plane

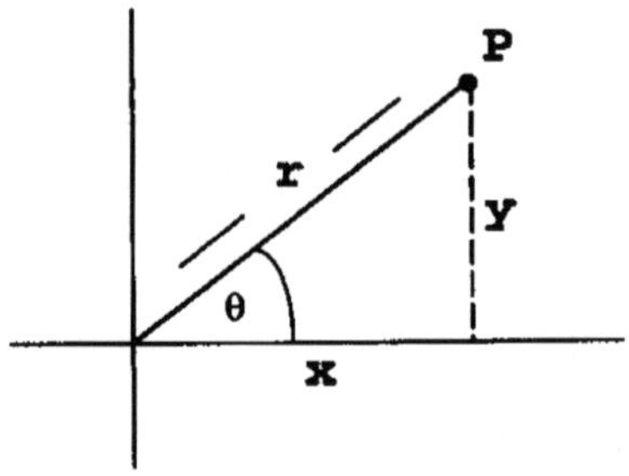

then we see that the point P with polar coordinates (r, θ) is also given by Cartesian coordinates (x, y) where

$$x = r\cos(\theta), \quad y = t\sin(\theta). \tag{21.4}$$

We can extend the definition of polar coordinates by defining $(-r, \theta)$ (with $r > 0$) to be the point $(r, \theta + \pi)$.

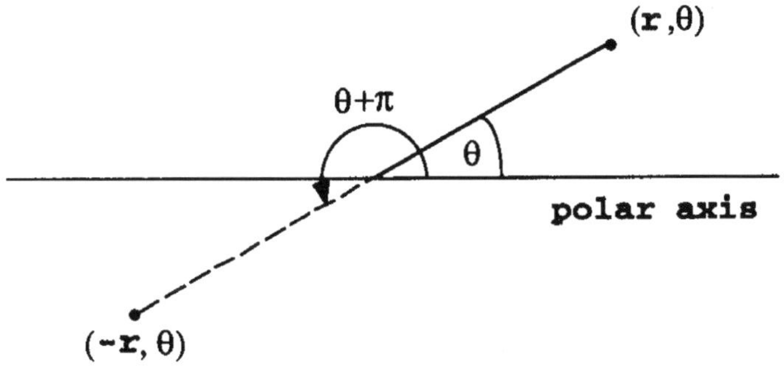

Example [21.15] Convert the point $(5, 2\pi/3)$ to Cartesian coordinates.

Applying equation (21.4) we have

$$x = 5\cos(2\pi/3) = -5/2,$$

and

$$y = 5\sin(2\pi/3) = 5\sqrt{3}/2.$$

Hence the Cartesian coordinates are $(-5/2, 5\sqrt{3}/2)$.

In Cartesian coordinates we considered and analyzed functions $y = f(x)$ and their graphs. We can develop an analogous theory when we consider functions $r = f(\theta)$ and their graphs. The transformations (21.4) allow us to move freely between the two coordinate systems.

Example [21.16] Express the function $y = x^2$ as a function in polar coordinates.

Using the transformations (21.4) we can express $y = x^2$ as

$$r\sin(\theta) = r^2\cos^2(\theta).$$

Hence the graph of this basic parabola is described by the function

$$r = \frac{\sin(\theta)}{\cos^2(\theta)}.$$

Question: What are the most elementary polar functions.

Answer: The simplest polar functions take the form $r = c$, where c is a constant. The graph of $r = c$ is simply a circle of radius c. Notice that this circle cannot be described as the graph of a function in Cartesian coordinates. The Cartesian form of $r = c$ leads to an equation as follows: since $x = r\cos(\theta)$ and $y = r\sin(\theta)$ we see that

$$x^2 + y^2 = r^2.$$

Hence the equation $r = c$ is equivalent to the equation $x^2 + y^2 = r^2$.

Example [21.17] (Four–leaved Rose) The polar function $r = \cos(2\theta)$ has as its graph the **four–leaved rose** below.

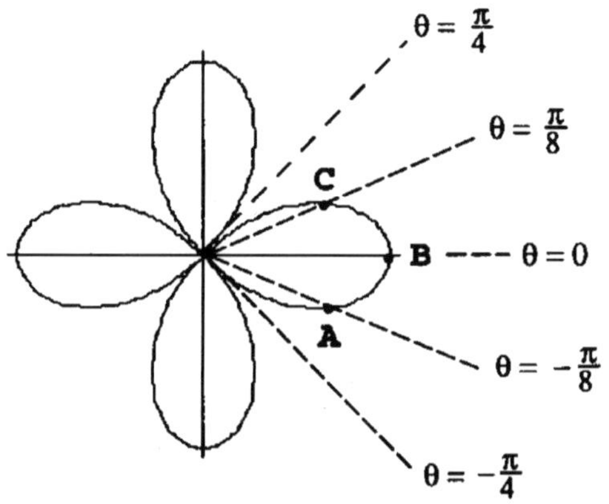

As an illustration, let's see how the rightmost leaf is graphed. When $\theta = \pm - \frac{\pi}{4}$, $r = 0$. When $\theta = \pm\frac{\pi}{8}$, $r = \frac{\sqrt{2}}{2}$. When $\theta = 0$, $r = 1$. We have graphed the indicated points A, B, C on the leaf. Notice that we began at the origin $\left(-\frac{\pi}{4}, 0\right)$ and ended at the origin $\left(\frac{\pi}{4}, 0\right)$.

Example [21.18] (The Cardioid) Consider the function $r = a(1 - \sin(\theta)$, where a is a constant. The graph is called the **cardioid**.

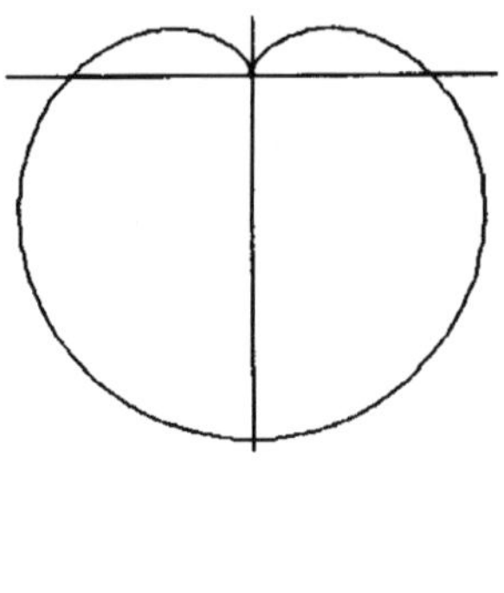

To find the Cartesian form of this equation we can write $y = r\sin(\theta)$, which implies $r = a\left(1 - \frac{y}{r}\right)$ and also $r^2 = a(r - y)$. Thus $r^2 + ay = ar$, which in combination with the fact $r^2 = x^2 + y^2$, yields $x^2 + y^2 + ay = a\sqrt{x^2 + y^2}$ or $(x^2 + y^2 + ay)^2 = a^2(x^2 + y^2)$.

Example [21.19] (Logarithmic Spiral) Consider the function $r = e^{a\theta}$, where $a > 0$ is a constant. The graph is a spiral.

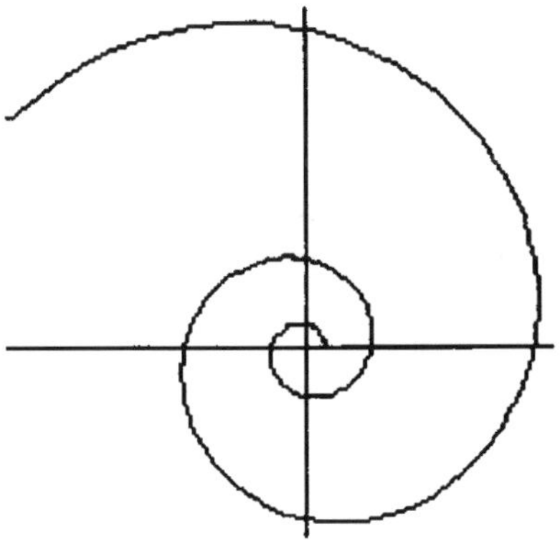

It is clear that as the angle θ increases the value of r grows exponentially.

Cylindrical Coordinates

Having seen how we can describe two–dimensional figures we previously could not handle easily it is natural to try to extend polar coordinates to three dimensional space. Thus, let $P = (x, y, z)$ be a point in Cartesian coordinates.

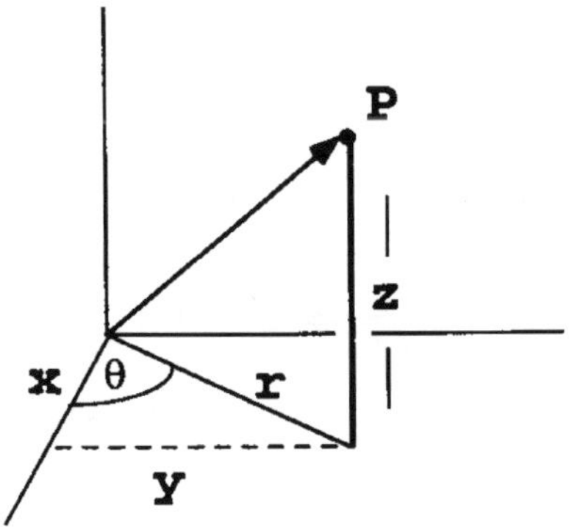

Perform the transformation

$$x = r\cos(\theta), \quad y = r\sin(\theta), \quad z = z, \tag{21.5}$$

we obtain the point in cylindrical coordinates (r, θ, z).

Example [21.20] Plot the point $\left(3, \frac{3\pi}{4}, 2\right)$ with cylindrical coordinates and convert it to rectangular coordinates.

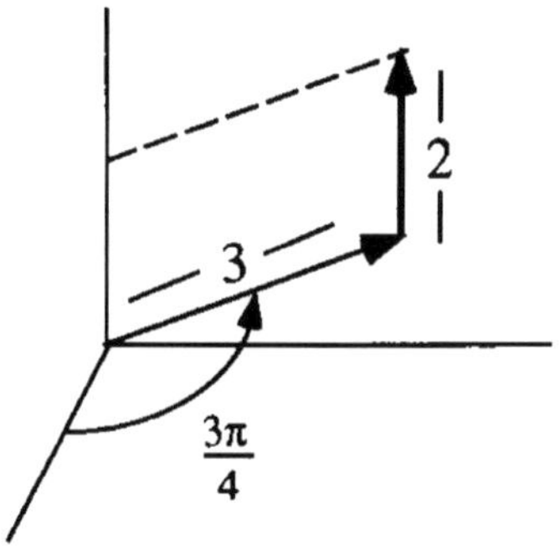

To convert to rectangular coordinates we utilize (21.5):

$$x = 3\cos\left(\frac{3\pi}{4}\right) = -\frac{3\sqrt{2}}{2}, \quad y = 3\sin\left(\frac{3\pi}{4}\right) = \frac{3\sqrt{2}}{2}, \quad z = z.$$

Exercises for §21.4

In exercises 1–6 plot the points with given polar coordinates and plot a point in which r has opposite sign.

(1) $r = 4,\ \theta = \frac{\pi}{3}$. (3) $r = -3,\ \theta = \pi$. (5) $r = -2,\ \theta = \frac{\pi}{6}$.

(2) $r = 2,\ \theta = \frac{3\pi}{4}$. (4) $r = 2,\ \theta = -\frac{\pi}{6}$. (6) $r = -5,\ \theta = -\frac{2\pi}{3}$.

(7) In converting from polar coordinates (r, θ) to rectangular coordinates (x, y), show that

$$\cos(\theta) = \frac{x}{\sqrt{x^2 + y^2}}, \quad \sin(\theta) = \frac{y}{\sqrt{x^2 + y^2}}.$$

(8) Show that the distance d between two points (r_1, θ_1) and (r_2, θ_2) in polar coordinates is given by the formula

$$d = \sqrt{r_1^2 + r_2^2 - 2r_1r_2\cos(\theta_2 - \theta_1)},$$

(9) Sketch the polar curve $r = 3(1 - 2\sin(\theta))$ and find its Cartesian equation.

(10) Sketch the curve $r = 2\sin(3\theta)$ and find its Cartesian equation.

(11) Consider the equation of a line $y = mx + b$ in Cartesian coordinates. Convert this to a polar equation.

(12) Sketch the polar curve $r = \frac{5}{\sin(\theta)}$ for $0 < \theta < \pi$ and find its Cartesian equation.

(13) Find the points of intersect of the two polar curves $r = 2(\cos(\theta))^2$ and $r = 2 - \sin(\theta)$.

(14) Find the polar equation for the curve $(x^2 + y^2)(x^2 + y^2 - 3) = y^2$.

(15) Let A, B be constants such that $AB \neq 0$. Show that the polar equation $r = A\sin(\theta) + B\cos(\theta)$ always represents a circle.

The following points are given in cylindrical coordinates. Plot then and convert them to rectangular coordinates.

(16) $(2, \frac{2\pi}{3}, 5)$.

(17) $(1, -\frac{\pi}{4}, -2)$.

(18) $(5, 0, 1)$.

(19) $(-2, \frac{\pi}{2}, 1)$.

(20) $(-3, -\frac{3\pi}{2}, 2)$.

(21) $(2, \frac{4\pi}{3}, -6)$.

21.5 Converting Polar Functions to Vector Valued Functions

Let $y = \phi(x)$ denote a curve in the $x\,y$–plane.

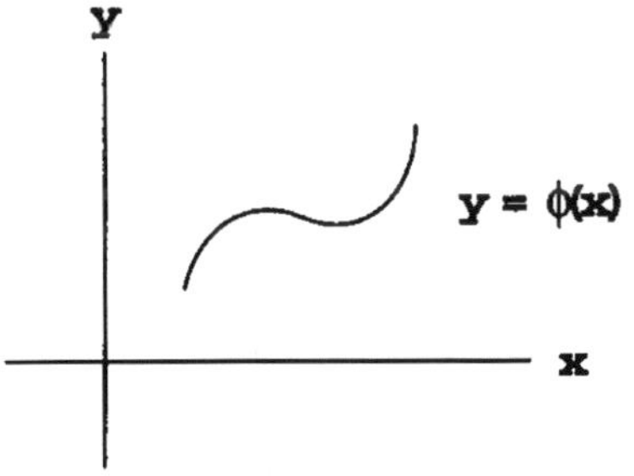

We can convert this to a space curve by parameterizing with the variable t, i.e., by setting $x = t$ and $y = \phi(t)$. Then the point (x, y) in the Cartesian plane can be realized as the endpoint of the vector $t\,\vec{i} + \phi(t)\,\vec{j}$ assuming we anchor the vector at the origin O.

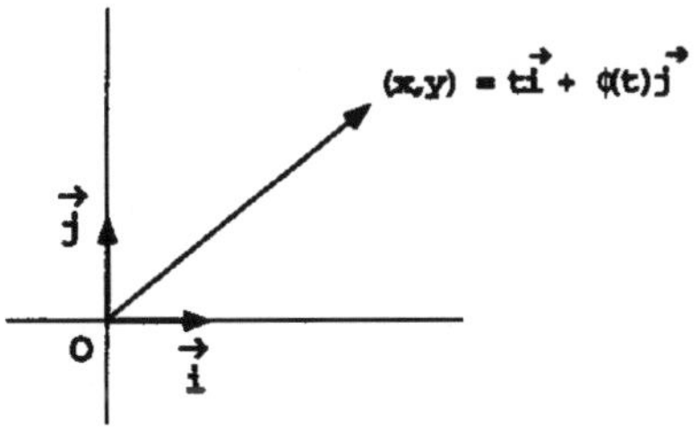

Question: If we are given a function $r = f(\theta)$ in polar form, how can we realize it as a space curve associated to a vector valued function?

Answer: Recall that a point (r, θ) in polar coordinates can be converted to rectangular coordinates by setting $x = r\cos(\theta)$ and $y = r\sin(\theta)$.

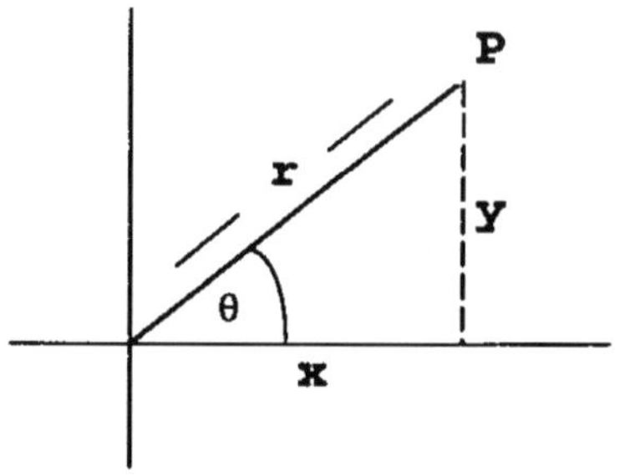

Notice that the arrow emanating from the origin O and ending at the point $P = (r, \theta)$ is simply the vector $\overrightarrow{OP} = r\cos(\theta)\,\vec{\mathrm{i}} + r\sin(\theta)\,\vec{\mathrm{j}}$.

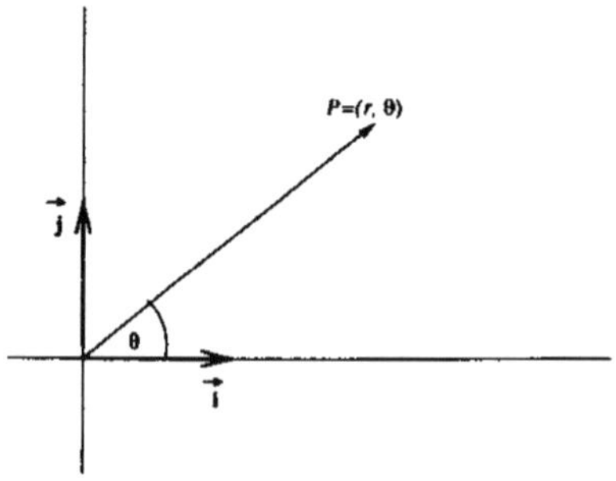

Hence the polar function $r = f(\theta)$ can be realized as the space curve

$$\begin{aligned}\overrightarrow{g(\theta)} &= r\cos(\theta)\,\vec{\mathrm{i}} + r\sin(\theta)\,\vec{\mathrm{j}} \\ &= f(\theta)\cos(\theta)\,\vec{\mathrm{i}} + f(\theta)\sin(\theta)\,\vec{\mathrm{j}}.\end{aligned}$$

Example [21.21] Convert the four leaved rose of Example [17] to a space curve.

The four leaved rose is given by the equation $r = \cos(2\theta)$, and thus the associated vector valued function is simply $\overrightarrow{g(\theta)} = \cos(2\theta)\cos(\theta)\,\vec{\mathrm{i}} + \cos(2\theta)\sin(\theta)\,\vec{\mathrm{j}}$.

Example [21.22] Convert the three leaved rose $r = \cos(3\theta)$ to a space curve in the $y\,z$–plane.

To place the 3–leaved rose in the yz–plane set $y = \cos(3\theta)\cos(\theta)$ and $z = \cos(3\theta)\sin(\theta)$. Then

$$\overrightarrow{g(\theta)} = \cos(3\theta)\cos(\theta)\,\vec{\mathbf{j}} + \cos(3\theta)\sin(\theta)\,\vec{\mathbf{k}}$$

will do the trick.

Example [21.23] Place the 3–leaved rose in the plane $x = 5$ centered at the point $(5, 0, 0)$.

The vector valued function $\overrightarrow{f(\theta)} = \cos(3\theta)\cos(\theta)\,\vec{\mathbf{j}} + \cos(3\theta)\sin(\theta)\,\vec{\mathbf{k}}$ yields a 3–leaved rose in the plane $x = 0$. In order to obtain the desired location of the 3–leaved rose we must translate (shift) by the vector $\overrightarrow{(5,0,0)}$. The resulting vector valued function is

$$\overrightarrow{(5,0,0)} + \overrightarrow{f(\theta)} = 5\,\vec{\mathbf{i}} + \cos(3\theta)\cos(\theta)\,\vec{\mathbf{j}} + \cos(3\theta)\sin(\theta)\,\vec{\mathbf{k}}.$$

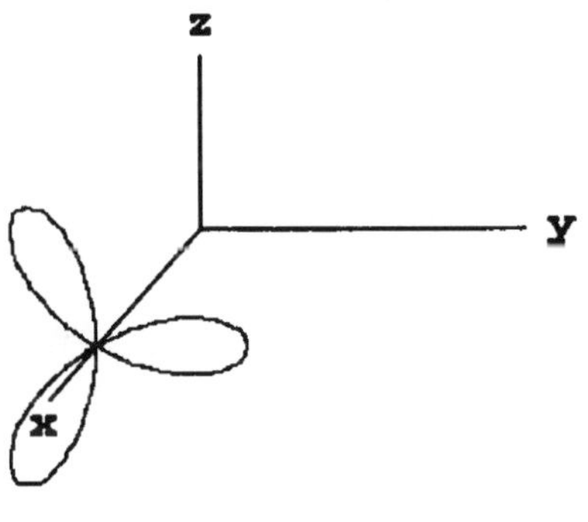

Example [21.24] Find the vector valued function which rotates the graph of the 3–leaved rose of the previous example by 15°in a clockwise direction in the plane $x = 5$.

Since $15° = \frac{\pi}{12}$radians, the desired rotation is simply the transformation $\theta \to \theta + \frac{\pi}{12}$. The resulting vector valued function is

$$5\,\vec{\mathbf{i}} + \cos\left(3\left(\theta + \frac{\pi}{12}\right)\right)\cos\left(\left(\theta + \frac{\pi}{12}\right)\right)\vec{\mathbf{j}} + \cos\left(3\left(\theta + \frac{\pi}{12}\right)\right)\sin\left(\left(\theta + \frac{\pi}{12}\right)\right)\vec{\mathbf{k}}.$$

Example [21.25] (The cycloid) Imagine a circle (lying in the plane $x = 0$) which rolls on the y–axis without slipping or sliding.

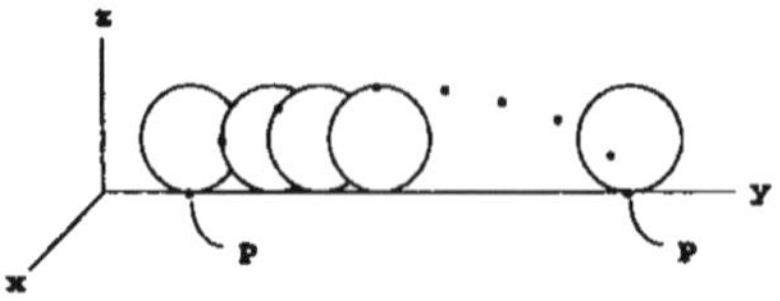

Fix a point P on the circle and consider the path (termed a **cycloid**) the point will traverse as the circle rolls. Assuming the circle has radius $a > 0$ find a vector valued function whose graph is the cycloid.

Lets assume the circle is initially positioned so that P is at the origin O. Suppose that the circle rolls some distance $d \le 2\pi a$.

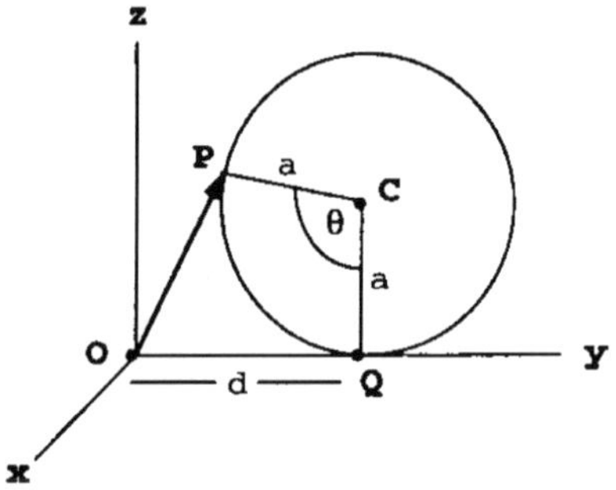

Notice that, by definition, the distance d is the same as the length of the arc QP, and hence $d = a\theta$. Thus the position of P after a roll of distance d is determined by the angle θ.

The function we are searching for must input the angle associated to the distance the circle has moved and yield a vector emanating from the origin and terminating at the position the point P is located at, i.e., $\overrightarrow{f(\theta)} = \overrightarrow{OP}$. To compute $\overrightarrow{f(\theta)}$ explicitly we use the vector identity

$$\overrightarrow{f(\theta)} = \overrightarrow{OP} = \overrightarrow{OQ} + \overrightarrow{QC} + \overrightarrow{CP}. \tag{21.6}$$

The first two vectors which appear in this sum are easily identified:

$$\overrightarrow{OQ} = a\vec{\mathbf{j}} = d\theta\,\vec{\mathbf{j}}, \quad \overrightarrow{QC} = a\vec{\mathbf{k}}.$$

To compute the vector $\overrightarrow{CP}$ in terms of θ we utilize the auxiliary triangle

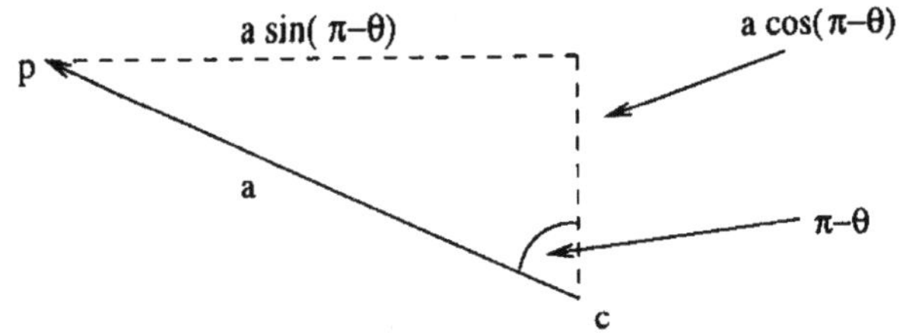

from which we learn that $\overrightarrow{CP} = -a\sin(\theta)\,\vec{\mathbf{j}} - a\cos(\theta)\,\vec{\mathbf{k}}$. Thus (21.6) takes the form

$$\overrightarrow{f(\theta)} = a(\theta - \sin(\theta))\,\vec{\mathbf{j}} + a(1 - \cos(\theta))\,\vec{\mathbf{k}}.$$

Exercises for §21.5

In the following exercises plot all vector valued functions with your CAS.

(1) Consider the polar curve $r = 4\sin(5\theta)$ which is a five–leaved rose. Find a vector valued function whose graph places this rose in the plane $z = -3$ centered at $(0, 0, -3)$.

(2) Find the vector valued function which rotates the graph of the five–leaved rose of exercise (1) by 45° in a counterclockwise direction in the plane $z = -3$.

(3) Consider the polar curve termed the spiral which is given by $r = \left(\frac{\theta}{3}\right)^2$ for $0 \leq \theta \leq 4\pi$. Find a vector valued function whose graph places this spiral in the plane $x = 1$ so that the polar point $r = 0, \theta = 0$ is translated to the point $(1, 6, 6)$.

(4) Find the vector valued function which rotates the graph of the spiral of exercise (3) by 60° in a clockwise direction in the plane $x = 1$.

(5) Find the vector valued function which translates the graph of the spiral of exercise (3) by the vector $\overrightarrow{(-1, -6, 1)}$.

(6) A circle rolls along the x–axis without slipping or sliding. Find the vector valued function whose graph is the cycloid traced out by a fixed point P on the circle. You may assume the initial position of P is at the origin.

Additional exercises for Chapter XXI

(1) Find a vector $\vec{\mathbf{v}}$ of length 4, perpendicular to the vectors $\overrightarrow{(2, 2, 2)}$ and and $\overrightarrow{(1, 0, -1)}$.

(2) It is seen that if the equation of a plane P is $ax + by + cz + d = 0$ then the vector $\overrightarrow{(a, b, c)}$ is normal to the plane P. Its length is $\sqrt{a^2 + b^2 + c^2}$. The equation of the plane P can be

normalized by requiring the normal vector to be of unit length. Let $\vec{\mathbf{n}}$ be such a vector, hence $\vec{\mathbf{n}} = \frac{\overrightarrow{(a,b,c)}}{\sqrt{a^2+b^2+c^2}}$. The normalized equation of the plane P is expressed as $\vec{\mathbf{n}} \cdot \overrightarrow{(x, y, z)} = e$. Prove that if $\vec{\mathbf{n}} \cdot \overrightarrow{(x, y, z)} = e$ is the normalized equation of a plane P and M is any point, then the distance from M to P is $|e - \vec{\mathbf{n}} \cdot \overrightarrow{(x, y, z)}|$.

(3) Find the distance from the following points to the indicated planes or lines:

(a) $(1, 0, -1)$; $\overrightarrow{(1, 1, 1)} \cdot \overrightarrow{(x, y, z)} = 1$,

(b) $(1, 0, -1)$; $2x + 3y + z = 1$,

(c) $(1, 2)$; $\overrightarrow{(3, 4)} \cdot \overrightarrow{(x, y)} = 0$.

(4) Determine whether the planes $2x + 3y + z = 0$ and $6x + 9y + 3z = 0$ are parallel.

(5) Determine whether the line

$$x = 2 + 5t$$

$$y = 1 + 2t$$

$$z = 4 + t$$

is parallel to the plane $x + 2y - 9z = 10$.

(6) Find parametric equations for the intersection of the planes $x+3y+2z = 6$ and $3x+y-z = 6$ and find the angle between these planes.

(7) On your CAS graph the function $\overrightarrow{f(t)} = 4\cos(t)\vec{\mathbf{i}} + 4\sin(t)\vec{\mathbf{j}}$ for $0 \leq t \leq \pi$. What is the functional form of one coordinate in terms of the other?

(8) Repeat problem 7 with $\overrightarrow{f(t)} = (4t - t^2)\vec{\mathbf{i}} + (4t^2 - t^3)\vec{\mathbf{j}}$.

(9) On your CAS, graph the function $\overrightarrow{f(u)} = \frac{1}{1+u}\vec{\mathbf{i}} + (\sqrt{u} + 1)\vec{\mathbf{j}}$. Find its rectangular equation.

(10) Repeat problem 9 with $\overrightarrow{f(u)} = \frac{1}{1+u}\vec{\mathbf{i}} + \frac{u}{1-u^2}\vec{\mathbf{j}}$.

(11) Convert from cylindrical coordinates to rectangular coordinates:

(a) $(3, \pi/6, 4)$,

(b) $(5, 5\pi/3, -1)$,

(c) $(5, 5\pi/3, -1)$,

(d) $(1, 5\pi/3, 0)$.

(12) On your CAS find the regions determined by

(a) $0 \leq \theta \leq \pi/3, \ 0 \leq r \leq \cos(\theta), \ 0 \leq z \leq 5$,

(b) $r = 4\sin(\theta), \quad 0 \leq z \leq 5,$

(c) $\theta = \pi/3, \quad 0 \leq z \leq 1.$

(13) On your CAS find the regions described by the following equations:

(a) $\overrightarrow{f(t)} = (\cos(t))\vec{\mathbf{i}} + (\cos(t))\vec{\mathbf{j}} + 3\vec{\mathbf{k}}, 0 \leq t \leq 2\pi,$

(b) $\overrightarrow{f(t)} = (t\cos(t))\vec{\mathbf{i}} + (t\cos(t))\vec{\mathbf{j}} + t\vec{\mathbf{k}}, t \geq 0.$

On your CAS, sketch the graqh of $\overrightarrow{f(t)}$ and indicate the direction as t increases:

(14) $\overrightarrow{f(t)} = 3t\vec{\mathbf{i}} + t\vec{\mathbf{j}} + 4t\vec{\mathbf{k}}.$

(15) $\overrightarrow{f(t)} = 3\cosh(t)\vec{\mathbf{i}} + 3\sinh(t)\vec{\mathbf{j}}.$

(16) $\overrightarrow{f(t)} = 3t\exp(-t)\vec{\mathbf{i}} + t\vec{\mathbf{k}}.$

(17) $\overrightarrow{f(t)} = 3\cos(t)\vec{\mathbf{i}} + 2\vec{\mathbf{j}} + 3\cos(t)\vec{\mathbf{k}}.$

(18) $\overrightarrow{f(t)} = 2\cos(2t)\vec{\mathbf{i}} + 2\sin(2t)\vec{\mathbf{j}} + 3\vec{\mathbf{k}}.$

Chapter XXII

Calculus of Vector Valued Functions

22.1 Derivatives of Vector Valued Functions

Let $\overrightarrow{f(t)} = x(t)\vec{i} + y(t)\vec{j} + z(t)\vec{k}$ be a vector valued function where $x(t), y(t)$, and $z(t)$ are real valued functions of the variable t.

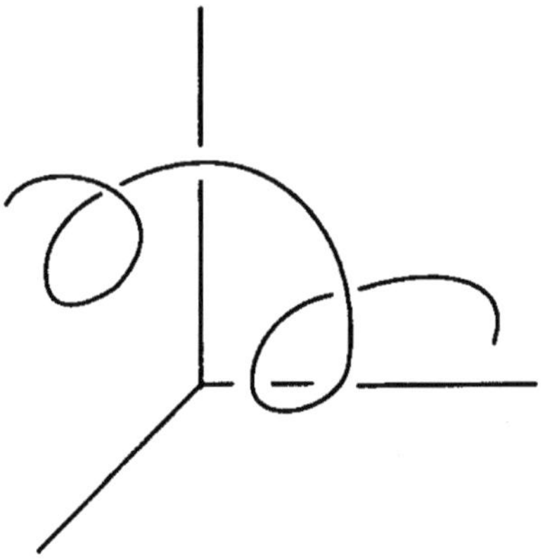

Question: How can we define the derivative $\frac{d}{dt}\overrightarrow{f(t)}$ (which is also denoted $\overrightarrow{f'(t)}$)?

Answer: Intuitively there is only one possible way to define this derivative. Provided the limit exists, we define the **derivative of the vector valued function** to be

$$\frac{d}{dt}\overrightarrow{f(t)} = \lim_{h\to 0} \frac{\overrightarrow{f(t+h)} - \overrightarrow{f(t)}}{h}.$$

Notice that, by definition, the derivative of $\overrightarrow{f(t)}$ can be expressed in terms of the derivatives of $x(t), y(t)$, and $z(t)$ in the following simple manner:

$$\begin{aligned}\frac{d}{dt}\overrightarrow{f(t)} &= \lim_{h\to 0}\frac{x(t+h)-x(t)}{h}\vec{\mathbf{i}} + \lim_{h\to 0}\frac{y(t+h)-y(t)}{h}\vec{\mathbf{j}} \\ &\quad + \lim_{h\to 0}\frac{z(t+h)-z(t)}{h}\vec{\mathbf{k}} \qquad (22.1)\\ &= x'(t)\vec{\mathbf{i}} + y'(t)\vec{\mathbf{j}} + z'(t)\vec{\mathbf{k}}.\end{aligned}$$

Example [22.1] Compute the derivative of the space curve

$$\overrightarrow{f(t)} = \cos(t)\vec{\mathbf{i}} + (t^5 - 3t^2 + 2)\vec{\mathbf{j}} + e^{t^2}\vec{\mathbf{k}}.$$

Following (22.1) we see that

$$\overrightarrow{f'(t)} = -\sin(t)\vec{\mathbf{i}} + (5t^4 - 6t)\vec{\mathbf{j}} + 2te^{t^2}\vec{\mathbf{k}}.$$

Question: How can we interpret the derivative of a space curve. Is it the slope of a tangent line (in some general sense)?

Answer: The concept of the slope of the tangent line does not exist in 3–dimensional space. To see this recall that in 2–dimensional space the slope of the tangent line is the ratio of the vertical distance to the horizontal distance $\frac{V}{H}$ as one travels from a point P to a point Q on the line.

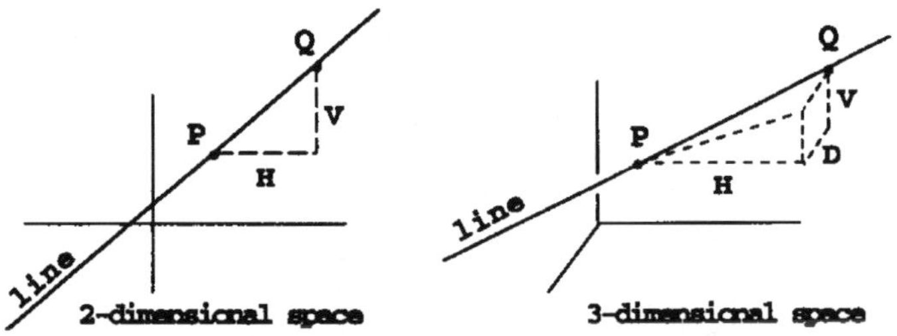

In 3–dimensional space, if one travels from P to Q, not only does one travel a horizontal distance H and a vertical distance V, but also a distance D in depth. Thus there is no natural analog of the slope for dimensions higher than 2.

Nevertheless, the derivative $\overrightarrow{f'(t)}$, turns out to be a vector which is itself tangent to the curve $\overrightarrow{f(t)}$ at the point t. This can be seen by examining the diagram below.

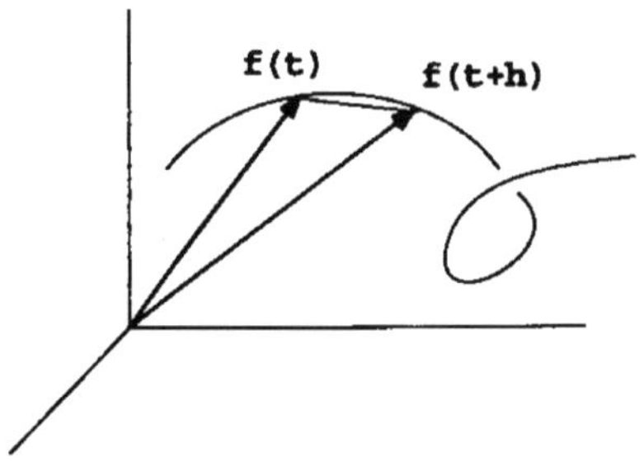

The vector $\overrightarrow{f(t+h)} - \overrightarrow{f(t)}$ is depicted below.

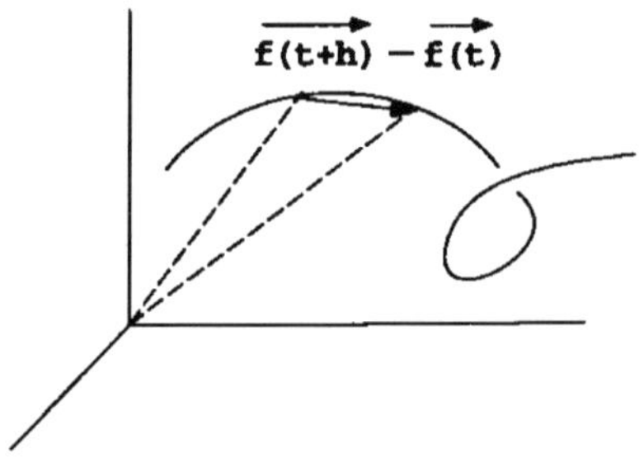

As $h \to 0$ the direction of the vector $\overrightarrow{f(t+h)} - \overrightarrow{f(t)}$ approaches the direction of the tangent line to the curve, and since dividing by the scalar h alters the length and not the direction of the vector, the limit

$$\lim_{h \to 0} \frac{\overrightarrow{f(t+h)} - \overrightarrow{f(t)}}{h}$$

results in a vector which is tangent to the space curve $\overrightarrow{f(t)}$ at the point t.

Question: Given a differentiable vector valued function $\overrightarrow{f(t)}$, does the length of the derivative, $|\overrightarrow{f'(t)}|$ have a meaning?

Answer: To approach this question let us suppose the variable t denotes time and that a particle is moving along the space curve. The position of the particle at time t is simply $(x(t), y(t), z(t))$. Then the length $|\overrightarrow{f(t+h)} - \overrightarrow{f(t)}|$ is the distance between the position of the particle at time $t+h$ and at time t.

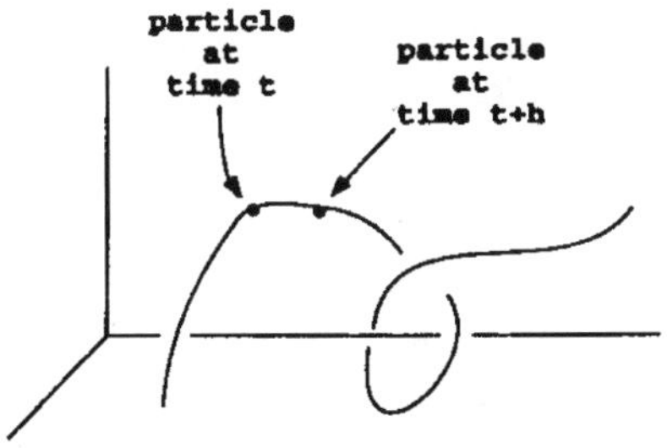

Upon dividing by h we obtain the ratio of the change in distance to the change in time, i.e., we obtain the average velocity of the particle. Thus $|\overrightarrow{f'(t)}|$ is the instantaneous velocity of the particle at the time t.

Example [22.2] Consider the space curve

$$\overrightarrow{f(t)} = \overrightarrow{(t^2, 1+t, 2-3t)}.$$

Compute $\overrightarrow{f'(1)}$ and $|\overrightarrow{f'(1)}|$. Find the equation of the tangent line at $t = 1$.

Since $\overrightarrow{f'(t)} = \overrightarrow{(2t, 1, -3)}$, we see that $\overrightarrow{f'(1)} = \overrightarrow{(2, 1, -3)}$ and thus

$$|\overrightarrow{f'(1)}| = \sqrt{4+1+9} = \sqrt{14}.$$

The tangent line in question must pass through the point $(1, 2, -1)$ and is parallel to the vector $\overrightarrow{f'(1)} = \overrightarrow{(2, 1, -3)}$. Thus the parametric form of the tangent line is simply

$$x(t) = 1 + 2t, \quad y(t) = 2 + t, \quad z = -1 - 3t.$$

The rules of differentiation we encountered in chapter 7 rapidly generalize to vector valued functions. To be specific, suppose $\overrightarrow{f(t)}, \overrightarrow{g(t)}$ are differentiable vector valued functions and $u(t)$ is a differentiable real valued function of t. The following assertions can be verified with the CAS and are left to the interested reader.

$$\frac{d}{dt}\left(\overrightarrow{f(t)} \pm \overrightarrow{g(t)}\right) = \overrightarrow{f'(t)} \pm \overrightarrow{g'(t)}$$

$$\frac{d}{dt}\left(u(t)\, \overrightarrow{f(t)}\right) = u'(t)\, \overrightarrow{f(t)} + u(t)\, \overrightarrow{f'(t)}$$

$$\frac{d}{dt}\left(\overrightarrow{f(t)} \cdot \overrightarrow{g(t)}\right) = \overrightarrow{f'(t)} \cdot \overrightarrow{g(t)} + \overrightarrow{f(t)} \cdot \overrightarrow{g'(t)}$$

$$\frac{d}{dt}\left(\overrightarrow{f(t)} \times \overrightarrow{g(t)}\right) = \overrightarrow{f'(t)} \times \overrightarrow{g(t)} + \overrightarrow{f(t)} \times \overrightarrow{g'(t)}$$

$$\frac{d}{dt}\overrightarrow{f(u(t))} = u'(t)\, \overrightarrow{f'(u(t))}.$$

Exercises for §22.1

For each of the following vector valued functions $\overrightarrow{f(t)}$, find the equation of the tangent line at the indicated point. Using your CAS graph the space curve and the tangent line.

In exercises (1)–(6) find the equation of a normal line (not uniquely defined) to the curve passing through the indicated point (i.e., a line which is perpendicular to the tangent line at the indicated point). Add to the graph of the space curve and the tangent line the graph of the normal line.

(1) $\overrightarrow{f(t)} = \overrightarrow{(1+5t^2, 2-t, 1+7t)}, t = 0.$

(2) $\overrightarrow{f(t)} = \overrightarrow{(t^3, -3+t^2, 1+t)}, t = 2.$

(3) $\overrightarrow{f(t)} = \overrightarrow{(3, 2\cos(t), 2\sin(t))}, t = \frac{\pi}{4}.$

(4) $\overrightarrow{f(t)} = \overrightarrow{(3\cos(t), 7, 2\sin(t))}, t = \frac{3\pi}{2}.$

(5) $\overrightarrow{f(t)} = \overrightarrow{(\sin(t), \cos(t), e^{\frac{t}{5}})}, t = 4\pi.$

(6) $\overrightarrow{f(t)} = \overrightarrow{(t^2, \sin(2t), e^{\frac{t}{4}})}, t = 3\pi.$

22.2 Integration and Arclength

Having defined the derivative of a vector valued function

$$\overrightarrow{f(t)} = x(t)\,\vec{\mathbf{i}} + y(t)\,\vec{\mathbf{j}} + z(t)\,\vec{\mathbf{k}}$$

we now define the indefinite and the definite integrals in an analogous manner:

$$\int \overrightarrow{f(t)}\,dt = \left(\int x(t)\,dt\right)\vec{\mathbf{i}} + \left(\int y(t)\,dt\right)\vec{\mathbf{j}} + \left(\int z(t)\,dt\right)\vec{\mathbf{k}},$$

and

$$\int_a^b \overrightarrow{f(t)}\,dt = \left(\int_a^b x(t)\,dt\right)\vec{\mathbf{i}} + \left(\int_a^b y(t)\,dt\right)\vec{\mathbf{j}} + \left(\int_a^b z(t)\,dt\right)\vec{\mathbf{k}},$$

Example [22.3] Evaluate $\int_0^2 (3t^2\vec{\mathbf{i}} + (t-2)\vec{\mathbf{j}} + e^t\vec{\mathbf{k}})\,dt$.

Following our definition the integral is evaluated as follows:

$$\begin{aligned} \int_0^2 3t^2\,dt\,\vec{\mathbf{i}} + \int_0^2 (t-2)\,dt\,\vec{\mathbf{j}} + \int_0^2 e^t\,dt\,\vec{\mathbf{k}} &= t^3\,\vec{\mathbf{i}} + \left(\frac{t^2}{2} - 2t\right)\vec{\mathbf{j}} + e^t\,\vec{\mathbf{k}}\Big]_0^2 \\ &= 8\vec{\mathbf{i}} - 2\vec{\mathbf{j}} + (e^2 - 1)\vec{\mathbf{k}}. \end{aligned}$$

Let $\overrightarrow{f(t)} = x(t)\vec{\mathbf{i}} + y(t)\vec{\mathbf{j}} + z(t)\vec{\mathbf{k}}$ be an arbitrary continuous space curve whose graph is depicted below.

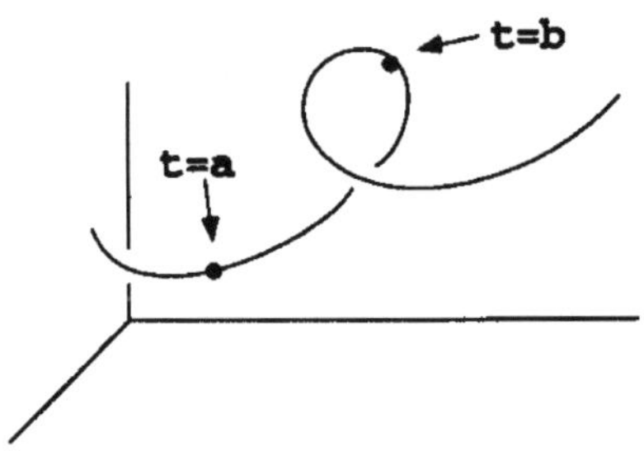

It is natural to ask, how can we compute the arclength L of the portion of the curve between $t = a$ and $t = b$. It is important to realize that since we are working in 3–dimensional space the space curve may be traversed more than once as t increases from a to b (notice that this cannot happen with the graph of a function in 2–dimensional space). Thus care must be taken to restrict our attention to the cases where the curve is traversed only once on the interval $a \leq t \leq b$.

Proposition [22.4] *Let $\overrightarrow{f(t)} = x(t)\,\vec{\mathbf{i}} + y(t)\,\vec{\mathbf{j}} + z(t)\,\vec{\mathbf{k}}$ be a continuous space curve so that the curve is traversed only once on the interval $a \le t \le b$. Then the length L of the curve on this interval is given by*

$$\begin{aligned} L &= \int_a^b |\overrightarrow{f'(t)}|\,dt \\ &= \int_a^b \sqrt{x'(t)^2 + y'(t)^2 + z'(t)^2}\,dt. \end{aligned} \tag{22.2}$$

Before verifying this result there are examples and applications worth spending time on.

Example [22.5] Find the arc length of the helix $\overrightarrow{f(t)} = \cos(t)\,\vec{\mathbf{i}} + \sin(t)\,\vec{\mathbf{j}} + t\,\vec{\mathbf{k}}$ for $0 \le t \le 5$.

Since $\overrightarrow{f'(t)} = -\sin(t)\,\vec{\mathbf{i}} + \cos(t)\,\vec{\mathbf{j}} + \vec{\mathbf{k}}$ its length is

$$|\overrightarrow{f'(t)}| = \sqrt{(\sin(t))^2 + (\cos(t))^2 + 1} = \sqrt{2},$$

and the desired arc length is given by

$$L = \int_0^5 \sqrt{2}\,dt = 5\sqrt{2}.$$

Proposition [22.4] actually remains valid in dimensions other than 3. For example, consider a vector valued function in 2–dimensional space

$$\overrightarrow{f(t)} = x(t)\,\vec{\mathbf{i}} + y(t)\,\vec{\mathbf{j}}.$$

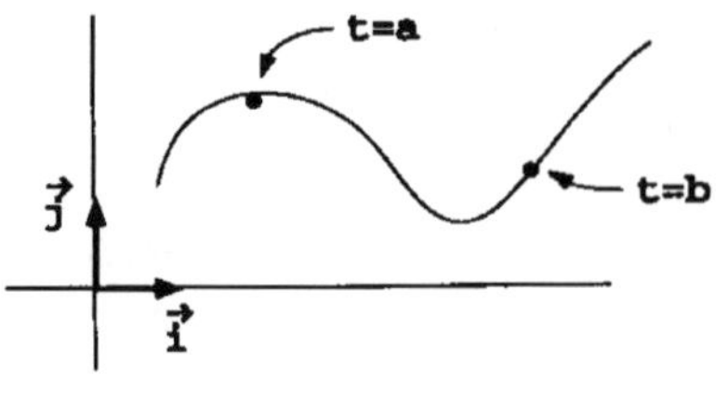

The arclength L of the piece of the curve as t ranges over the interval $a \le t \le b$ is given by

$$L = \int_a^b |\overrightarrow{f'(t)}|\,dt. \tag{22.3}$$

Example [22.6] Compute the circumference C of a circle of radius R.

As we have seen on various occasions, there are several ways to obtain the value of C. Here we begin by representing the circle as the space curve realized by the vector valued function $\overrightarrow{f(t)} = R\cos(t)\,\vec{\imath} + R\sin(t)\,\vec{\jmath}$ at which point the circumference can be readily computed by our proposition:

$$\begin{aligned} C &= \int_0^{2\pi} \sqrt{R^2(\sin(t))^2 + R^2(\cos(t))^2}\,dt \\ &= \int_0^{2\pi} R\,dt \;=\; 2\pi R. \end{aligned}$$

Example [22.7] Let $y = \phi(x)$ be a real valued continuous function of x. Demonstrate that the arclength L of $\phi(x)$ on the interval $a \le x \le b$ is given by the formula

$$L \;=\; \int_a^b \sqrt{1 + (\phi'(x))^2}\,dx.$$

This formula for arc length is precisely formula (16.4) and was verified in Chapter 16 with some measurable effort. Using the theory of vector valued functions we can derive (16.4) from (22.2) immediately once the original function is put into parametric form. In fact, the function $y = \phi(x)$

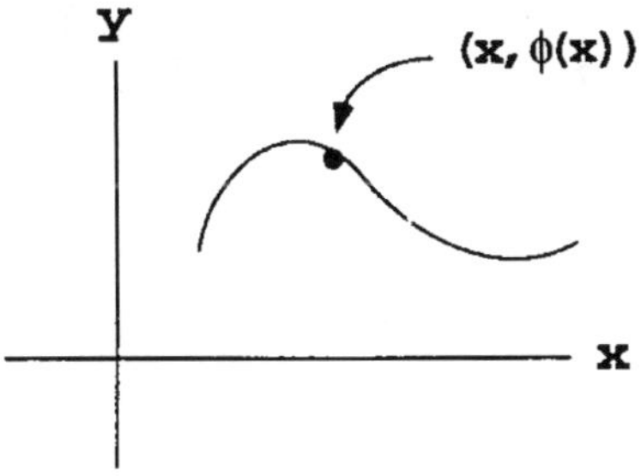

can be realized as the space curve $\overrightarrow{f(t)} = \overrightarrow{(t, \phi(t))}$.

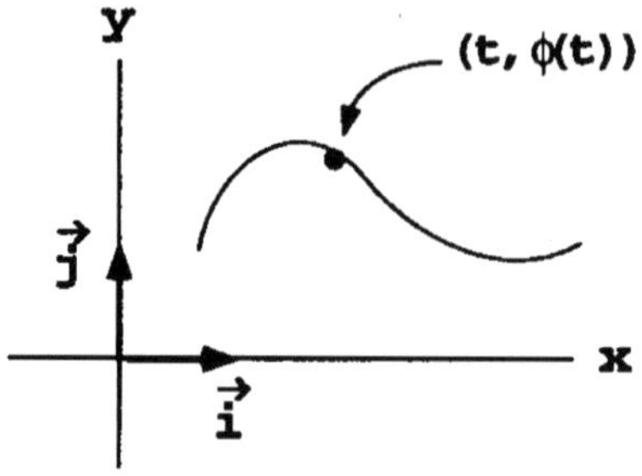

Noting that $\overrightarrow{f'(t)} = \overrightarrow{(1, \phi'(t))}$, and $|\overrightarrow{f'(t)}| = \sqrt{1 + (\phi'(t))^2}$, the formula (22.3) becomes obvious.

Derivation of the Arclength Formula

Fix a basepoint $t = a$ and let $L(t)$ denote the length of the piece of the space curve between $t = a$ and $t = T$.

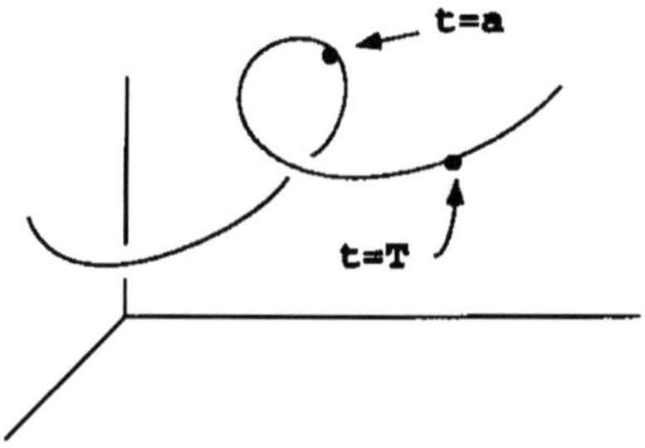

Then $L(t)$ is the length of the arc depicted below.

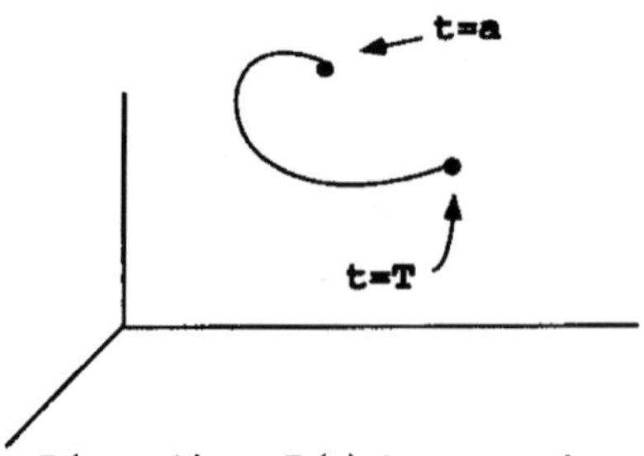

Note that $L(a) = 0$ and that $L(t + h) - L(t)$ is approximately the length of the vector $\overrightarrow{f(t+h)} - \overrightarrow{f(t)}$. Thus when we divide by h we obtain the quotients

$$\frac{L(t+h)-L(t)}{h} \approx \left|\frac{\overrightarrow{f(t+h)} - \overrightarrow{f(t)}}{h}\right|,$$

and letting $h \to 0$ on both sides we see that

$$L'(t) = |\overrightarrow{f'(t)}|.$$

Integrating both sides yields the desired formula

$$L(b) = L(b) - L(a)$$

$$= \int_a^b L'(t)\,dt = \int_a^b |\overrightarrow{f'(t)}|\,dt.$$

Exercises for §22.2

Compute the following integrals.

(1) $\int \left((t^3-1)\vec{\mathbf{i}} + 2t\vec{\mathbf{j}} + (e^t-1)\vec{\mathbf{k}}\right) dt.$

(2) $\int (\cos(t)\vec{\mathbf{i}} + \ln(t)\vec{\mathbf{j}} + e^t\vec{\mathbf{k}})\,dt.$

(3) $\int (te^{-t})\,\vec{\mathbf{i}} + \sqrt{t+1}\vec{\mathbf{j}} + t^4\vec{\mathbf{k}}\,dt.$

Compute the lengths of the following space curves on the indicated intervals. Graph the curves with your CAS.

(4) $\overrightarrow{f(t)} = t\vec{\mathbf{i}} + 2t - 1\vec{\mathbf{j}} + 1 - t\vec{\mathbf{k}}, (-1 \le t \le 2).$

(5) $\overrightarrow{f(t)} = (\frac{t^3}{6} + \frac{1}{2t})\vec{\mathbf{i}} + t\vec{\mathbf{j}},\ (1 \le t \le 3).$

(6) $\overrightarrow{f(t)} = 5\sin(t)\vec{\mathbf{i}} + 7t\vec{\mathbf{j}} + 5\cos(t)\vec{\mathbf{k}},\ (0 \le t \le 10).$

(7) $\overrightarrow{f(t)} = \ln(t)\vec{\mathbf{i}} + 2t\vec{\mathbf{j}} + t^2\vec{\mathbf{k}},\ (2 \le t \le 5).$

(8) $\overrightarrow{f(t)} = (t + \frac{t^3}{3})\vec{\mathbf{i}} + (t - \frac{t^3}{3})\vec{\mathbf{j}} + t^2\vec{\mathbf{k}},\ (0 \le t \le 1).$

(9) $\overrightarrow{f(t)} = 2\sin(t)\vec{\mathbf{i}} + 2\ln(\cos(t))\vec{\mathbf{j}} + 2\cos(t)\vec{\mathbf{k}},\ (0 \le t \le \frac{\pi}{4}).$

For each of the following real valued functions, find a space curve that represents them and compute the arclength on the indicated interval.

(10) $y = (x-1)^{\frac{3}{2}},\ (2 \le x \le 4).$

(11) $y = \frac{x^3}{6} + \frac{1}{2x},\ (1 \le x \le 3).$

(12) $y = e^x,\ (0 \le x \le 2).$

22.3 Arclength and Area in Polar Coordinates

Let $r = f(\theta)$ be a function given in polar form.

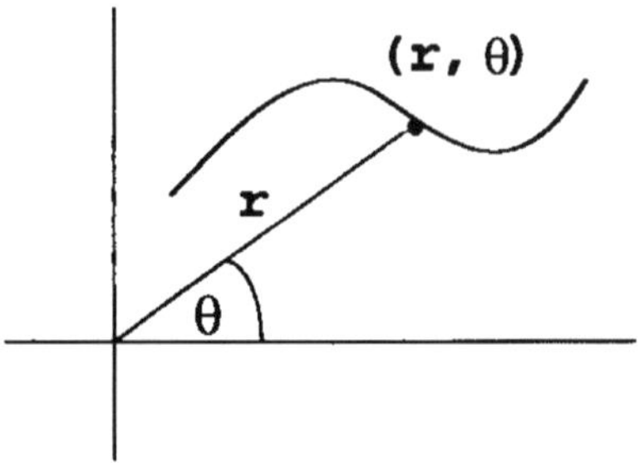

Let L denote the arclength of the piece of the curve between $\theta = \alpha$ and $\theta = \beta$.

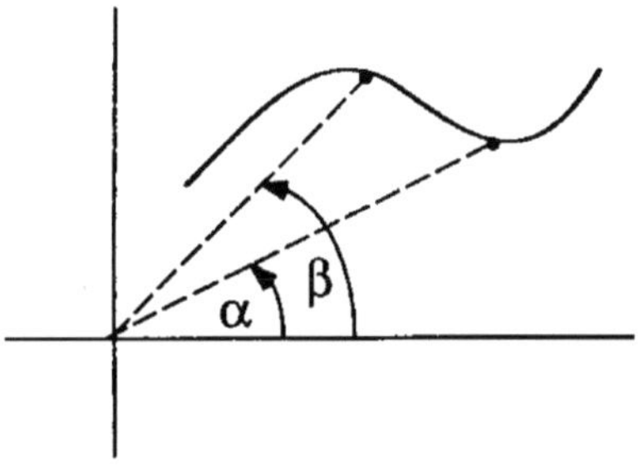

By converting $f(\theta)$ from polar form to the vector valued function

$$\overrightarrow{g(\theta)} = f(\theta)\cos(\theta)\,\vec{\mathbf{i}} + f(\theta)\sin(\theta)\,\vec{\mathbf{j}},$$

we can compute L using Proposition [22.4]:

$$L = \int_a^b |\overrightarrow{g'(\theta)}|\, d\theta.$$

Now

$$\begin{aligned}\overrightarrow{g'(\theta)} &= (f'(\theta)\cos(\theta) - f(\theta)\sin(\theta))\,\vec{\mathbf{i}} + (f'(\theta)\sin(\theta) + f(\theta)\cos(\theta))\,\vec{\mathbf{j}} \\ &= (r'\cos(\theta) - r\sin(\theta))\,\vec{\mathbf{i}} + (r'\sin(\theta) + r\cos(\theta))\,\vec{\mathbf{j}},\end{aligned}$$

and hence

$$\begin{aligned}|\overrightarrow{g'(\theta)}|^2 &= (r'\cos(\theta) - r\sin(\theta))^2 + (r'\sin(\theta + r\cos(\theta))^2 \\ &= (r')^2 + r^2.\end{aligned}$$

In summary we have proved,

Proposition [22.8] *The length L of a curve with polar equation $r = f(\theta)$ for $\alpha \leq \theta \leq \beta$ is given by*

$$L = \int_{\alpha}^{\beta} \sqrt{r^2 + \left(\frac{dr}{d\theta}\right)^2}\, d\theta.$$

Example [22.9] Find the length of one leaf of the four–leaved rose $r = \cos(2\theta)$.

We can plot one leaf of this rose by letting $-\frac{\pi}{4} \leq \theta \leq \frac{\pi}{4}$. Thus the arclength L is given by the integral

$$L = \int_{-\frac{\pi}{4}}^{\frac{\pi}{4}} \sqrt{(\cos(2\theta))^2 + 4(\sin(2\theta))^2}\, d\theta.$$

Recall that we have shown that the area of a circle of radius R is πR^2. Similarly, the area of a semicircle is $\frac{1}{2}\pi R^2$, and the area of a quarter circle is $\frac{1}{4}\pi R^2$. More generally, the area of a circular sector of radius R and angle α with $0 \leq \alpha \leq 2\pi$ is given by $\frac{1}{2}\alpha R^2$.

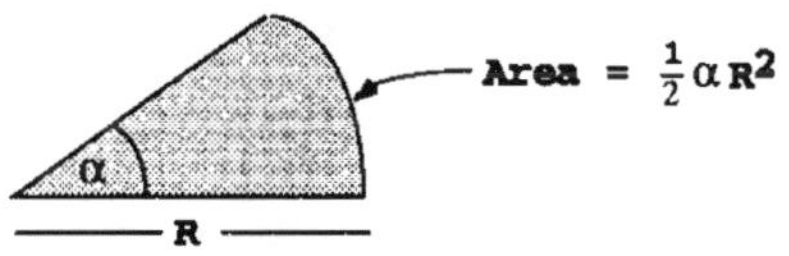

Warning: It is *critical* not to omit the $\frac{1}{2}$ in the formula for the area of a circular sector.

Returning again to our polar curve $r = f(\theta)$, let A denote the area illustrated below.

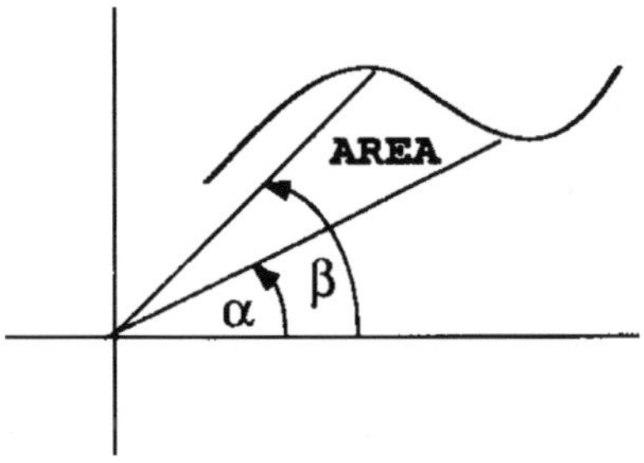

We claim that this area is given by the integral

$$A = \int_{\alpha}^{\beta} \frac{1}{2} f(\theta)^2 \, d\theta. \tag{22.4}$$

Example [22.10] Compute the area A of one leaf of the three–leaved rose $r = \sin(3\theta)$.

A single leaf of the rose appears when we plot the function on the interval $0 \leq \theta \leq \frac{2\pi}{3}$. Thus the desired area is given by

$$A = \int_{0}^{\frac{2\pi}{3}} \frac{1}{2} (\sin(3\theta))^2 \, d\theta.$$

To verify (22.4) we begin by fixing α and let $A(\theta)$ denote the area depicted below.

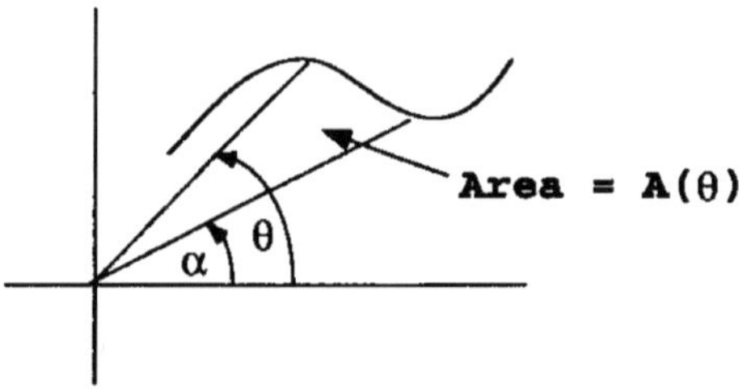

Clearly $A(\alpha) = 0$. Viewing θ as a variable consider for h sufficiently small, the difference $A(\theta + h) - A(\theta)$. By definition this difference represents the area of the thin wedge below,

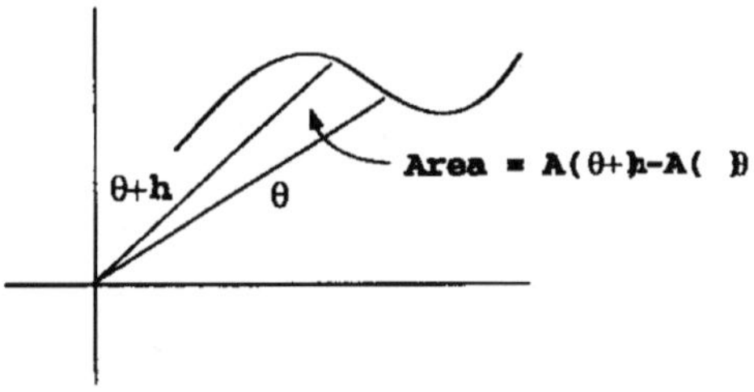

which is approximately a circular sector whose radius is $r = f(\theta)$ and whose angle is h. Thus

$$A(\theta + h) - A(\theta) \approx \frac{1}{2} h f(\theta)^2.$$

Upon dividing by h and letting $h \to 0$ we obtain the derivative of A explicitly:

$$A'(\theta) = \lim_{h\to 0} \frac{A(\theta+h) - A(\theta)}{h}$$

$$= \frac{1}{2} f(\theta)^2.$$

Equation (22.4) follows by integrating both sides of this equation with respect to θ.

Exercises for §22.3

Plot one leaf of the following roses with your CAS. In each case express the area and the arclength of the leaf by an integral. Evaluate these integrals with your CAS.

(1) $r = 4\cos(4\theta)$.

(2) $r = 2\cos(8\theta)$.

(3) $r = 7\cos(5\theta)$.

(4) $r = 2\cos(3\theta + \frac{\pi}{2})$.

(5) $r = 4\cos(2\theta + \frac{\pi}{9})$.

(6) $r = 3\cos(-7\theta)$.

In exercises (7)–(9) plot the arcs with your CAS. Compute the arclength in each case.

(7) $r = e^{\frac{1}{3}\theta}$ for $-10 \le \theta \le 10$.

(8) $r = 5\left(\sin\left(\frac{\pi}{2}\right)\right)^2$ for $0 \le \theta \le \pi$.

(9) $r = (\tan(\theta))^2$ for $-\frac{\pi}{3} \le \theta \le \frac{\pi}{3}$.

(10) Find the area bounded by $r^2 = 9\left(1 + 3(\sin(\theta))^2\right)$ for $-\pi \le \theta \le \pi$.

(11) Find the area bounded by the parabola $r = \frac{3}{1-\cos(\theta)}$ for $0 \le \theta \le 2\pi$.and the line $\theta = \frac{\pi}{3}$.

(12) Find the area of the smaller loop of the limacon $r = \frac{1}{2} + \cos(\theta)$.

(13) Find the area of the region that lies inside both curves $r = \sin(2\theta)$ and $r = \sin(\theta)$.

(14) Find the area inside the curve $r = 3(1 + \cos(\theta))$ for $-\pi \le \theta \le \pi$ but outside $r^2 = 4(1 + \cos(\theta))$ for $-\frac{\pi}{2} \le \theta \le \frac{\pi}{2}$.

22.4 Direction and Curvature

Let us return now to considering a particle traveling through 3–dimensional space along a space curve which does not traverse any part of itself more than once.

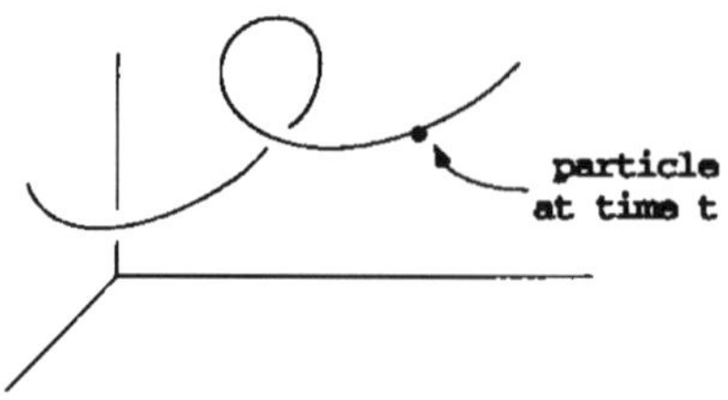

Assume the curve is the graph of a **smooth** vector valued function

$$\overrightarrow{f(t)} = x(t)\,\vec{\mathbf{i}} + y(t)\,\vec{\mathbf{j}} + z(t)\,\vec{\mathbf{k}}$$

(i.e., $\overrightarrow{f(t)}$ is differentiable infinitely often and non–zero except possibly at the end-points of a given interval being considered). We have already shown that $\overrightarrow{f'(t)}$ is a tangent vector to the curve whose length is the velocity of the particle at the time t. The assumption $\overrightarrow{f'(t)} \neq 0$ allows us to consider the vector

$$\overrightarrow{T(t)} = \frac{\overrightarrow{f'(t)}}{|\overrightarrow{f'(t)}|} = \frac{\overrightarrow{f'(t)}}{v(t)},$$

which we term the **unit tangent vector**. The term *unit* arises from the observation that $\overrightarrow{T(t)}$ has length 1, i.e., the dot product of $\overrightarrow{T(t)}$ with itself is one:

$$\overrightarrow{T(t)} \cdot \overrightarrow{T(t)} = 1. \tag{22.5}$$

Question: What properties does $\overrightarrow{T'(t)}$ have?

Answer: We claim that the vector, $\overrightarrow{T'(t)}$, is always perpendicular to $\overrightarrow{T(t)}$ and is thus termed a **normal vector**. To see that this is the case it is sufficient to show that the dot product

$$\overrightarrow{T'(t)} \cdot \overrightarrow{T(t)} = 0.$$

This identity follows from (22.5) when we differentiate both sides of (22.5) with respect to t:

$$\frac{d}{dt}\left(\overrightarrow{T(t)} \cdot \overrightarrow{T(t)}\right) = \overrightarrow{T'(t)} \cdot \overrightarrow{T(t)} + \overrightarrow{T(t)} \cdot \overrightarrow{T'(t)} = 0,$$

and hence

$$2\overrightarrow{T'(t)} \cdot \overrightarrow{T(t)} = 0.$$

Having found a normal vector we now wish to produce one whose length is 1. Following the method used to produce a tangent vector of length 1 we define

$$\overrightarrow{N(t)} = \frac{\overrightarrow{T'(t)}}{|\overrightarrow{T'(t)}|}$$

to be the **unit normal vector**. Visually, the vectors $\overrightarrow{T(t)}$ and $\overrightarrow{N(t)}$ are perpendicular vectors of length one on our space curve.

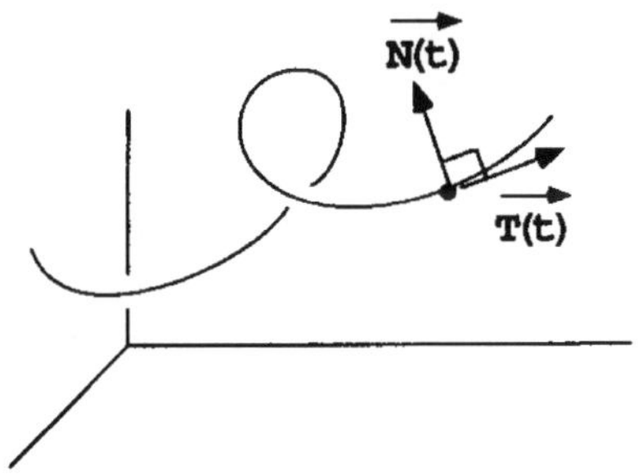

Example [22.11] Let $\overrightarrow{f(t)} = \cos(t)\,\vec{\mathbf{i}} + 2t\,\vec{\mathbf{j}} + \sin(t)\,\vec{\mathbf{k}}$ be a space curve. Find $\overrightarrow{T(t)}$ and $\overrightarrow{N(t)}$.

We begin by computing

$$\overrightarrow{f'(t)} = -\sin(t)\,\vec{\mathbf{i}} + 2\,\vec{\mathbf{j}} + \cos(t)\,\vec{\mathbf{k}}$$

at which point we see that

$$|\overrightarrow{f'(t)}| = \sqrt{(\sin(t))^2 + 4 + (\cos(t))^2} = \sqrt{5}.$$

By definition

$$\overrightarrow{T(t)} = -\frac{\sin(t)}{\sqrt{5}}\,\vec{\mathbf{i}} + \frac{2}{\sqrt{5}}\,\vec{\mathbf{j}} + \frac{\cos(t)}{\sqrt{5}}\,\vec{\mathbf{k}}.$$

Now to compute $\overrightarrow{N(t)}$ we first differentiate $\overrightarrow{T(t)}$:

$$\overrightarrow{T'(t)} = -\frac{\cos(t)}{\sqrt{5}}\,\vec{\mathbf{i}} - \frac{\sin(t)}{\sqrt{5}}\,\vec{\mathbf{k}}.$$

Then $\overrightarrow{T'(t)}$ has length $\frac{1}{\sqrt{5}}$ and

$$\overrightarrow{N(t)} = -\cos(t)\,\vec{\mathbf{i}} - \sin(t)\,\vec{\mathbf{k}}.$$

Curvature:

The concept of curvature is a formalization of a basic and intuitive perception. Any object which is in motion (a car, bicycle, airplane, etc.) experiences a variety of turns. Some turns are sharp, others are barely noticeable. In the abstract setting of the space curve below we can see that at the points C, D the curvature (i.e., amount of turning) is substantial, while at the points P, Q, R the curve is actually fairly straight.

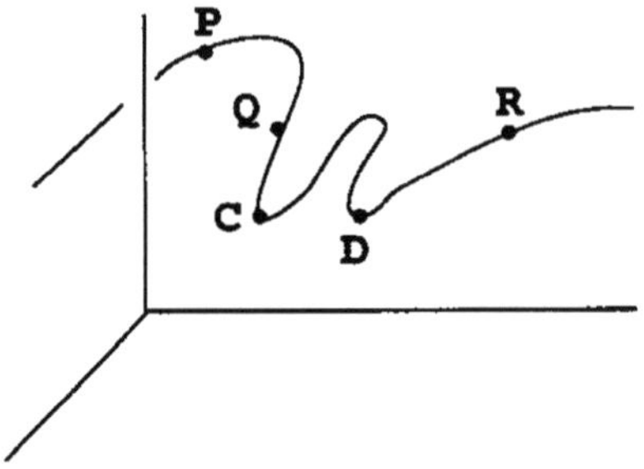

Question: How can we measure curvature at a point?

One way of assessing the curvature at a point is to look at the unit tangent vectors of nearby points. For example, the unit tangent vectors of points near P are all approximately the same. On the other hand, when we focus on points near C, it is clear that the direction of the tangent vectors change frequently even when we move only slightly along the curve.

Thus to measure the curvature we must measure the change in the direction of the unit tangent vector as we move along the curve. To make this observation precise we begin by fixing a **base point** $t = a$ on the smooth space curve

$$\overrightarrow{f(t)} = x(t)\,\vec{\mathbf{i}} + y(t)\,\vec{\mathbf{j}} + z(t)\,\vec{\mathbf{k}}.$$

Let $s = s(t)$ denote the arclength between a and t. Then $s(t)$ is a one–to–one function (since the distance between a and t can only have one value) which is given by the integral

$$s(t) = \int_a^t |\overrightarrow{f'(t)}|\, dt = \int_a^t v(t)\, dt.$$

Since $s(t)$ is one–to–one it has an inverse which we shall denote $t = \phi(s)$. Thus we may write

$$\overrightarrow{f(t)} = \overrightarrow{f(\phi(s))},$$

which is termed parameterizing $\vec{f}$ with respect to arclength. Having done this we can now discuss the rate of change (i.e., the derivative) of $\vec{f}$ with respect to the arclength variable s.

Definition: *Let* $\overrightarrow{f(t)} = x(t)\,\vec{\mathbf{i}} + y(t)\,\vec{\mathbf{j}} + z(t)\,\vec{\mathbf{k}}$ *be a smooth space curve which does not traverse any part of itself more than once. The* ***curvature*** $\kappa(t)$ *at t is defined to be*

$$\left|\frac{d\overrightarrow{T(t)}}{ds}\right|.$$

In order to compute the curvature easily we first apply the chain rule to see that

$$\frac{d\overrightarrow{T(t)}}{dt} = \frac{d\overrightarrow{T(t)}}{ds} \cdot \frac{ds}{dt}.$$

Furthermore, since $s(t) = \int_a^t |\overrightarrow{f'(t)}|\,dt$, the Fundamental theorem of calculus gives us

$$\frac{ds}{dt} = |\overrightarrow{f'(t)}| = v(t).$$

Thus we have obtained an alternative (and useful form) for the curvature:

$$\kappa(t) = \frac{|\overrightarrow{T'(t)}|}{|v(t)|}. \tag{22.6}$$

Example [22.12] Show that the curvature of a straight line is 0.

The space curve whose graph is the most general line which passes through a fixed point (x_0, y_0, z_0) is

$$\overrightarrow{f(t)} = (x_0 + at)\,\vec{\mathbf{i}} + (y_0 + bt)\,\vec{\mathbf{j}} + (z_0 + ct)\,\vec{\mathbf{k}},$$

where a, b, c are constants. Thus $\overrightarrow{f'(t)} = a\,\vec{\mathbf{i}} + b\,\vec{\mathbf{j}} + c\,\vec{\mathbf{k}}$, and the unit tangent vector at any point t is

$$\overrightarrow{T(t)} = \frac{1}{\sqrt{a^2+b^2+c^2}}(a\,\vec{\mathbf{i}} + b\,\vec{\mathbf{j}} + c\,\vec{\mathbf{k}}).$$

Since $\overrightarrow{T'(t)} = \vec{0}$ the curvature is zero.

Example [22.13] Show that the curvature of a circle of radius R is $\frac{1}{R}$.

A circle of radius R centered at the origin in the $x\,y$–plane can be realized as the space curve

$$\overrightarrow{f(t)} = R\cos(t)\,\vec{\mathbf{i}} + R\sin(t)\,\vec{\mathbf{j}}.$$

Thus $\overrightarrow{f'(t)} = -R\sin(t)\,\vec{\mathbf{i}} + R\cos(t)\,\vec{\mathbf{j}}$ is a tangent vector whose length is R. The unit tangent vector at a point t is given by

$$\overrightarrow{T(t)} = -\sin(t)\,\vec{\mathbf{i}} + \cos(t)\,\vec{\mathbf{j}},$$

and thus $\overrightarrow{T'(t)} = -\cos(t)\,\vec{\mathbf{i}} - \sin(t)\,\vec{\mathbf{j}}$, also has length one. Equation (22.6) then dictates that the curvature is given by

$$\kappa(t) = \frac{|\overrightarrow{T'(t)}|}{|v(t)|} = \frac{1}{R}.$$

Exercises for §22.4

Find $\overrightarrow{T(t)}$, $\overrightarrow{N(t)}$, and $\overrightarrow{\kappa(t)}$ for the following space curves. Plot the space curves as well as the vectors $\overrightarrow{T(t)}$ and $\overrightarrow{N(t)}$ which emanate from them with the CAS. Visually confirm that these vectors are perpendicular.

(1) $\overrightarrow{f(t)} = \sin(2t)\,\vec{\mathbf{i}} + 5t\,\vec{\mathbf{j}} + \cos(2t)\,\vec{\mathbf{k}}.$

(2) $\overrightarrow{f(t)} = e^t\,\vec{\mathbf{i}} + e^{-t}\,\vec{\mathbf{j}} + \sqrt{2}t\,\vec{\mathbf{k}}.$

(3) $\overrightarrow{f(t)} = \ln(\cos(t))\,\vec{\mathbf{i}} + \cos(t)\,\vec{\mathbf{j}} + \sin(t)\,\vec{\mathbf{k}}.$

(4) $\overrightarrow{f(t)} = \left(t - \frac{t^3}{3}\right)\vec{\mathbf{i}} + t^2\,\vec{\mathbf{j}} + \left(t + \frac{t^3}{3}\right)\vec{\mathbf{k}}.$

(5) Find the curvature $\kappa(0)$ (at $t = 0$) for the space curve of exercise (4). The points $A = \overrightarrow{f(\frac{1}{2})} = \overrightarrow{(\frac{11}{24}, \frac{1}{4}, \frac{13}{24})}$ and $B = \overrightarrow{f(-\frac{1}{2})} = \overrightarrow{(-\frac{11}{24}, \frac{1}{4}, -\frac{13}{24})}$ are close to $\overrightarrow{f(0)} = \overrightarrow{(0,0,0)}$. Find the equation of a circle of radius $\frac{1}{\kappa(0)}$ which passes through the points A, B. Graph this circle and the space curve with your CAS. Do the curvatures match near $t = 0$?

(6) Show with your CAS that the curvature $\kappa(t)$ of the space curve $\overrightarrow{f(t)}$ is given by the formula

$$\kappa(t) = \frac{|\overrightarrow{f'(t)} \times \overrightarrow{f''(t)}|}{|\overrightarrow{f'(t)}|^3}.$$

(7) Show that in the special case of a plane curve $y = f(x)$, the curvature $\kappa(x)$ is given by

$$\kappa(x) = \frac{|f''(x)|}{(1 + (f'(x))^2)^{\frac{3}{2}}}.$$

(8) Compute the curvature of $y = e^x$ at a general point (x, y). For which value(s) of x is the curvature maximized?

(9) Let $f(t) > 0$ and $g(t)$ be a differentiable real-valued function of t. Show that the space curve

$$\overrightarrow{f(t)} = \int f(t)\sin(t)\,dt\,\vec{\mathbf{i}} + \int f(t)\cos(t)\,dt\,\vec{\mathbf{j}} + \int f(t)\,g(t)\,dt\,\vec{\mathbf{k}}$$

has curvature

$$\kappa(t) = \frac{1}{f(t)}\sqrt{\frac{1+g(t)^2+(g'(t))^2}{(1+g(t)^2)^3}}.$$

22.5 Velocity and Acceleration

When driving a car around a curve, if one wants to maintain a constant speed, it is necessary to accelerate while simultaneously turning the car in a direction perpendicular to the tangent vector to the curve. This basic observation has a mathematical model which we shall now develop. Let $\overrightarrow{f(t)} = x(t)\,\vec{\mathbf{i}} + y(t)\,\vec{\mathbf{j}} + z(t)\,\vec{\mathbf{k}}$ be a smooth space curve whose graph is traced out by a moving particle. Let $\overrightarrow{T(t)}$ and $\overrightarrow{N(t)}$ be the unit tangent and normal vectors of the particle at time t.

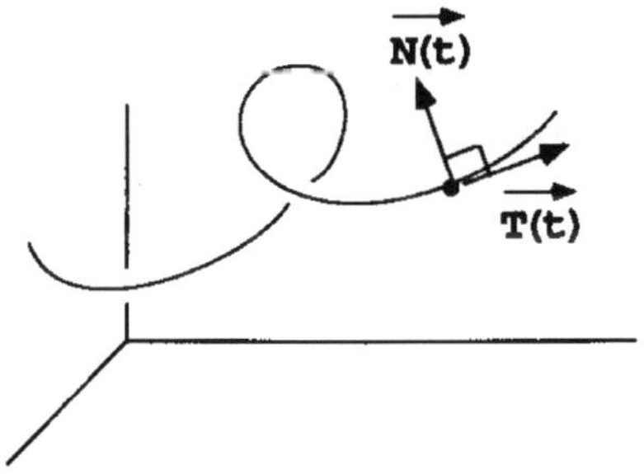

As we have seen, the velocity vector is simply

$$\overrightarrow{\mathbf{v}(t)} = \overrightarrow{f'(t)} = v(t)\overrightarrow{T(t)}$$

where $v(t)$ is the instantaneous velocity. The natural definition for the acceleration vector is the derivative of the velocity vector,

$$\overrightarrow{\mathbf{a}(t)} = \overrightarrow{\mathbf{v}'(t)} = v'(t)\overrightarrow{T(t)} + v(t)\overrightarrow{T'(t)}. \qquad (22.7)$$

Intuitively the acceleration should depend, in part, on the curvature.

Proposition [22.15] *Let $a(t) = v'(t)$. Then the acceleration vector can be expressed as the sum*

$$\overrightarrow{\mathbf{a}(t)} = a(t)\,\overrightarrow{T(t)} + \kappa(t)\,v(t)^2\,\overrightarrow{N(t)}.$$

Proof: By combining the definition of the unit normal vector, $\overrightarrow{T'(t)} = |\overrightarrow{T'(t)}|\,\overrightarrow{N(t)}$, with equation (22.6), $|\overrightarrow{T'(t)}| = \kappa(t)\,v(t)$, we obtain the identity

$$\overrightarrow{T'(t)} = \kappa(t)\,v(t)\overrightarrow{N(t)}.$$

But then (22.7) becomes

$$\overrightarrow{\mathbf{a}(t)} = v'(t)\,\overrightarrow{T(t)} + \kappa(t)\,v(t)^2\,\overrightarrow{N(t)}.$$

Upon replacing $v'(t)$ by $a(t)$ (which is termed the instantaneous acceleration) the proposition is proved.

Remark: A common way of succinctly stating Proposition [22.15] is to say that $\overrightarrow{\mathbf{a}(t)}$ has the component $a(t)$ in the $\overrightarrow{T(t)}$–direction (or the **tangential component**) and the component $\kappa(t)v(t)^2$ in the $\overrightarrow{N(t)}$–direction (or the **normal direction**).

Example [22.16] A bicycle is turning around a circle of radius R which is represented by the space curve

$$\overrightarrow{f(t)} = R\sin(t)\,\vec{\mathbf{i}} + R\cos(t)\,\vec{\mathbf{j}}.$$

Compute $\overrightarrow{T(t)}$, $\overrightarrow{N(t)}$, $\overrightarrow{\mathbf{v}(t)}$, $\overrightarrow{\mathbf{a}(t)}$, and the components of acceleration.

To begin, we see that

$$\overrightarrow{\mathbf{v}(t)} = \frac{d\overrightarrow{f(t)}}{dt} = R\cos(t)\,\vec{\mathbf{i}} - R\sin(t)\,\vec{\mathbf{j}},$$

and thus

$$v(t) = |\overrightarrow{\mathbf{v}(t)}| = R,$$

$$a(t) = v'(t) = 0.$$

Furthermore,

$$\overrightarrow{T(t)} = \cos(t)\,\vec{\mathbf{i}} - \sin(t)\,\vec{\mathbf{j}},$$

$$\overrightarrow{T'(t)} = -\sin(t)\,\vec{\mathbf{i}} - \cos(t)\,\vec{\mathbf{j}},$$

and

$$\overrightarrow{N(t)} = -\sin(t)\,\vec{\mathbf{i}} - \cos(t)\,\vec{\mathbf{j}}.$$

Recalling from Example [22.13] that $\kappa(t) = \frac{1}{R}$, we see that

$$\overrightarrow{\mathbf{a}(t)} \;=\; a(t)\,\overrightarrow{T(t)} + \kappa(t)\,v(t)^2\,\overrightarrow{N(t)} \;=\; R\overrightarrow{N(t)}.$$

Thus there is no tangential acceleration when turning around the circle. The only acceleration is in the direction $\overrightarrow{N(t)}$ which points toward the center of the circle.

As a conclusion to this chapter we restate the various derived formulae. This list is useful for reference as well as for gaining a perspective on the material.

Summary of Basic Formulae

$$\overrightarrow{\mathbf{v}(t)} = \frac{d\overrightarrow{f(t)}}{dt}, \qquad v(t) = |\overrightarrow{\mathbf{v}(t)}|,$$

$$\overrightarrow{T(t)} = \frac{\overrightarrow{\mathbf{v}(t)}}{v(t)}, \qquad \overrightarrow{N(t)} = \frac{\overrightarrow{T'(t)}}{|\overrightarrow{T'(t)}|},$$

$$\overrightarrow{\mathbf{a}(t)} = \frac{d\overrightarrow{\mathbf{v}(t)}}{dt}, \qquad a(t) = \frac{dv}{dt},$$

$$s(t) = \int v(t)\,dt, \qquad \kappa(t) = \frac{|T'(t)|}{dt},$$

$$\overrightarrow{\mathbf{a}(t)} = a(t)\overrightarrow{T(t)} + \kappa(t)v(t)^2\overrightarrow{N(t)}.$$

Exercises for §22.5

In exercises (1)–(4) plot the space curve $\overrightarrow{f(t)}$ with your CAS. Compute the tangential and normal components of the acceleration vector.

(1) $\overrightarrow{f(t)} = 8\vec{\mathbf{i}} + (2t^2 - 3)\vec{\mathbf{j}} + (t-2)\vec{\mathbf{k}}.$

(2) $\overrightarrow{f(t)} = 3\sin(t)\vec{\mathbf{i}} - 5t\vec{\mathbf{j}} + 3\cos(t)\vec{\mathbf{k}}.$

(3) $\overrightarrow{f(t)} = \sqrt{2}t\vec{\mathbf{i}} + e^{-t}\vec{\mathbf{j}} + e^{t}\vec{\mathbf{k}}.$

(4) $\overrightarrow{f(t)} = \frac{1}{2}t^2\vec{\mathbf{i}} + \frac{2\sqrt{2}}{3}t^{\frac{3}{2}}\vec{\mathbf{j}} + t\vec{\mathbf{k}}.$

(5) Assume that the earth is located at the origin $(0,0,0)$. Suppose a telecommunications satellite is traveling around the earth in an elliptical orbit. Its path is the space curve

$$\overrightarrow{f(t)} = 3\sqrt{2}\cos(\pi t)\vec{\mathbf{j}} + 2\sqrt{2}\sin(\pi t)\vec{\mathbf{k}}.$$

Now suppose a meteor is approaching earth whose path is given by the space curve

$$\overrightarrow{g(t)} = \left(t - \frac{5}{4}\right)\vec{\mathbf{i}} + \left(t^2 - 3t - \frac{13}{16}\right)\vec{\mathbf{j}} - \left(t + \frac{3}{4}\right)\vec{\mathbf{k}}.$$

Plot the earth and the paths of the satellite and the meteor for $0 \le t \le 2\pi$. What do you observe? At which time t does the meteor collide with the satellite? At the moment of collision, compute the following:

(a) the angle between the paths of the meteor and the satellite,

(b) the curvature of the meteor,

(c) the velocity of the satellite and the meteor,

(d) the acceleration vector of the satellite and the meteor.

(6) Assume again that the earth is a point which is fixed at the origin $(0,0,0)$. A reconnaissance satellite belonging to country A is traveling in the elliptical orbit around the earth given by the space curve

$$\overrightarrow{f(t)} = 5\cos\left(\frac{t\pi}{12}\right)\vec{\mathbf{i}} + 3\sin\left(\frac{t\pi}{12}\right)\vec{\mathbf{k}}.$$

How long does it take the satellite to make one full rotation around the earth? Suppose country Z, an enemy of country A, wishes to destroy the satellite by launching a missile (travelling at constant speed, starting at $t = 0$) which, when it hits the satellite, will set off an explosive device. The path of the missile must be a straight line. Find the equation of the line (as a space curve) which will result in a direct hit after exactly 3 hours. Plot the space curves for the satellite and the missile with your CAS. Compute the following (at the moment of collision):

(a) the angle between the paths of the satellite and the missile,

(b) the curvature of the satellite,

(c) the velocity of the satellite and the missile,

(d) the acceleration vector of the satellite and the missile.

(7) Assume a particle is moving with constant velocity. Show the the direction of the acceleration vector must be in the direction of the unit normal vector $\overrightarrow{N(t)}$.

(8) Show the the magnitude of the acceleration vector is $\sqrt{v'(t)^2 + \kappa(t)^2 v(t)^4}$.

Additional exercises for Chapter XXII

(1) Let $\overrightarrow{f(t)} = \overrightarrow{(t, t^2, t^3)}, 0 \leq t \leq 1$. On your CAS sketch the curve described by $\vec{f}$ and the tangent line at $(1/2, 1/4, 1/8)$. Next find $\overrightarrow{|f'(t)|}$. Finally, suppose t runs through $\mathbf{R}$. Find all the points of the curve described by $\vec{f}$ at which the tangent vector is perpendicular to the vector $\overrightarrow{(2, 2, 1)}$.

(2) Prove that if the vector valued, differentiable function $\vec{f}$ is never zero for $u \leq t \leq v$ then:

(a) $\vec{f} \cdot \frac{d\vec{f}}{dt} = |\vec{f}| \frac{d|\vec{f}|}{dt}$,

(b) $\vec{f}$ is constant if and only if $\vec{f} \cdot \vec{f'} = 0$,

(c) $\vec{f}(t)$ has a constant direction if and only if $\vec{f} \times \vec{f'} \equiv 0$.

(3) On your CAS, sketch the curve represented by $(x, y) = (t^3, t^5)$ and show that this parameterization does not allow for a tangent vector at the origin. Find an alternative parameterization that allows for a tangent vector at the origin.

(4) Prove that if $\vec{g}(t)$ takes values in $\mathbf{R}^3$, with $t \in \mathbf{R}$, and if $\vec{g'}(t) = 0$ for $a \leq t \leq b$, then $\vec{g}(t)$ is a constant vector on that interval. (**Hint:** apply the mean value theorem to each coordinate function).

(5) Let $\vec{f}(t) = \overrightarrow{(\cos(t), \sin(t), 1)}$. On your CAS, sketch the graph of the curve and describe the tangent vector to the curve, what can you say about its direction with respect to the vector $\overrightarrow{(\cos(t), \sin(t), 0)}$?

Find the length of the curves given by the following parametric equation:

(6) $\vec{f}(t) = \overrightarrow{(t^3, \frac{2}{3}t^3 - \frac{3}{2}t)}, 0 \leq t \leq 3$.

(7) $\vec{f}(t) = \overrightarrow{(6t^2, \sqrt{2}t^3, t^4)}, -1 \leq t \leq 2$.

(8) $\vec{f}(t) = \overrightarrow{(\cos(t), \sin(t), t)}, 0 \leq t \leq \pi$.

(9) $y = x^{2/3}, 0 \leq x \leq 5$.

(10) $r = \exp(2t), \theta = t, z = \exp(2t); 0 \leq t \leq \ln(2)$.

(11) Find the parametric equations of the circle using arclength s as the variable (recall that the equation of the circle is $x = a\cos(t), y = a\sin(t), 0 \leq t \leq 2\pi$

(12) Find the parametric equation of the curve given by $\vec{f}(t) = \overrightarrow{(a\cos(t), a\sin(t), ct)}, t \geq 0$ using arclength s as the variable

(13) Find the parametric equation of the curve given by $\vec{f}(t) = \overrightarrow{(\sin(\exp(2t)), \cos(\exp(2t)),}$ $\sqrt{3}\exp(2t), t \geq 0$ using arclength s as the variable

(14) Find the area which is outside the circle $r = 2\cos(\theta)$ and inside the cardioid $r = 2+2\cos(\theta)$.

(15) Find the area which is enclosed by the curves $3\sin(\theta)$ and $3\cos(\theta)$.

(16) Prove that the curvature for the curve described by $\vec{f}(t) = \overrightarrow{(x(t), y(t), z(t))}$ can be expressed by $\kappa(t) = \frac{|\vec{f}'(t)\times\vec{f}''(t)|}{|\vec{f}'(t)|^3}$.

(17) On your CAS, graph the circular helix defined by $\vec{f}(t) = \overrightarrow{(a\cos(t), a\sin(t), ct)}, a > 0$ and find its curvature $\kappa(t)$ using the result of the previous problem.

Chapter XXIII
Functions of Several Variables

23.1 Functions of Several Variables

When functions appear in nature they generally involve many variables. For example, the price of gold depends on a variety of factors including the number of people wanting to buy or sell gold, the amount of gold being mined, unpredictable political events, etc. An accurate mathematical model for the price of gold is probably impossible to obtain. Nevertheless, there are many instances where mathematical models involving functions of several variables have been developed and successfully implemented. An illustration of such an achievement involves the flight of a commercial airplane which is controlled mainly by a computer. The computer outputs a flight path (i.e., a space curve) which is a function of many variables such as atmospheric pressure, windspeed and direction, rotation of the earth, etc. It is plainly evident that a study of functions of several variables and the development of a multidimensional calculus is essential. This is the goal of the remainder of this book.

> **Definition:** *A* ***real valued function of*** n ***variables****, denoted* $f(x_1, \ldots, x_n)$*, is a rule which assigns to every* n***–tuple*** *of real numbers,* $x_1, \ldots, x_n$*, in a domain of* n*–tuples, a unique real number,* $f(x_1, \ldots, x_n)$*.*

Remark: When the concept of function (along with the notion of domain and range) was introduced in Chapter 1 the presentation was sufficiently general that it carries over to this discussion. Thus, in the definition above, the domain referred to is familiar.

Example [23.1] (Area of a rectangle) Perhaps the simplest example of a function of two variables is the function which determines the area of a rectangle of length L and width W.

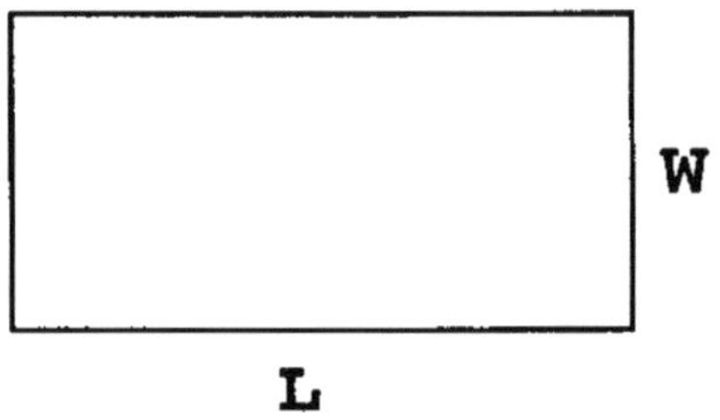

Its area $A(L, W)$ is a function of both L and W and is given by

$$A(L, W) = L \cdot W.$$

Notice that the function A is defined on all 2–tuples (L, W) where both L and W are non–negative.

Example [23.2] (Additive Mean Function) Let $x_1, \ldots, x_n$ be any n–tuple of numbers. The additive mean function, $M(x_1, \ldots, x_n)$ is a function of n variables and is defined to be

$$M(x_1, \ldots, x_n) = \frac{1}{n}(x_1 + x_2 + \cdots + x_n).$$

For example, when $n = 3$ the additive mean of $2, 5, 8$ is

$$M(2, 5, 8) = \frac{1}{3}(2 + 5 + 8) = 5.$$

Example [23.3] (The Pizza Function) A pizzeria offers a small cheese and tomato pizza for \$5.50. There are four different toppings available each of which cost extra. Assume that peppers are an additional 40¢, onions are 50¢, mushrooms are 75¢, and eggplant is \$1.00 per topping. Let P, O, M, and E denote the number of additional toppings of peppers, onions, mushrooms, and eggplant ordered respectively by a customer. The cost of the pizza $C(P, O, M, E)$ is a function of the 4 variables P, O, M, and E and is given by the formula

$$C(P, O, M, E) = 5.50 + (.4)P + (.5)O + (.75)M + E.$$

Example [23.4] (Distance Function) Let (x, y, z) be a point in three dimensional space. Then the distance $D(x, y, z)$ from (x, y, z) to the origin $(0, 0, 0)$ is given by the formula

$$D(x, y, z) = \sqrt{x^2 + y^2 + z^2}.$$

The distance function is a function of three variables x, y, z and is defined on every 3–tuple.

Remarks: When we identify each triple (x, y, z) with a point in 3–dimensional space the distance function can be viewed as a function of points in space. This direct connection is valid in general and any function of three variables is thus related to the theory of vectors. Functions of n variables can similarly be viewed as functions of points in n–dimensional space. Notice that our humble pizza function (Example [23.3]) is in fact a function on a space of dimension 4!

Exercises for §23.1

Construct the following real valued functions of several variables with your CAS. Determine the domain and range of the given functions. You can check whether or not a value of the variable is in the domain by evaluating the function at that point. An error message or an imaginary number will appear if the point was not in the domain.

(1) $f(x, y) = x^2 + y - 2.$

(2) $f(x, y) = x^2 + y^2 - 3.$

(3) $f(x, y) = \frac{x}{y}.$

(4) $f(x, y) = \frac{3x^2 - 7y^3}{x^4 + y^4}.$

(5) $f(x, y, z) = \sqrt{x^2 + y^2 + z^2 - 11}.$

(6) $f(x, y, z) = \frac{\sqrt{x}}{yz}.$

(7) $f(x, y, z) = \ln(9 - 3x^2 - y^2 - 3z^2).$

(8) $f(x, y, z, w) = \frac{\sqrt{xy}}{z^2 + w^2}.$

23.2 Graphical Display

Recall that a function $f(x)$ of one variable is graphically displayed by plotting the points $(x, f(x))$ in 2–dimensional space. Since we have only **one degree of freedom** (the x–variable), the graphical display is a one–dimensional curve. When we label the two axes of 2–dimensional space in the usual manner, i.e., with x, y, we can write our function in the form $y = f(x)$. The point in the graphical display, $(x, f(x))$, is thus (x, y).

Example [23.5] Let $f(x) = x^2$. Then $(1, 1)$, $(2, 4)$, $(-3, 9)$, and $(5, 25)$ are examples of points on the graph where the general point takes the form (x, x^2) for $x \in \mathbb{R}$.

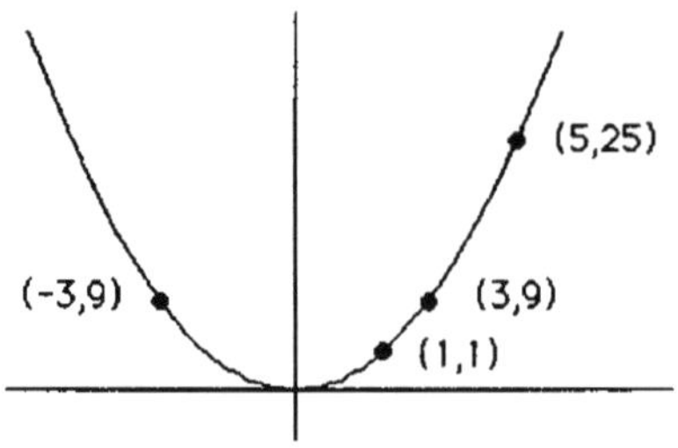

Question: How do we obtain the graphical display of a function $f(x, y)$ of 2 variables x, y?

Answer: The natural analogy with the above discussion is to graphically display the function $f(x, y)$ by plotting the points of the form $(x, y, f(x, y))$ in 3–dimensional space. By labeling the axes of 3–dimensional space in the customary manner, x, y, z, we can rewrite our function in the form

$$z = f(x, y).$$

Example [23.6] Characterize the graphical display of the function $f(x, y) = 2x - 3y$.

If we graph a few random points of the form $(x, y, 2x - 3y)$ in hopes of seeing what is going on, the resulting picture is quite confusing (in the figure below we have graphed the points $(0, 0, 0)$, $(0, 1, -3)$, $(1, 0, 2)$, $(1, 1, -1)$, $(3, -1, 9)$).

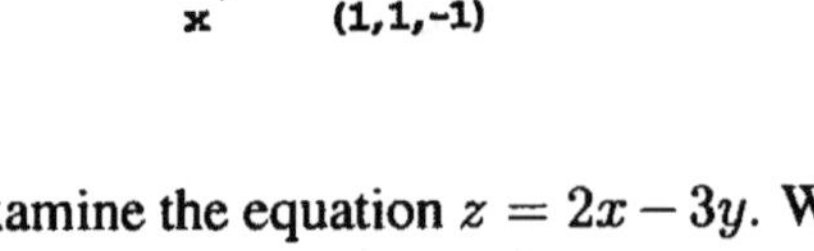

A better approach is to examine the equation $z = 2x - 3y$. We showed in §21.2 this is in fact the equation of a plane.

Example [23.7] Describe the graphical display of the function $f(x, y) = \sqrt{x^2 + y^2}$.

This function is more complex than our previous example. Observe first that when we write $z = \sqrt{x^2 + y^2}$, since the square root cannot be negative, $z \geq 0$. A basic method of obtaining the graph of a function of two variables is to fix a value of z, say $z = 1$, and then plot the points $(x, y, 1)$ which lie on our graph. Now

$(x, y, 1)$ lies on our graph provided $\sqrt{x^2 + y^2} = 1$, which is the (familiar) equation of a circle. More generally, for any fixed $z = z_0 > 0$ the points (x, y, z_0) where $\sqrt{x^2 + y^2} = z_0$ must lie on a circle of radius z_0 which lies in the plane $z = z_0$. By visualizing the collection of circles in concert we see that the graph of $f(x, y) = \sqrt{x^2 + y^2}$ is simply the cone with vertex at $(0, 0, 0)$.

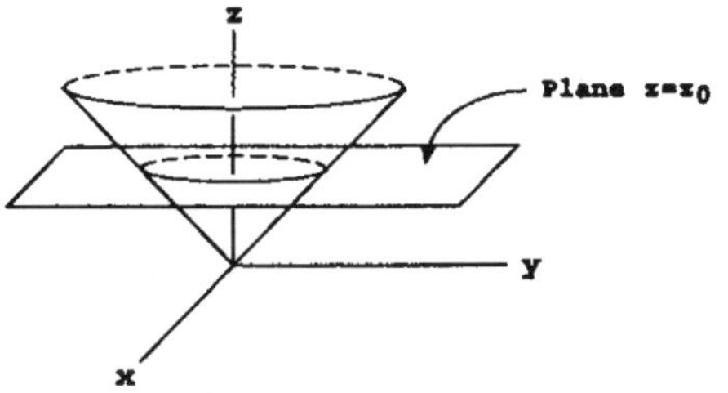

Example [23.8] Describe the graphical display of $f(x, y) = 100 - 3x^2 - 2y^4$.

Letting $z = 100 - 3x^2 - 2y^4$ it is clear that, since $3x^2 + 2y^4$ is always positive, the highest z–value (and thus the highest point on the z–axis the graph attains) is $z = 100$. For each fixed $z = z_0$ we consider the set of points (x, y, z_0) which lie both on our graph and on the plane $z = z_0$. Amalgamating these curves we obtain the graphical display of a surface whose shape is that of a mountain.

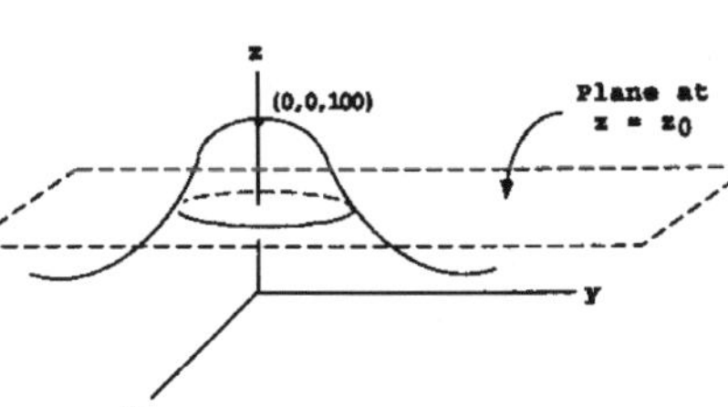

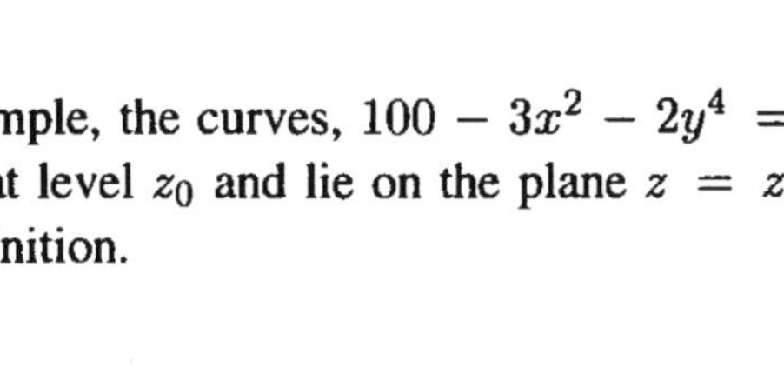

Remark: In this example, the curves, $100 - 3x^2 - 2y^4 = z_0$ are termed **level curves**. They occur at level z_0 and lie on the plane $z = z_0$. This motivates the following general definition.

Definition: *Let $z = f(x, y)$ be a function of two variables. For each fixed number z_0, the equation $f(x, y) = z_0$ is the equation of a curve in the plane $z = z_0$. This curve is termed the* ***level curve*** *of f in the plane $z = z_0$.*

When we examine an arbitrary function $z = f(x, y)$ of two variables, the graph is a 2–dimensional surface which twists and turns through 3–dimensional space. Some examples are depicted below.

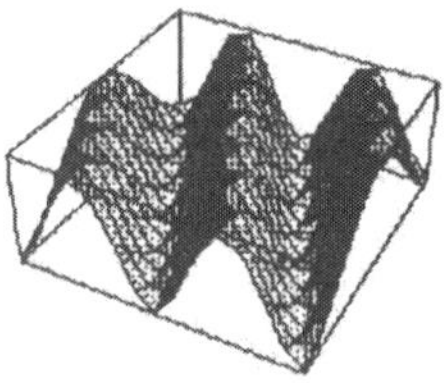

z = cos(x+y)

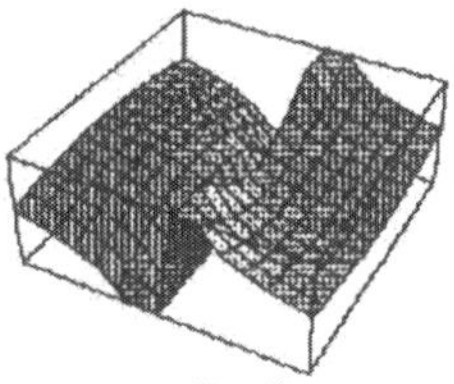

z = 1-x²+y²

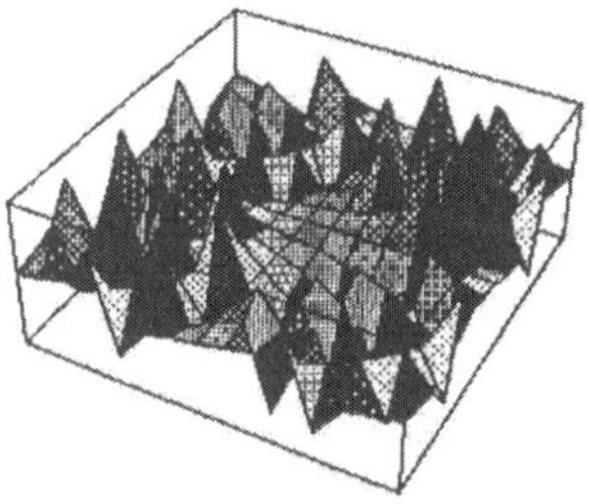

z = tan(xy)

If all the level curves $f(x, y) = z_0$ are projected (plotted) on the $x\,y$–plane we obtain a **contour map** of the surface.

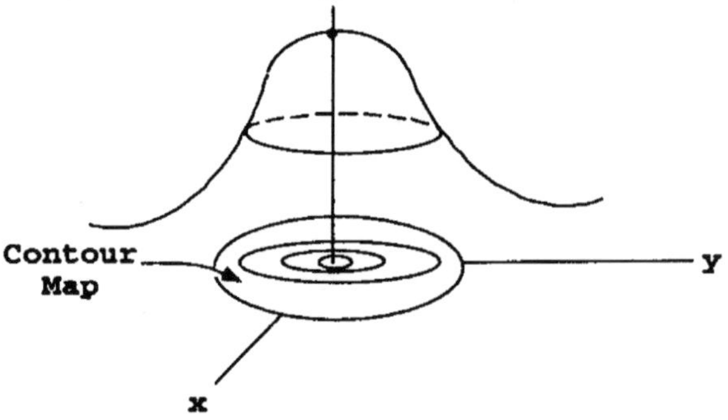

When functions of more than 2 variables are considered it is not possible to obtain a graphical display in 3–dimensional space. We can, however, abstractly consider 4 mutually perpendicular unit vectors $\vec{\mathbf{i}}$, $\vec{\mathbf{j}}$, $\vec{\mathbf{k}}$, and $\vec{\mathbf{l}}$ forming 4–dimensional space. A point in 4–dimensional space is a 4–tuple (x, y, z, w) of real numbers and vectors take the form $a\vec{i}+b\vec{j}+c\vec{k}+d\vec{l}$. Despite our inability to visualize four dimensional space it is possible to deceive the eye. Just as a three dimensional image can be drawn on a sheet of paper (which is 2–dimensional) and appear to have length, breadth, and width (using perspective), a four dimensional image can be approximated by holograms in 3–dimensional space.

The graphical display of a function

$$w = f(x, y, z)$$

is, by definition, the set of all points in 4–dimensional space which are of the form (x, y, z, w), where $w = f(x, y, z)$. If we fix $w = w_0$ we can look at the **level surfaces** $f(x, y, z) = w_0$ which are analogous to the level curves considered for surfaces in 3–dimensional space.

Example [23.9] (The 4–dimensional sphere) Consider the function of 3 variables

$$w = \sqrt{x^2 + y^2 + z^2}.$$

The level surface at $w = w_0$ is the sphere

$$w_0 = \sqrt{x^2 + y^2 + z^2}.$$

This is the reflection of the 4–dimensional sphere in a 3–dimensional space of level w_0.

Exercises for §23.2

Plot the following surfaces with your CAS. Determine the points where the surface has a singularity (i.e., it shoots off to $\pm\infty$). To see a particular part of the surface you will have to choose a suitable plot range.

(1) $z = 5x - 3y.$

(2) $z = x^2 + y^2.$

(3) $z = \frac{1}{3x^2 + 2y^2}.$

(4) $z = \cos(x) \cdot \cos(y).$

(5) $z = \frac{20 - x^2 - y^2}{(x - 5)}.$

(6) $z = \frac{x^2 + y^2}{x + y}.$

Draw a contour map for the following surfaces. Use your CAS to show a few level curves in each case. Graph the contour map with the surface plot of your CAS.

(7) $z = 5x^2 + 3y^2.$

(8) $z = 5x^2 - 3y^2.$

(9) $z = 1 - xy.$

(10) $z = 2y - \cos(x).$

(11) $z = (x - 2)^2 - (y - 4)^2.$

(12) $z = \frac{x^2}{9} + \frac{y^2}{16}.$

23.3 Partial Derivatives and the Gradient

Let $f(x, y)$ be a real valued function which is defined for all pairs of real numbers x, y.

Question: How should we define the derivative of such a function?

One natural approach to this question is to try to define the derivative as a double limit: i.e.,

$$f'(x,y) = \lim_{h\to 0}\lim_{k\to 0}\frac{f(x+h,y+k)-f(x,y)}{hk}.$$

Unfortunately this doesn't work. The difficulty with this approach can be seen even when one attempts to compute the double limit for a simple function. For example, when $f(x,y) = xy$, (a function one would expect to be able to differentiate) the double limit becomes

$$\begin{aligned}\lim_{h\to 0}\lim_{k\to 0}\frac{f(x+h,y+k)-f(x,y)}{hk} &= \lim_{h\to 0}\lim_{k\to 0}\frac{(x+h)(y+k)-xy}{hk}\\ &= \lim_{h\to 0}\lim_{k\to 0}\frac{hy+kx+hk}{hk}\\ &= \lim_{h\to 0}\lim_{k\to 0}\frac{y}{k}+\frac{x}{h}+1.\end{aligned}$$

Clearly this limit does not exist. Despite this difficulty it is possible to define **partial derivatives** of $f(x, y)$ with respect to each of the variables x and y. in the following manner.

Definition: *Fix the variable y. The* ***partial derivative with respect to*** x *of $f(x, y)$, denoted $\frac{\partial f}{\partial x}$ (also denoted $\frac{\partial f}{\partial x}(x, y)$) is defined to be*

$$\lim_{h\to 0}\frac{f(x+h,y)-f(x,y)}{h}$$

provided the limit exists.

Definition: *Fix the variable x. The* ***partial derivative with respect to*** y *of $f(x, y)$, denoted $\frac{\partial f}{\partial y}$ (also denoted $\frac{\partial f}{\partial y}(x, y)$) is defined to be*

$$\lim_{k\to 0}\frac{f(x,y+k)-f(x,y)}{k}$$

provided the limit exists.

Example [23.10] Evaluate $\frac{\partial}{\partial x}(x^2y)$.

In this example $f(x,y) = x^2y$ and we have

$$\begin{aligned}\frac{\partial}{\partial x}(x^2y) &= \lim_{h\to 0}\frac{(x+h)^2y - x^2y}{h}\\ &= y\lim_{h\to 0}\frac{(x+h)^2 - x^2}{h}\\ &= y\cdot 2x.\end{aligned}$$

Partial derivatives can be easily computed without recourse to limits. To evaluate

$$\frac{\partial}{\partial x}f(x,y)$$

simply treat the variable y as a constant, and then differentiate with respect to x.

Example [23.11] Evaluate $\frac{\partial}{\partial x}(x^4 + 2x^3y^2 + x^2y - y^4)$.

Treating y as a constant we have

$$\frac{\partial}{\partial x}(x^4 + 2x^3y^2 + x^2y - y^4) = 4x^3 + 6x^2y^2 + 2xy.$$

Example [23.12] Evaluate $\frac{\partial}{\partial y}(x^3y^2 + \cos(xy^2) + 3x - 4)$.

We have

$$\frac{\partial}{\partial y}(x^3y^2 + \cos(xy^2) + 3x - 4) = 2x^3y - 2xy\sin(xy^2).$$

Warning: The partial derivatives do behave differently from derivatives of functions of one variable. For example,

$$\frac{\partial}{\partial y}(3x-4) = 0$$

even though $3x - 4$ is not a constant function. In fact, if $f(x,y)$ is any function involving only x, (e.g. $f(x,y) = x^4 + 3x + 1$) then

$$\frac{\partial}{\partial y}f(x,y) = \lim_{h\to 0}\frac{f(x,y+h) - f(x,y)}{h} = 0,$$

since $f(x, y+h) = f(x, y)$ for all x.

Given a real valued function of n variables, $f(x_1, x_2, \ldots, x_n)$, we can define the jth **partial derivative** to be

$$\begin{aligned}
&\frac{\partial}{\partial x_j} f(x_1, x_2, \ldots, x_n) \\
&= \lim_{h \to 0} \frac{f(x_1, x_2, \ldots, x_j + h, \ldots x_n) - f(x_1, x_2, \ldots, x_j, \ldots, x_n)}{h}.
\end{aligned}$$

Example [23.13] Evaluate

$$\frac{\partial}{\partial z}(3x^2y^4z^3 + 2xy^2 + xz).$$

Viewing x, y as constants, we have

$$\frac{\partial}{\partial z}(3x^2y^4z^3 + 2xy^2 + xz) = 9x^2y^4z^2 + x.$$

Second Order Partial Derivatives:

When we take the partial derivative of a function $f(x, y)$, with respect to either x or y we obtain another function of two variables. We can then take the partial derivatives of these new functions. This process will yield **four second partial derivatives**:

$$\frac{\partial^2 f}{\partial x^2}, \ \frac{\partial^2 f}{\partial y^2},$$

$$\frac{\partial^2 f}{\partial x \partial y}, \ \frac{\partial^2 f}{\partial y \partial x}.$$

Example [23.14] Let $f(x, y) = x^2 + x^3y^2 + y^3$. Compute all four second partial derivatives.

The initial partial derivatives are

$$\frac{\partial f}{\partial x} = 2x + 3x^2y^2, \quad \frac{\partial f}{\partial y} = 2x^3y + 3y^2.$$

Thus

$$\frac{\partial^2 f}{\partial x^2} = 2 + 6xy^2, \quad \frac{\partial^2 f}{\partial y^2} = 2x^3 + 6y,$$

and

$$\frac{\partial^2 f}{\partial x \partial y} = 6x^2 y, \qquad \frac{\partial^2 f}{\partial y \partial x} = 6x^2 y.$$

Remark: The phenomena we observed in the previous example,

$$\frac{\partial^2 f}{\partial x \partial y} = \frac{\partial^2 f}{\partial y \partial x},$$

is not an accident. This identity can be verified in general by carefully examining the definition of the second partial derivative as a limit.

The Gradient Vector

Let $f(x, y)$ be a function of two variables which is defined on the domain $a \leq x \leq b$, $c \leq y \leq d$ where $a < b$ and $c < d$ are fixed numbers. Suppose that the partial derivatives $\frac{\partial f}{\partial x}$ and $\frac{\partial f}{\partial y}$ exist on the domain.

Definition: *The* ***gradient*** *of f, denoted $\overrightarrow{\nabla f}$, is the vector*

$$\overrightarrow{\nabla f(x, y)} = \frac{\partial f}{\partial x}\vec{\mathbf{i}} + \frac{\partial f}{\partial y}\vec{\mathbf{j}}.$$

Similarly, given a function $f(x, y, z)$ of 3 variables we define

Definition: *The* ***gradient*** *of f, denoted $\overrightarrow{\nabla f}$, is the vector*

$$\overrightarrow{\nabla f(x, y, z)} = \frac{\partial f}{\partial x}\vec{\mathbf{i}} + \frac{\partial f}{\partial y}\vec{\mathbf{j}} + \frac{\partial f}{\partial z}\vec{\mathbf{k}}.$$

Example [23.15] Let $f(x, y) = x^3 y^2$ and $g(x, y, z) = \cos(xy^2 z^3) + y^4$. Compute $\overrightarrow{\nabla f}$ and $\overrightarrow{\nabla g}$.

Following the definition we have

$$\overrightarrow{\nabla f} = 3x^2 y^2\,\vec{\mathbf{i}} + 2x^3 y\,\vec{\mathbf{j}},$$

and

$$\overrightarrow{\nabla g} = -y^2 z^3 \sin(xy^2 z^3)\,\vec{\mathbf{i}} - \left(2xyz^3 \sin(xy^2 z^3) + 4y^3\right)\vec{\mathbf{j}} - 3xy^2 z^2 \sin(xy^2 z^3)\,\vec{\mathbf{k}}.$$

The gradient vector we have just defined, will play a major role in the remainder of this chapter when we attempt to isolate the concept of the derivative of a function of two variables.

Exercises for §23.3

Compute the partial derivatives $\frac{\partial f}{\partial x}, \frac{\partial f}{\partial y}, \frac{\partial f}{\partial z}$ for the following functions f. Confirm your answers with your CAS.

(1)	x^4y^2+7.	**(4)**	$\cos(2y^3z)-x^2$.	**(7)**	x^{yz}.
(2)	$x^4y^2+3x^5$.	**(5)**	$e^{x^2z}+3y^2z^4$.	**(8)**	$xy+zx+y^3$.
(3)	$\ln(xy^2z)$.	**(6)**	$\tan(xyz)$.	**(9)**	$\cosh(x-5x+z)$.

Compute $\overrightarrow{\nabla f}$ and $\overrightarrow{\nabla f(1,2,3)}$ for the following functions f. Confirm your answers with your CAS.

(10)	xy^2.	**(12)**	$\cos(xz^3)+y^3$.	**(14)**	$\tan(y^3z^3)+x^2$.
(11)	$x^3y^2z^4$.	**(13)**	e^{x^2z}.	**(15)**	$\ln(x^2+y^2)$.

Find all three second partial derivatives $\frac{\partial^2 f}{\partial x^2}$, $\frac{\partial^2 f}{\partial y^2}$, $\frac{\partial^2 f}{\partial x \partial y}$ for the following functions f. Confirm your answers with your CAS.

(16)	xy.	**(18)**	$\sqrt{x^2+y^2}$.	**(20)**	x^y.
(17)	x^3y^2.	**(19)**	$\ln(x+y)$.	**(21)**	$\cos(x^2+y^2)$.

23.4 The Total Derivative

The partial derivatives of a function $f(x,y)$ defined in the previous section were obtained by holding one of the two variables constant, which amounts to looking at a function of one variable. The question as to whether the theory of differentiation can truly be generalized to two dimensions remained unanswered until 1909, when the British mathematician William H. Young introduced the concept of the **total derivative.**

In §23.3 we attempted, unsuccessfully, to define the derivative of $f(x,y)$ as a double limit. As it turned out, even the tamest functions were not differentiable (when computed by the double limit), leading us to the opinion that the proposed definition was not a good one. Yet intuitively a double limit does seem to be appropriate in this context, and a clever modification will make it work.

In order to view differentiation in two dimensions in the proper perspective we must first look back at the one dimensional situation. Let $f(x)$ be a function of one variable. The derivative of f was defined to be

$$\lim_{h\to 0}\frac{f(x+h)-f(x)}{h}.$$

Notice that the existence of this limit is equivalent to the statement that when h is very small,

$$f(x+h) \approx f(x) + hf'(x). \tag{23.1}$$

By definition the approximation symbol in (23.1) indicates that

$$f(x+h) = f(x) + hf'(x) + \varepsilon(h),$$

where $\varepsilon(h)$ is a function such that

$$\lim_{h \to 0} \frac{\varepsilon(h)}{h} = 0.$$

For example, consider $f(x) = x^2$. Then $f(x+h) = x^2 + 3xh + h^2$, $f'(x) = 2x$, and

$$(x+h)^2 \approx x^2 + h \cdot 2x.$$

This approximation can be replaced by the exact identity

$$(x+h)^2 = x^2 + h \cdot 2x + \varepsilon(h),$$

where $\varepsilon(h) = h^2$. It is remarkable that within this basic example lies the key to formulating the concept of the total derivative.

Definition: *Let $f(x,y)$ be a real valued function defined on the domain $a \le x \le b$, $c \le y \le d$. We say f is* ***differentiable*** *at (x_0, y_0) if both $\frac{\partial f}{\partial x}(x_0, y_0)$ and $\frac{\partial f}{\partial y}(x_0, y_0)$ exist and there exists a function $\varepsilon(h,k)$ such that*

$$f(x_0+h, y_0+k) = f(x_0,y_0) + \frac{\partial f}{\partial x}(x_0,y_0)h + \frac{\partial f}{\partial y}(x_0,y_0)k + \varepsilon(h,k), \tag{23.2}$$

where

$$\lim_{h \to 0} \lim_{k \to 0} \frac{\varepsilon(h,k)}{\sqrt{h^2+k^2}} = 0.$$

Remark: Notice that the partial derivatives, $\frac{\partial f}{\partial x}$ and $\frac{\partial f}{\partial y}$, appear in this definition. This was to be expected in that fixing one of the variables would reduce us to taking the partial derivative.

If we let $H = (h, k)$, and $\overrightarrow{H} = \overrightarrow{(h,k)} = h\vec{\mathbf{i}} + k\vec{\mathbf{j}}$, then we can rewrite (23.2) in the compact form

$$f(x+h, y+k) - f(x,y) = \overrightarrow{\nabla f} \cdot \overrightarrow{H} + \varepsilon(H).$$

Definition: *Let f be a real valued differentiable function. The* ***total derivative****, denoted $\overrightarrow{f'(x,y)}$, is the vector of partial derivatives*

$$\overrightarrow{f'(x,y)} = \frac{\partial f}{\partial x}\vec{\mathbf{i}} + \frac{\partial f}{\partial y}\vec{\mathbf{j}}.$$

Remark: The definition of the total derivative is best viewed in the natural progression. For a function of one variable, $f(x)$, we have

$$f(x+h) - f(x) \approx f'(x) \cdot h.$$

For a function of two variables, $f(x,y)$, the analogous approximation is

$$f(x+h, y+k) - f(x,y) \approx \overrightarrow{f'(x,y)} \cdot \overrightarrow{(h,k)}.$$

A function of three variables can also be handled in a similar manner: given a function of three variables, $f(x,y,z)$, define the total derivative to be the gradient vector,

$$\begin{aligned}\overrightarrow{f'(x,y,z)} &= \overrightarrow{\nabla f} \\ &= \frac{\partial f}{\partial x}\vec{\mathbf{i}} + \frac{\partial f}{\partial y}\vec{\mathbf{j}} + \frac{\partial f}{\partial z}\vec{\mathbf{k}},\end{aligned}$$

the analogous approximation is

$$f(x+h, y+k, z+\ell) - f(x,y,z) \approx \overrightarrow{f'(x,y,z)} \cdot \overrightarrow{(h,k,\ell)}.$$

Example [23.16] Compute the total derivative of $f(x,y,z) = x^2y^3z + 2x$.

In this case we have

$$\begin{aligned}\overrightarrow{f'(x,y,z)} &= \overrightarrow{\nabla f} \\ &= (2xy^3z + 2)\vec{\mathbf{i}} + 3x^2y^2z\vec{\mathbf{j}} + x^2y^3\vec{\mathbf{k}}.\end{aligned}$$

Warning: The total derivative is a vector not a scalar.

Directional Derivatives

The motivation for defining the directional derivative lies in the fact that, since the total derivative is a vector, it should be possible to measure the rate of change of our function in the direction of an arbitrary (unit) vector. It is the projection of the total derivative onto the unit vector that captures this intuition.

> **Definition:** *Let f be a differentiable real valued function. Let $\overrightarrow{u}$ be a unit vector. The* ***directional derivative*** *of f in the direction of $\overrightarrow{u}$, denoted $\frac{\partial f}{\partial \overrightarrow{u}}$, is defined to be*
>
> $$\frac{\partial f}{\partial \overrightarrow{u}} = \overrightarrow{\nabla f} \cdot \overrightarrow{u}.$$

Example [23.17] Let $f(x, y, z) = x^2y^3z^4$. Find $\frac{\partial f}{\partial \overrightarrow{u}}$ when $\overrightarrow{u} = \frac{\vec{\mathbf{i}}+\vec{\mathbf{j}}+\vec{\mathbf{k}}}{\sqrt{3}}$.

The total derivative of f is given by

$$\begin{aligned}
\overrightarrow{f(x,y,z)} &= \overrightarrow{\nabla f} \\
&= 2xy^3z^4\vec{\mathbf{i}} + 3x^2y^2z^4\vec{\mathbf{j}} + 4x^2y^3z^3\vec{\mathbf{k}},
\end{aligned}$$

and hence the desired directional derivative is

$$\begin{aligned}
\frac{\partial f}{\partial \overrightarrow{u}} &= \overrightarrow{\nabla f} \cdot \overrightarrow{u} \\
&= \overrightarrow{\nabla f} \cdot \frac{\vec{\mathbf{i}}+\vec{\mathbf{j}}+\vec{\mathbf{k}}}{\sqrt{3}} \\
&= \frac{1}{\sqrt{3}}(2xy^3z^4 + 3x^2y^2z^4 + 4x^2y^3z^3).
\end{aligned}$$

There is an alternate form of the directional derivative which arises naturally in application (and thus is frequently given) which we mention for completeness. Let $\overrightarrow{u} = \overrightarrow{(u_1, u_2, u_3)}$ be a unit vector. Then it can be shown that

$$\frac{\partial f}{\partial \overrightarrow{u}} = \lim_{h \to 0} \frac{f(x + hu_1, y + hu_2, z + hu_3) - f(x, y, z)}{h}.$$

Exercises for §23.4

Compute the total derivative and the directional derivative for the following functions f in the given direction $\vec{u}$.

(1) $f(x, y) = x^2 + 3y^2$, $\vec{u} = \frac{3}{5}\vec{i} + \frac{4}{5}\vec{j}$.

(2) $f(x, y) = x^2y^4 + 3y$, $\vec{u} = \frac{3}{5}\vec{i} - \frac{4}{5}\vec{j}$.

(3) $f(x, y) = x^2yz^4 + 2\sin(xy)$, $\vec{u} = \frac{3\vec{i} - 2\vec{j} + \vec{k}}{\sqrt{14}}$.

(4) $f(x, y, z) = x^3 + y^2 - xz$, $\vec{u} = \overrightarrow{\left(\frac{1}{3}, \frac{2}{3}, \frac{2}{3}\right)}$.

(5) $f(x, y) = z^4e^{3x^2y}$, $\vec{u} = \overrightarrow{\left(\frac{2}{3}, \frac{1}{3}, \frac{2}{3}\right)}$.

(6) $f(x, y) = x^{yz}$, $\vec{u} = \frac{\vec{i} - 2\vec{j} + 3\vec{k}}{\sqrt{14}}$.

(7) Compute $\frac{\partial f}{\partial \vec{u}}(1, -1, 2)$ for $f(x, y, z) = x^2 - 2y^2 + z^3$ and $\vec{u} = \frac{\vec{i} - \vec{j}}{\sqrt{2}}$.

(8) Compute $\frac{\partial f}{\partial \vec{u}}(-1, 2, 1)$ for $f(x, y, z) = x^2y^3z$ and $\vec{u} = \vec{k}$.

23.5 The Chain Rule

One of the most prominent results on differentiation we encountered was the chain rule (see §7.4). The chain rule dictates that the derivative of the composite of two real valued functions $f(x)$ and $g(x)$ is given by

$$\frac{d}{dx} f\left(g(x)\right) = f'\left(g(x)\right) \cdot g'(x).$$

When we move to functions of two variables $f(x, y)$, $u(x, y)$, and $v(x, y)$, the situation becomes more complex. Two questions arise: **(1)** how can we compute

$$\frac{\partial}{\partial x} f\left(u(x, y), v(x, y)\right), \qquad \frac{\partial}{\partial y} f\left(u(x, y), v(x, y)\right),$$

(2) given two functions $g(t)$ and $h(t)$ of one variable t, how we can compute the derivative of $f(g(t), h(t))$.

In order to develop some intuition about these questions (and how we might answer them) lets look at some examples.

Example [23.18] Let $f(x, y) = x^2y^3$, $g(t) = \sqrt{t}$, and $h(t) = t^2$. Evaluate $\frac{d}{dt} f(g(t), h(t))$.

The composite in this case is given by

$$\begin{aligned} f(g(t),h(t)) &= g(t)^2h(t)^3 \\ &= \sqrt{t}^2\cdot(t^2)^3 \\ &= t^7, \end{aligned}$$

and thus $\frac{d}{dt}f(g(t),h(t)) = 7t^6$.

We now state the chain rule for functions of the form $f(g(t),h(t))$ which will allow us to perform the computation in Example [23.18] with much less fuss.

Proposition [23.19] *Let $f(x,y)$ be a real valued differentiable function of two variables x,y. Let $g(t)$ and $h(t)$ be differentiable functions of another variable t. Assuming that $f(g(t),h(t))$ is itself differentiable, we have*

$$\frac{d}{dt}f(g(t),h(t)) = \frac{\partial f}{\partial g}\frac{dg}{dt} + \frac{\partial f}{\partial h}\frac{dh}{dt}. \tag{23.3}$$

Example [23.20] Recompute the derivative of Example [23.18] using Proposition [23.19].

Since $f(g,h) = g^2\cdot h^3$ the partial derivatives are

$$\frac{\partial f}{\partial g} = 2gh^3, \quad \frac{\partial f}{\partial h} = 3g^2h^2.$$

In addition, since $g(t) = \sqrt{t}$, $\frac{dg}{dt} = \frac{1}{2}t^{-\frac{1}{2}}$, and $h(t) = t^2$, $\frac{dh}{dt} = 2t$. Plugging these derivatives into the chain rule yields

$$\begin{aligned} \frac{d}{dt}f(g(t),h(t)) &= \overbrace{2gh^3}^{\frac{\partial f}{\partial g}}\cdot\overbrace{\frac{1}{2}t^{-\frac{1}{2}}}^{\frac{dg}{dt}} + \overbrace{3g^2h^2}^{\frac{\partial f}{\partial h}}\cdot\overbrace{2t}^{\frac{dh}{dt}} \\ &= 2\sqrt{t}\cdot t^6\cdot\frac{1}{2}t^{-\frac{1}{2}} + 3(\sqrt{t})^2t^4\cdot 2t \\ &= 7t^6. \end{aligned}$$

Having considered functions of the form $f(g(t), h(t))$, we now state the more general chain rule for functions of two variables.

Proposition [23.21] *Let $f(x, y)$ be a real valued differentiable function of two variables x, y. Let $u(s,t)$ and $v(s,t)$ be real valued differentiable functions of two other variables s, t. Assuming $f(u(s,t), v(s,t))$ is differentiable we have*

$$\frac{\partial}{\partial s} f(u(s,t), v(s,t)) = \frac{\partial f}{\partial u}\frac{\partial u}{\partial s} + \frac{\partial f}{\partial v}\frac{\partial v}{\partial s}$$

$$\frac{\partial}{\partial t} f(u(s,t), v(s,t)) = \frac{\partial f}{\partial u}\frac{\partial u}{\partial t} + \frac{\partial f}{\partial v}\frac{\partial v}{\partial t}.$$

Example [23.22] Let $f(x, y) = x^2y^3$, $u(s,t) = s^2+t^2$, and $v(s,t) = st$. Compute $\frac{\partial}{\partial s} f(u, v)$.

Proposition [23.21] allows us to compute the partial derivative easily as seen below:

$$\begin{aligned}\frac{\partial}{\partial s} f(u, v) &= \frac{\partial f}{\partial u}\frac{\partial u}{\partial s} + \frac{\partial f}{\partial v}\frac{\partial v}{\partial s} \\ &= 2uv^3 \cdot 2s + 3u^2v^2 \cdot t \\ &= 2(s^2+t^2)(st)^3 \cdot 2s + 3(s^2+t^2)^2(st)^2 \cdot t.\end{aligned}$$

When considering composites of functions of three or more variables there are even more possibilities. Rather than working with functions of three variables (or any other particular number of variables), we state the most general result.

Proposition [23.23] (General Chain Rule) *Let $f(x_1, \ldots, x_n)$ be a real valued differentiable function of n variables $x_1, x_2, \ldots, x_n$ and let $u_1(t_1, \ldots, t_m)$, $u_2(t_1, \ldots, t_m)$, $\ldots$, $u_n(t_1, \ldots, t_m)$ be real valued functions of m variables $t_1, \ldots, t_m$. Assuming $f(u_1, \ldots u_n)$ is differentiable we have*

$$\frac{\partial}{\partial t_i} f(u_1, \ldots, u_n) = \frac{\partial f}{\partial u_1}\frac{\partial u_1}{\partial t_i} + \frac{\partial f}{\partial u_2}\frac{\partial u_2}{\partial t_i} + \cdots + \frac{\partial f}{\partial u_n}\frac{\partial u_n}{\partial t_i},$$

for every $i = 1, 2, \ldots, m$.

Example [23.24] Compute the partial derivatives of $f(u, v, w)$ where $u = u(s, t)$, $v = v(s, t)$, and $w = w(s, t)$.

The general chain rule yields the following formulae:

$$\frac{\partial f}{\partial s} = \frac{\partial f}{\partial u}\frac{\partial u}{\partial s} + \frac{\partial f}{\partial v}\frac{\partial v}{\partial s} + \frac{\partial f}{\partial w}\frac{\partial w}{\partial s}$$

$$\frac{\partial f}{\partial t} = \frac{\partial f}{\partial u}\frac{\partial u}{\partial t} + \frac{\partial f}{\partial v}\frac{\partial v}{\partial t} + \frac{\partial f}{\partial w}\frac{\partial w}{\partial t}.$$

Example [23.25] Let $f(x, y, z) = xy^2 + z^3$, $u(s, t) = s^2 + t^2$, $v(s, t) = t^3$, and $w(s, t) = s^2 t$. Compute

$$\frac{\partial}{\partial t} f(u, v, w).$$

Since $f(u, v, w) = uv^2 + w^3$

$$\frac{\partial f}{\partial u} = v^2, \quad \frac{\partial f}{\partial v} = 2uv, \quad \frac{\partial f}{\partial w} = 3w^2.$$

Applying the chain rule we obtain the desired derivative:

$$\frac{\partial f}{\partial t} = \frac{\partial f}{\partial u}\frac{\partial u}{\partial t} + \frac{\partial f}{\partial v}\frac{\partial v}{\partial t} + \frac{\partial f}{\partial w}\frac{\partial w}{\partial t}$$

$$= v^2 \cdot 2t + 2uv \cdot 3t^2 + 3w^2 \cdot s^2$$

$$= 2t^7 + 6(s^2 + t^2)t^5 + 3s^6 t^2.$$

Exercises for §23.5

Use the general chain rule to compute $\frac{df}{dt}$ or $\frac{\partial g}{\partial s}$ for the following functions f, g. Confirm your answers by using your CAS to compose the required functions and differentiate them directly.

(1) $f(x, y) = 2x^2 - 3y^2, x = t^4, y = t^2.$

(2) $f(x, y) = (x - y^2)^{10}, x = \cos(t), y = t^3.$

(3) $f(x, y, z) = \ln(x^2 + y^2 + z^4), x = 3t^2, y = t^5, z = \sqrt{t}.$

(4) $f(x, y, z) = e^{xyz}, x = \ln(t), y = t^2, z = t^3.$

(5) $f(x, y, z, w) = \cos(x + y + z + w^2), x = t^2, y = t, z = \ln(t), w = t^3.$

(6) $g(x, y) = 2x^2 + y^2, x = s^2 + t^2, y = s^2 - t^2.$

(7) $g(x, y) = (x + y^3)^5, x = st, y = \frac{s}{t}.$

(8) $g(x, y, z) = (x^2 + y^2 + z^2)^4, x = s + t, y = s - t, z = t^3.$

(9) $g(x, y, z) = \cos(xyz), x = s + t + u, y = s^2u^3, z = u^4.$

(10) $g(x, y, z, w) = \ln(x + y + z + w), x = stu, y = \frac{s-t}{u}, z = s^3.$

23.6 Tangent Planes

When the derivative of a function of one variable was defined we considered the tangent line to the curve at a point P in great detail. We conclude this chapter by developing the analogous theory for differentiable functions of two variables. Since the graph of the function $z = f(x, y)$ is a 2–dimensional surface in three dimensional space, at each point P on the surface there will be a tangent plane.

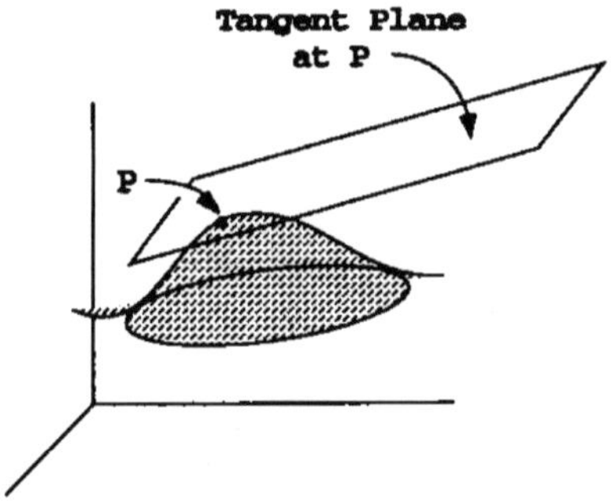

Question: Given a differentiable function $z = f(x, y)$ and a point $P = (x_0, y_0, z_0)$ on the surface, how do we determine the equation of the tangent plane at P?

As we showed in §21.2, the equation of the tangent plane at the point P can be computed provided we can find a vector $\vec{n}$ which is perpendicular to the surface at the point P. Once this vector $\vec{n}$ is found the equation of the tangent plane is simply

$$\overrightarrow{(x - x_0, y - y_0, z - z_0)} \cdot \vec{n} = 0,$$

since the vector $\overrightarrow{(x - x_0, y - y_0, z - z_0)}$ would always be perpendicular to $\vec{n}$ whenever (x, y, z) is on the plane.

Method to find a Normal Vector

Beginning again with the differentiable function $z = f(x, y)$ which defines our surface, consider the function of three variables

$$F(x, y, z) = f(x, y) - z = 0.$$

Then F is a function of x, y, z and the total derivative of F is

$$\begin{aligned}\overrightarrow{\nabla F} &= \frac{\partial F}{\partial x}\vec{\mathbf{i}} + \frac{\partial F}{\partial y}\vec{\mathbf{j}} + \frac{\partial F}{\partial z}\vec{\mathbf{k}} \\ &= \frac{\partial f}{\partial x}\vec{\mathbf{i}} + \frac{\partial f}{\partial y}\vec{\mathbf{j}} - \vec{\mathbf{k}}.\end{aligned}$$

Claim: The vector $\overrightarrow{\nabla F}(x_0, y_0, z_0)$ is normal to the surface $z = f(x, y)$ at the point (x_0, y_0, z_0).

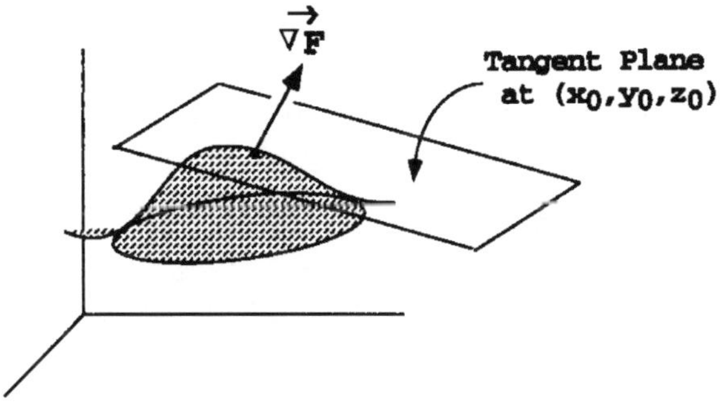

Proof: Let $\overrightarrow{g(t)} = x(t)\,\vec{\mathbf{i}} + y(t)\,\vec{\mathbf{j}} + z(t)\,\vec{\mathbf{k}}$ be any space curve lying on the surface $F(x, y, z) = 0$ which passes through (x_0, y_0, z_0). Since the space curve passes through the point (x_0, y_0, z_0), there is some t_0 such that $\overrightarrow{g(t_0)} = \overrightarrow{(x_0, y_0, z_0)}$. Now we know that for all t, $F(x(t), y(t), z(t)) = 0$, and in addition, the chain rule tells us that

$$\begin{aligned}\frac{d}{dt}F(x(t), y(t), z(t)) &= \overrightarrow{\nabla F}\cdot\left(x'(t)\,\vec{\mathbf{i}} + y'(t)\,\vec{\mathbf{j}} + z'(t)\,\vec{\mathbf{k}}\right) \\ &= \overrightarrow{\nabla F}\cdot\overrightarrow{g'(t)} \\ &= 0.\end{aligned}$$

In particular, at $t = t_0$, $\overrightarrow{\nabla F}\cdot\overrightarrow{g'(t_0)} = 0$. Since $\overrightarrow{g'(t_0)}$ is a tangent vector to the curve at the point (x_0, y_0, z_0) we have demonstrated that $\overrightarrow{\nabla F}(x_0, y_0, z_0)$ is perpendicular to this tangent vector. Recalling that $\overrightarrow{g(t)}$ was an arbitrary space curve on the surface which passes through (x_0, y_0, z_0), we see that $\overrightarrow{\nabla F}(x_0, y_0, z_0)$ is perpendicular to the tangent vector of every space curve on the surface which passes through (x_0, y_0, z_0), and thus $\overrightarrow{\nabla F}(x_0, y_0, z_0)$ must be perpendicular to the surface at (x_0, y_0, z_0).

Example [23.26] Find the equation of the tangent plane to the surface $z = x^2 + y^2$ at the point $(1, 2, 5)$.

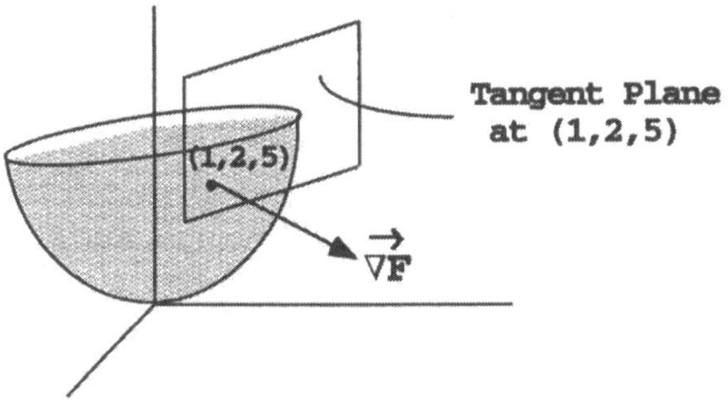

Defining $F(x, y, z) = x^2 + y^2 - z$ we have $\overrightarrow{\nabla F} = 2x\,\vec{\mathbf{i}} + 2y\,\vec{\mathbf{j}} - \vec{\mathbf{k}}$, and the vector

$$\overrightarrow{\nabla F}(1, 2, 5) = 2\vec{\mathbf{i}} + 4\vec{\mathbf{j}} - \vec{\mathbf{k}}$$

is a normal vector at $(1, 2, 5)$. The equation of the tangent plane is given by the dot product

$$\overrightarrow{(x-1, y-2, z-5)} \cdot \overrightarrow{(2, 4, -1)} = 0,$$

or equivalently,

$$2x + 4y - z = 5.$$

The method we have just developed which uses the gradient vector to find the tangent plane to a surface, actually remains valid in any dimension. We illustrate this by returning to a simple example of a curve and its tangent line in two dimensions.

Example [23.27] Find the equation of the tangent line to the curve $y = x^2$ at the point $x = 2$.

Let $F(x, y) = x^2 - y = 0$ be the equation of the curve. Then

$$\overrightarrow{\nabla F}(x, y) = 2x\,\vec{\mathbf{i}} - \vec{\mathbf{j}},$$

and this vector is normal to the curve. Now $\overrightarrow{\nabla F}(2,4) = 4\vec{\mathbf{i}} - \vec{\mathbf{j}}$, and if (x, y) is a general point on the tangent line

$$\overrightarrow{(x-2, y-4)} \cdot \overrightarrow{(4,-1)} = 0.$$

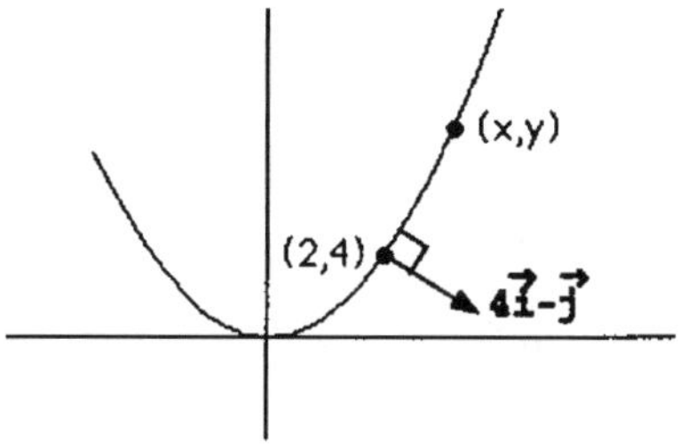

Thus the equation of the tangent line is

$$4x - y = 4.$$

Exercises for §23.6

In the following exercises, find the equation of the tangent plane to the given surface at the given point.

(1) $z = 3x^2 + 2y^2$, $(1, 2, 11)$.

(2) $z = 3x^2 - 2y^2$, $(1, 2, -5)$.

(3) $z = \cos(xy)$, $(\pi, \frac{1}{2}, 0)$.

(4) $z = x^2 y$, $(-2, 3, 12)$.

(5) $z = x^y$, $(2, 3, 8)$.

(6) $z = e^{-x^2 - y^2}$, $(10, 10, e^{-200})$.

(7) $z = \frac{1}{x^2 + 3y^2}$, $(2, 1, \frac{1}{7})$.

(8) $z = \sqrt{x - 2y}$, $(12, 4, 2)$.

(9) Plot the curve $y = e^{-x^2}$ with your CAS. Find the normal vector to this curve at the point $x = 1$ using the gradient. Graph this gradient vector (anchored at $(1, e^{-1})$) using your CAS. Find the equation of the tangent line at $x = 1$.

(10) Find the equation of the tangent line to the curve $y = x\ln(x)$ at the point $x = e$ using the gradient method.

Additional exercises for Chapter XXIII

---

(1) On your CAS, sketch the graph of the function $f(x, y) = \sqrt{1 - x^2 - y^2}$.

(2) On your CAS, sketch the graph of the function $f(x, y) = -\sqrt{1 - x^2 - y^2}$.

(3) For problems 1 and 2, use your CAS to draw level curves of the surfaces.

(4) Use your CAS to plot the three dimensional surface $f(x, y) = \exp(x + y)$.

Find an equation of the level curve that passes through the point:

(a) (ln(3),ln(3)), **(b)** (0,0), **(c)** (ln(2),0).

(5) On your CAS, plot the graph of the function $f(x, y) = \frac{x^2y^2}{1-xy}$. Find the domain of f. Is f continuous everywhere? Is it differentiable?

(6) Let $f(x, y) = \frac{x^2y^2}{x^4+y^4}$ for $(x, y) \neq (0, 0)$ and $f(0, 0) = 0$. Find the limit of $f(x, y)$ as $(x, y) \to (0, 0)$ along

(a) the x-axis, **(b)** the y-axis, **(c)** the parabola $y = x^2$, **(d)** the line $y = x$. What do you conclude about the continuity of f at $(0, 0)$?

(7) Find the equation of the tangent plane to the given surface at the given point. Graph the surface and the tangent plane with your CAS.

(a) $z = x^2 + 3y^2 - x + 3y - 2$, $(1, 3, 34)$.

(b) $z = e^{-(x^2+y^2-2x)}$, $(1, 0, e^1)$.

(c) $z = xy^2 - yx^2 - 2xy$, $(1, -1, 4)$.

Chapter XXIV

Multidimensional Optimization

24.1 The Method of Steepest Descent

The physical motivation for studying the method of steepest descent lies in observations of the following type. After a heavy rain, water which will meander down a mountain (forming a stream) will naturally choose the path of steepest descent. Another example of this nature involves ski trails. Ski trails vary in difficulty, the most challenging ones will have the steepest descent.

Problem: Given a surface $z = f(x, y)$ and a point (x_0, y_0, z_0) in the surface, how do we find the path of steepest descent to a lower point (x_1, y_1, z_1) (where by lower we mean $z_1 < z_0$) on the surface.

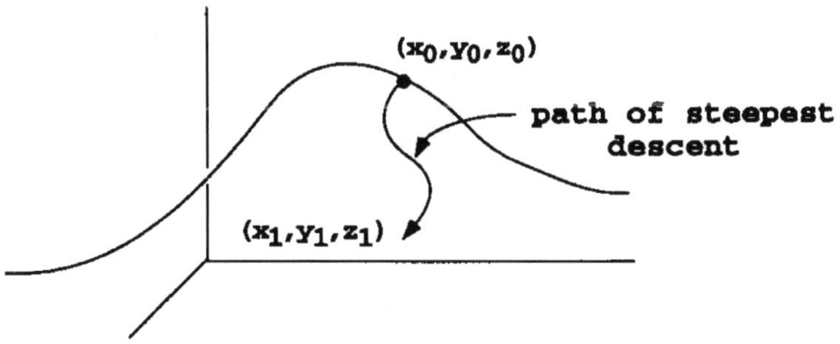

This problem is not entirely new to us. We encountered the one variable analogue in Chapter 9 where we saw that the steepness of the curve $y = f(x)$ at a point

(x_0, y_0) is measured by the slope of the curve, i.e., the derivative $f'(x_0)$. The larger $f'(x_0)$ is, the steeper the curve is at that point.

To begin our analysis of the two variable situation, lets imagine that we are at the point (x_0, y_0, z_0) on the surface. In order to move down the surface we must choose a direction, which in mathematical terms amounts to choosing a unit vector $\vec{\mathbf{u}} = \overrightarrow{(u_1, u_2)}$ on the plane $z = z_0$ and then begin traveling down the surface using $\vec{\mathbf{u}}$ as a guide (the way one would use a compass).

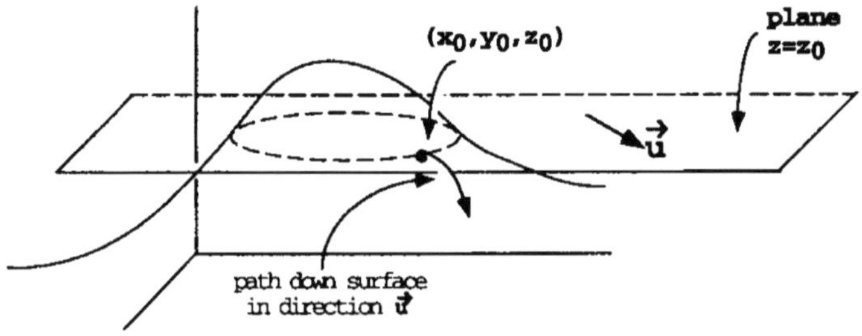

In order to describe the path we have chosen (and its slope), let P denote the plane which

(1) contains the line passing through (x_0, y_0, z_0) which is parallel to $\vec{\mathbf{u}}$ and,

(2) is perpendicular to the plane $z = z_0$.

Then the intersection of the surface with the plane P is precisely the path we take moving down the surface in the $\vec{\mathbf{u}}$ direction.

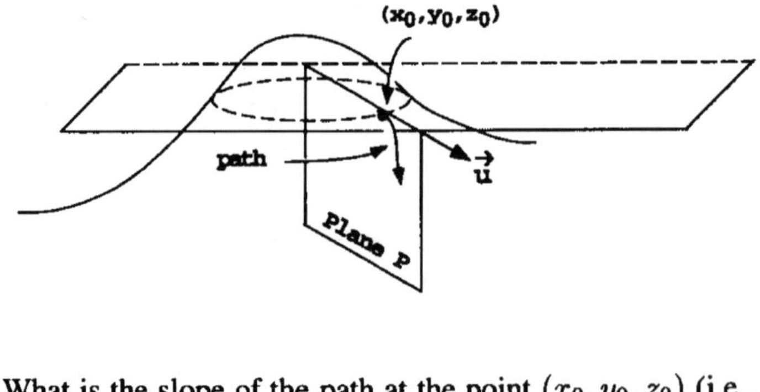

Question: What is the slope of the path at the point (x_0, y_0, z_0) (i.e., the slope of the tangent line at (x_0, y_0, z_0))?

In general, the slope of a line in a given plane is the ratio of vertical to horizontal distance $\frac{V}{H}$ (we saw this in §3.1 in the Cartesian plane). Notice that the larger the value of the slope the steeper the incline of the line.

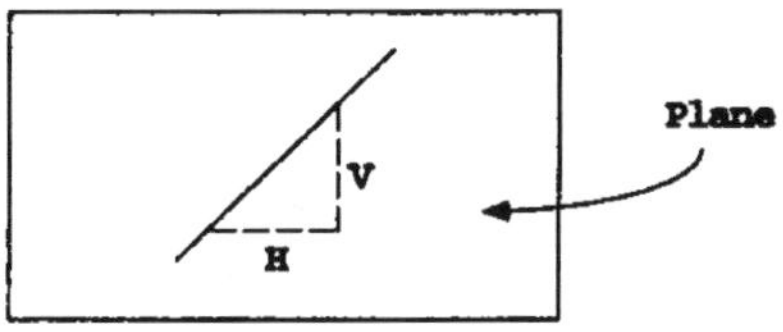

The tangent line to our path at the point (x_0, y_0, z_0) is simply the intersection of the tangent plane to the surface at (x_0, y_0, z_0) with the plane P. Thus the slope of the path is given by the directional derivative:

$$\begin{aligned}\frac{\partial f}{\partial \vec{\mathbf{u}}} &= \lim_{h \to 0} \frac{f(x + u_1 h, y + u_2 h) - f(x, y)}{h} \\ &= \vec{\nabla f} \cdot \vec{\mathbf{u}}.\end{aligned} \tag{24.1}$$

Since $\vec{\mathbf{u}}$ is a unit vector (which indicates that $|\vec{\mathbf{u}}| = 1$), we can evaluate the directional derivative (24.1) using the dot product formula from Theorem (20.8):

$$\begin{aligned}\frac{\partial f}{\partial \vec{\mathbf{u}}} &= |\vec{\nabla f}| \cdot |\vec{\mathbf{u}}| \cos(\theta) \\ &= |\vec{\nabla f}| \cos(\theta),\end{aligned}$$

where θ is the angle between the gradient vector $\vec{\nabla f}$ and $\vec{\mathbf{u}}$. Since $-1 \le \cos(\theta) \le 1$, we see that $\frac{\partial f}{\partial \vec{\mathbf{u}}}$ is maximized when $\cos(\theta) = 1$ and minimized when $\cos(\theta) = -1$. Thus $\frac{\partial f}{\partial \vec{\mathbf{u}}}$ is maximized when the vectors $\vec{\nabla f}$ and $\vec{\mathbf{u}}$ point in the same direction and minimized when they point in opposite directions. We state our conclusions in the following proposition.

Proposition [24.1] *Let $z = f(x, y)$ be a differentiable function. The direction of the gradient vector $\vec{\nabla f}$ is that of maximal increase of the function f.*

Example [24.2] Suppose the graph of $z = 8{,}000 - 2x^2 - 5y^2$ on the domain $-50 \le x \le 50$, $-25 \le y \le 25$ yields the surface of a mountain. A skier is at the point $(5, 10, 7450)$. In what direction should he push off to have the steepest descent down the mountain?

Let $f(x,y) = 8,000 - 2x^2 - 5y^2$. Then $\overrightarrow{\nabla f} = -4x\,\vec{\mathbf{i}} - 10y\,\vec{\mathbf{j}}$, and

$$\overrightarrow{\nabla f}(5,10) = -20\,\vec{\mathbf{i}} - 100\,\vec{\mathbf{j}}.$$

This is the direction of maximal increase of f. The steepest descent will occur in the opposite direction.

Example [24.3] A hiker is near the bottom of the mountain described in Example [24.2] at coordinates $(50, 20, 1000)$. In which direction should he start to climb to have the steepest ascent?

Here $\overrightarrow{\nabla f}(50,20) = -200\,\vec{\mathbf{i}} - 200\,\vec{\mathbf{j}}$ is the direction of steepest ascent.

Example [24.4] Referring again to the mountain of Example [24.2], suppose a water pipe on the mountain side bursts and water gushes out. The break occurs at the coordinates $(10, -20, 5800)$. Draw a contour map of the mountain clearly marking the point $(10, -20, 5800)$. If there are no obstacles the water will flow down the mountain along the path of steepest descent. Sketch an approximation of the path of steepest descent on the contour map.

We begin by graphing the level curves $z_0 = 8000 - 2x^2 - 5y^2$ with the following values of $z_0 = 0,\ 1000,\ 2000,\ 3000,\ 4000,\ 5000,\ 5800,\ 6500,\ 7000,\ 7500.$ The point $(10, -20, 5800)$ is marked.

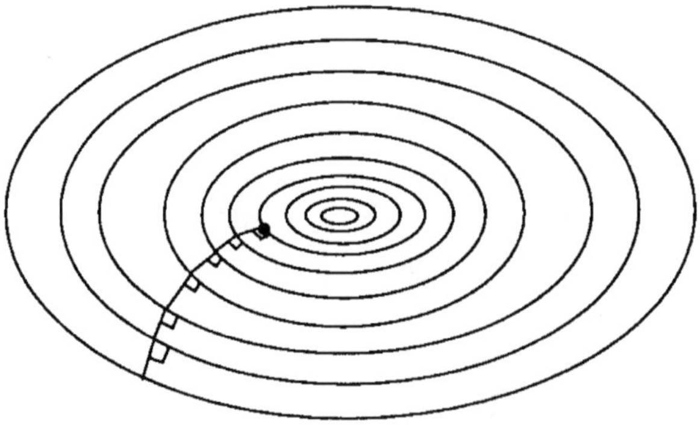

For each of the above values of z_0 let $F(x,y) = f(x,y) - z_0 = 0$. Then $\overrightarrow{\nabla F} = \overrightarrow{\nabla f}$, and as we showed previously (see §23.6) $\overrightarrow{\nabla F}$ is perpendicular to the level curve. Since the path of steepest descent is always in the direction of $\overrightarrow{\nabla f}$, it follows that we can approximate this path on the contour map by drawing in short perpendicular lines between level curves.

Exercises for §24.1

(1) Let $z = 10000 - 3x^2 - y^4$ denote the surface of a mountain. What is the highest point on the mountain. A hiker is at coordinates $(10, 5, 9075)$. In which direction should he begin going down in order to have the steepest descent?

(2) Using your CAS draw several level curves for the surface $z = 10000 - 3x^2 - y^4$ and clearly mark the hiker's coordinates $(10, 5, 9075)$. Sketch in by hand the path of steepest descent.

(3) Let $z = 5x^2 + xy + y^3$. Find the direction in which f is increasing most rapidly at the point $(1, -2)$.

(4) Suppose you are located at a point with coordinates $x = 1.5$, $y = 2$ in a region where the altitude is given by the function $a(x, y) = \tan\left(2\pi x + \frac{\pi}{3} y\right)$. In which direction (in the $x\,y$–plane) should you travel to maximize the increase of elevation? In which direction should you travel to stay at the same elevation? **Hint:** The directional derivative of a is zero in a direction perpendicular to $\overrightarrow{\nabla a}$.

(5) Sonic experimentation has determined an enormous underground cave in a certain country. The depth in feet of the cave is given by the function $d(x, y) = 150 - .01x^2 - .006y^2$. A scientist is at coordinates $(10, 20)$ with a sonar based devise for measuring the depth of the cave.

  (a) What depth reading will he observe at coordinate $(10, 20)$?

  (b) In which direction should he walk to observe the most rapid increase of depth?

  (c) Sketch a contour map of the cave showing its depth.

(6) The equations $z = f(x, y)$ and $F(x, y, z) = f(x, y) - z = 0$ are actually the same. Their graphs will be identical surfaces. Explain why $\overrightarrow{\nabla f}$ and $\overrightarrow{\nabla F}$ are not the same vectors. Does $\overrightarrow{\nabla F}$ have anything to do with the steepest descent on the surface?

24.2 The Method of Critical Points

An important application of the derivative of a function of one variable was the ability to find maximum and minimum values of the function. When we move to differentiable functions of two variables, $z = f(x, y)$, the graph may exhibit various types of extrema such as peaks of hills, bottoms of valleys, and saddle points. We illustrate these three possibilities by examples.

Example [24.5] (Peak of a Hill) Let $z = 100 - x^2 - y^2$. Then the point $(0, 0, 100)$ is the peak of the hill.

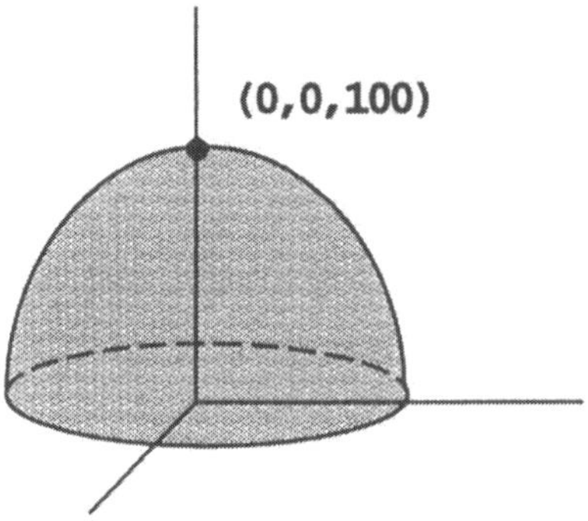

Example [24.6] (Bottom of the Valley) Let $z = x^2 + y^2 - 100$. Then the point $(0, 0, -100)$ is the bottom of the valley.

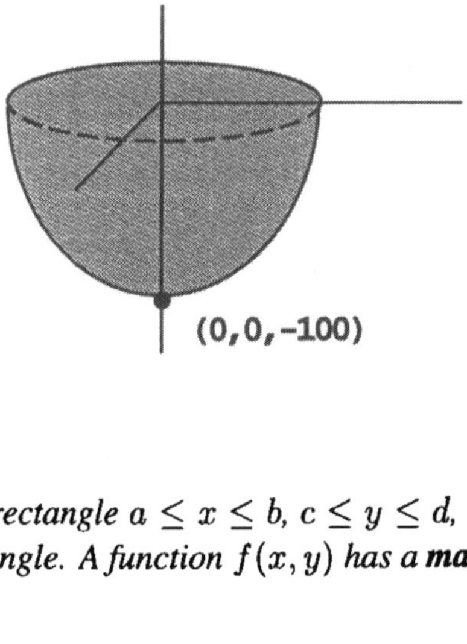

Definition: *Fix a rectangle $a \le x \le b$, $c \le y \le d$, and let (x_0, y_0) be a point in the rectangle. A function $f(x, y)$ has a **maximum** at (x_0, y_0) provided*

$$f(x, y) \le f(x_0, y_0),$$

for every (x, y) in the rectangle.

Remark: By reversing inequalities we can define a **minimum value.**

It is not unusual to find that the maximum or minimum values of a function on a rectangle $a \le x \le b$, $c \le y \le d$ occur on the boundary of the rectangle. For example, the function $f(x, y) = x^2 + y^2$ when considered on the square $1 \le x \le 2$, $1 \le y \le 2$ has its maximum at $(2, 2)$, namely $f(2, 2) = 8$. Clearly, $f(x, y) \le 8$ for every point in the square.

In Example [24.1], the point $(0, 0, 100)$ is a maximum on any rectangle which includes the point $(0, 0, 100)$, while in Example [24.2], the point $(0, 0, -100)$ is a minimum on any rectangle which includes $(0, 0, -100)$. The following example illustrates yet a third possibility.

Example [24.7] (Saddle Point) Let $z = y^2 - x^2$. The graph of this function is displayed below.

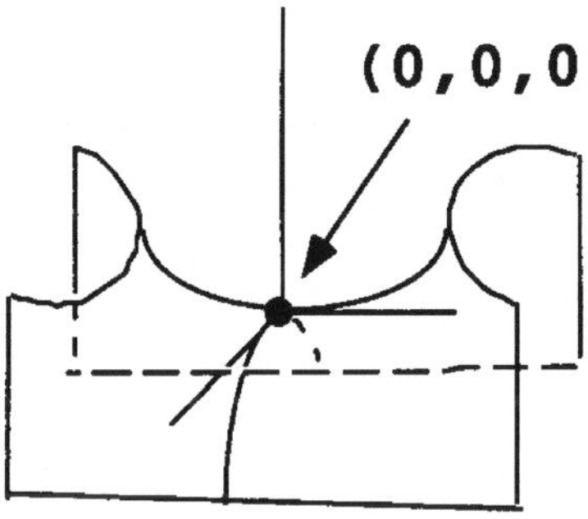

The point $(0, 0, 0)$ is termed a saddle point. Note that it is neither a maximum or a minimum.

A common feature of these three examples is that the tangent plane at the maximum, minimum, or saddle points must be horizontal (i.e., perpendicular to the z–axis).

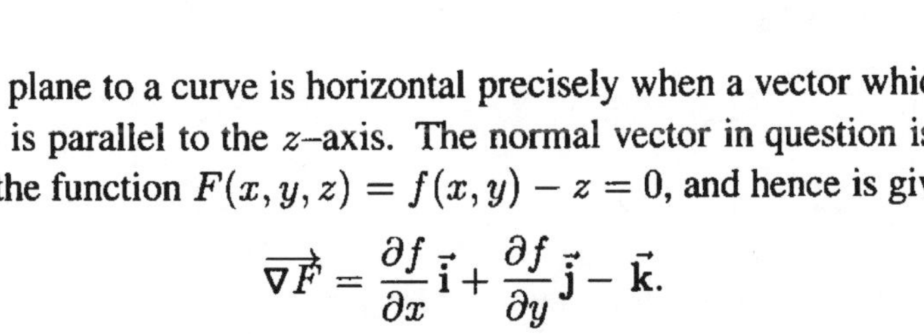

The tangent plane to a curve is horizontal precisely when a vector which is normal to the plane is parallel to the z–axis. The normal vector in question is simply the gradient of the function $F(x, y, z) = f(x, y) - z = 0$, and hence is given by

$$\overrightarrow{\nabla F} = \frac{\partial f}{\partial x}\vec{\mathbf{i}} + \frac{\partial f}{\partial y}\vec{\mathbf{j}} - \vec{\mathbf{k}}.$$

This vector is parallel to the z–axis if and only if

$$\frac{\partial f}{\partial x} = 0, \quad \text{and} \quad \frac{\partial f}{\partial y} = 0.$$

This discussion motivates the important definition.

Definition: *Let $f(x, y)$ be a function of two variables. A point (x_0, y_0) is termed a **critical point** of f if*

$$\frac{\partial f}{\partial x}(x_0, y_0) = 0, \quad \text{and} \quad \frac{\partial f}{\partial y}(x_0, y_0) = 0$$

or one of these partial derivatives does not exist.

Remark: With this terminology in place we see that, if the partial derivatives exist, then a critical point is a point where the tangent plane to the surface is horizontal.

Example [24.8] Find the critical points of the function $f(x, y) = 2x^2 + 8x + y^2 - 3y + 11$.

The critical points occur when

$$\frac{\partial f}{\partial x} = 4x + 8 = 0, \qquad \frac{\partial f}{\partial y} = 2y - 3 = 0.$$

Thus $(-2, \frac{3}{2})$ is the only critical point on the surface.

Example [24.9] Find the critical points of the function $f(x, y) = \cos(x + y)$ in the region $0 \le x \le \pi, 0 \le y \le \pi$.

In this case

$$\frac{\partial f}{\partial x} = \frac{\partial f}{\partial y} = -\sin(x + y),$$

and the critical points must occur when $\sin(x+y) = 0$ that is to say when $x+y = \pi$. Thus there are infinitely many critical points each of which takes the form $(x, \pi - x)$ where $0 \le x \le \pi$. These points from a line at the bottom of the wave and have a common tangent plane.

The Three Step Method for Finding Maxima and Minima

Since a maximum or minimum value of a function $f(x, y)$ defined on a rectangle can only occur at a critical point or a boundary point of the rectangle, we immediately obtain a 3–step method for finding maxima and minima which generalizes the 3-step method of §10.1.

Step 1: Find all critical points (x_0, y_0) within the rectangle where $\frac{\partial f}{\partial x}(x_0, y_0) = \frac{\partial f}{\partial y}(x_0, y_0) = 0$.

Step 2: Determine the largest and smallest values the function attains at these critical points (if the set of critical points forms a line use the method in §10.1 to accomplish this task).

Step 3: Find the maximum or minimum value of the function $f(x, y)$ on the boundary of the rectangle (this may again require the use of methods of §10.1). Determine if this is larger or smaller than the value $f(x_0, y_0)$ obtained in step 2.

Example [24.10] Find the maximum and minimum values of the function $f(x, y) = xy^2$ on the rectangle below.

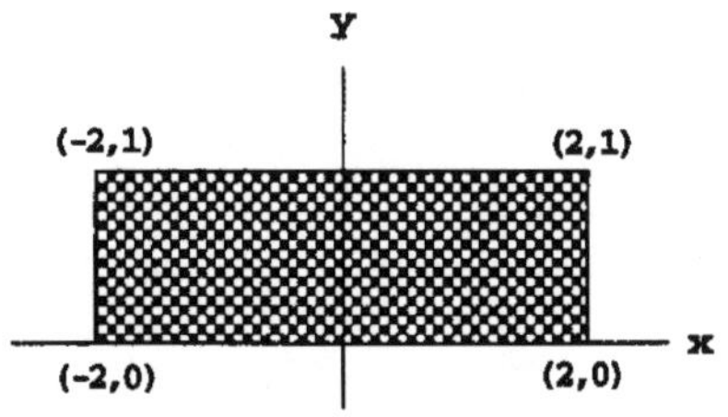

Step 1. The partial derivatives are $\frac{\partial f}{\partial x} = y^2$ and $\frac{\partial f}{\partial y} = 2xy$. By setting these derivatives equal to zero we see that we have a critical point when $y = 0$, and thus the critical points are on the boundary.

Step 2. For an arbitrary critical point $(x, 0)$ the function takes the value $f(x, 0) = 0$.

Step 3. We can complete out analysis by examining $f(x, y)$ on the remainder of the boundary. Consider first the line segment consisting of the points $(x, 1)$ where $-2 \leq x \leq 2$, i.e., the segment between the points $(-2, 1)$ and $(2, 1)$. The value of f on the line is $f(x, 1) = x$ and thus the maximum occurs at $(2, 1)$ and the minimum occurs at $(-2, 1)$. Next we must consider the line segment consisting of the points $(2, y)$ where $0 \leq y \leq 1$. Here $f(2, y) = 2y^2$ which is maximized at $(2, 1)$. Similarly on the final line segment which consists of the points $(-2, y)$ where $0 \leq y \leq 1$, $f(-2, y) = -2y^2$ which is minimized at $(-2, 1)$ and maximized at $(-2, 0)$.

We conclude that $f(x, y)$ is maximized at $(2, 1)$ with the value $f(2, 1) = 2$ and is minimized at $(-2, 1)$ with the value $f(-2, 1) = -2$.

Example [24.11] Find the shortest distance between the point $(1, 2, 0)$ and the plane $x + y - z = 0$.

Let (x, y, z) be an arbitrary point on the plane. Its distance to the point $(1, 2, 0)$ is

$$\sqrt{(x-1)^2 + (y-2)^2 + z^2} = \sqrt{(x-1)^2 + (y-2)^2 + (x+y)^2}.$$

This distance will be minimized if and only if the square of the distance is minimized (the latter function being easier to work with). Thus we must find the critical points of the functions

$$s(x,y) = (x-1)^2 + (y-2)^2 + (x+y)^2.$$

Setting the partial derivatives equal to zero gives us the equations

$$\frac{\partial s}{\partial x} = 2(x-1) + 2(x+y) = 0$$

$$\frac{\partial s}{\partial y} = 2(y-2) + 2(x+y) = 0,$$

which are equivalent to the equations

$$4x + 2y = 2$$

$$2x + 4y = 4.$$

Thus $(0, 1)$ is a critical point and at this critical point $s(x, y)$ takes the value $s(0, 1) = 1 + 1 + 1 = 3$.

Question: Is $(0, 1)$ a maximum, a minimum, or a saddle point of the function $s(x, y)$?

Answer: We claim that $(0, 1)$ is in fact a minimum and hence the solution to our original problem is $\sqrt{3}$. Intuitively we can verify our claim by graphing the function $z = s(x, y)$ in three dimensional space.

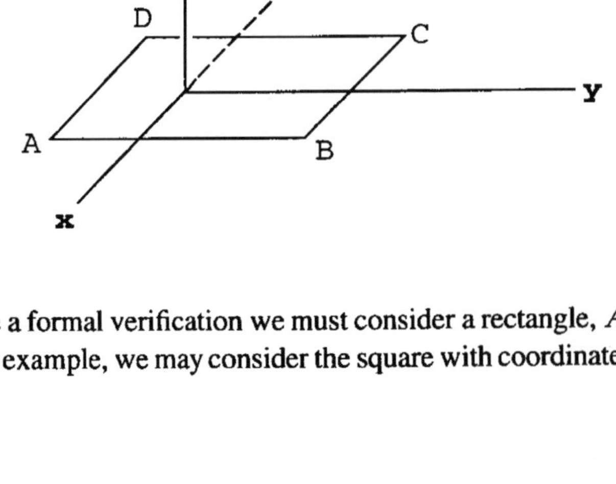

In order to give a formal verification we must consider a rectangle, $ABCD$ which is very large. For example, we may consider the square with coordinates $A = (50, -50)$,

$B = (50,50)$, $C = (-50,50)$, and $D = (-50,-50)$. Our three step algorithm dictates that the minimum for $s(x,y)$ on this square must occur either at the critical point $(0,1)$ or on the boundary of the square $ABCD$. But on this boundary either $x = \pm 50$ or $y = \pm 50$, hence $s(x,y) \geq (48)^2$ (which is certainly larger than 3). Notice that this argument remains valid as the length of the side of the square tends to infinity. We conclude that $s(x,y) \geq 3$ for all (x,y) in the xy–plane.

Example [24.12] Find the maximum value of the function $f(x,y) = \frac{1}{x^2+y^2}$ on the rectangle $-1 \leq x \leq 1, -1 \leq y \leq 1$;

We begin by taking the partial derivatives of $f(x,y)$:

$$\frac{\partial f}{\partial x} = \frac{-2x}{(x^2+y^2)^2}, \quad \frac{\partial f}{\partial y} = \frac{-2y}{(x^2+y^2)^2}.$$

Setting these derivatives equal to zero yield a single critical point, $(0,0)$. But notice that the partial derivatives of $f(x,y)$ are not defined at $(0,0)$. In fact the function shoots up to infinity as we approach $(0,0)$ from any direction, and thus the function has no maximum value.

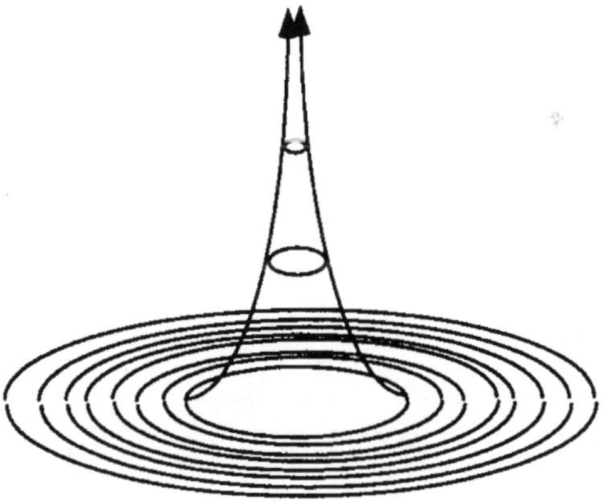

Remark: This last example illustrates that the critical point may occur where a partial derivative does not exist.

Exercises for §24.2

In exercises (1)–(5) find the maximum and minimum values of $f(x,y)$ on the given rectangle. Confirm your answers visually by plotting the given function with your CAS.

(1) $f(x,y) = 20 - 2x^2 + 4x - y^2 + 4y$ for $0 \leq x \leq 3, 0 \leq y \leq 4$.

(2) $f(x,y) = x^4 + 3xy$ for $0 \leq x \leq 1, -1 \leq y \leq 1$.

(3) $f(x,y) = x^2 - xy - y^2 + 5y - 3$ for $-1 \leq x \leq 2, 0 \leq y \leq 3$.

(4) $f(x,y) = x^3y^2(3 - x - y)$ for $0 \leq x \leq 10, 0 \leq y \leq 10$.

(5) $f(x,y) = y^2 + \cos(\pi x) - y$ for $-\frac{1}{2} \leq x \leq \frac{1}{2}, 0 \leq y \leq 1$.

(6) A shipping company wants to make rectangular cardboard boxes (without lids). If the boxes are to have a volume of 4 cubic feet each, what dimensions should the boxes be to minimize the amount of cardboard used.

(7) A crating company is producing rectangular plywood boxes with lids. Assuming each box is to be made with 24 square feet of plywood, what dimensions should the boxes be to maximize the volume.

(8) Find the point on the line $x = t, y = 3t - 1, z = 2t + 3$ which is closest to the point $(1, 0, -1)$.

(9) Find the point on the plane $2x - 5y + z = 2$ which is closest to the point $(1, 0, -1)$

(10) A long piece of wire (240 inches long) is cut into three or fewer pieces. If each piece is bent into the shape of a square, how should the cutting be done so that the sum of the areas of the squares is minimized?

(11) Consider the surface $z = \frac{2}{xy}$. What are the critical points? For which values of x, y in the rectangle $-1 \leq x \leq 1, -1 \leq y \leq 1$ is z maximized or minimized? Visually confirm your answers by plotting with your CAS.

(12) Find the point on the surface $z = \frac{2}{xy}$ which is closest to the origin $(0, 0, 0)$.

24.3 Taylor Series and the Classification of Critical Points

We begin this section by recalling the Taylor series about the point $x = 0$ (also termed the Maclaurin series) for a differentiable function $f(x)$ of one variable,

$$f(x) = f(0) + f'(0)x + \frac{f''(0)}{2!}x^2 + \frac{f^{(3)}(0)}{3!}x^3 + \cdots$$

which at the point $x = a$ yields

$$f(a) = f(0) + f'(0)a + \frac{f''(0)}{2!}a^2 + \frac{f^{(3)}(0)}{3!}a^3 + \cdots \tag{24.2}$$

We would like to generalize this formula to differentiable functions $z = f(x,y)$ of two variables. We shall show that

$$f(a,b) = f(0,0) + \frac{\partial f}{\partial x}(0,0)a + \frac{\partial f}{\partial y}(0,0)b + \cdots$$

where the higher terms (not indicated in the above expansion) involve higher order partial derivatives. The required expansion can be obtained with what can only be called a very clever trick which will reduce the problem to a one dimensional Taylor series.

Fixing numbers a, b let us define the function of one variable $F(t)$ by

$$F(t) = f(at, bt).$$

Notice that $F(1) = f(a, b)$. When we replace a by t in the expansion (24.2), we obtain the expansion

$$F(t) \;=\; F(0) \;+\; F'(0)t \;+\; \frac{F''(0)}{2!}t^2 + \frac{F^{(3)}(0)}{3!}t^3 \;+\; \cdots \tag{24.3}$$

We claim that by computing $F(0)$, $F'(0)$, $F''(0)$, etc. and then setting $t = 1$ we will arrive at the desired expansion.

Having described the essence of the plan we now move to the (somewhat lengthy) computations. The first term in (24.3) is $F(0)$ which is simply $F(0) = f(0,0)$. The first derivative of F is obtained by the chain rule:

$$\begin{aligned} \frac{d}{dt}F(t) &= \frac{d}{dt}f(at, bt) \\ &= a\frac{\partial f}{\partial x}(at, bt) + b\frac{\partial f}{\partial y}(at, bt), \end{aligned}$$

and thus

$$F'(0) = a\frac{\partial f}{\partial x}(0,0) + b\frac{\partial f}{\partial y}(0,0).$$

The second derivative is a bit more involved:

$$\begin{aligned} F''(t) &= \frac{d}{dt}F'(t) \\ &= \frac{d}{dt}\left(a\frac{\partial f}{\partial x}(at, bt) + b\frac{\partial f}{\partial y}(at, bt)\right) \\ &= a^2\frac{\partial^2 f}{\partial x^2}(at, bt) + 2ab\frac{\partial^2 f}{\partial x \partial y}(at, bt) + b^2\frac{\partial^2 f}{\partial y^2}(at, bt), \end{aligned}$$

hence

$$F''(0) = a^2\frac{\partial^2 f}{\partial x^2}(0,0) + 2ab\frac{\partial^2 f}{\partial x \partial y}(0,0) + b^2\frac{\partial^2 f}{\partial y^2}(0,0).$$

At this point it is clear how to proceed further (we omit these computations since they are cumbersome). Combining the computations above with equation (24.3) yields

$$\begin{aligned} F(t) &= f(at, bt) \\ &= f(0,0) + \left(a\frac{\partial f}{\partial x}(0,0) + b\frac{\partial f}{\partial y}(0,0)\right)t \\ &\quad + \frac{1}{2!}\left(a^2\frac{\partial^2 f}{\partial x^2}(0,0) + 2ab\frac{\partial^2 f}{\partial x \partial y}(0,0) + b^2\frac{\partial^2 f}{\partial y^2}(0,0)\right)t^2 + \cdots \end{aligned}$$

Choosing $t = 1$ gives us the expansion we initially desired:

$$\begin{aligned} f(a,b) &= f(0,0) + \left(a\frac{\partial f}{\partial x}(0,0) + b\frac{\partial f}{\partial y}(0,0)\right) \\ &\quad + \frac{1}{2}\left(a^2\frac{\partial^2 f}{\partial x^2}(0,0) + 2ab\frac{\partial^2 f}{\partial x \partial y}(0,0) + b^2\frac{\partial^2 f}{\partial y^2}(0,0)\right) \\ &\quad + \cdots \qquad (24.4) \end{aligned}$$

The terms exhibited on the right hand side of (24.4) are called the **terms of degree at most two** in the Taylor expansion.

While we have not chosen to dwell on questions of convergence they essentially reduce to our previous study since the expansion was obtained from the expansion of a function of a single variable.

Example [24.13] Compute the terms of degree at most two in the Taylor series for the function $f(x, y) = e^{xy}$.

The computations here require the following derivatives:

$$f(0,0) = e^0 = 1,$$

$$\frac{\partial f}{\partial x}(0,0) = ye^{xy}\Big]_{x=0,\ y=0} = 0, \quad \frac{\partial f}{\partial y}(0,0) = xe^{xy}\Big]_{x=0,\ y=0} = 0,$$

$$\frac{\partial^2 f}{\partial x^2}(0,0) = y^2e^{xy}\Big]_{x=0,\ y=0} = 0, \quad \frac{\partial^2 f}{\partial y^2}(0,0) = x^2e^{xy}\Big]_{x=0,\ y=0} = 0,$$

$$\frac{\partial^2 f}{\partial x \partial y}(0,0) = e^{xy} + xye^{xy}\Big]_{x=0,\ y=0} = 1.$$

Plugging into (24.2), we obtain the (seemingly familiar) expansion

$$e^{ab} = 1 + ab + \cdots$$

which exhibits the terms of degree at most two in the Taylor series of e^{xy}. Can you conjecture what the higher terms might be? A little more work shows that

$$e^{ab} = 1 + ab + \frac{(ab)^2}{2!} + \frac{(ab)^3}{3!} + \cdots$$

Classification of Critical Points

Let $z = f(x, y)$ be a differentiable function of two variables x, y. We shall say that f has a **local minimum** at a point (x_0, y_0) if

$$f(x_0 + h, y_0 + k) \geq f(x_0, y_0)$$

for all sufficiently small $h, k \geq 0$, and f has a **local maximum** at (x_0, y_0) if

$$f(x_0 + h, y_0 + k) \leq f(x_0, y_0)$$

for all sufficiently small $h, k \geq 0$.

We have already demonstrated that a necessary condition for f to have a local maximum or minimum at a point (x_0, y_0) is that the tangent plane to the surface (at the point (x_0, y_0)) is horizontal, i.e., (x_0, y_0) is a critical point. What remains to be developed is a test which will determine, given a critical point, whether it is a local maximum, a local minimum, or a saddle point. The **critical point test** which we will state satisfies this need, and is a natural generalization of the second derivative test (see §9.3) to function of two variables. A verification of this test can be obtained by carefully examining the terms of degree two in the Taylor expansion of the function in question. In the following statement, $f(x, y)$ will denote a differentiable function with a critical point (x_0, y_0). When we evaluate the various derivatives at (x_0, y_0), for example $\frac{\partial f}{\partial x}(x_0, y_0)$, we shall, for the sake of notational convenience, simply write $\frac{\partial f}{\partial x}$.

The Critical Point Test

Let (x_0, y_0) be a critical point of $f(x, y)$. Defining

$$D = \frac{\partial^2 f}{\partial x^2} \cdot \frac{\partial^2 f}{\partial y^2} - \left(\frac{\partial^2 f}{\partial x \partial y} \right)^2,$$

then the following statements hold:

(1) (x_0, y_0) *is a local maximum if* $\frac{\partial^2 f}{\partial x^2} < 0$ *and* $D > 0$.

(2) (x_0, y_0) *is a local minimum if* $\frac{\partial^2 f}{\partial x^2} > 0$ *and* $D > 0$.

(3) (x_0, y_0) *is a saddle point if* $D < 0$.

(4) *If* $D = 0$ *the test provides no information.*

Example [24.14] Find and classify the critical points of the following functions:

$$(i)\ f(x,y) = 1 - x^2 - y^2, \quad (ii)\ f(x,y) = x^2 + y^2, \quad (iii)\ f(x,y) = xy.$$

In all three cases the point $(0,0)$ is the only critical point. Applying the critical point test to each of these functions demonstrates their distinct behavior around the origin.

(i) $D = (-2) \cdot (-2) - 0 = 4,\ \frac{\partial^2 f}{\partial x^2} = -2 < 0$ and hence $(0,0)$ is a local maximum.

(ii) $D = 4,\ \frac{\partial^2 f}{\partial x^2} = 2 > 0$ indicating that $(0,0)$ is local minimum.

(iii) $D = 0 - 1 < 0$ and thus $(0,0)$ is a saddle point.

Example [24.15] Find and classify the critical points of the function

$$f(x,y) = e^{-x^2-y^2}.$$

In this case the partial derivatives are given by

$$\frac{\partial f}{\partial x} = -2xe^{-x^2-y^2}, \quad \frac{\partial f}{\partial y} = -2ye^{-x^2-y^2},$$

and thus here too $(0,0)$ is the only critical point. Now the second partial derivatives are

$$\begin{aligned} \frac{\partial^2 f}{\partial x^2} &= (-2 + 4x^2)e^{-x^2-y^2}, \\ \frac{\partial^2 f}{\partial y^2} &= (-2 + 4y^2)e^{-x^2-y^2}, \\ \frac{\partial^2 f}{\partial x \partial y} &= 4xye^{-x^2-y^2}, \end{aligned}$$

and thus

$$D = \frac{\partial^2 f}{\partial x^2} \cdot \frac{\partial^2 f}{\partial y^2} - \left(\frac{\partial^2 f}{\partial x \partial y}\right)^2 \Bigg]_{x=0\ y=0} = 4.$$

By evaluating the second partial derivative with respect to x at the origin, we obtain $\frac{\partial^2 f}{\partial x^2}(0,0) = -2 < 0$. W see that $(0,0)$ is a local maximum.

Exercises for §24.3

Compute the terms of degree at most 2 in the Taylor expansion for the following functions.

(1) $f(x,y) = (x - y)^2$.

(2) $f(x,y) = (x + 2y)^3$.

(3) $f(x,y) = \ln(2 + x + y)$.

(4) $f(x,y) = e^{x - 3y}$.

(5) $f(x,y) = \cos(x + xy)$.

(6) $f(x,y) = \sin(y - 2xy)$.

Find and classify the critical points of each of the following functions. Visually confirm your answer by graphing with your CAS.

(7) $f(x,y) = 1 - x^2 - y^2 + 2x + 6y$.

(8) $f(x,y) = 6 + x^2 + y^2 + 8x - 4y$.

(9) $f(x,y) = xy - 3x - 3y + 9$.

(10) $f(x,y) = 2x - x^3 - y^3$.

(11) $f(x,y) = 5xy - x^3 - y^3$.

(12) $f(x,y) = 6x + 12y - x^3 - y^3$.

(13) $f(x,y) = 2x^4 + y^4 - 2x^2 - 2y^2 + 8$.

(14) Show that each of the functions

$$1 - x^2y^2, \quad x^2y^2, \quad x^3y^2$$

has a critical point at $(0,0)$ which is a maximum, minimum, and saddle point, respectively, by graphing these functions with your CAS. Can you show this using the critical point test?

24.4 The Method of Lagrange Multipliers

We conclude this chapter by presenting an immensely powerful method for finding maxima and minima of functions of several variables which was discovered by Joseph Lagrange (1736–1813). The method is applicable to problems of the following form:

Problem: *Maximize or minimize a function* $f(x_1, x_2, \ldots, x_n)$ *of* n *variables* $x_1, x_2 \ldots, x_n$ *subject to the* ***constraint*** *(i.e., condition)* $g(x_1, x_2, \ldots, x_n) = 0$ *where* g *is some other function.*

A wide array of problems which arise in science and engineering can be mathematically modeled in this form. We examine two examples which will help us get a feel for the approach.

Example [24.16] Find the shortest distance between a point (x_0, y_0, z_0) and a plane $ax + by + cz = d$.

We have already encountered and solved this problem on more than one occasion. We now transform this geometric problem into one which takes the form of minimization subject to a constraint. Let (x, y, z) be an arbitrary point on the plane. The square of the distance from (x, y, z) to the point (x_0, y_0, z_0) is given by

$$f(x, y, z) = (x - x_0)^2 + (y - y_0)^2 + (z - z_0)^2.$$

Define $g(x, y, z) = ax + by + cz - d$. With these functions in place our problem can now be restated in the desired form: minimize $f(x, y, z)$ subject to the constraint $g(x, y, z) = 0$.

Example [24.17] Find the largest possible rectangle which can be inscribed in a circle of radius R.

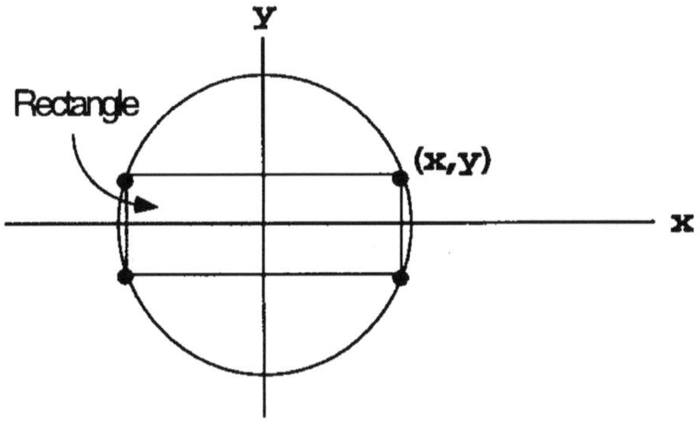

The question posed here can be restated as follows: maximize the area function $f(x, y) = 4xy$ subject to the constraint $g(x, y) = x^2 + y^2 - R^2 = 0$.

The following exposition will be simplified greatly by introducing the following notation: given a point $P = (\alpha_1, \alpha_2, \ldots, \alpha_n)$ in n–dimensional space, and $\phi = \phi(x_1, x_2, \ldots, x_n)$ a function of n variables, let $\phi(P)$ denote $\phi(\alpha_1, \alpha_2, \ldots, \alpha_n)$.

Lagrange's solution to the general problem of maximizing (or minimizing) a function subject to constraints is both commanding and direct. It is based on the following proposition in which a constant, λ (which is now called the **Lagrange multiplier**), is introduced.

Proposition [24.18] *Let*

$$f(x_1, \ldots, x_n), \quad g(x_1, \ldots, x_n)$$

be two real valued functions of n variables $x_1, \ldots, x_n$ which have partial derivatives of all orders. Assume that the maximum or minimum value of f subject to the constraint $g = 0$ occurs at a point

$$P = (\alpha_1, \alpha_2, \ldots, \alpha_n)$$

where $\nabla g(P) \neq 0$. Then

$$\vec{\nabla} f(P) = \lambda \vec{\nabla} g(P) \tag{24.5}$$

for some constant λ.

It is remarkable that the simple identity (24.5) encodes all the information needed to solve our general max–min problem.

Example [24.19] Minimize the function $f(x, y) = x^2 + 2y^2$ subject to the constraint $2x + y = 3$.

We begin by setting $g(x, y) = 2x + y - 3 = 0$. The gradient vectors are

$$\vec{\nabla} f = 2x\,\vec{\mathbf{i}} + 4y\,\vec{\mathbf{j}},$$

and

$$\vec{\nabla} g = 2\vec{\mathbf{i}} + \vec{\mathbf{j}},$$

and thus (24.5) becomes

$$2x\,\vec{\mathbf{i}} + 4y\,\vec{\mathbf{j}} = \lambda(2\vec{\mathbf{i}} + \vec{\mathbf{j}}).$$

This identity yields the equations

$$2x = 2\lambda$$

$$4y = \lambda,$$

whose solutions are

$$x = \lambda = 4y.$$

Plugging $x = 4y$ into the constraint equation $2x + y = 3$ yields $9y = 3$, and thus $y = \frac{1}{3}$. We conclude that $\left(\frac{4}{3}, \frac{1}{3}\right)$ is the required solution and the minimum value of $f(x, y) = x^2 + 2y^2$ is

$$f\left(\frac{4}{3}, \frac{1}{3}\right) = \frac{18}{9} = 2.$$

Example [24.20] Find the points on the surface $z^2 - 2xy = 3$ closest to the origin.

Let (x, y, z) be an arbitrary point on the surface. The distance of this point to the origin is $\sqrt{x^2 + y^2 + z^2}$. Since it is sufficient to minimize $x^2 + y^2 + z^2$ (the square of the distance) our problem can be restated as follows: minimize $f(x, y, z) = x^2 + y^2 + z^2$ subject to the constraint $g(x, y, z) = z^2 - 2xy - 3 = 0$. The gradients in this case are

$$\vec{\nabla} f = 2x\,\vec{\mathbf{i}} + 2y\,\vec{\mathbf{j}} + 2z\,\vec{\mathbf{k}},$$

$$\vec{\nabla} g = -2y\,\vec{\mathbf{i}} - 2x\vec{\mathbf{j}} + 2z\,\vec{\mathbf{k}},$$

and thus (24.5) yields

$$2x = -2y\lambda, \qquad 2y = -2x\lambda, \qquad 2z = 2z\lambda.$$

We conclude that $\lambda = 1$ and $x = -y$. When we plug this solution into our constraint equation we obtain

$$z^2 + 2x^2 = 3.$$

Thus we see that the points on the surface $z^2 - 2xy = 3$ which are closest to the origin must take the form $(x, -x, \sqrt{3 - 2x^2})$ where $\frac{-\sqrt{3}}{2} \le x \le \frac{\sqrt{3}}{2}$. These points form a space curve on our surface.

·

Outline of the Proof of Proposition [24.18]

All of the essential ideas involved in this proof appear when we focus on functions of three variables and hence we will discuss the three variable case exclusively. Thus we are given a function $f(x, y, z)$ which we must maximize (or minimize) subject to the constraint $g(x, y, z) = 0$.

The set of points (x, y, z) which satisfy the constraint equation $g(x, y, z) = 0$ all lie on a surface which we refer to as S. Let $P = (x_0, y_0, z_0) \in S$ be the point on the surface S where $f(x, y, z)$ is maximized (or minimized). Since $\vec{\nabla} g(P)$ is assumed to be nonzero, it must be perpendicular to the surface at the point P.

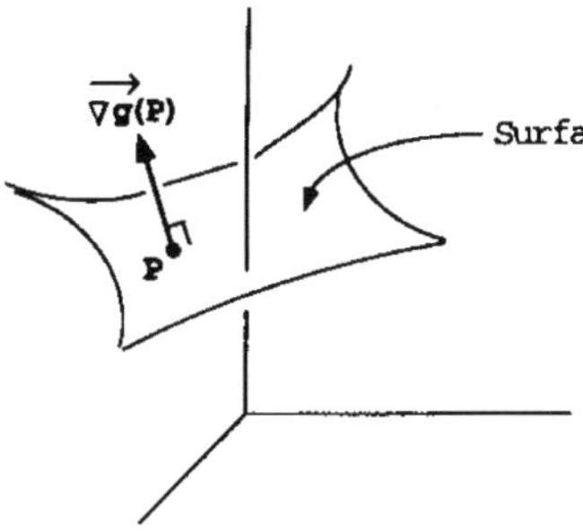

Choose an arbitrary space curve

$$\overrightarrow{r(t)} = \overrightarrow{(x(t), y(t), z(t))}$$

which lies on the surface S and passes through the point P

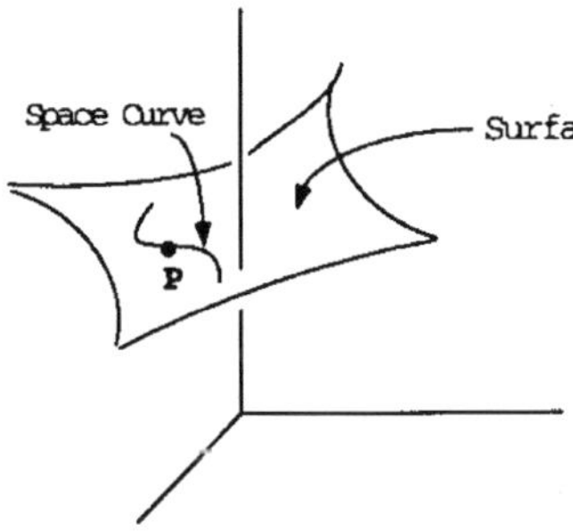

at some point $t = t_0$, i.e., $(x(t_0), y(t_0), z(t_0)) = (x_0, y_0, z_0)$.

Now, by definition, $f(x(t), y(t), z(t))$ is a function of a single variable t which attains a maximum at $t = t_0$, and thus

$$\frac{d}{dt} f(x(t), y(t), z(t))\bigg]_{t=t_0} = 0.$$

Computing this derivative with the chain rule we find that

$$\begin{aligned} \frac{d}{dt} f(x(t), y(t), z(t)) &= \frac{\partial f}{\partial x}\frac{dx}{dt} + \frac{\partial f}{\partial y}\frac{dy}{dt} + \frac{\partial f}{\partial z}\frac{dz}{dt} \\ &= \vec{\nabla} f \cdot \frac{d\vec{r}}{dt}, \end{aligned}$$

and evaluating this at t_0 we see that

$$\vec{\nabla} f(x_0, y_0, z_0) \cdot \overrightarrow{r'(t_0)} = 0.$$

This final identity implies that $\vec{\nabla} f(P)$ is perpendicular to every space curve on S which passes through the point P. We deduce that $\vec{\nabla} f(P)$ is in fact perpendicular to the tangent plane of S at P.

Since $\vec{\nabla} g(P)$ is also perpendicular to the tangent plane at P, it follows that $\vec{\nabla} f(P)$ and $\vec{\nabla} g(P)$ are *parallel* vectors, i.e., there exists a real number λ such that

$$\vec{\nabla} f(P) = \lambda \vec{\nabla} g(P).$$

The Case of Several Constraints

Having seen how useful the Lagrange multiplier is in the case of one constraint, we are motivated to generalize the method. Thus suppose we wish to maximize a function $f(x_1, x_2, \ldots, x_n)$ subject to several constraints,

$$g_1(x_1, x_2, \ldots, x_n) = 0$$

$$g_2(x_1, x_2, \ldots, x_n) = 0$$

$$\vdots$$

$$g_k(x_1, x_2, \ldots, x_n) = 0$$

This situation will lead to a *set* of k Lagrange multipliers, $\lambda_1, \lambda_2, \ldots, \lambda_k$ and the system of equations

$$\vec{\nabla} f = \lambda_1 \vec{\nabla} g_1 + \lambda_1 \vec{\nabla} g_2 + \cdot + \lambda_1 \vec{\nabla} g_k$$

$$g_1 = 0, g_2 = 0, \ldots, g_k = 0$$

which we must solve.

Example [24.21] Maximize the function $f(x, y, z) = 2x - y + z$ subject to the constraints that (x, y, z) lie simultaneously on the plane $x + y + z = 1$ and on the cylinder $x^2 + y^2 = 4$ (z arbitrary).

In this case we have two constraints

$$g_1(x, y, z) = x + y + z - 1 = 0$$

$$g_2(x, y, z) = x^2 + y^2 - 4 = 0.$$

The various gradients in this final example are

$$\vec{\nabla} f = 2\vec{\mathbf{i}} - \vec{\mathbf{j}} + \vec{\mathbf{k}}$$

$$\vec{\nabla} g_1 = \vec{\mathbf{i}} + \vec{\mathbf{j}} + \vec{\mathbf{k}}$$

$$\vec{\nabla} g_2 = 2x\vec{\mathbf{i}} + 2y\vec{\mathbf{j}},$$

and hence the equation $\vec{\nabla} f = \lambda_1 \vec{\nabla} g_1 + \lambda_2 \vec{\nabla} g_2$ becomes

$$2\vec{\mathbf{i}} - \vec{\mathbf{j}} + \vec{\mathbf{k}} = (\lambda_1 + 2x\lambda_2)\vec{\mathbf{i}} + (\lambda_1 + 2y\lambda_2)\vec{\mathbf{j}} + \lambda_1 \vec{\mathbf{k}}.$$

Thus $\lambda_1 = 1$,

$$2 = 1 + 2x\lambda_2 \implies \lambda_2 = \frac{1}{2x}$$

and

$$-1 = 1 + 2y\lambda_2 \implies -2 = \frac{y}{x}.$$

We conclude that $y = -2x$ which upon plugging this into the second constraint $x^2 + y^2 - 4 = 0$ yields the equation of one variable

$$x^2 + 4x^2 = 4.$$

The solutions of this equation are $x = \pm\frac{2}{\sqrt{5}}$ and hence $y = \mp\frac{4}{\sqrt{5}}$. Plugging these values into the first constraint $x+y+z = 1$ yields $z = 1 \pm \frac{2}{\sqrt{5}}$. The maximum value of $2x - y + z$ subject to the two constraints occurs at the point $\left(\frac{2}{\sqrt{5}}, \frac{-4}{\sqrt{5}}, 1 + \frac{2}{\sqrt{5}}\right)$ and the maximum value is $\frac{10}{\sqrt{5}} + 1 = 1 + 2\sqrt{5}$. Notice that the minimum value occurs at the other point $\left(-\frac{2}{\sqrt{5}}, \frac{4}{\sqrt{5}}, 1 - \frac{2}{\sqrt{5}}\right)$.

Exercises for §24.4

Solve all of the following optimization problems using the method of Lagrange multipliers.

(1) Show that the largest possible rectangle which can be inscribed in a circle is a square.

(2) Find the shortest distance between the point $(1, 1, 2)$ and the plane $2x - y + 3z = 2$.

(3) Minimize the function $2x^2 + 3y^2$ subject to the constraint $3x + 4y = 1$.

(4) Find the maximum and minimum values of the function $3x - y + 4z$ on the surface of the sphere $x^2 + y^2 + z^2 = 4$.

(5) Which point on the hyperbola $xy = 5$ minimizes the function $3x + 2y$?

(6) Find the point on the sphere $(x - 1)^2 + y^2 + (z + 1)^2 = 1$ which is furthest from the point $(1, 2, 3)$.

(7) Using your CAS, graph the ellipsoid $\frac{x^2}{4} + \frac{y^2}{9} + z^2 = 1$. Which point on the ellipsoid is furthest from the origin?

(8) Minimize the function $x^2 - y + 2z^2 - 1$ subject to the constraint $z^2 = x^4 + y^4$.

(9) An airline accepts carry on baggage in the shape of a box whose length plus width plus depth is at most $100''$. Find the largest volume such a box can have.

(10) Minimize $2x^2 + y^2 + 3z^2$ subject to the constraint $xyz = 1$.

(11) Find the points which lie on the intersection of the plane $x + y = 2$ with the sphere $x^2 + (y-3)^2 + z^2 = 1$ which are nearest to and furthest from the origin.

(12) Maximize the mean value $\frac{1}{K}(x_1 + x_2 + \cdots + x_K)$ of K variables subject to the constraint $x_1^2 + x^2 + \cdots + x_K^2 = 1$.

(13) Find a formula for the shortest distance between a point (x_0, y_0, z_0) and a plane $ax + by + cz = d$.

(14) Find the points on the surface $z = 2 - \sin(x) - \cos(y)$ which are closest to the origin.

Additional exercises for Chapter XXIV

Find the direction in which f is increasing most rapidly at the point P. Find the directional derivative in the direction $\vec{\mathbf{a}}$. Graph the functions with your CAS.

(1) $f(x,y) = 7x^3y^2 - 3;\ P = (2,3), \vec{\mathbf{a}} = 4\vec{\mathbf{i}} - 3\vec{\mathbf{j}}$.

(2) $f(x,y) = xe^y - ye^x;\ P = (2,3),\ \vec{\mathbf{a}} = 2\vec{\mathbf{i}} + \vec{\mathbf{j}}$.

(3) $f(x,y) = \frac{x}{x+y};\ P = (1,0),\ \vec{\mathbf{a}} = \vec{\mathbf{i}}$.

(4) $f(x,y) = \ln(1 + x^2 + y);\ P = (1,2),\ \vec{\mathbf{a}} = \vec{\mathbf{i}} + \vec{\mathbf{j}}$.

(5) $f(x,y) = \sqrt{(1-x)y};\ P = (-2,3),\ \vec{\mathbf{a}} = \vec{\mathbf{i}}$.

(6) Let $f(x,y) = yx^3 + 4y^2$. Graph the surface with your CAS and find all points where $|\vec{\nabla} f| = 10$.

(7) On a certain mountain, the elevation above sea level ($z = 0$), is given by $f(x,y) = 3000 - x^2 - 3y^2$ feet. The positive x-axis points east and the positive y-axis north. A climber is at the point (2,10,2696). If the climber walks due west, will he be ascending or descending? What if he walks northeast? Can he take a heading to walk a level path? Graph with your CAS to assist you.

Find and classify all critical points of $f(x,y)$ on the given region. Confirm your answers visually by plotting the given function with your CAS.

(8) $f(x,y) = x^2 + y^2 + \frac{2}{xy}$ for $-5 \le x \le -1/2,\ -5 \le y \le -1/2$.

(9) $f(x,y) = \sin(x) + \sin(y) + \sin(x - 2y)$ for $-\pi \le x, y \le \pi$.

(10) $f(x,y) = x^2 + 2y^2 - x^2y$ for $-2 \le x \le 5,\ -5 \le y \le 1$.

(11) $f(x,y) = abxy + \frac{a}{x^2+a^2} + \frac{b}{y^2+b^2}$. for all x,y.

(12) $f(x,y) = e^{-(x^2+y^2-2x+3y+2)}$ for all x,y.

(13) $f(x,y) = xy^2$, for x, y in the first quadrant and $x^2 + y^2 \leq 1$.

(14) Find three positive numbers whose sum is 100 and whose product is as large as possible.

(15) Find three positive numbers whose sum is 100 and the sum of their squares is as small as possible.

(16) Maximize $f(x,y) = xy^2$ subject to the constraint $4x^2 + 8y^2 = 16$.

(17) Find the point on the parabaloid $f(x,y) = x^2 + y^2 + 2$ closest to the plane $x + y + z = 0$.

Chapter XXV Double Integrals

25.1 Review of One Variable Integration

As a motivation and guide to our discussion of multiple integration, we begin with a brief review of the single variable case. Thus we consider, $y = f(x)$, an arbitrary continuous function of one variable x whose graphical display is depicted below.

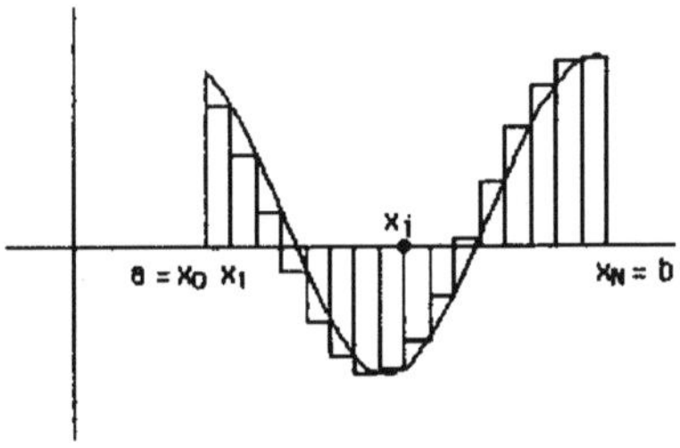

By breaking the interval $a \leq x \leq b$ into N pieces (where N is large), then we have defined

$$\int_a^b f(x)\,dx = \lim_{N\to\infty} \sum_{i=1}^{N} f(x_i)\,\frac{b-a}{N}.$$

The integral turns out to give the area between the curve and the x–axis where the area of regions below the x–axis are taken negatively. In fact, each term

$$f(x_i)\,\frac{b-a}{N} \qquad (i = 1, 2, \ldots, N)$$

in the above sum is the approximate area of the ith rectangular type region shaded below.

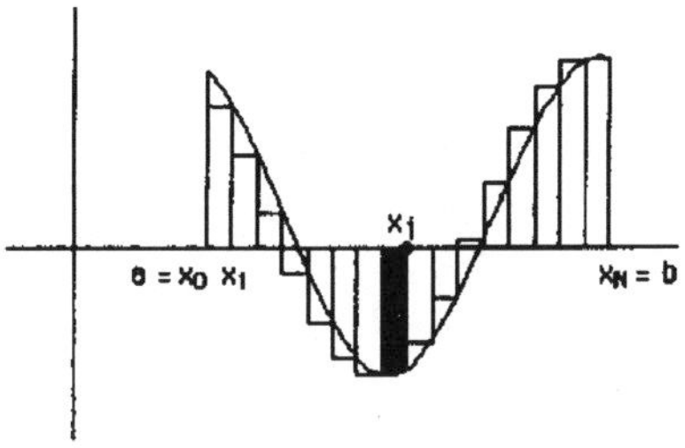

We now extend these notions to higher dimensional spaces and functions of several variables.

25.2 Double Integrals

Consider a continuous function $z = f(x, y)$ of two variables x, y with the following graphical display.

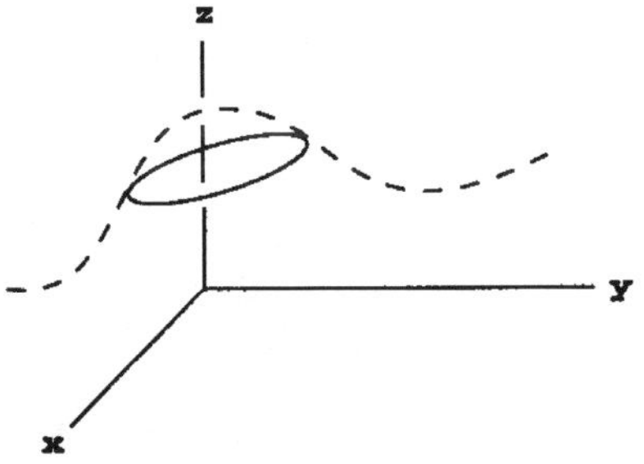

Let $\mathcal{R}$ be a two–dimensional region in the $x\,y$–plane as depicted below.

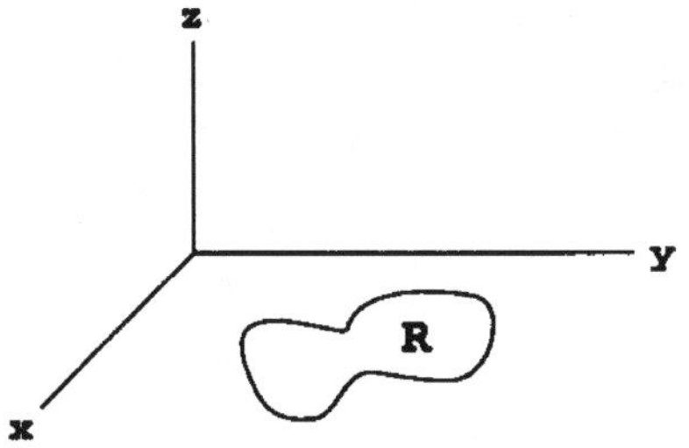

In order to analyze the function $f(x, y)$ over the region $\mathcal{R}$ we need to make some assumptions to insure that the region $\mathcal{R}$ is reasonable, i.e., is as *tame* as in the above figure. The first assumption we make is that $\mathcal{R}$ is **bounded**. This simply means that $\mathcal{R}$ lies in a finite rectangle (which will be referred to as a bounding rectangle).

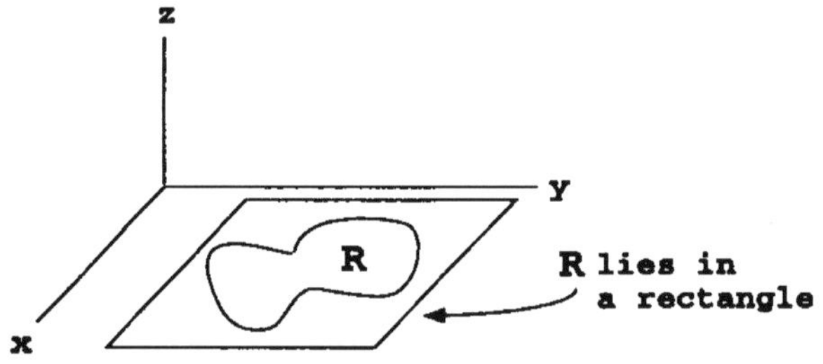

The second assumption we shall make is that the boundary of $\mathcal{R}$ is a continuous curve in the $x\,y$–plane which does not intersect itself.

Under the assumption that the graphical display of a function $z = f(x, y)$ lies above the $x\,y$–plane (i.e., $f(x, y) > 0$) we intuitively define

$$\iint_{\mathcal{R}} f(x,y)\,dx\,dy$$

as the volume of the solid above the region $\mathcal{R}$.

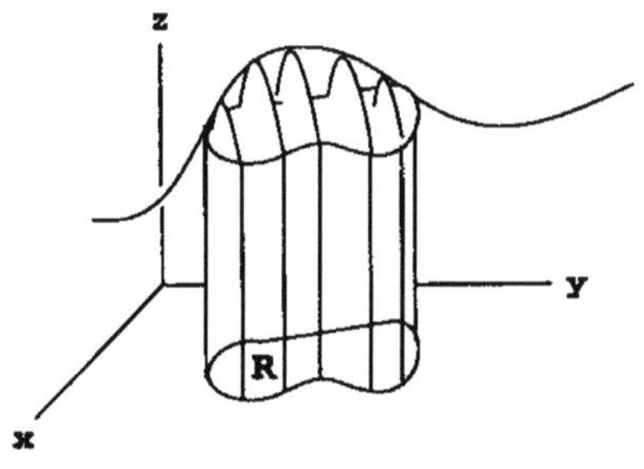

We now formally define the double integral as a limit of a sum in a manner analogous to the definition for a single integral. Assume that our bounding rectangle consists of all points (x, y) such that $a \le x \le b$, $c \le y \le d$.

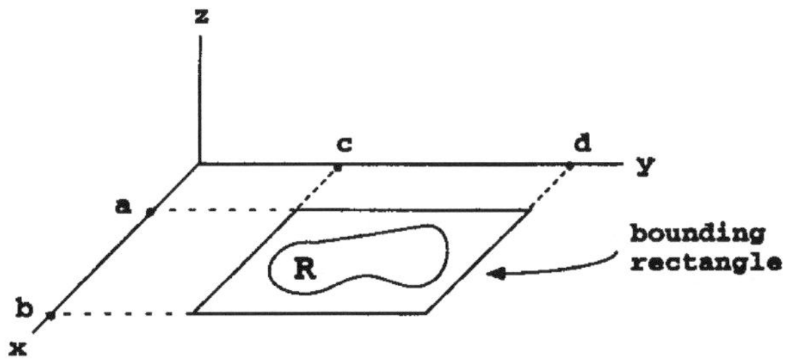

Superimpose a rectangular grid on our bounding rectangle by dividing each side of the rectangle into N equal pieces, i.e., $a < x_1 < x_2 < \cdots < x_N = b$, $c < y_1 < y_2 < \cdots < y_N = d$.

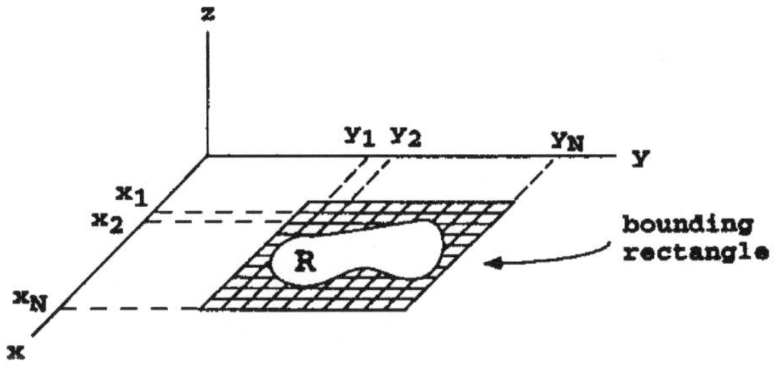

Notice that our bounding rectangle is thus divided into N^2 small rectangles each of which has area $\frac{b-a}{N} \cdot \frac{d-c}{N}$. In order to distinguish points in $\mathcal{R}$ from the entire bounding rectangle we introduce the function $\delta_{\mathcal{R}}(x, y)$ whose value is 1 or 0 depending on whether or not the point in question is in $\mathcal{R}$ or not, i.e., $\delta_{\mathcal{R}}(x, y)$ is defined by

$$\delta_{\mathcal{R}}(x, y) = \begin{cases} 1, & \text{if } (x, y) \in \mathcal{R} \\ 0, & \text{if } (x, y) \notin \mathcal{R}. \end{cases}$$

With this done, we formally define the **double integral** to be

$$\iint\limits_{\mathcal{R}} f(x, y)\, dx\, dy = \lim_{N \to \infty} \sum_{i=1}^{N} \sum_{j=1}^{N} \delta_{\mathcal{R}}(x_i, y_j) f(x_i, y_j) \cdot \frac{(b-a)(d-c)}{N^2}. \tag{25.1}$$

To see that this definition does in fact capture the volume of the solid above the region $\mathcal{R}$ (in the case $f(x, y) > 0$), notice that

$$f(x_i, y_j) \cdot \frac{(b-a)(d-c)}{N^2}$$

is approximately the volume of the ij^{th} column in the figure below.

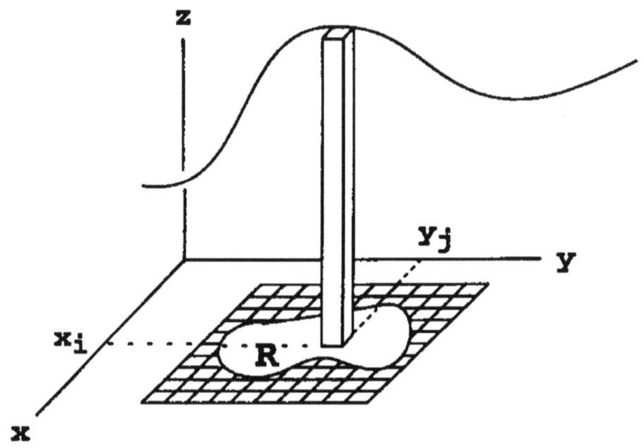

By summing these volumes over all (x_i, y_j) which lie in the region $\mathcal{R}$ and taking the limit, we obtain the volume of the solid above the region $\mathcal{R}$.

When we cease to restrict our attention to functions whose graphs lie above the $x\,y$–plane, the value of the integral behaves in a manner which is analogous to the single variable case: the integral is given by the sum of the volumes of solids taken positively or negatively according as the graphical display of $z = f(x, y)$ lies above or below the $x\,y$–plane. Our first example of a double integral demonstrates that in the case of a rectangular region the double integral amounts to iterating the single integral.

Example [25.1] (The region $\mathcal{R}$ is a rectangle) Suppose the region $\mathcal{R}$ is given by $\mathcal{R} = \{(x, y) \,|\, a \leq x \leq b, c \leq y \leq d\}$ and $z = f(x, y)$ is a continuous function. Compute the integral of $f(x, y)$ over $\mathcal{R}$.

If we fix a value of y with $c \leq y \leq d$ and let $A(y)$ denote the area of the slice depicted below

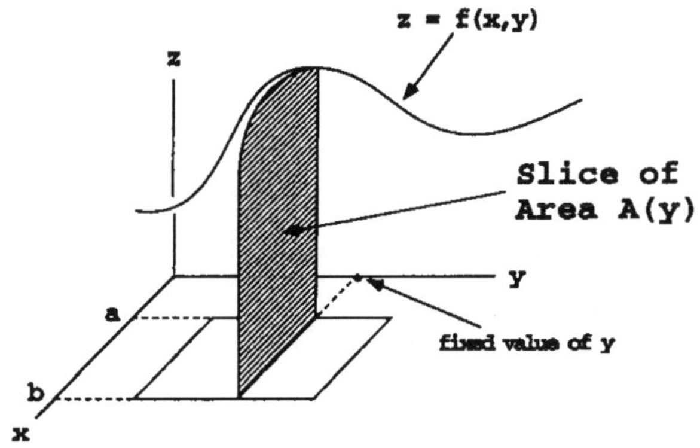

then

$$A(y) = \int_a^b f(x,y)\,dx.$$

The volume of the solid region between the surface $z = f(x,y)$ and the rectangle $\mathcal{R}$ is thus the limit of the sum (i.e., integral) of the areas $A(y)$ (see §16.3, Volume as Summation of Cross-sectional Areas) and thus:

$$\begin{aligned}\iint_{\mathcal{R}} f(x,y)\,dx\,dy &= \int_c^d A(y)\,dy \\ &= \int_c^d \left(\int_a^b f(x,y)\,dx\right) dy.\end{aligned} \tag{25.2}$$

Similarly, we can fix a value of x in the interval $a \le x \le b$, and let $A_1(x)$ denote the area of the region formed by intersecting the solid with the plane at the x (which is parallel to the $y\,z$–plane). As before

$$A_1(x) \;=\; \int_c^d f(x,y)\,dy, \tag{25.3}$$

and applying again the method from §16.3 we obtain

$$\begin{aligned}\iint_{\mathcal{R}} f(x,y)\,dx\,dy &= \int_a^b A_1(x)\,dx \\ &= \int_a^b \left(\int_c^d f(x,y)\,dy\right) dx.\end{aligned} \tag{25.4}$$

Combining equations (25.3) and (25.4) we obtain **Fubini's Theorem**:

$$\int_a^b \left(\int_c^d f(x,y)\, dy \right) dx = \int_c^d \left(\int_a^b f(x,y)\, dx \right) dy. \qquad (25.5)$$

It should be noted that there are occasions, when integrating over a rectangle, where switching the order of integration allows for a substantially easier computation.

Example [25.2] Let $f(x,y) = 10 - x^2 - 3xy$ and

$$\mathcal{R} = \{(x,y) \,|\, 0 \le x \le 1, 2 \le y \le 3\}.$$

Evaluate $\iint_{\mathcal{R}} f(x,y)\, dx\, dy$.

We have

$$\begin{aligned}
\iint_{\mathcal{R}} f(x,y)\, dx\, dy &= \int_2^3 \left(\int_0^1 (10 - x^2 - 3xy)\, dx \right) dy \\
&= \int_2^3 10x - \frac{x^3}{3} - \frac{3x^2y}{2} \Big]_0^1 dy \\
&= \int_2^3 \left(10 - \frac{1}{3} - \frac{3y}{2}\right) dy \\
&= \frac{29}{3}y - \frac{3y^2}{4} \Big]_2^3 = \frac{29}{3} - \frac{27}{4} + 3 = \frac{71}{12}.
\end{aligned}$$

Exercises for §25.2

Compute the following integrals over the indicated rectangles. Use Fubini's Theorem to change the order of integration when this seems convenient. Interpret the integrals as volumes by graphing the volumes in question with the CAS.

(1) $\int_1^3 \int_0^1 (x^2y - y^2 + y)\, dx\, dy$.

(2) $\int_{\frac{1}{2}}^1 \int_1^3 (1 + \frac{y}{x})\, dx\, dy$.

(3) $\int_0^\pi \int_0^1 y \sin(xy)\, dx\, dy$.

(4) $\int_0^\pi \int_0^1 (e^x \cos(y))\, dx\, dy$.

(5) Find the volume of the solid which lies under the paraboloid $z = 16 - x^2 - y^2$ and over the rectangle $\mathcal{R} = \{0 \le x \le 2, 0 \le y \le 1\}$. Graph the solid with your CAS.

(6) Verify that if f_1 and f_2 are continuous functions of the region $\mathcal{R}$, then

$$\iint_{\mathcal{R}} (f_1 + f_2)\, dx\, dy = \iint_{\mathcal{R}} f_1\, dx\, dy + \iint_{\mathcal{R}} f_2\, dx\, dy.$$

(7) Verify that if f is a continuous function on the region $\mathcal{R}$ and c is a constant then

$$\iint_{\mathcal{R}} c \cdot f \, dx \, dy = c \cdot \iint_{\mathcal{R}} f \, dx \, dy.$$

(8) Suppose that the regions $\mathcal{R}_1$ and $\mathcal{R}_2$ do not overlap and that f is a continuous function on their union $\mathcal{R} = \mathcal{R}_1 \cup \mathcal{R}_2$. Verify that

$$\iint_{\mathcal{R}} f \, dx \, dy = \iint_{\mathcal{R}_1} f \, dx \, dy + \iint_{\mathcal{R}_2} f \, dx \, dy.$$

(9) Use **(8)** to compute the volume of the solid which lies below $f(x, y) = 1$ and above the unit squares centered at $(\frac{3}{2}, 0, 0)$ and $(-\frac{1}{2}, 0, 0)$. Graph the solid with the CAS.

(10) Let $h(x, y) = f(x) \cdot g(y)$ where $f(x)$ and $g(y)$ are continuous functions on the intervals $a \le x \le b$ and $c \le y \le d$ respectively. Let $\mathcal{R}$ be the rectangle $\{(x, y) \,|\, a \le x \le b, \; c \le y \le d\}$. Prove that

$$\iint_{\mathcal{R}} h(x, y) \, dx \, dy = \left(\int_a^b f(x) \, dx \right) \cdot \left(\int_c^d g(y) \, dy \right).$$

Use **(10)** above to compute the following integrals.

(11) $\int_0^{\frac{\pi}{4}} \int_0^{\frac{\pi}{4}} (\cos(x)) \cdot (\sin(y)) \, dx \, dy.$

(12) $\int_0^1 \int_1^e (y \cdot (\ln(x)) \, dx \, dy.$

(13) Approximate the double integral

$$\int_0^2 \int_0^1 (x - 2y) \, dx \, dy$$

by graphing $f(x, y) = x - 2y$ over the rectangle

$$\mathcal{R} = \{(x, y) \,|\, 0 \le x \le 1, \; 0 \le y \le 2\},$$

breaking $\mathcal{R}$ into 18 smaller squares, and using the CAS to compute the appropriate sum. Compare your answer to the actual value of the integral.

(14) Repeat (13) above with the function $f(x, y) = x^2 - y^2$.

25.3 Evaluation of Double Integrals

Let $z = f(x, y)$ be a continuous function. Consider the region $\mathcal{R}$ in the xy–plane (which is shaded in the figure below), whose boundary consists of known continuous functions $g_1(x), g_2(x)$ and the sides of a rectangle

$$\left\{ (x, y) \;\middle|\; a \le x \le b, \quad c \le y \le d \right\}.$$

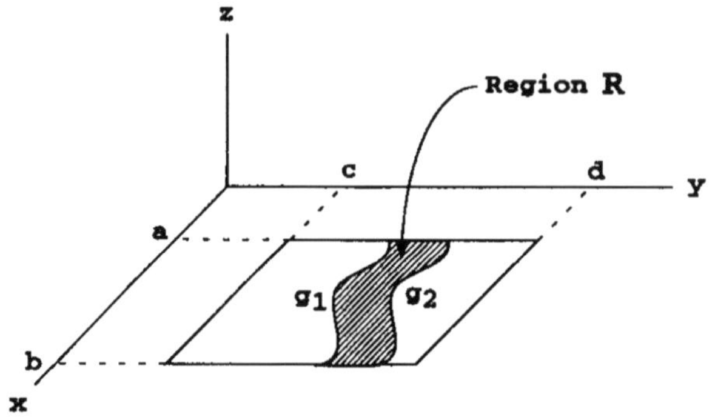

The double integral over such a region,

$$\iint\limits_{\mathcal{R}} f(x,y)\,dx\,dy$$

occurs frequently and the following proposition gives a method for its computation.

Proposition [25.3] *Let $g_1(x)$, $g_2(x)$ be continuous functions defined on an interval $a \le x \le b$. Let*

$$\mathcal{R} = \Big\{(x,y)\,\Big|\, a \le x \le b,\ g_1(x) \le y \le g_2(x)\Big\}.$$

If $f(x,y)$ is continuous on $\mathcal{R}$ then

$$\iint\limits_{\mathcal{R}} f(x,y)\,dx\,dy \;=\; \int_a^b \left(\int_{g_1(x)}^{g_2(x)} f(x,y)\,dy\right)dx. \qquad (25.6)$$

Proof: For a fixed value of x in the interval $a \le x \le b$, let $A(x)$ denote the area of the slice depicted below.

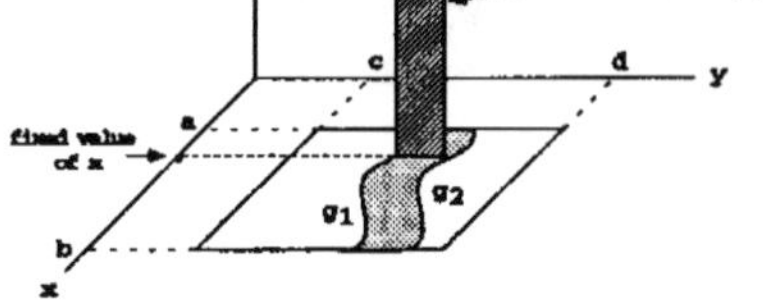

By definition $A(x)$ is given by the integral

$$A(x) = \int_{g_1(x)}^{g_2(x)} f(x,y)\, dy.$$

The volume of the solid region between $\mathcal{R}$ and the surface $z = f(x,y)$ is thus the integral of $A(x)$ and we have demonstrated the proposition:

$$\iint_{\mathcal{R}} f(x,y)\, dx\, dy = \int_a^b A(x)\, dx$$

$$= \int_a^b \left(\int_{g_1(x)}^{g_2(x)} f(x,y)\, dy \right) dx.$$

Proposition [25.3] indicates that many double integrals can be converted into repeated single integrals. In light of our previously developed technique (for example Chapter XVII), we have in Proposition [25.3] a powerful method for evaluating double integrals. We illustrate this with several examples.

Example [25.4] (Triangular Region) Consider the region $\mathcal{R}$ bounded by the x–axis, the line $y = 3x$ (where $0 \le x \le 1$), and the line $x = 1$. Evaluate

$$\iint_{\mathcal{R}} (x^2 + y^2)\, dx\, dy.$$

We first sketch the region $\mathcal{R}$.

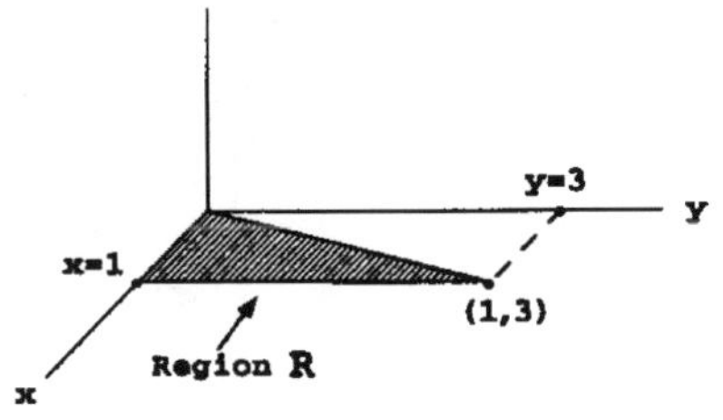

In this case $g_1(x) = 0$ and $g_2(x) = 3x$, and we may compute the integral as follows:

$$\begin{aligned}
\iint_{\mathcal{R}} (x^2+y^2)\,dx\,dy &= \int_0^1 \left(\int_0^{3x} (x^2+y^2)\,dy \right) dx \\
&= \int_0^1 \left(x^2 y + \frac{y^3}{3} \Big]_{y=0}^{y=3x} \right) dx \\
&= \int_0^1 \left(3x^3 + \frac{27x^3}{3} \right) dx \\
&= \frac{3x^4}{4} + \frac{9x^4}{4} \Big]_0^1 \\
&= 3.
\end{aligned}$$

Example [25.5] Let $\mathcal{R}$ be the region bounded by the curves $y = x^2+1$, $y = x^3+2$, and the lines $x = 0$, $x = 1$. Evaluate

$$\iint_{\mathcal{R}} xy\,dx\,dy.$$

In this case the region is sketched below,

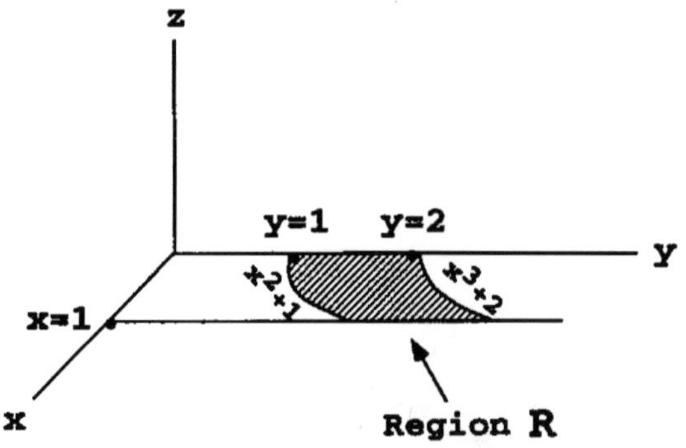

and the integral is computed as follows:

$$\begin{aligned}
\iint\limits_{\mathcal{R}} xy\,dx\,dy &= \int_0^1 \left(\int_{x^2+1}^{x^3+2} xy\;dy \right) dx \\
&= \int_0^1 \frac{xy^2}{2} \bigg]_{y=x^2+1}^{y=x^3+2} dx \\
&= \int_0^1 \frac{x\Big((x^3+2)^2 - (x^2+1)^2 \Big)}{2}\,dx \\
&= \int_0^1 \frac{x(x^6+4x^3+4) - (x^4+2x^2+1)}{2}\,dx \\
&= \frac{x^8}{16} + \frac{2x^5}{5} + x^2 - \frac{x^5}{10} - \frac{x^3}{3} - \frac{x}{2} \bigg]_0^1 \\
&= \frac{1}{16} + \frac{2}{5} + 1 - \frac{1}{10} - \frac{1}{3} - \frac{1}{2}.
\end{aligned}$$

Remark: There are occasions when the most convenient description for a region $\mathcal{R}$ is of the form

$$\mathcal{R} = \Big\{ (x,y) \,\Big|\, h_1(y) \le x \le h_2(y),\ c \le y \le d \Big\},$$

i.e., $\mathcal{R}$ is bounded by the curves $x = h_1(y)$ and $x = h_2(y)$ and the lines $y = c,\ y = d$. In this case we have an analogue to Proposition [25.3]: the integral of $f(x,y)$ over $\mathcal{R}$ is given by

$$\iint\limits_{\mathcal{R}} f(x,y)\,dx\,dy = \int_c^d \left(\int_{h_1(y)}^{h_2(y)} f(x,y)\,dx \right) dy. \tag{25.7}$$

Example [25.6] Evaluate $\iint_{\mathcal{R}} (x+y)\,dx\,dy$ where $\mathcal{R}$ is the region bounded by the curves $x = y^2,\ x = y^2+1$, and the lines $y = 0,\ y = 1$.

The region $\mathcal{R}$ is sketched below:

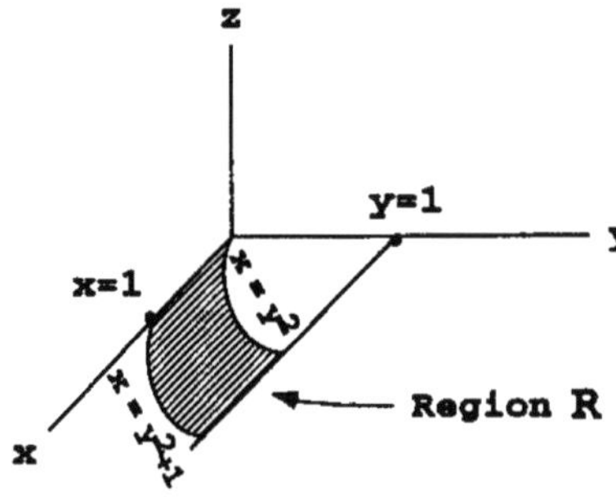

The integral is computed as follows:

$$\begin{aligned}
\iint_{\mathcal{R}} (x+y)\,dx\,dy &= \int_0^1 \left(\int_{y^2}^{y^2+1} (x+y)\,dx \right) dy \\
&= \int_0^1 \frac{x^2}{2} + xy \Big]_{x=y^2}^{x=y^2+1} dy \\
&= \int_0^1 \left(\frac{(y^2+1)^2}{2} + y^3 + y - \frac{y^4}{2} - y^3 \right) dy \\
&= \int_0^1 \left(y^2 + \frac{1}{2} + y \right) dy \\
&= \frac{1}{3} + \frac{1}{2} + \frac{1}{2} \\
&= \frac{4}{3}.
\end{aligned}$$

Example [25.7] (Reversing the order of integration) Reverse the order of integration in the integral

$$\int_0^1 \left(\int_y^{y^2} (x+y)\,dx \right) dy.$$

The region we are integrating over in this example consists of the set of points (x, y) such that

$$0 \le y \le 1, \qquad y \le x \le y^2.$$

This may be rewritten as

$$0 \le y \le x \le y^2 \le 1,$$

or

$$0 \leq x \leq 1, \qquad \sqrt{x} \leq y \leq x.$$

Thus

$$\int_0^1 \left(\int_y^{y^2} (x+y)\, dx \right) dy = \int_0^1 \left(\int_{\sqrt{x}}^{x} (x+y)\, dy \right) dx.$$

Exercises for §25.3

In the following exercises evaluate the double integral $\iint\limits_{\mathcal{R}} f(x,y)\ dx\ dy$ over the given region $\mathcal{R}$. Use your CAS to obtain a sketch of $\mathcal{R}$ as well as the piece of the surface $z = f(x,y)$ over $\mathcal{R}$.

(1) $f(x,y) = xy$, where $\mathcal{R}$ is the triangular region bounded by the lines $x = 0$, $y = 3x$, and $y = 6$.

(2) $f(x,y) = x^2 + y^2$, where $\mathcal{R}$ is the triangular region bounded by the x–axis, and the lines $x = 2$, and $y = x + 1$.

(3) $f(x,y) = x + y$, where $\mathcal{R}$ is the region bounded by the curves $y = x^3$, $y = x^2 + 2$, and the lines $x = 0$ and $x = 1$.

(4) $f(x,y) = x^2 y - x$, where $\mathcal{R}$ is the region bounded by the curves $y = x^4$, $y = x^3 + 3$, and the lines $x = 0$ and $x = 1$.

(5) $f(x,y) = 2x^3 y + xy$, where $\mathcal{R}$ is the region bounded by the curves $x = y^2$, $x = y^2 + 2$, and the lines $y = 0$ and $y = 1$.

(6) $f(x,y) = 2x - y$, where $\mathcal{R}$ is the region bounded by the curves $x = y^3$, $x = y^4 + 3$, and the lines $y = 1$ and $y = 2$.

(7) $f(x,y) = 4y$, where $\mathcal{R}$ is the region bounded by the x–axis and the curve $y = \sin(x)$ for $0 \leq x \leq \pi$.

(8) $f(x,y) = x + y$, where $\mathcal{R}$ is the region bounded by the curves $y = x^2$, $y = x^3$, for $0 \leq x \leq 1$.

(9) $f(x,y) = x + y$, where $\mathcal{R}$ is the region bounded by the curves $x = y^2$, $x = y^3$, for $0 \leq y \leq 1$.

(10) $f(x,y) = 3x$, where $\mathcal{R}$ is the disk $x^2 + y^2 \leq 9$.

(11) $f(x,y) = 3y$, where $\mathcal{R}$ is the disk $x^2 + y^2 \leq 9$.

(12) $f(x,y) = x + y$, where $\mathcal{R}$ is the disk $x^2 + y^2 \leq 9$.

Reverse the order of integration in the remaining exercises. Check your answer by computing both double integrals with your CAS.

(13) $\int_0^1 \left(\int_0^{y^4} x\, dx \right) dy.$

(14) $\int_1^3 \left(\int_y^{y^2} (x^2 - y)\, dx \right) dy.$

(15) $\int_0^2 \left(\int_x^{2x^3} y\, dy \right) dx.$

(16) $\int_0^1 \left(\int_{2\sqrt{x}}^{x} dy \right) dx.$

25.4 Double Integrals in Polar Coordinates

It is clear that, regardless of our technical prowess in integration, double integrals can be quite challenging. The difficulty lies in the fact that the region we are integrating over may be given by a complex description (i.e., may be bounded by curves which are given by intricate functions). It turns out that our computations can be greatly simplified if we look at our region from a different geometric perspective, for example polar coordinates.

Recall that any point (x, y) in the $x\,y$–plane

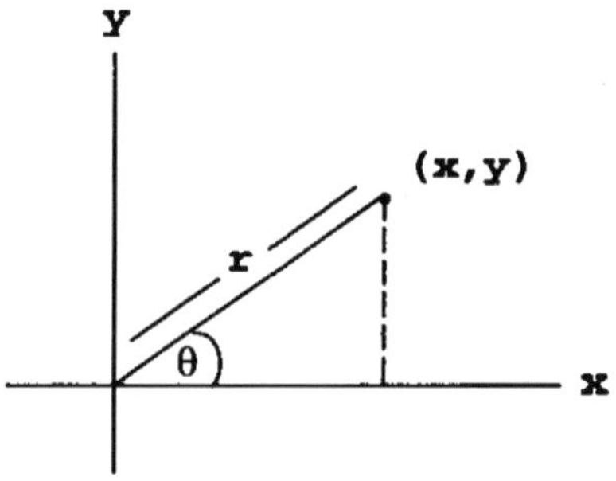

can be converted to polar coordinates (r, θ) by setting

$$x = r\cos(\theta), \qquad y = r\sin(\theta).$$

Further we may transform a function of two variables x, y,

$$z = f(x, y),$$

into a function of the polar coordinates (r, θ) by setting

$$z = f(r\cos(\theta), r\sin(\theta)).$$

Example [25.8] Convert the function $z = x^2 + 2x + y^2$ into a function of the two polar variables (r, θ).

Substituting $x = r\cos(\theta)$, $y = r\sin(\theta)$ into our function $z = x^2 + 2x + y^2$ yields

$$\begin{aligned} z &= r^2(\cos(\theta))^2 + 2r\cos(\theta) + r^2(\sin(\theta))^2 \\ &= r^2((\cos(\theta))^2 + (\sin(\theta))^2) + 2r\cos(\theta) \\ &= r^2 + 2r\cos(\theta). \end{aligned}$$

Thus $z = r^2 + 2r\cos(\theta)$ is the required polar function.

We now return to the general double integral

$$\iint_{\mathcal{R}} f(x,y)\,dx\,dy, \tag{25.8}$$

where $f(x,y)$ is a continuous function on a bounded region $\mathcal{R}$. If the region $\mathcal{R}$ is circular (or is visibly a polar region) it may be substantially easier to compute (25.3) by transforming the entire integral to an integral in polar coordinates. We now detail how this is done.

To begin this exposition, the concept of a **standard polar region** is needed. Fix real numbers $a < b$ and $\alpha < \beta$. A **standard polar region** is defined to be the set

$$\Big\{(x,y)\ \Big|\ x = r\cos(\theta),\ y = r\sin(\theta),\ a \le r \le b,\ \alpha \le \theta \le \beta\Big\}.$$

In polar notation the standard polar region takes on the familiar form

$$\Big\{(r,\theta)\ \Big|\ a \le r \le b,\ \alpha \le \theta \le \beta\Big\},$$

and is depicted below.

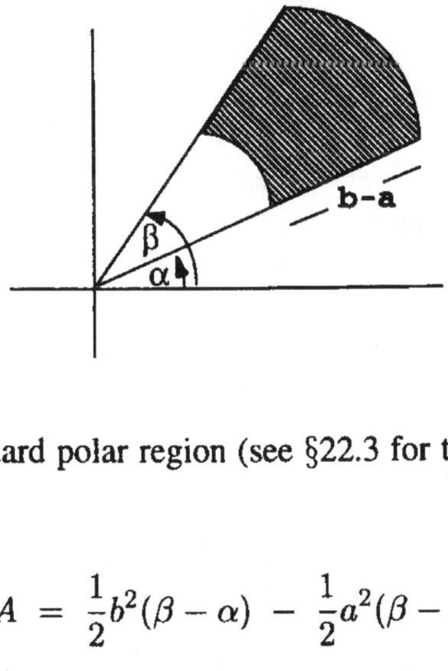

The area A of our standard polar region (see §22.3 for the derivation) is given by the formula

$$\begin{aligned} A &= \frac{1}{2}b^2(\beta-\alpha) - \frac{1}{2}a^2(\beta-\alpha) \\ &= \frac{1}{2}(b^2-a^2)(\beta-\alpha) \\ &= \frac{1}{2}(a+b)(b-a)(\beta-\alpha). \end{aligned} \tag{25.9}$$

Returning now to the given region $\mathcal{R}$ in the $x\,y$–plane,

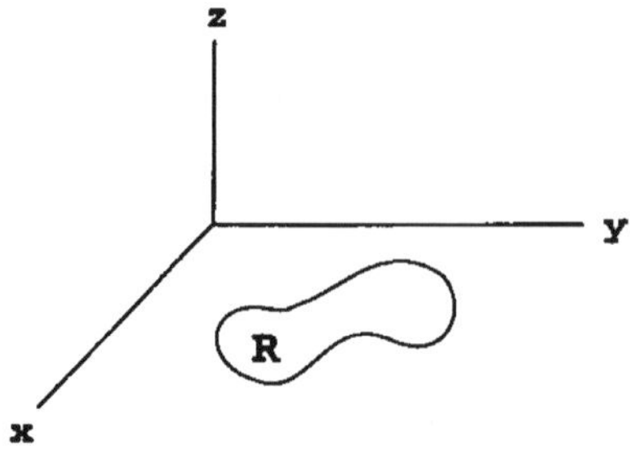

we propose to replace the previously considered rectangular grid (on the bounding rectangle) with a **polar grid** on a standard polar region ($ABCD$ in the figure below) which bounds our region.

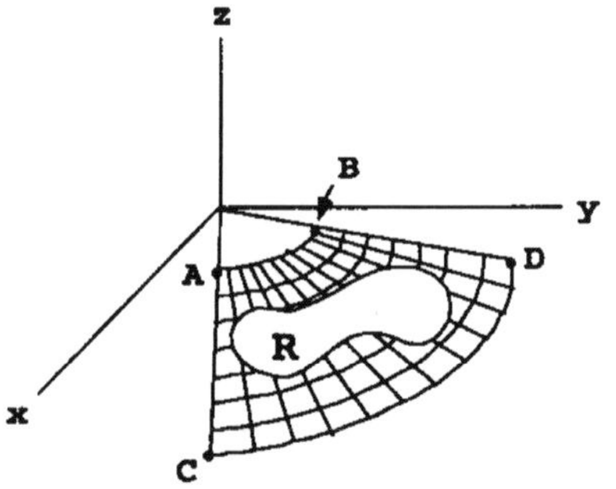

The polar grid is constructed by dividing each side of the bounding standard polar region into N equal pieces and yields a grid of N^2 smaller standard polar regions (where N will of course be a large positive integer).

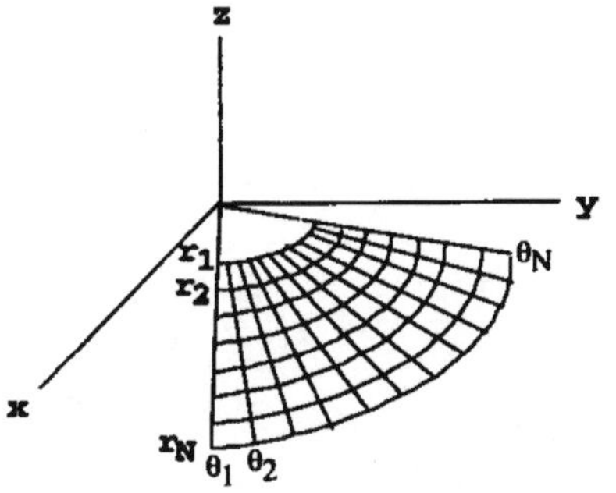

In order to make use of our grid, we focus on the small standard polar region at (r_i, θ_i) which is shaded below,

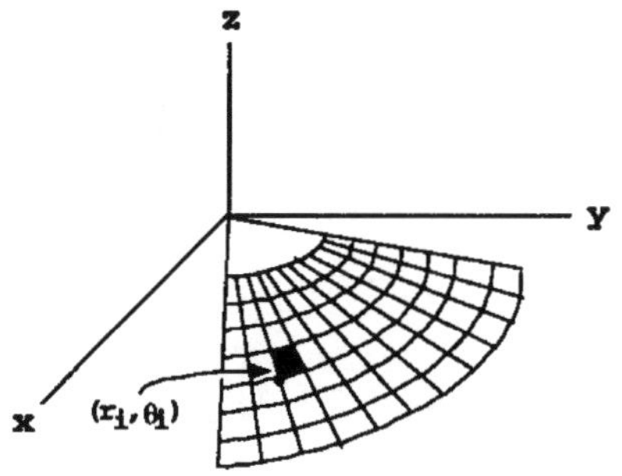

and try to compute its area. By putting this standard polar region under a microscope and labeling its endpoints we obtain,

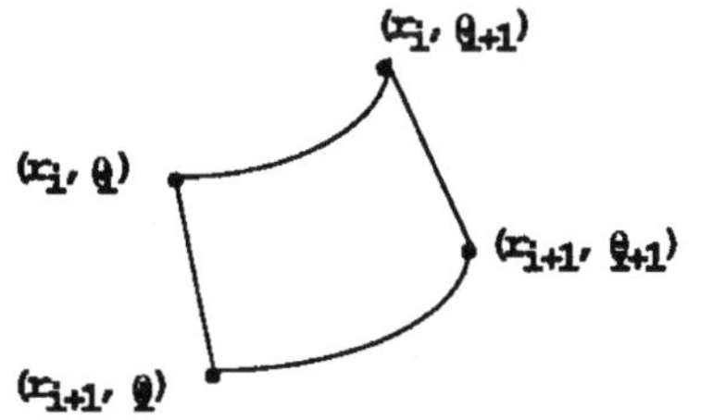

where $r_{i+1} - r_i = \frac{b-a}{N}$ and $\theta_{i+1} - \theta_i = \frac{\beta-\alpha}{N}$. Thus by (25.9) the area of the small standard polar region is

$$\frac{1}{2}(r_{i+1} + r_i)(r_{i+1} - r_i)(\theta_{i+1} - \theta_i) = \frac{1}{2}(r_{i+1} + r_i)\frac{(b-a)(\beta-\alpha)}{N^2}.$$

We are now in a position to compute our original integral in polar coordinates. Define the function

$$\delta_{\mathcal{R}}(r,\theta) = \begin{cases} 1, & \text{if } (r,\theta) \in \mathcal{R} \\ 0, & \text{if } (r,\theta) \notin \mathcal{R}, \end{cases}$$

which distinguished points in the region from those outside. The double integral (25.8) may now be expressed as the limit of the sum

$$\iint_{\mathcal{R}} f(x,y)\,dx\,dy$$

$$= \lim_{N\to\infty} \sum_{i=1}^{N}\sum_{j=1}^{N} \delta_{\mathcal{R}}(r,\theta)\cdot f\Big(r_i\cos(\theta_j), r_i\sin(\theta_j)\Big)\frac{1}{2}(r_i+r_{i+1})\frac{(b-a)(\beta-\alpha)}{N^2}. \tag{25.10}$$

Now recalling that $r_{i+1} - r_i = \frac{b-a}{N}$, we have

$$\frac{1}{2}\Big(r_i + r_{i+1}\Big) = r_i + \frac{1}{2}\frac{b-a}{N}.$$

For very large N, we see that

$$\frac{1}{2}(r_i + r_{i+1}) = \left(r_i + \frac{1}{2}\frac{b-a}{N}\right)$$

$$\approx r_i.$$

Substituting this into (25.10) yields

$$\iint_{\mathcal{R}} f(x,y)\,dx\,dy$$

$$= \lim_{N\to\infty} \sum_{i=1}^{N}\sum_{j=1}^{N} \delta_{\mathcal{R}}(r,\theta)\cdot f\Big(r_i\cos(\theta_j), r_i\sin(\theta_j)\Big)\, r_i \frac{(b-a)(\beta-\alpha)}{N^2},$$

and we claim that the expression on the right is simply the polar integral, i.e.,

$$\iint_{\mathcal{R}} f\Big(r\cos(\theta), r\sin(\theta)\Big) r\,dr\,d\theta$$

$$= \lim_{N\to\infty} \sum_{i=1}^{N}\sum_{j=1}^{N} \delta_{\mathcal{R}}(r,\theta)\cdot f\Big(r_i\cos(\theta_j), r_i\sin(\theta_j)\Big)\, r_i \frac{(b-a)(\beta-\alpha)}{N^2}. \tag{25.11}$$

To prove this identity recall that we have already shown in §25.2 that for any function g which is continuous on $\mathcal{R}$,

$$\iint_{\mathcal{R}} g(x,y)\,dx\,dy = \lim_{N\to\infty} \sum_{i=1}^{N}\sum_{j=1}^{N} \delta_{\mathcal{R}}(x_i,y_j) g(x_i,y_j)\cdot \frac{(b-a)(d-c)}{N^2}. \tag{25.12}$$

Furthermore, we are free to change the names of our variables and replace (x, y) by (r, θ) and (c, d) by (α, β). Hence we have

$$\iint_{\mathcal{R}} g(r,\theta)\,dr\,d\theta = \lim_{N\to\infty} \sum_{i=1}^{N}\sum_{j=1}^{N} \delta_{\mathcal{R}}(r_i,\theta_j) g(r_i,\theta_j)\cdot \frac{(b-a)(\beta-\alpha)}{N^2}.$$

Now choosing

$$g(r,\theta) = f(r\cos(\theta), r\sin(\theta))\cdot r,$$

we have a derivation of (25.11).

Our discussion has established the basic formula

$$\iint_{\mathcal{R}} f(x,y)\,dx\,dy \;=\; \iint_{\mathcal{R}} f\Big(r\cos(\theta), r\sin(\theta)\Big)\cdot r\,dr\,d\theta. \tag{25.13}$$

25.5 Evaluation of Double Integrals in Polar Coordinates

Consider a bounded region $\mathcal{R}$ in the $r\,\theta$–plane whose boundary consists of known functions $g_1(\theta)$, $g_2(\theta)$ and the lines $\theta = \alpha$, $\theta = \beta$.

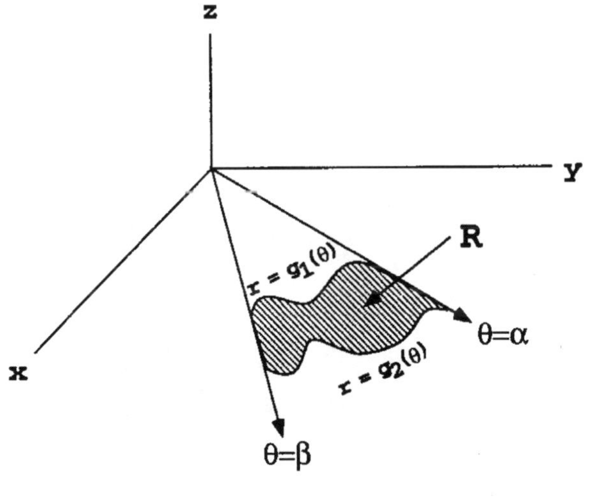

By definition the region $\mathcal{R}$ consists of the points (r, θ) which satisfy the conditions

$$\alpha \le \theta \le \beta, \qquad\qquad g_1(\theta) \le r \le g_2(\theta).$$

In this case (in analogy with Proposition [25.3]) we have the identity

$$\iint_{\mathcal{R}} f(x,y)\,dx\,dy \;=\; \int_{\alpha}^{\beta}\left(\int_{g_1(\theta)}^{g_2(\theta)} f\Big(r\cos(\theta), r\sin(\theta)\Big)\cdot r\,dr\right)d\theta. \tag{25.14}$$

Example [25.9] (The Disk at the Origin) Let $\mathcal{D}$ be the disk of radius a centered at the origin of the $x\,y$–plane. Evaluate the integral

$$\iint_{\mathcal{D}} (x^2 + y^2)\, dx\, dy.$$

The region $\mathcal{D}$ is a perfect candidate for integration with polar coordinates. We begin by describing $\mathcal{D}$ as a polar region: every point $(r, \theta) \in C$ satisfies $0 \le r \le a$ and $0 \le \theta \le 2\pi$. Thus

$$\begin{aligned}
\iint_{\mathcal{D}} (x^2 + y^2)\, dx\, dy &= \int_0^{2\pi} \left(\int_0^a \left((r\cos(\theta))^2 + (r\sin(\theta))^2 \right) \cdot r\, dr \right) d\theta \\
&= \int_0^{2\pi} \left(\int_0^a r^3\, dr \right) d\theta \\
&= \frac{2\pi a^4}{4}.
\end{aligned}$$

Example [25.10] (A Translated Disk) Let $\mathcal{D}$ be the disk of radius a centered at $x = a$, $y = 0$.

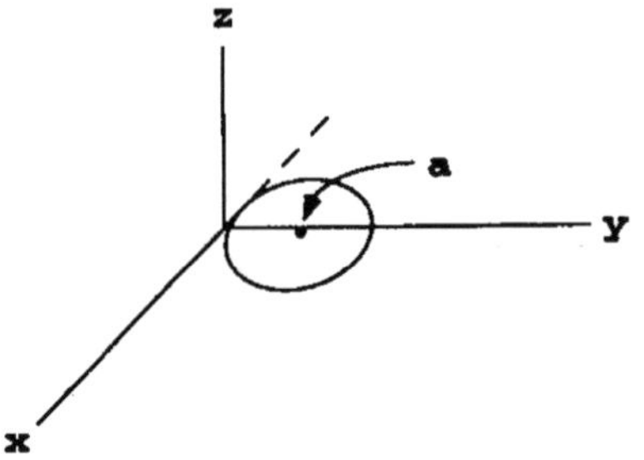

Evaluate the integral

$$\iint_{\mathcal{D}} x\, dx\, dy.$$

Again we must begin by expressing $\mathcal{D}$ using polar coordinates. We accomplish this by converting the equation for $\mathcal{D}$ in rectangular coordinates

$$(x - a)^2 + y^2 = a^2,$$

via the transformation $x = r\cos(\theta)$, $y = r\sin(\theta)$ to yield

$$(r\cos(\theta) - a)^2 + (r\sin(\theta))^2 = a^2.$$

Expanding this equation gives

$$r^2(\cos(\theta))^2 - 2ar\cos(\theta) + a^2 + r^2(\sin(\theta))^2 = a^2,$$

and hence

$$r = 2a\cos(\theta)$$

where $-\frac{\pi}{2} \leq \theta \leq \frac{\pi}{2}$. The integral in question may now be computed as follows:

$$\begin{aligned}
\iint\limits_{\mathcal{D}} x\,dx\,dy &= \int_{-\frac{\pi}{2}}^{\frac{\pi}{2}} \left(\int_0^{2a\cos(\theta)} r\cos(\theta)\cdot r\,dr \right) d\theta \\
&= \int_{-\frac{\pi}{2}}^{\frac{\pi}{2}} \cos(\theta) \left(\int_0^{2a\cos(\theta)} r^2\,dr \right) d\theta \\
&= \int_{-\frac{\pi}{2}}^{\frac{\pi}{2}} \cos(\theta)\,\frac{(2a\cos(\theta))^3}{3}\,d\theta \\
&= \frac{8a^3}{3} \int_{-\frac{\pi}{2}}^{\frac{\pi}{2}} (\cos(\theta))^4\,d\theta \\
&= \frac{8a^3}{3}\cdot\frac{3\pi}{8} = \pi a^3.
\end{aligned}$$

Exercises for §25.5

In the following exercises evaluate the double integral $\iint\limits_{\mathcal{R}} f(x,y)\,dx\,dy$ over the region $\mathcal{R}$ by converting to polar coordinates. Use your CAS to obtain a sketch of $\mathcal{R}$ as well as the piece of the surface $z = f(x,y)$ over $\mathcal{R}$.

(1) $f(x,y) = x^2 + y^2$, where $\mathcal{R}$ is the portion of the disk $x^2 + y^2 \leq 4$ in the second quadrant.

(2) $f(x,y) = y$, where $\mathcal{R}$ is the disk of radius 2 centered at $(2,0)$.

(3) $f(x,y) = xy$, where $\mathcal{R}$ is the disk $x^2 + y^2 \leq 9$.

(4) $f(x,y) = x$, where $\mathcal{R}$ is the region bounded by the lines $y = 0$, $y = \sqrt{3} \cdot x$ and the circle of radius 1.

(5) $f(x,y) = \sqrt{x^2 + y^2}$, where $\mathcal{R}$ is the region bounded by the cardioid $r = 1 + \sin(\theta)$, $0 \leq \theta \leq 2\pi$.

(6) $f(x,y) = y$, where $\mathcal{R}$ is the region bounded by the limacon $r = 3 + 2\cos(\theta)$, $0 \leq \theta \leq 2\pi$.

25.6 Computing Areas and Volumes with Double Integrals

In §25.2 we demonstrated that if $z = f(x,y)$ is a continuous and nonnegative function on a bounded region $\mathcal{R}$ in the xy–plane,

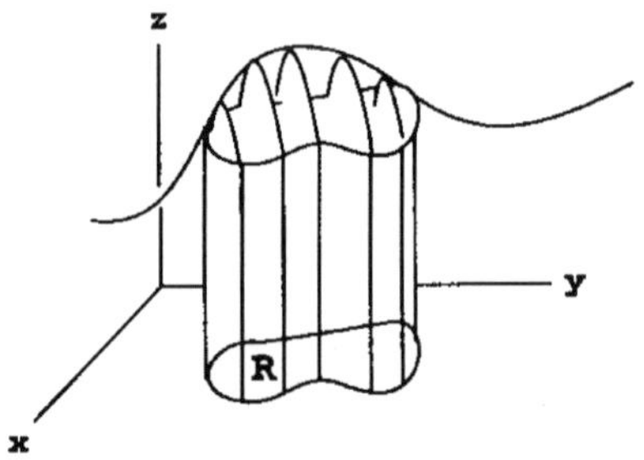

then the volume V of the solid that lies below the surface $z = f(x,y)$ and above $\mathcal{R}$ is given by the double integral

$$V = \iint_{\mathcal{R}} f(x,y)\, dx\, dy \tag{25.15}$$

provided the integral exists. The area of the region $\mathcal{R}$ can itself be recovered from the double integral if we set $f(x,y) = 1$ in (25.15):

$$A = \iint_{\mathcal{R}} dx\, dy. \tag{25.16}$$

We illustrate these formulae with several examples (which will demonstrate that double integrals allow for easier computation of volumes and areas than single integrals did).

Example [25.11] Find the volume V of the solid which lies between the surface $z = xy + e^x$ and the rectangle $\{(x,y) \mid 0 \le x \le 1, 0 \le y \le 2\}$ in the $x\,y$–plane.

Equation (25.15) dictates that

$$\begin{aligned} V &= \int_0^2 \left(\int_0^1 (xy + e^x)\, dx \right) dy \\ &= \int_0^2 \left(\frac{y}{2} + e - 1 \right) dy \\ &= 1 + 2e - 2 \\ &= 2e - 1. \end{aligned}$$

Example [25.12] Find the volume V of a sphere of radius R.

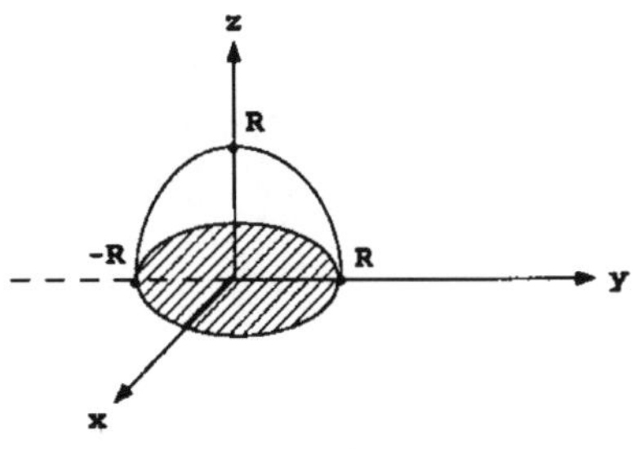

In this example the geometry clearly lends itself to polar coordinates. The shaded circle of radius R (in the $x\,y$–plane) consists of all points (r, θ) with $0 \le r \le R$, $0 \le \theta \le 2\pi$. Furthermore, we claim that the surface of the hemisphere is given by the polar function

$$z = \sqrt{R^2 - r^2}. \tag{25.17}$$

To verify this we note that in rectangular coordinates any point (x, y, z) on the hemisphere satisfies the equation $x^2 + y^2 + z^2 = R^2$ which implies

$$z = \sqrt{R^2 - x^2 - y^2}.$$

Converting to polar coordinates (by letting $x = r\cos(\theta)$ and $y = r\sin(\theta)$) yields (25.17).

The volume in question thus takes the form of a double integral in polar coordinates

$$V = 2 \cdot \int_0^{2\pi} \left(\int_0^R \sqrt{R^2 - r^2} \cdot r \, dr \right) d\theta.$$

To evaluate this inner integral we use the substitution

$$u = R^2 - r^2, \qquad du = -2r \, dr,$$

which allows us to readily compute the desired volume:

$$\begin{aligned} V &= 2 \cdot \int_0^{2\pi} \left(\int_{R^2}^0 \sqrt{u} \cdot -\frac{du}{2} \right) d\theta \\ &= 2 \cdot \int_0^{2\pi} \left(\int_0^{R^2} \sqrt{u} \cdot \frac{du}{2} \right) d\theta \\ &= 2 \cdot \frac{2\pi R^3}{3} = \frac{4\pi R^3}{3}. \end{aligned}$$

Example [25.13] Compute via double integration the area A of the region (in the $x\,y$–plane) bounded by the curves $y = x^2$ and $y = 3x$.

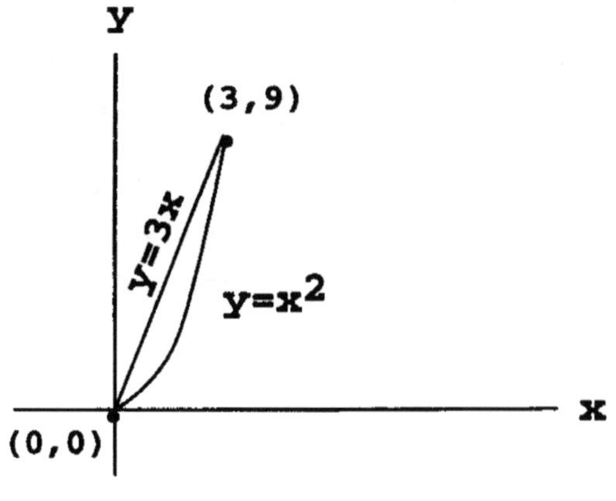

Following (25.16), the area of this region is captured by the double integral

$$A = \int_0^3 \left(\int_{x^2}^{3x} 1\, dy \right) dx$$

$$= \int_0^3 (3x - x^2)\, dx$$

$$= \frac{27}{2} - \frac{27}{3} = \frac{9}{2}.$$

Exercises for §25.6

Using double integration, find the area of the following regions $\mathcal{R}$ in the $x\,y$–plane. Graph the boundary of each $\mathcal{R}$ with your CAS.

(1) $\mathcal{R}$ is the region in the first quadrant bounded by the curves $y = x^2$ and $y = x^4$.

(2) $\mathcal{R}$ is the region in the first quadrant bounded by the curves $x = y^2$ and $x = y^4$.

(3) $\mathcal{R}$ is the region bounded by the curve $y = x^2$ and the line $y = 3x - 2$ for $-2 \leq x \leq -1$.

(4) $\mathcal{R}$ is the intersection of the two disks $x^2 + y^2 \leq 4$ and $x^2 + (y - 2)^2 \leq 4$.

(5) $\mathcal{R}$ is the region bounded by the line $y = x$ and the curve $x + y^2 = 2$.

(6) $\mathcal{R}$ is the region in the first quadrant bounded by the line $y = 1 - x$ and the curve $y = \frac{2}{x}$.

(7) $\mathcal{R}$ is a circle of radius r.

Find the volume of the following solids Ω.

(8) Ω is the solid region below the plane $2x + y - z = 2$ and above the rectangle $1 \leq x \leq 2, 0 \leq y \leq 1$.

(9) Ω is the solid region below the plane $x - 2y + 3z = 4$ and above the rectangle $-1 \leq x \leq 1, 0 \leq y \leq 2$.

(10) Ω is the solid region which lies between the surface $z = x^2 y + e^x$ and the rectangle $0 \leq x \leq 1, 0 \leq y \leq 1$.

(11) Ω is the solid region which lies between the surface $z = \sin(\pi(x + y))$ and the rectangle $0 \leq x \leq \frac{1}{2}, 0 \leq y \leq \frac{1}{2}$.

(12) Ω is the solid region below the ellipsoid $z = x^2 + 2y^2$ and above the triangle whose vertices are located at the points $(0,0,0)$, $(1,0,0)$, and $(0,1,0)$.

(13) Ω is the solid region which lies below the surface $z = x^2 y$ and lies above the region (in the $x\,y$-plane) which is bounded by the line $y = x$ and the curve $y = x^2$.

25.7 Changing Variables in Double Integrals

Let $z = f(x, y)$ be a continuous function defined on a bounded region $\mathcal{R}$. In §25.4 we demonstrated that by changing to polar coordinates (via $x = r\cos(\theta)$ and $y = r\sin(\theta)$ the double integral transforms to

$$\iint_{\mathcal{R}} f(x, y)\, dx\, dy = \iint_{\mathcal{R}} f(r\cos(\theta), r\sin(\theta)) \cdot r\, dr\, d\theta,$$

i.e., $dx\, dy$ is transformed into $r\ dr\, d\theta$. Based on this success we are motivated to ask the following more general question:

Question: If we consider an arbitrary change of variables

$$x = g_1(u, v), \qquad y = g_2(u, v)$$

with new variables u, v how do we transform a double integral

$$\iint_{\mathcal{R}} f(x, y)\, dx\, dy$$

into an integral involving the variables u and v?

Since the function $f(x, y)$ clearly transforms to $f(g_1(u, v), g_2(u, v))$ the question at hand amounts to, what happens to $dx\, dy$?

To approach this question we first review the one–variable situation. Consider a transformation

$$x = g(u)$$

where u is a new variable and g is a differentiable one–to–one function. Geometrically we have the following picture.

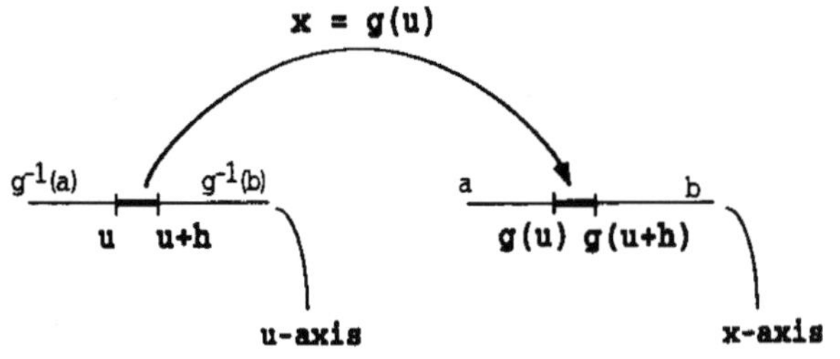

As $h \to 0$ the small interval $[u, u+h]$ on the u–axis is transformed into the interval $[g(u), g(u+h)]$ which is itself a small interval on the x–axis since g is by assumption differentiable (and hence continuous). Thus

$$\begin{aligned} dx &\approx g(u+h) - g(u) \\ &\approx g'(u)\, du, \end{aligned}$$

and an arbitrary integral

$$\int_a^b f(x)\, dx$$

becomes

$$\int_{g^{-1}(a)}^{g^{-1}(b)} f\Big(g(u)\Big) \cdot g'(u)\, du.$$

In the two variable situation when we are given a transformation

$$x = g_1(u,v), \qquad y = g_2(u,v)$$

where g_1, g_2 are differentiable one–to–one functions and u, v are new variables, define $g^{-1}(\mathcal{R})$ to be the set of all points (u,v) such that $(g_1(u,v), g_2(u,v)) \in \mathcal{R}$. We shall assume that each $(x,y) \in \mathcal{R}$ is the image of a unique point (u,v) in the uv–plane. Geometrically the effect of the transformation on a small rectangle is that of being pulled in four directions.

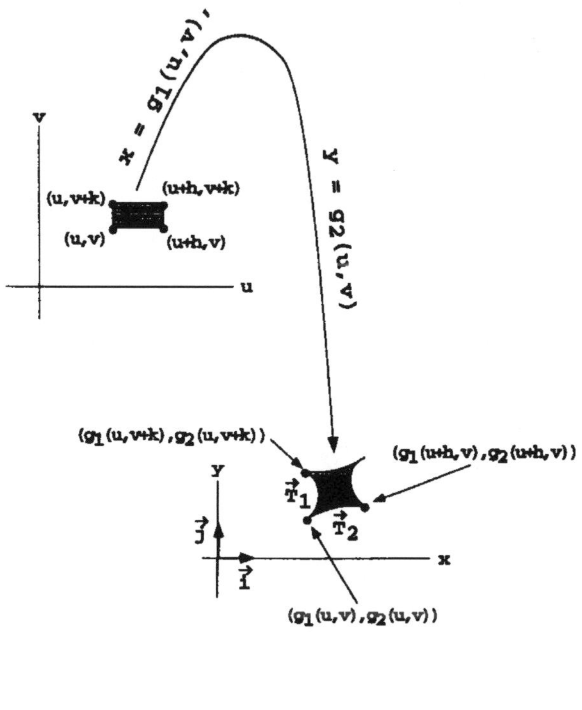

As $h, k \to 0$ a small rectangular increment of area in the $u\,v$–plane is transformed into a small increment of area in the $x\,y$–plane. Let $\vec{\mathbf{T}}_1$ and $\vec{\mathbf{T}}_2$ be vectors in the $x\,y$–plane which approximate the sides of increment of area in the $x\,y$–plane. By definition

$$\vec{\mathbf{T}}_1 = \Big(g_1(u, v+k) - g_1(u,v)\Big)\vec{\mathbf{i}} + \Big(g_2(u, v+k) - g_2(u,v)\Big)\vec{\mathbf{j}}$$

$$\vec{\mathbf{T}}_2 = \Big(g_1(u+h, v) - g_1(u,v)\Big)\vec{\mathbf{i}} + \Big(g_2(u+h, v) - g_2(u,v)\Big)\vec{\mathbf{j}},$$

and therefore

$$\vec{\mathbf{T}}_1 \approx \left(\frac{\partial g_1}{\partial v}\vec{\mathbf{i}} + \frac{\partial g_2}{\partial v}\vec{\mathbf{j}}\right) dv = \left(\frac{\partial x}{\partial v}\vec{\mathbf{i}} + \frac{\partial y}{\partial v}\vec{\mathbf{j}}\right) dv$$

$$\vec{\mathbf{T}}_2 \approx \left(\frac{\partial g_1}{\partial u}\vec{\mathbf{i}} + \frac{\partial g_2}{\partial u}\vec{\mathbf{j}}\right) du = \left(\frac{\partial x}{\partial u}\vec{\mathbf{i}} + \frac{\partial y}{\partial u}\vec{\mathbf{j}}\right) du.$$

But then the area of the region in the $u\,v$–plane is approximately given by the cross product of the two vectors $\vec{\mathbf{T}}_1$ and $\vec{\mathbf{T}}_2$ i.e.,

$$dx\,dy \approx |\vec{\mathbf{T}}_1 \times \vec{\mathbf{T}}_2| = \left|\frac{\partial x}{\partial u}\frac{\partial y}{\partial v} - \frac{\partial x}{\partial v}\frac{\partial y}{\partial u}\right| du\,dv.$$

The above computations motivate the following important definition.

Definition: *Let* $x = g_1(u,v)$, $y = g_2(u,v)$ *be differentiable functions which define a change of variables. The* ***Jacobian*** *of this transformation is defined to be*

$$\frac{\partial(x,y)}{\partial(u,v)} = \frac{\partial x}{\partial u}\frac{\partial y}{\partial v} - \frac{\partial x}{\partial v}\frac{\partial y}{\partial u}. \tag{25.18}$$

A careful elaboration of the previous discussion yields the following proposition.

Proposition [25.14] *Let $f(x, y)$ be a continuous function on a bounded region $\mathcal{R}$ in the $x\,y$–plane. Let*

$$x = g_1(u, v), \qquad y = g_2(u, v)$$

be a change of coordinates where the functions g_1 and g_2 are assumed to be one–to–one and differentiable. Then assuming each $(x, y) \in \mathcal{R}$ is the image of a unique point (u, v) (in the $u\,v$–plane) the following identity holds:

$$\iint\limits_{\mathcal{R}} f(x, y)\, dx\, dy = \iint\limits_{g^{-1}(\mathcal{R})} f\Big(g_1(u, v), g_2(u, v)\Big) \left|\frac{\partial(x, y)}{\partial(u, v)}\right| du\, dv \tag{25.19}$$

provided $\left|\frac{\partial(x,y)}{\partial(u,v)}\right| \neq 0$ for all $(u, v) \in g^{-1}(\mathcal{R})$.

Example [25.15] Compute the Jacobian of the polar coordinate transformation given by $x = r\cos(\theta)$, $y = r\sin(\theta)$.

By definition we have

$$\begin{aligned} \frac{\partial(x, y)}{\partial(r, \theta)} &= \frac{\partial x}{\partial r}\frac{\partial y}{\partial \theta} - \frac{\partial x}{\partial \theta}\frac{\partial y}{\partial r} \\ &= (\cos(\theta)) \cdot (r\cos(\theta)) - (-r\sin(\theta)) \cdot (\sin(\theta)) \\ &= r\Big((\cos(\theta))^2 + (\sin(\theta))^2\Big) \\ &= r. \end{aligned}$$

Example [25.16] Compute the Jacobian of the transformation $x = uv$ and $y = \frac{u}{v}$.

In this case we have

$$\begin{aligned} \frac{\partial(x, y)}{\partial(u, v)} &= \frac{\partial x}{\partial u}\frac{\partial y}{\partial v} - \frac{\partial x}{\partial v}\frac{\partial y}{\partial u} \\ &= v \cdot \left(\frac{-u}{v^2}\right) - \left(u \cdot \frac{1}{v}\right) \\ &= \frac{-2u}{v}, \end{aligned}$$

Example [25.17] Given numbers $a, b > 0$ show that

$$\iint\limits_{\frac{x^2}{a^2}+\frac{y^2}{b^2} \leq 1} dx\,dy \;=\; \pi ab.$$

To compute this integral we use a classical trick. The idea is to eliminate the denominators in the definition of the region $\mathcal{R}$. Let $x = au$ and $y = bv$. Then

$$\frac{\partial(x,y)}{\partial(u,v)} \;=\; \frac{\partial x}{\partial u}\frac{\partial y}{\partial v} - \frac{\partial x}{\partial v}\frac{\partial y}{\partial u} \;=\; ab,$$

and our integral becomes

$$\iint\limits_{u^2+v^2 \leq 1} ab\,du\,dv \;=\; \pi ab,$$

since the unit circle has area π.

Example [25.18] Transform the double integral

$$\iint\limits_{\substack{xy+\frac{x}{y} \leq 1 \\ x,y>0}} y^{-1}e^{x}\,dx\,dy$$

into a $u\,v$–integral using the transformation $x = uv$, $y = \frac{u}{v}$.

Just as in Example [25.167], the change of variables suggested here is motivated by the desire to convert our original region into a region we can handle more easily (in this case, a circle). The Jacobian for this transformation was computed in Example [25.16],

$$\frac{\partial(x,y)}{\partial(u,v)} \;=\; \frac{-2u}{v},$$

and thus

$$\iint\limits_{\substack{xy+\frac{x}{y} \leq 1 \\ x,y>0}} y^{-1}e^{x}\,dx\,dy \;=\; \iint\limits_{\substack{u^2+v^2 \leq 1 \\ uv>0}} \frac{v}{u}e^{uv}\left|\frac{2u}{v}\right|du\,dv$$

$$=\; 2\iint\limits_{\substack{u^2+v^2 \leq 1 \\ uv>0}} e^{uv}\,du\,dv.$$

Exercises for §25.7

Evaluate the Jacobian of the following transformations using your CAS.

(1) $x = r\sin(\theta), y = r\cos(\theta)$.

(2) $x = 3u + 2v, y = u - v$.

(3) $x = ve^u, y = ue^v$.

(4) $x = \cos(u + v), y = \sin(u - v)$.

Evaluate the following double integrals with the suggested change of coordinates.

(5)

$$\iint_{\substack{0 \le x - y \le 1 \\ 0 \le 2x+y \le 2}} 2x\, dx\, dy,$$

Set $u = x - y, v = 2x + y$, i.e., $x = \frac{u+v}{3}$, and $y = \frac{v-2u}{3}$.

(6)

$$\iint_{\substack{0 \le x + 2y \le 2 \\ -1 \le x-3y \le 1}} (x + y)\, dx\, dy,$$

Set $u = x + 2y, v = x - 3y$, i.e., $x = \frac{3u+2v}{5}$, and $y = \frac{u-v}{5}$.

(7)

$$\iint_{\substack{1 \le 3x - 2y \le 2 \\ 2 \le x+3y \le 3}} xy\, dx\, dy,$$

Set $u = 3x - 2y, v = x + 3y$.

(8)

$$\iint_{\substack{0 \le x - 2y \le 2 \\ 0 \le 2x+y \le 1}} e^{2x+y}\, dx\, dy.$$

(9)

$$\iint_{\substack{1 \le x(1-y) \le 3 \\ 1 \le xy \le 2}} 3x\, dx\, dy,$$

Set $x = u + v, y = \frac{v}{u+v}$.

(10)

$$\iint_{\frac{x^2}{4} + \frac{y^2}{9} \le 1} y^2\, dx\, dy,$$

Set $x = 2r\cos(\theta), y = 3r\sin(\theta)$.

Additional exercises for Chapter XXV

Evaluate the double integral of the function $f(x,y)$ over the given region, $\mathcal{R}$.

(1) $f(x,y) = 3x^2y - 4yx^2,\ \mathcal{R} = \{(x,y) \mid -1 \le x \le 1, -2 \le y \le 2\}$.

(2) $f(x,y) = \frac{xy}{\sqrt{x^2+y^2+2}},\ \mathcal{R} = \{(x,y) \mid 0 \le x \le 2, 0 \le y \le 3\}$.

(3) $f(x,y) = \ln(xy) + x^2$, $\mathcal{R}$ is the region bounded by the curves $x = 1$, $y = 0$ and $y = x^2$.

(4) $f(x,y) = x^2 - xy^3$, where $\mathcal{R}$ is the region bounded by $y = x$ and $y = 3x - x^2$.

(5) $f(x,y) = x^2y^3$, where $\mathcal{R}$ is the region bounded by the curves $xy = 1$, $y = x$ and $y = 2x$, $x \ge 0$.

Reverse the order of integration in exercises 6,7,8. Check your answer by computing both double integrals with your CAS.

(6) $\int_{-2}^{2}\left(\int_{-\sqrt{4-x^2}}^{\sqrt{4-x^2}}(4-y)\,dy\right)dx.$

(7) $\int_{1/4}^{1}\left(\int_{x^2}^{x}\sqrt{\frac{x}{y}}\,dy\right)dx.$

(8) $\int_0^{\pi/2}\left(\int_0^{y^2}\exp\left(\frac{x}{y^2}\right)\,dx\right)dy.$

Use your CAS to graph the following regions R. Then compute the double integrals of $f(x,y)$ over the region R.

(9) $R = \{r = 6\sin(3\theta)\}$ (three-petaled rose). $f(x,y) = 1$.

(10) R is the three-petaled rose and $f(x,y) = \frac{1}{1+x^2+y^2}$.

(11) $R = \{r = 7\sin(5\theta)\}$ (five-petaled rose). $f(x,y) = x^2y$.

(12) $R = \{3r = 2\cos(4\theta)\}$. $f(x,y) = \frac{x^2}{1+y^2}$.

Find the volume of the following solids Ω. Obtain the 3-dimensional plot with your CAS.

(13) $f(x,y) = 1$ for $x^2 + y^2 \le 16$.

(14) $f(x,y) = x^4 + y^4$ for $o \le x \le 1,\ 0 \le y \le 1$.

Evaluate the Jacobian of the following transformations using your CAS.

(15) $x = ue^v$, $y = u^2 - v^2$.

(16) $x = \frac{1}{2}(u-v)\ y = \frac{1}{2}(u+v)$.

Evaluate the following double integrals

(17)

$$\iint_{\substack{1 \le x(2-y) \le 4,\\ 1 \le 2xy \le 3}} 4y \, dx \, dy.$$

(18)

$$\iint_{\substack{0 \le x-3y \le 4,\\ 0 \le 2x+1 \le 4}} \exp(3x + 2y) \, dx \, dy.$$

Chapter XXVI
Triple Integrals

26.1 Triple Integrals and the Fourth Dimension

Let $w = f(x, y, z)$ be a continuous function of three variables x, y, z. Let Ω be a 3–dimensional solid region.

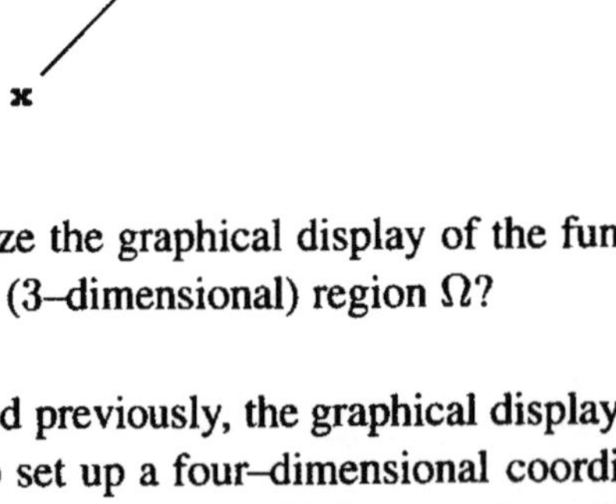

Question: Can we realize the graphical display of the function $w = f(x, y, z)$ as (x, y, z) ranges over the (3–dimensional) region Ω?

As we have discussed previously, the graphical display referred to in this question would require us to set up a four–dimensional coordinate system (x, y, z, w) with four mutually perpendicular axes. Unfortunately, at this point it is impossible to even simulate the drawing of the four axes and our discussion will have to be based on a combination of vivid imagination and (careful) formal generalization.

Return for a moment to the one and two dimensional situations depicted below.

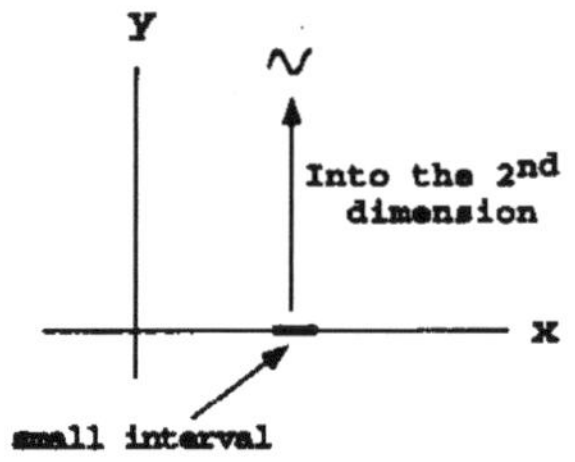

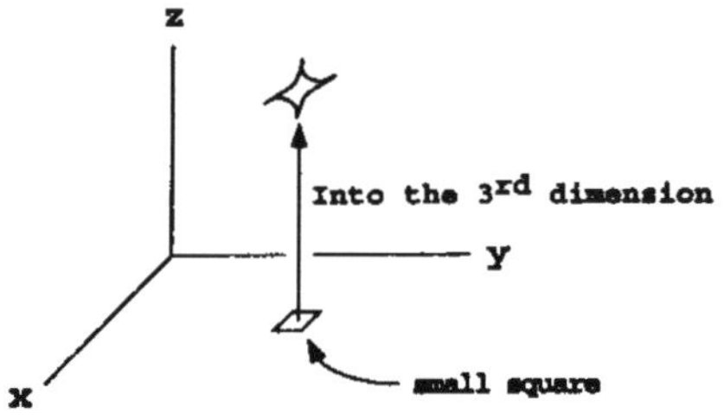

In the graph of a continuous function, $y = f(x)$, defined on the x–axis, a small interval maps to a small arc in two–dimensional space. Similarly, in the graph of a function, $z = f(x, y)$, on the $x\,y$–plane, a small square maps to a small surface in 3–dimensional space.

Now consider a continuous function, $w = f(x, y, z)$, on three–dimensional space. In this situation a small cube will be mapped to a small three–dimensional solid region. This solid region will be projected a distance $w = f(x, y, z)$ into the fourth dimension.

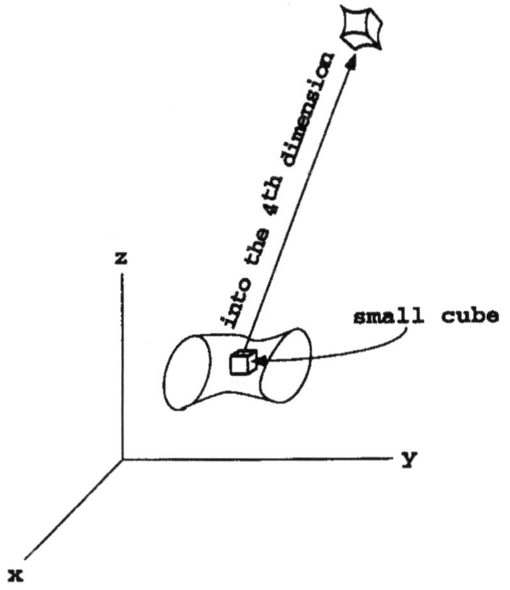

As we traverse through our solid region Ω, the graphical display of $w = f(x, y, z)$ will be a three dimensional object Ω' which twists and turns in four–dimensional space.

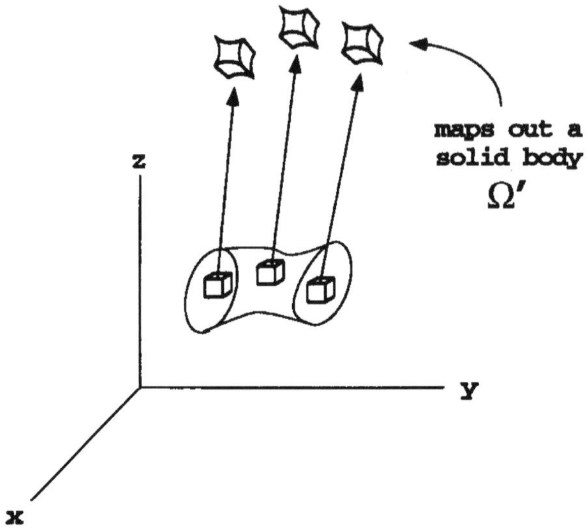

Assuming that the graphical display of $w = f(x,y,z)$ lies *above* the $x\,y\,z$ –plane, i.e., $w = f(x,y,z) > 0$ (alternately Ω' is projected into the positive w direction), we may intuitively define

$$\iiint\limits_{\Omega} f(x,y,z)\,dx\,dy\,dz \tag{26.1}$$

as the volume bounded by Ω' and $x\,y\,z$–space.

In the special case where $f(x,y,z) = 1$ then our triple integral simply yields the volume $V(\Omega)$ of the body Ω,

$$\iiint\limits_{\Omega} dx\,dy\,dz = V(\Omega). \tag{26.2}$$

The precise definition of the integral of $w = f(x,y,z)$ is the formal generalization of the definition of the double integral given in §25.2. We begin by assuming that Ω is **bounded**, i.e., there exists a rectangular box B such that Ω lies in B.

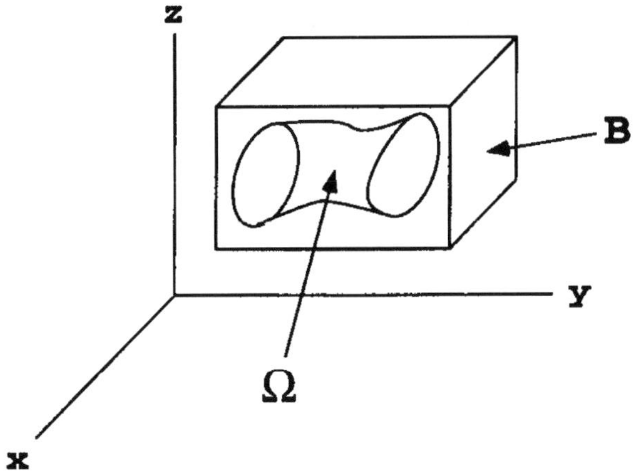

By definition B takes the form

$$B = \left\{ (x, y, z) \;\middle|\; a \le x \le b,\ c \le y \le d, e \le z \le f \right\},$$

and in order to distinguish points in Ω from those outside Ω we define (just as we did in the previous chapter)

$$\delta_\Omega(x, y, z) = \begin{cases} 1, & \text{if } (x, y, z) \in \Omega \\ 0, & \text{if } (x, y, z) \notin \Omega. \end{cases}$$

If we assume in addition that the surface of Ω does not intersect itself then we may define

$$\iiint_\Omega f(x, y, z)\, dx\, dy\, dz$$

$$= \lim_{N\to\infty} \sum_{i=1}^{N} \sum_{j=1}^{N} \sum_{k=1}^{N} \delta_\Omega(x_i, y_j, z_k) f(x_i, y_j, z_k) \cdot \frac{(b-a)(d-c)(f-e)}{N^3}, \tag{26.3}$$

where $a < x_1 < x_2 < \cdots < x_N = b$, $c < y_1 < y_2 < \cdots < y_N = d$, and $e < z_1 < z_2 < \cdots < z_N = f$ are equally spaced points.

When we focus attention on the special case $\Omega = B$ then $\delta_\Omega(x, y, z) = 1$ whenever $(x, y, z) \in B$ and our somewhat fierce looking triple integral reduces to iterated single integrals, i.e.,

$$\iiint_B f(x, y, z)\, dx\, dy\, dz = \int_e^f \left(\int_c^d \left(\int_a^b f(x, y, z)\, dx \right) dy \right) dz. \tag{26.4}$$

Of course, the above iterated integral can be evaluate in any order.

Example [26.1] Let $B = \{(x,y,z)\,|\,0 \le x \le 1, 1 \le y \le 2, -1 \le z \le 0\}$. Evaluate

$$\iiint_B (x^2yz + ye^{x+z})\,dx\,dy\,dz.$$

The iterated integral gives us

$$\begin{aligned}
\iiint_B (x^2yz + ye^{x+z})\,dx\,dy\,dz &= \int_{-1}^{0}\left(\int_1^2\left(\int_0^1 (x^2yz + ye^{x+z})\,dx\right)dy\right)dz \\
&= \int_{-1}^{0}\left(\int_1^2\left(\frac{1}{3}yz + ye^z(e-1)\right)dy\right)dz \\
&= \int_{-1}^{0}\frac{3}{2}\left(\frac{z}{3} + e^z(e-1)\right)\,dz \\
&= -\frac{1}{4} + (e-1)(1-e^{-1}).
\end{aligned}$$

Exercises for §26.1

Evaluate the following integrals. Check your computations with your CAS.

(1) $\int_0^1\int_0^1\int_0^1 xy^2z\,dx\,dy\,dz.$

(2) $\int_0^1\int_{-1}^0\int_{-1}^2 (x^2+y^2+z^2)\,dx\,dy\,dz.$

(3) $\int_{-1}^1\int_2^3\int_0^1 (x^2y + xz^3)\,dx\,dy\,dz.$

(4) $\int_0^1\int_1^3\int_{-1}^1 e^x(y+z)\,dx\,dy\,dz.$

(5) $\int_3^4\int_2^3\int_0^1 \frac{y^2+z}{\sqrt{x}}\,dx\,dy\,dz.$

(6) $\int_0^2\int_0^1\int_{-2}^1 yz\cos(\pi xy/2)\,dx\,dy\,dz.$

(7) Suppose $0 \le f(x,y,z) \le M$ for all (x,y,z) in the solid region Ω. Show that

$$\iiint_\Omega f(x,y,z)\,dx\,dy\,dz \le M\cdot \text{Volume}(\Omega).$$

(8) Suppose $0 \le f(x,y,z) \le g(x,y,z)$ for all (x,y,z) in the solid region Ω. Show that

$$\iiint_{\Omega} f(x,y,z)\,dx\,dy\,dz \;\le\; \iiint_{\Omega} g(x,y,z)\,dx\,dy\,dz.$$

(9) Let $p(x,y,z) = f(x)\cdot g(y)\cdot h(z)$. Suppose that $f(x)$, $g(y)$, and $h(z)$ are continuous on the intervals $a \le x \le b, c \le y \le d, e \le z \le f$, respectively. Letting Ω denote the box

$$\Omega = \{(x,y,z) \mid a \le x \le b, c \le y \le d, e \le z \le f\}$$

show that

$$\iiint_{\Omega} p(x,y,z)\,dx\,dy\,dz = \left(\int_a^b f(x)\,dx\right)\left(\int_c^d g(y)\,dy\right)\left(\int_e^f h(z)\,dz\right).$$

26.2 Evaluation of Triple Integrals

In this section we shall develop a technique for integrating over solids Ω which are bounded by two surfaces $z = s_1(x,y)$ and $z = s_2(x,y)$,

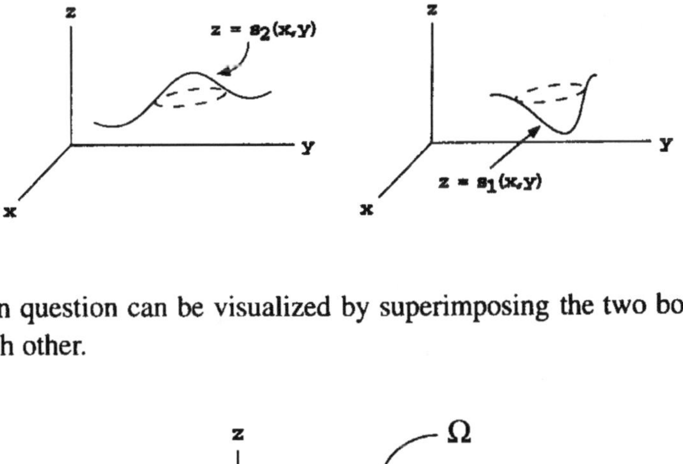

The solids in question can be visualized by superimposing the two bounding surfaces on each other.

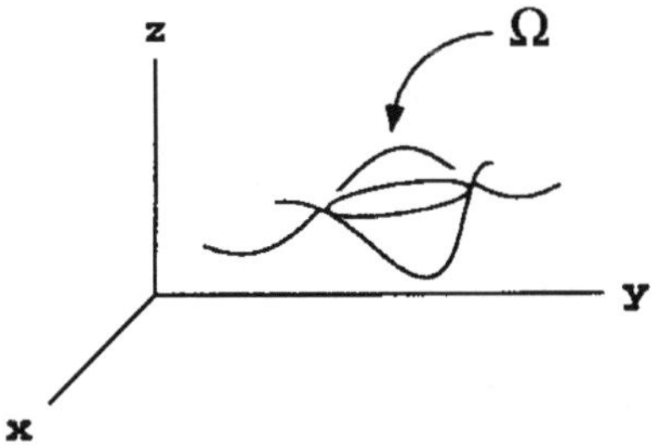

Imagine for a moment that a bright light is shining parallel to the z–axis high above and directly on Ω. A shadow of Ω will immediately appear on the xy–plane. The

shadow is itself a region $\mathcal{R}$ in the $x\,y$–plane: it is the perpendicular projection of Ω onto the $x\,y$–plane.

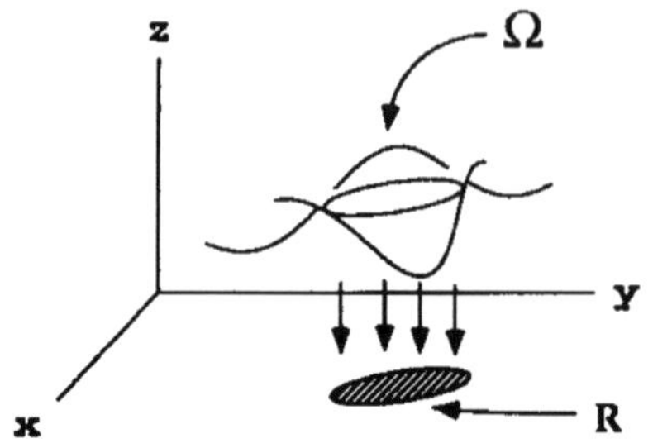

Every point $(x, y, z) \in \Omega$ satisfies the properties

$$(x, y) \in \mathcal{R},$$

and

$$s_1(x, y) \leq z \leq s_2(x, y).$$

It can be shown (though we will not dwell on the details here the argument is similar to that given in §25.3) that the triple integral can evaluated by first performing a single integral and then a double integral,

$$\iiint\limits_{\Omega} f(x, y, z)\, dx\, dy\, dz = \iint\limits_{\mathcal{R}} \left(\int_{s_1(x,y)}^{s_2(x,y)} f(x, y, z)\, dz \right) dx\, dy. \tag{26.5}$$

Example [26.2] Evaluate the integral

$$\iiint\limits_{\Omega} xz\, dx\, dy\, dz$$

where Ω is the triangular wedge depicted below.

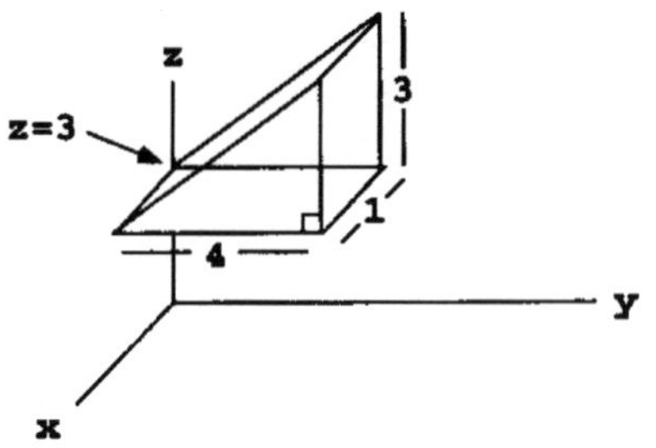

In this problem the surface $z = s_1(x, y)$ is the plane $z = 3$,

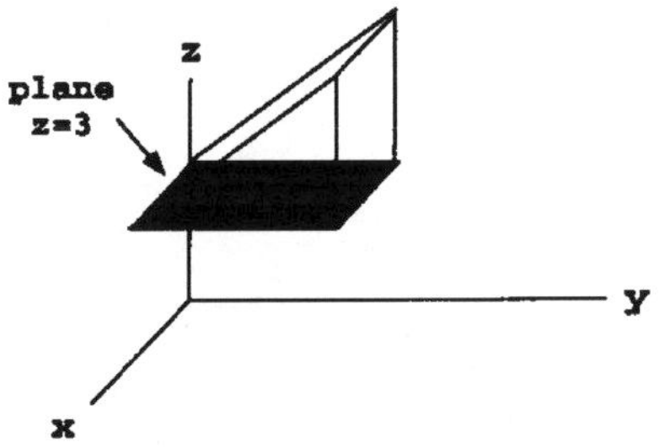

while the surface $z = s_2(x, y)$ is the plane which passes through the points $(0, 0, 3)$, $(1, 0, 3)$, $(0, 4, 6)$. The equation of the latter plane can be obtained by the methods of §21.2 and is given by

$$4z \; - \; 3y = 12.$$

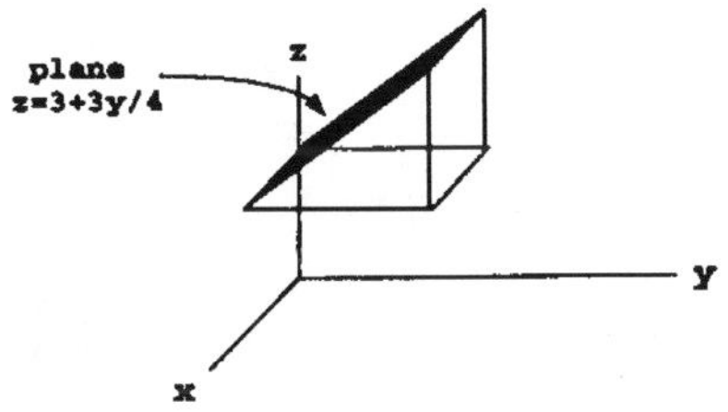

The perpendicular projection of Ω onto the $x\,y$–plane is the rectangle $\mathcal{R}$, $0 \le x \le 1$, $0 \le y \le 4$,

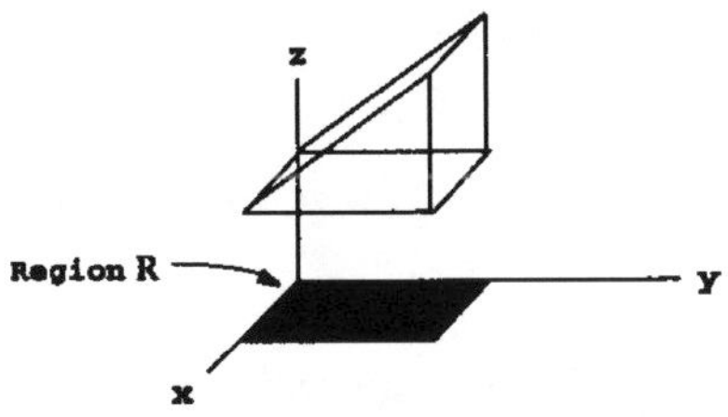

and thus

$$\begin{aligned}
\iiint\limits_{\Omega} xz\,dx\,dy\,dz &= \int_0^4\int_0^1\left(\int_3^{3+\frac{3}{4}y} xz\,dz\right)dx\,dy \\
&= \int_0^4\int_0^1 \frac{xz^2}{2}\Bigg]_{z=3}^{z=3+\frac{3}{4}y} dx\,dy \\
&= \int_0^4\int_0^1\left(\frac{x(9+\frac{9}{2}y+\frac{9}{16}y^2)}{2}-\frac{9x}{2}\right)dx\,dy \\
&= \int_0^4\left(\frac{(9+\frac{9}{2}y+\frac{9}{16}y^2)}{4}-\frac{9}{4}\right)dy \\
&= \int_0^4\left(\frac{9}{8}y+\frac{9}{64}y^2\right)dy \\
&= 9+3 \;=\; 12.
\end{aligned}$$

Remark: In this section we focused on integrating over solids which are bounded by surfaces which are bounded by $z = s_1(x,y)$ and $z = s_2(x,y)$. It is possible to consider solids Ω which are bounded by functions

$$y = t_1(x,z),\quad y = t_2(x,z),\quad \text{or}\quad x = u_1(y,z),\quad x = u_2(y,z).$$

The triple integral of a continuous function $f(x,y,z)$ over Ω will be given by, respectively

$$\begin{aligned}
&\iint\limits_{\mathcal{R}}\left(\int_{t_1(x,z)}^{t_2(x,z)} f(x,y,z)\,dy\right)dx\,dz, \\
&\iint\limits_{\mathcal{R}}\left(\int_{u_1(y,z)}^{u_2(y,z)} f(x,y,z)\,dx\right)dy\,dz,
\end{aligned} \tag{26.6}$$

where the region under the double integrals is the projection of Ω onto the $x\,z$–plane or the $y\,z$–plane respectively.

Example [26.3] Evaluate the volume $V(\Omega)$ of the triangular wedge Ω of Example [26.2].

In this case the integration is somewhat simpler:

$$V(\Omega) = \iiint\limits_{\Omega} 1\, dx\, dy\, dz$$

$$= \int_0^4 \int_0^1 \left(\int_3^{3+\frac{3}{4}y} 1 \cdot dz \right) dx\, dy$$

$$= \int_0^4 \int_0^1 \left(\frac{3}{4} y \right) dx\, dy$$

$$= \int_0^4 \frac{3}{4} y dy = 6.$$

Exercises for §26.2

In the following exercises compute the value of the triple integral $\iiint\limits_{\Omega} f(x, y, z)\, dx\, dy\, dz$. Use your CAS to help sketch the solid region Ω if it is not already depicted below.

(1) $f(x, y, z) = (x^2 + y) \cdot z^2$, where Ω is the unit cube centered at the origin.

(2) $f(x, y, z) = (x - y) \cdot \sin(z)$, where Ω is the rectangular box $1 \le x \le 3$, $0 \le y \le 1$, $0 \le z \le \pi$.

(3) $f(x, y, z) = 2x$, where Ω is the triangular wedge depicted below.

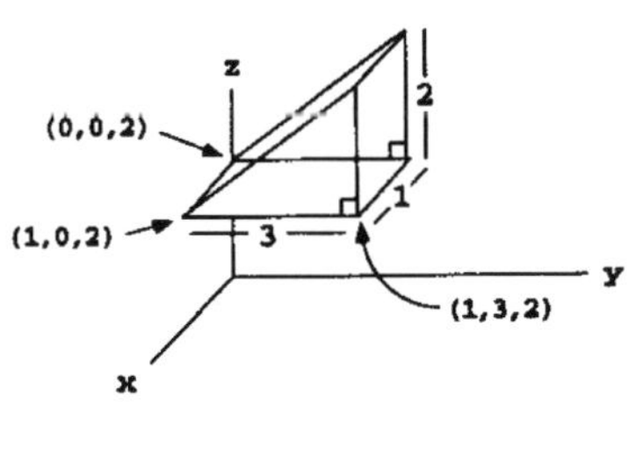

(4) $f(x, y, z) = xyz$, where Ω is the solid region which lies below the surface $z = x^2 + 1$ and above the rectangle $-1 \le x \le 1$, $0 \le y \le 1$.

(5) $f(x, y, z) = y + 2z$, where Ω is the solid region which lies below the surface $z = x^2 + y^2$ and above the rectangle $-1 \le x \le 2$, $2 \le y \le 3$.

(6) $f(x, y, z) = yz$, where Ω is the triangular wedge depicted below.

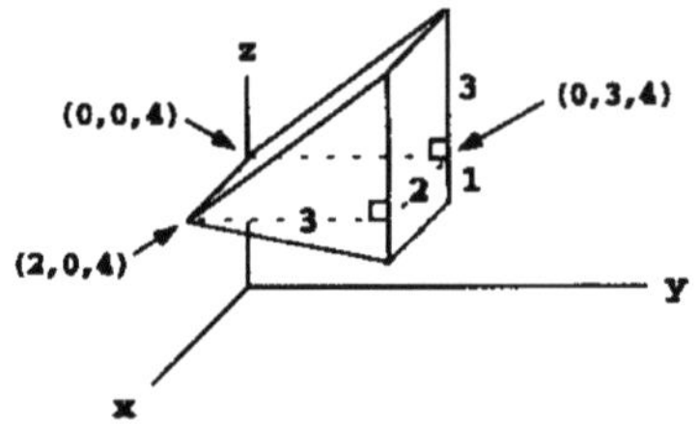

(7) $f(x, y, z) = 2$, where Ω is depicted below.

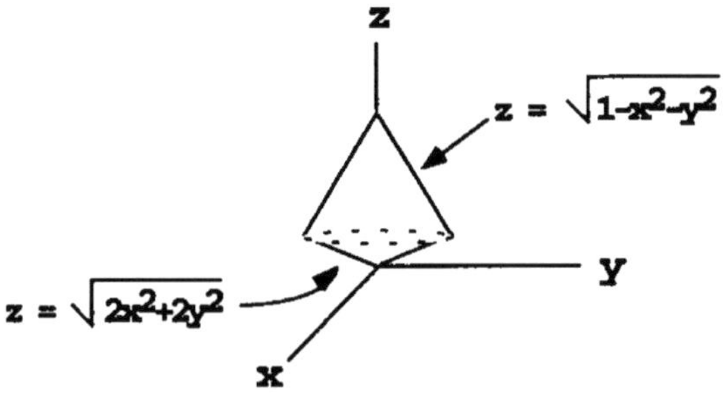

Hint: Show that the equation of the dotted circle is $x^2 + y^2 = \frac{1}{3}$.

(8) $f(x, y, z) = 3$, where Ω is the solid region bounded by the surfaces $z = x^2 + y^2$ and $z = 2 - x^2 - y^2$.

(9) $f(x, y, z) = 3$, where Ω is the solid tetrahedron formed by the three coordinate planes and the plane $x + y + z = 1$.

(10) Compute the volumes of the solid regions Ω in all of the previous problems.

26.3 Changing Coordinates in Triple Integrals

Given a triple integral

$$\iiint_{\Omega} f(x, y, z)\, dx\, dy\, dz,$$

let

$$x = g_1(u, v, w), \quad y = g_2(u, v, w), \quad z = g_3(u, v, w)$$

be a change of variables where u, v, w are new variables, and g_i are differentiable one–to–one functions. Recalling the method we used to approximate the image of an element of area under the change of variables in two dimensions (see §25.7) we define

$$\vec{\mathbf{U}} \approx \left(\frac{\partial x}{\partial u}\vec{\mathbf{i}} + \frac{\partial y}{\partial u}\vec{\mathbf{j}} + \frac{\partial z}{\partial u}\vec{\mathbf{k}}\right) du$$

$$\vec{\mathbf{V}} \approx \left(\frac{\partial x}{\partial v}\vec{\mathbf{i}} + \frac{\partial y}{\partial v}\vec{\mathbf{j}} + \frac{\partial z}{\partial v}\vec{\mathbf{k}}\right) dv$$

$$\vec{\mathbf{W}} \approx \left(\frac{\partial x}{\partial v}\vec{\mathbf{i}} + \frac{\partial y}{\partial v}\vec{\mathbf{j}} + \frac{\partial z}{\partial v}\vec{\mathbf{k}}\right) dw$$

as approximations of the images of the sides of an increment of volume.

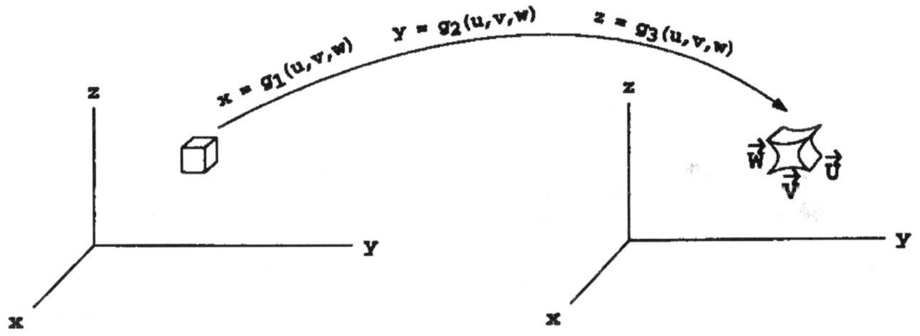

The volume element in the new coordinate system is given by

$$dx\,dy\,dz = \vec{\mathbf{U}} \cdot (\vec{\mathbf{V}} \times \vec{\mathbf{W}})$$

$$= \left|\frac{\partial(x,y,z)}{\partial(u,v,w)}\right| du\ dv\ dw,$$

where

$$\frac{\partial(x,y,z)}{\partial(u,v,w)} = \begin{vmatrix} \frac{\partial x}{\partial u} & \frac{\partial y}{\partial u} & \frac{\partial z}{\partial u} \\ \frac{\partial x}{\partial v} & \frac{\partial y}{\partial v} & \frac{\partial z}{\partial v} \\ \frac{\partial x}{\partial w} & \frac{\partial y}{\partial w} & \frac{\partial z}{\partial w} \end{vmatrix} \tag{26.7}$$

is the Jacobian of the transformation.

Having defined the Jacobian for the three dimensional change of variables, we are now in a position to state our change of variables formula:

$$\iiint\limits_{\Omega} f(x,y,z)\,dx\,dy\,dz = \iiint\limits_{\Omega'} f^*(u,v,w)\left|\frac{\partial(x,y,z)}{\partial(u,v,w)}\right| du\,dv\,dw \tag{26.8}$$

where

$$f^*(u,v,w) = f\Big(g_1(u,v,w), g_2(u,v,w), g_3(u,v,w)\Big),$$

and Ω' is the set of points (u,v,w) such that

$$\Big(g_1(u,v,w), g_2(u,v,w), g_3(u,v,w)\Big) \in \Omega.$$

Determining which change of variables is convenient in a given situation can be delicate. In the case of double integrals we found that polar coordinates were appropriate provided the region we were integrating over could easily be described as a polar region. In general, our goal when we change variables is to somehow obtain a simpler region over which to integrate (and hope we can complete the computation). In the next section two of the most important variable changes are introduced: the Cylindrical Coordinates and the Spherical Coordinates.

Exercises for §26.3

Using your CAS, compute the Jacobian for the following three dimensional change of variables.

(1) $x = u + v + w, y = u - v + 2w, z = 2u + v - 3w$

(2) $x = -3u + v - 2w, y = u + 5v - 2w, z = -2u - 3v + 4w$

(3) $x = a_1u + b_1v + c_1w, y = a_2u + b_2v + c_2w, z = a_3u + b_3v + c_3w$

(4) $x = u^2 + v^2, y = v^2 - w^2, z = uvw.$

(5) $x = u + v, y = uw^2, z = w - v.$

Evaluate the following integrals with the suggested change of coordinates.

(6) Set $x = 2u, y = 5v$, and $z = 3w$ in

$$\iiint\limits_{\frac{x^2}{4}+\frac{y^2}{25}+\frac{z^2}{9}\leq 1} dx\,dy\,dz.$$

(7) Let Ω be the solid defined by $\{0 \leq x, y \leq 2\ 0 \leq x + y + z \leq 1\}$. Set $u = x + y + z, v = x$, and $w = y$, (i.e., $x = v, y = w$, and $z = u - v - w$) in $\iiint\limits_{\Omega} xyz\,dx\,dy\,dz$.

(8) Let Ω be the solid defined by $\{-1 \leq z \leq 0,\ 0 \leq x + 2y \leq 1,\ 1 \leq 2x + y - z \leq 2\}$. Set $u = 2x + y - z, v = x + 2y$, and $w = z$, (i.e., $x = \frac{2u+2w-v}{3}, y = \frac{2v-u-w}{3}$, and $z = w$) in the triple integral $\iiint\limits_{\Omega} x\,dx\,dy\,dz$.

26.4 Cylindrical and Spherical Coordinates

Cylindrical coordinates,

$$x = r\cos(\theta), \quad y = r\sin(\theta), \quad z = z$$

were introduced in §21.4 as the natural extension of polar coordinates to three dimensions. Recall that every point $p = (x, y, z)$ can be realized as a point (r, θ, z) in cylindrical coordinates.

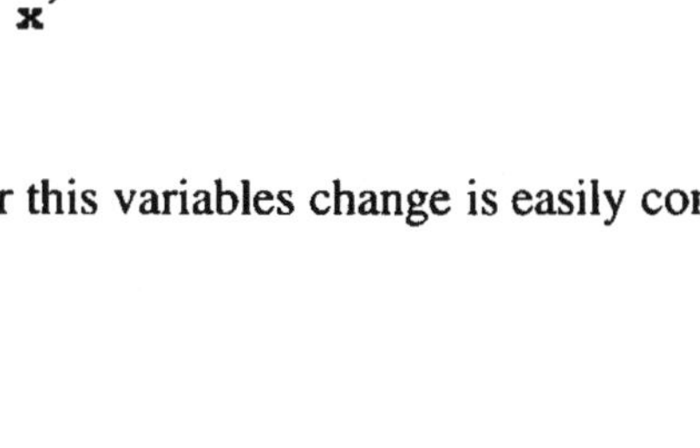

The Jacobian $\frac{\partial(x,y,z)}{\partial(r,\theta,z)}$ for this variables change is easily computed:

$$\frac{\partial(x,y,z)}{\partial(r,\theta,z)} = \begin{vmatrix} \frac{\partial x}{\partial r} & \frac{\partial y}{\partial r} & \frac{\partial z}{\partial r} \\ \frac{\partial x}{\partial \theta} & \frac{\partial y}{\partial \theta} & \frac{\partial z}{\partial \theta} \\ \frac{\partial x}{\partial z} & \frac{\partial y}{\partial z} & \frac{\partial z}{\partial z} \end{vmatrix} = \begin{vmatrix} \cos(\theta) & \sin(\theta) & 0 \\ -r\sin(\theta) & r\cos(\theta) & 0 \\ 0 & 0 & 1 \end{vmatrix} \tag{26.9}$$

$$= r(\cos^2\theta + \sin^2\theta)$$

$$= r.$$

Hence equation (26.8) becomes

$$\iiint\limits_{\Omega} f(x,y,z)\,dx\,dy\,dz = \iiint\limits_{\Omega'} f^*(r,\theta,z)\cdot r\,dr\,d\theta\,dz \tag{26.10}$$

where $f^*(r,\theta,z) = f(r\cos(\theta), r\sin(\theta), z)$.

Example [26.4] Evaluate the integral

$$\iiint\limits_{\Omega} xz\,dx\,dy\,dz$$

where Ω is one quarter of the cylinder of radius $\frac{1}{2}$ and height 5 as in the figure below.

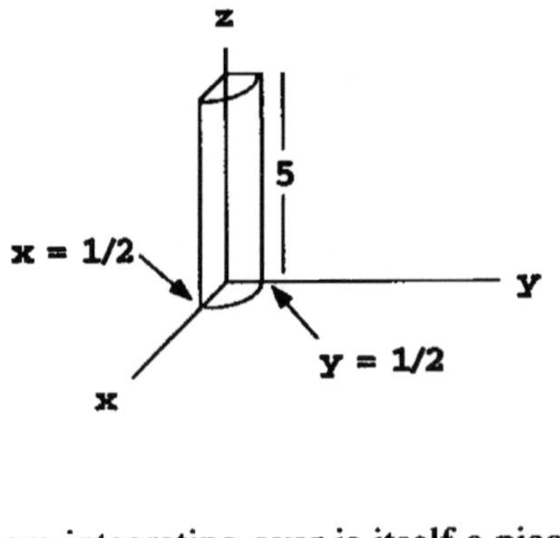

Since the region we are integrating over is itself a piece of a cylinder it can be easily described by cylindrical coordinates (as one would expect) : Ω' consists of the points (r, θ, z) with $0 \le r \le \frac{1}{2}$, $0 \le \theta \le \frac{\pi}{2}$, and $0 \le z \le 5$. Thus we have

$$\iiint_{\Omega} f(x, y, z)\, dx\, dy\, dz = \int_0^5 \int_0^{\frac{\pi}{2}} \int_0^{\frac{1}{2}} r(\cos(\theta))z \cdot r\, dr\, d\theta\, dz$$

$$= \int_0^5 \int_0^{\frac{\pi}{2}} (\cos(\theta))z \frac{r^3}{3}\Big]_{r=0}^{r=\frac{1}{2}} d\theta\, dz$$

$$= \frac{1}{24} \int_0^5 z \cdot \sin(\theta)\Big]_{\theta=0}^{\theta=\frac{\pi}{2}} dz$$

$$= \frac{1}{24} \frac{z^2}{2}\Big]_{z=0}^{z=5}$$

$$= \frac{25}{48}.$$

Spherical Coordinates:

Every point $P = (x, y, z)$ can be specified by three angles (ρ, ϕ, θ) where $0 \le \rho$, $0 \le \phi \le \pi$, and $0 \le \theta \le 2\pi$.

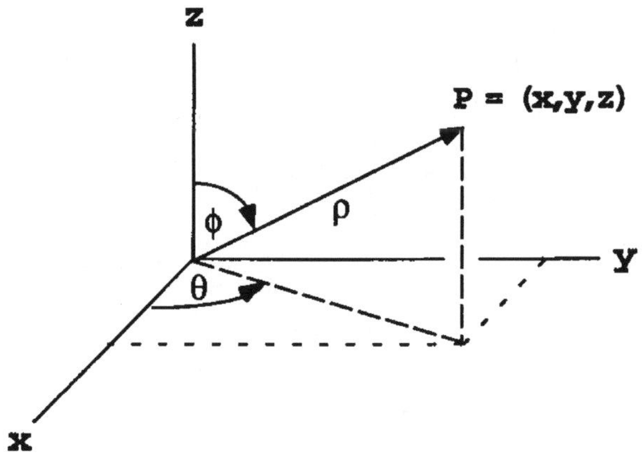

The precise formulae for this representation, which we shall term **spherical coordinates**, is given by

$$\begin{aligned} x &= \rho\sin(\phi)\cos(\theta) \\ y &= \rho\sin(\phi)\sin(\theta) \\ z &= \rho\cos(\phi). \end{aligned} \tag{26.11}$$

In order to visualize a surface which is given in spherical coordinates we will, in general, have to do some manipulation to bring it to rectangular form. The identities

$$\begin{aligned} \rho &= \sqrt{x^2+y^2+z^2} \\ \cos(\phi) &= \frac{z}{\rho} \\ \sin(\phi) &= \frac{\sqrt{x^2+y^2}}{\rho} \\ \cos(\theta) &= \frac{x}{\sqrt{x^2+y^2}} \\ \sin(\theta) &= \frac{y}{\sqrt{x^2+y^2}}. \end{aligned} \tag{26.12}$$

are very useful for this conversion.

Example [26.5] Identify the surface $\rho = \cos(\phi)$ given in spherical coordinates.

By definition,

$$\rho = \cos(\phi) = \frac{z}{\rho},$$

and further

$$\rho^2 = x^2 + y^2 + z^2.$$

We conclude that

$$x^2 + y^2 + z^2 = z,$$

and, hence, by completing the square

$$x^2 + y^2 + \left(z - \frac{1}{2}\right)^2 = \left(\frac{1}{2}\right)^2$$

our surface is seen to be the sphere of radius 2 centered at $\left(0, 0, \frac{1}{2}\right)$.

Example [26.6] Compute the Jacobian $\frac{\partial(x,y,z)}{\partial(\rho,\phi,\theta)}$ for the spherical transformation.

By definition the Jacobian is given by the somewhat cumbersome derivative

$$\frac{\partial(x,y,z)}{\partial(\rho,\phi,\theta)} = \begin{vmatrix} \frac{\partial x}{\partial \rho} & \frac{\partial y}{\partial \rho} & \frac{\partial z}{\partial \rho} \\ \frac{\partial x}{\partial \phi} & \frac{\partial y}{\partial \phi} & \frac{\partial z}{\partial \phi} \\ \frac{\partial x}{\partial \theta} & \frac{\partial y}{\partial \theta} & \frac{\partial z}{\partial \theta} \end{vmatrix}$$

$$= \begin{vmatrix} \sin(\phi)\cos(\theta) & \sin(\phi)\sin(\theta) & \cos(\phi) \\ \rho\cos(\phi)\cos(\theta) & \rho\cos(\phi)\sin(\theta) & -\rho\sin(\phi) \\ -\rho\sin(\phi)\sin(\theta) & \rho\sin(\phi)\cos(\theta) & 0 \end{vmatrix} \qquad (26.13)$$

$$= \rho^2 \sin(\phi).$$

We leave some details as an exercise.

The computation in Example [26.6] allows us to write down the change of coordinates (26.4) for spherical coordinates:

$$\iiint\limits_{\Omega} f(x,y,z)\, dx\, dy\, dz = \iiint\limits_{\Omega'} f^*(\rho,\phi,\theta) \cdot \rho^2 \sin(\phi)\, d\rho\, d\phi\, d\theta. \qquad (26.14)$$

Example [26.7] (Volume of the Sphere) Compute the volume of a sphere of radius R.

The solid sphere of radius R (centered at the origin) consists of the points

$$\Omega_{\mathcal{R}} = \left\{(x,y,z) \,\middle|\, x^2 + y^2 + z^2 \le R\right\}.$$

Spherical coordinates are entirely natural in this case:

$$\Omega_{\mathcal{R}} = \left\{(\rho,\phi,\theta) \,\middle|\, 0 \le \rho \le R,\ 0 \le \phi \le \pi,\ 0 \le \theta \le 2\pi\right\}.$$

Applying equation (26.14) we obtain the classical formula for the volume,

$$\iiint_{\Omega} dx\,dy\,dz = \int_0^{2\pi}\int_0^{\pi}\int_0^{R} \rho^2 \sin(\phi)\,d\rho\,d\phi\,d\theta$$

$$= \frac{R^3}{3}\int_0^{2\pi}\left(\int_0^{\pi} \sin(\phi)\,d\phi\right)d\theta$$

$$= \frac{4\pi R^3}{3}.$$

Exercises for §26.4

(1) Compute the integral

$$\iiint_{\substack{1 \le x^2+y^2 \le 4\\ 2 \le z \le 3}} x^2 z\,dx\,dy\,dz$$

using cylindrical coordinates.

(2) Compute the integral

$$\iiint_{\substack{0 \le y^2+z^2 \le 1\\ -1 \le x \le 1}} z\,dx\,dy\,dz$$

by setting $x = x, y = r\cos(\theta)$, and $z = r\sin(\theta)$.

(3) Compute the integral

$$\iiint_{1 \le x^2+y^2+z^2 \le 2} dx\,dy\,dz$$

by converting to spherical coordinates.

(4) Compute the integral

$$\iiint_{1 \le x^2+y^2+z^2 \le 2} xyz\,dx\,dy\,dz$$

by converting to spherical coordinates.

Identify the following surfaces given in cylindrical coordinates. Check your answer by graphing with the CAS.

(5) $r = 2, 0 \leq \theta \leq 2\pi,$
$0 \leq z \leq 2.$

(6) $r = 3, 0 \leq \theta \leq \frac{\pi}{2},$
$-1 \leq z \leq 0.$

(7) $r(\cos(\theta) + \sin(\theta)) + z = 1.$

(8) $r^2 = z, 0 \leq z \leq 4.$

(9) $r^2 = z, -1 \leq z \leq 1.$

(10) $r\cos(\theta) = 1.$

Identify the following surfaces given in spherical coordinates. Check your answer by graphing with the CAS.

(11) $\rho = 1, 0 \leq \phi \leq \pi,$
$0 \leq \theta \leq 2\pi.$

(12) $\rho = 3, 0 \leq \phi \leq 2\pi,$
$0 \leq \theta \leq \frac{\pi}{2}.$

(13) $0 \leq \rho \leq 1, \phi = \frac{\pi}{4},$
$0 \leq \theta \leq 2\pi.$

(14) $0 \leq \rho \leq 1, \phi \leq \frac{3\pi}{4}, 0 \leq \theta \leq 2\pi.$

(15) $\rho = \frac{1}{\cos(\phi)}.$

(16) $\rho = 2\cos(\phi).$

Additional exercises for Chapter XXVI

Evaluate the following integrals. Check your computations with your CAS

(1) $\int_{-1}^{1} \int_{0}^{1} \int_{-1}^{1} x \frac{\cos^2(\pi z y)}{x^2+y^2} \, dx \, dy \, dz.$

(2) $\int_{1}^{3} \int_{x}^{x^2} \int_{0}^{\ln(x)} x e^y dy \, dz \, dx.$

(3) $\int_{0}^{1} \int_{0}^{1} \int_{1}^{e^y} z \cdot \ln(x) \, dx \, dy \, dz.$

In the following exercises find the volume of the the solid region Ω. Use your CAS to help sketch the solid region.

(4) Ω is the wedge in the first octant bounded by $2y^2 + 2z^2 = 1,$ and the planes $y = x$ and $x = 0.$

(5) Ω is the solid region bounded by the parabaloids $z = 4x^2 + 4y^2$ and $z = 6 - 7x^2 - y^2.$

(6) Ω is the solid region bounded by the parabaloids $z = 4x^2 + y^2$ and $z = 4 - 3y^2.$

(7) Ω is the solid enclosed by the parabaloid $z = 9 - x^2 - y^2$ and $z = 0.$ (**Hint:** use cylindrical coordinates)

(8) Ω is the solid enclosed by the parabaloid $z = x^2 + y^2$ and $z = 10$. (**Hint:** use cylindrical coordinates)

(9) Ω is the solid bounded by the surface $r^2 + 2z^2 = 20$ and the surface $z = r^2$. (**Hint:** use cylindrical coordinates).

(10) Ω is the solid bounded by the cone $r = \frac{5}{a}$ and $z = 5$

(11) Ω is the solid in the first octant bounded by the sphere $\rho = 4$, the coordinate planes, and the cones $\phi = \frac{\pi}{4}$ and $\phi = \frac{\pi}{3}$. (**Hint:** use polar coordinates).

(12) Ω is the solid bounded by the sphere $\rho = 4a^2$ and the planes $z = -1$ and $z = a$.

The **centroid** of a solid region Ω, is $(\bar{x}, \bar{y}, \bar{z})$, where

$$\bar{x} = \frac{1}{\text{Volume}(\Omega)} \iint_{\Omega} x\, dx\, dy\, dz. \quad (26.15)$$

$$\bar{y} = \frac{1}{\text{Volume}(\Omega)} \iint_{\Omega} y\, dx\, dy\, dz. \quad (26.16)$$

$$\bar{z} = \frac{1}{\text{Volume}(\Omega)} \iint_{\Omega} z\, dx\, dy\, dz. \quad (26.17)$$

Roughly speaking, if the solid region Ω has its mass evenly distributed, than one can "balance" the solid at the centroid. Find the centroid of the given solid region Ω. Graph the solid and plot the centroid with your CAS.

(13) Ω is the solid region bounded by the parabaloids $z = 4x^2 + 4y^2$ and $z = 6 - 7x^2 - y^2$.

(14) Ω is the solid enclosed by the parabaloid $z = 9 - x^2 - y^2$ and $z = 0$. (**Hint:** use cylindrical coordinates).

(15) Ω is the solid in the first octant bounded by the sphere $\rho = 4$, the coordinate planes and the cones $\phi = \frac{\pi}{4}$ and $\phi = \frac{\pi}{3}$. (**Hint:** use polar coordinates).

(16) Ω is the solid bounded by the surface $r^2 + 2z^2 = 20$ and the surface $z = r^2$. (**Hint:** use cylindrical coordinates).

(17) Ω is the solid region bounded by the cone $z = \frac{h}{a}r$ and $z = h$.

Chapter XXVII

Vector Fields and Line Integrals

27.1 Vector Fields

A **vector field** is a vector valued function on three dimensional space, i.e., a vector field assigns to every point (x, y, z) in 3–dimensional space, a unique vector $\vec{\mathbf{v}} = \vec{\mathbf{f}}(x, y, z)$. The graphical displays of vector fields are often visually dramatic. As can be seen in the example below, displays of vector fields give the viewer the feeling of motion or perhaps a physical force.

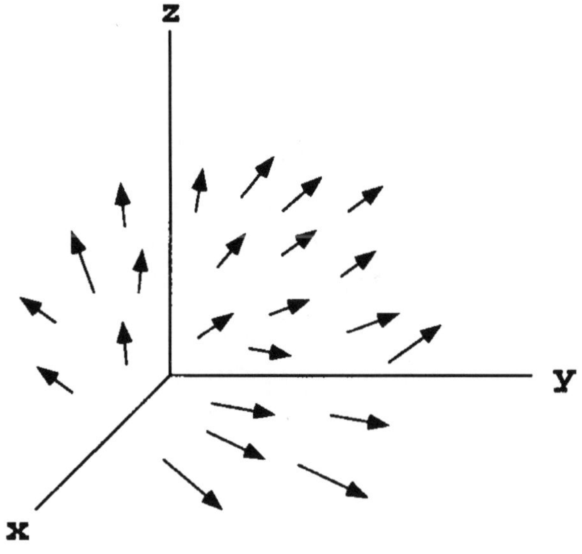

In order to develop some intuition on these functions we need to look at some examples.

Example [27.1] (The Constant Vector Field) Let $\vec{\mathbf{f}}(x, y, z) = \vec{\mathbf{i}} + \vec{\mathbf{j}} + \vec{\mathbf{k}}$. Since the value of $\vec{\mathbf{f}}(x, y, z)$ is independent of the point in question, all the vectors are the same and equal $\vec{\mathbf{i}} + \vec{\mathbf{j}} + \vec{\mathbf{k}}$.

The display is given below.

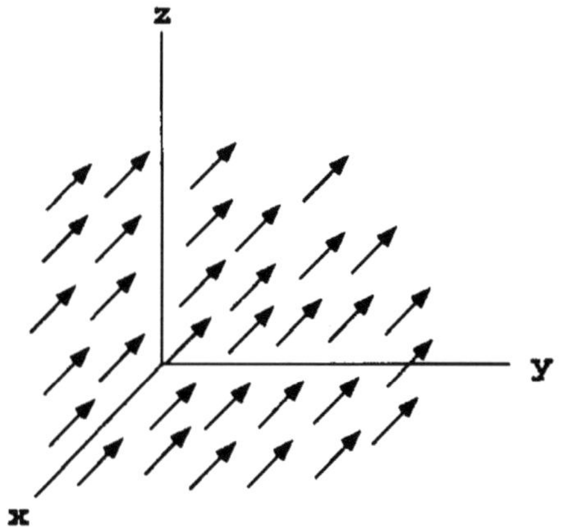

Example [27.2] (Position Vector Field) Consider the vector field $\vec{\mathbf{f}}(x, y, z) = x\vec{\mathbf{i}} + y\vec{\mathbf{j}} + z\vec{\mathbf{k}}$. Observe that the vectors grow larger as we move away from the origin, and all the vectors point away from the origin.

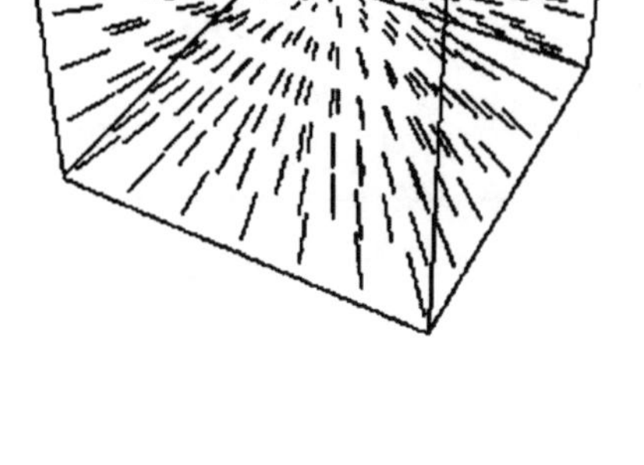

Example [27.3] In the case of the vector field

$$\vec{\mathbf{f}}(x, y, z) = \frac{(x\vec{\mathbf{i}} + y\vec{\mathbf{j}} + z\vec{\mathbf{k}})}{\sqrt{x^2 + y^2 + z^2}},$$

(where $(x, y, z) \neq (0, 0, 0)$), the vectors all have unit length (i.e., length one), and point away from the origin.

Returning to the position vector field

$$\vec{\mathbf{f}}(x, y, z) = x\vec{\mathbf{i}} + y\vec{\mathbf{j}} + z\vec{\mathbf{k}},$$

focus for a moment on a particular point (x, y, z), and its position vector $x\vec{\mathbf{i}}+y\vec{\mathbf{j}}+z\vec{\mathbf{k}}$ (depicted below).

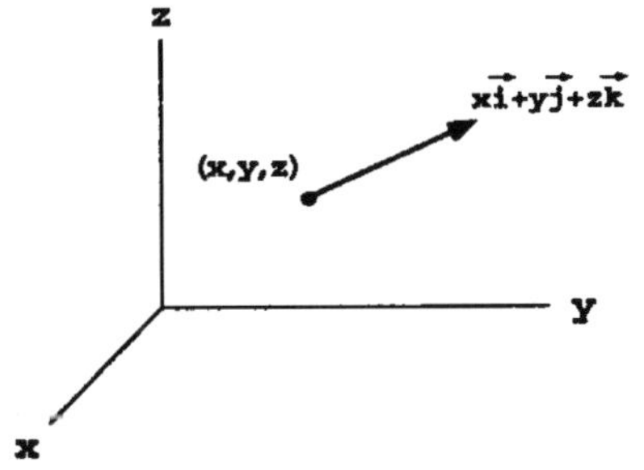

Consider now the vector $-y\vec{\mathbf{i}} + x\vec{\mathbf{j}}$ which lies in the $x\,y$–plane. Since the dot product

$$(-y\vec{\mathbf{i}} + x\vec{\mathbf{j}}) \cdot (x\vec{\mathbf{i}} + y\vec{\mathbf{j}} + z\vec{\mathbf{k}}) \;=\; 0,$$

we deduce that the two vectors are perpendicular. When we display the new vector field

$$\vec{\mathbf{g}} \;=\; -y\vec{\mathbf{i}} + x\vec{\mathbf{j}}$$

(in the $x\,y$–plane)

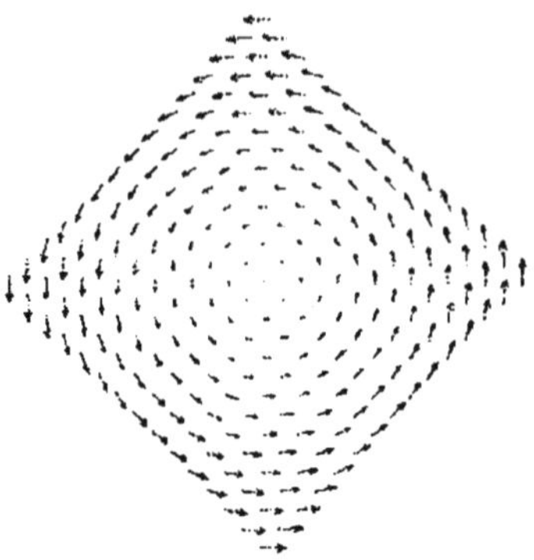

a definite turning is visible. This is an example of a spin vector field.

In general, it is difficult to envisage the graphical display of a vector field. Even in the simple examples considered thus far, the behavior varies substantially. There is, however, an important class of vector fields, termed **gradient vector fields**, whose graphical display are well understood and shall be discussed in the remainder of this section.

Recall that the **gradient** of a differentiable function $\phi = \phi(x, y, z)$ (see §23.3) is given by

$$\vec{\nabla}\phi(x, y, z) \;;= \frac{\partial \phi}{\partial x}\vec{\mathbf{i}} + \frac{\partial \phi}{\partial y}\vec{\mathbf{j}} + \frac{\partial \phi}{\partial z}\vec{\mathbf{k}}.$$

Definition: *A vector field* $\vec{\mathbf{v}} = \vec{\mathbf{f}}(x, y, z)$ *is a* ***gradient vector field*** *if there exists a differentiable function* $\phi(x, y, z)$ *such that*

$$\vec{\mathbf{v}} = \vec{\mathbf{f}}(x, y, z) = \vec{\nabla}\phi(x, y, z).$$

We shall term ϕ *a* ***potential function*** *for* $\vec{\mathbf{f}}$.

In the midst of the proof of Proposition [24.18], we observed that the vector $\vec{\nabla}\phi$ is perpendicular to the level surface $\phi(x, y, z) = c$, i.e., at a point $P = (x_0, y_0, z_0)$,

$$\vec{\nabla}\phi(x_0, y_0, z_0)$$

is perpendicular to the tangent plane to the surface $\phi(x, y, z) = c$ at P.

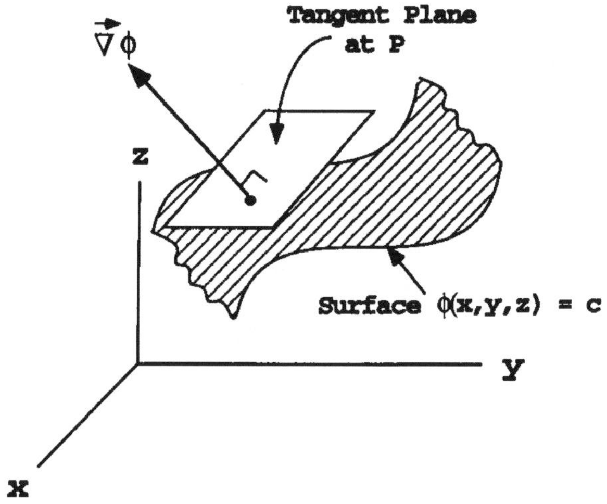

When we view $\vec{\nabla}\phi$ as a vector field, its graphical display can be readily described (using the observations above).

Corollary [27.4] *Let $\vec{v} = \vec{\nabla}\phi(x, y, z)$ be a gradient vector field with potential function ϕ. The graphical display of $\vec{\nabla}\phi$ consists of vectors which are perpendicular to the level surfaces*

$$\phi(x, y, z) = c$$

where the constants c vary over the real numbers.

Example [27.5] (Ellipsoid) Let d, e, and f be constants. The surface

$$\frac{x^2}{d^2} + \frac{y^2}{e^2} + \frac{z^2}{f^2} = 1$$

will be an ellipsoid

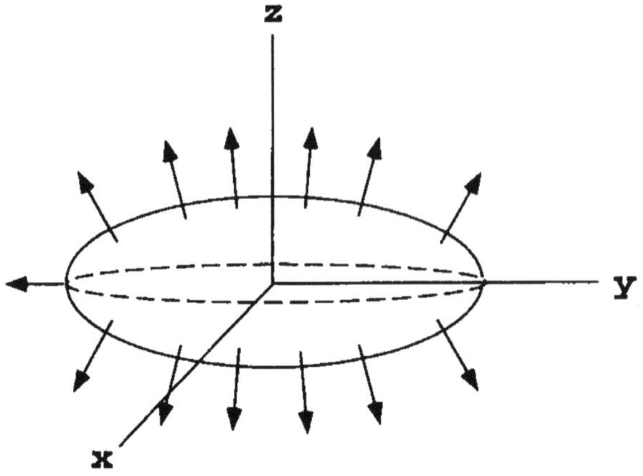

which intercepts the three coordinate axes at the points

$$(\pm d, 0, 0), \quad (0, \pm e, 0), \quad (0, 0, \pm f).$$

The graphical display of the gradient vector field

$$\begin{aligned} \vec{\mathbf{v}} &= \vec{\nabla}\left(\frac{x^2}{d^2} + \frac{y^2}{e^2} + \frac{z^2}{f^2}\right) \\ &= \frac{2x}{d^2}\vec{\mathbf{i}} + \frac{2y}{e^2}\vec{\mathbf{j}} + \frac{2z}{f^2}\vec{\mathbf{k}} \end{aligned}$$

consists of the vectors which are perpendicular to the ellipsoids

$$\frac{x^2}{d^2} + \frac{y^2}{e^2} + \frac{z^2}{f^2} = c$$

as c ranges over all positive constants.

In our next example we examine an example of Corollary [27.4] in two dimensions

Example [27.6] (Parabolic Potential) Let $\phi(x, y) = y - x^2$. The curves $\phi(x, y) = c$ are the easily sketched parabolas $y = x^2 + c$.

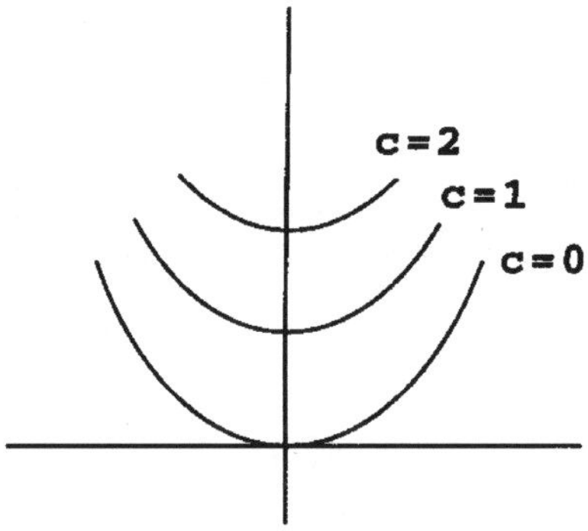

The gradient vector field $\vec{\mathbf{v}} = \vec{\nabla}\phi = -2x\vec{\mathbf{i}} + \vec{\mathbf{j}}$ has for its graphical display all of the vectors which are perpendicular to the level curves $y - x^2 = c$.

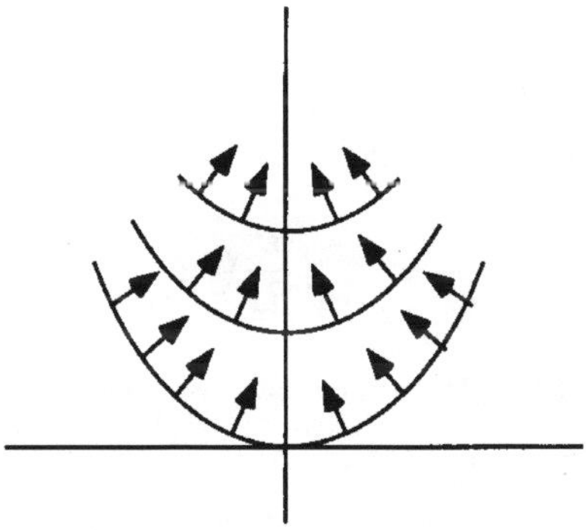

Our success in describing the displays of gradient vector fields leads us naturally to ask:

Question: Given a 2 or 3–dimensional vector field $\vec{\mathbf{v}} = \vec{\mathbf{f}}(x, y)$ how can we determine whether or not it is a gradient vector field?

We begin this discussion with the 2–dimensional case in that it is somewhat less complex. By definition $\vec{\mathbf{v}} = \vec{\mathbf{f}}(x, y)$ takes the form

$$\vec{\mathbf{v}} = f_1(x, y)\vec{\mathbf{i}} + f_2(x, y)\vec{\mathbf{j}}.$$

If $\vec{\mathbf{v}}$ is in fact a gradient vector field there exists a potential function $\phi(x, y)$ such that $\vec{\mathbf{v}} = \vec{\nabla}\phi = \frac{\partial\phi}{\partial x}\vec{\mathbf{i}} + \frac{\partial\phi}{\partial y}\vec{\mathbf{j}}$. Thus we may infer that

$$\begin{aligned} f_1(x,y) &= \frac{\partial \phi}{\partial x} \\ f_2(x,y) &= \frac{\partial \phi}{\partial y}. \end{aligned} \tag{27.1}$$

Recalling that the mixed partial derivatives always coincide (see §23.3), upon differentiating (27.1) we see that

$$\frac{\partial f_1}{\partial y} = \frac{\partial^2 \phi}{\partial x \partial y} = \frac{\partial f_2}{\partial x}. \tag{27.2}$$

Condition (27.2) provides an elegant test to see if $\vec{\mathbf{v}}$ is a gradient vector field. It can be shown that (27.2) is *necessary and sufficient* for $\vec{\mathbf{v}}$ to be a gradient vector field provided we are working over a **simply connected** region (by definition a region $\mathcal{R}$ is simply connected if any closed curve in $\mathcal{R}$ can be shrunk to a point without leaving $\mathcal{R}$). Furthermore, in order to solve for the potential function ϕ we need only integrate (27.1) with respect to x and differentiate the result with respect to y.

Example [27.7] Demonstrate that $\vec{\mathbf{v}} = y^3\vec{\mathbf{i}} + 3xy^2\vec{\mathbf{j}}$ is a gradient vector field and find its potential function ϕ.

In this example equation (27.2) is readily verified,

$$\frac{\partial}{\partial y} y^3 = \frac{\partial}{\partial x}(3xy^2),$$

and we deduce that $\vec{\mathbf{v}}$ is a gradient vector field. To compute a potential function we begin with the equation (27.1)

$$y^3 = \frac{\partial \phi}{\partial x},$$

which upon integrating both sides with respect to x yields

$$\phi(x,y) = \int \frac{\partial \phi}{\partial x}\, dx = \int y^3\, dx = y^3 x + C(y),$$

where $C(y)$ is a function involving only y. Differentiating with respect to y we then obtain

$$\frac{\partial \phi}{\partial y} = 3xy^2 + C'(y).$$

Applying (27.1) once again we see that

$$\frac{\partial \phi}{\partial y} = 3xy^2,$$

and we conclude that $C'(y) = 0$, hence $C(y) = C_1$ a constant, and

$$\phi(x, y) = y^3 x + C_1$$

is the required potential function.

Example [27.8] Is $\vec{\mathbf{v}} = y^4\vec{\mathbf{i}} + 2xy\vec{\mathbf{j}}$ a gradient vector field?

The vector field in this example cannot be a gradient vector field in that it fails condition (27.2):

$$\frac{\partial}{\partial y} y^4 \neq \frac{\partial}{\partial x}(2xy).$$

We now move on to finding conditions which insure that a 3–dimensional vector field is a gradient vector field. Consider the 3–dimensional vector field

$$\vec{\mathbf{v}} = f_1(x, y, z)\vec{\mathbf{i}} + f_2(x, y, z)\vec{\mathbf{j}} + f_3(x, y, z)\vec{\mathbf{k}},$$

where f_1, f_2, f_3 are differentiable functions. If a potential function $\phi(x, y, z)$ exists then, again by definition, we have the identities

$$\frac{\partial \phi}{\partial x} = f_1, \qquad \frac{\partial \phi}{\partial y} = f_2, \qquad \frac{\partial \phi}{\partial z} = f_3.$$

When we take further derivatives (and assume our region is simply connected), we arrive at the the necessary and sufficient conditions:

$$\begin{aligned} \frac{\partial^2 \phi}{\partial x \partial y} &= \frac{\partial f_1}{\partial y} = \frac{\partial f_2}{\partial x} \\ \frac{\partial^2 \phi}{\partial x \partial z} &= \frac{\partial f_1}{\partial z} = \frac{\partial f_3}{\partial x} \\ \frac{\partial^2 \phi}{\partial y \partial z} &= \frac{\partial f_2}{\partial z} = \frac{\partial f_3}{\partial y}. \end{aligned} \tag{27.3}$$

Example [27.9] Is $\vec{\mathbf{v}} = 2xyz\vec{\mathbf{i}} + x^2z\vec{\mathbf{j}} + x^2y\vec{\mathbf{k}}$ a gradient vector field?

The answer to this question can be seen to be affirmative by verifying equations (27.3):

$$\begin{aligned} \frac{\partial}{\partial y}(2xyz) &= \frac{\partial}{\partial x}(x^2 z) \\ \frac{\partial}{\partial z}(2xyz) &= \frac{\partial}{\partial x}(x^2 y) \\ \frac{\partial}{\partial z}(x^2 z) &= \frac{\partial}{\partial y}(x^2 y). \end{aligned}$$

Equations (27.3) are not easily committed to memory and don't seem particularly natural. To remedy this situation we shall introduce (though not very formally) the concept of an **operator on a set of functions.**

An **operator** is basically a function whose domain and range are themselves sets of functions (possibly real–valued, vector valued, etc.). We have already seen one important example of an operator: the derivative acting on the set of real–valued differentiable functions of one variable. There are numerous examples of operators which appear in various contexts. In our situation we shall consider the **gradient** operator, denoted by

$$\vec{\nabla} = \frac{\partial}{\partial x}\vec{\mathbf{i}} + \frac{\partial}{\partial y}\vec{\mathbf{j}} + \frac{\partial}{\partial z}\vec{\mathbf{k}}.$$

By definition the gradient takes real–valued differentiable functions (on 3–dimensional space), $\psi(x, y, z)$, to their gradient (which is a 3–dimensional vector field),

$$\vec{\nabla}\psi = \frac{\partial\psi}{\partial x}\vec{\mathbf{i}} + \frac{\partial\psi}{\partial y}\vec{\mathbf{j}} + \frac{\partial\psi}{\partial z}\vec{\mathbf{k}}.$$

When we view the gradient operator as a vector we can define the **curl** of a vector

$$\vec{\mathbf{v}} = f_1(x, y, z)\vec{\mathbf{i}} + f_2(x, y, z)\vec{\mathbf{j}} + f_3(x, y, z)\vec{\mathbf{k}},$$

by

$$\mathrm{curl}(\vec{\mathbf{v}}) = \vec{\nabla} \times \vec{\mathbf{v}}.$$

Explicitly, the curl is given by

$$\vec{\nabla} \times \vec{\mathbf{v}} = \begin{vmatrix} \vec{\mathbf{i}} & \vec{\mathbf{j}} & \vec{\mathbf{k}} \\ \frac{\partial}{\partial x} & \frac{\partial}{\partial y} & \frac{\partial}{\partial z} \\ f_1 & f_2 & f_3 \end{vmatrix}$$

$$= \left(\frac{\partial f_3}{\partial y} - \frac{\partial f_2}{\partial z}\right)\vec{\mathbf{i}} + \left(\frac{\partial f_1}{\partial z} - \frac{\partial f_3}{\partial x}\right)\vec{\mathbf{j}} + \left(\frac{\partial f_2}{\partial x} - \frac{\partial f_1}{\partial y}\right)\vec{\mathbf{k}}$$

A quick inspection shows that (27.3) holds if and only if

$$\mathrm{curl}(\vec{\mathbf{v}}) = 0. \tag{27.4}$$

Furthermore, it can be shown that if $\vec{\mathbf{v}}$ is a vector field and $\text{curl}(\vec{\mathbf{v}}) = 0$ in a simply connected 3–dimensional region (the definition here is analogous to the 2–dimensional one), then $\vec{\mathbf{v}}$ must in fact be a gradient vector field.

Example [27.10] Find the potential function $\phi(x, y, z)$ for the gradient vector field $\vec{\mathbf{v}} = 2xyz\vec{\mathbf{i}} + x^2z\vec{\mathbf{j}} + x^2y\vec{\mathbf{k}}$ of example [27.9].

A simple computation verifies that (27.4) holds:

$$\text{curl}(\vec{\mathbf{v}}) = \begin{vmatrix} \vec{\mathbf{i}} & \vec{\mathbf{j}} & \vec{\mathbf{k}} \\ \frac{\partial}{\partial x} & \frac{\partial}{\partial y} & \frac{\partial}{\partial z} \\ 2xyz & x^2z & x^2y \end{vmatrix} = 0.$$

We now focus on finding ϕ. To begin with we know that $\frac{\partial \phi}{\partial x} = 2xyz$, and hence

$$\phi(x, y, z) = x^2yz + C(y, z),$$

where $C(y, z)$ depends only on y, z. Next we use the identity $\frac{\partial \phi}{\partial y} = x^2z$ to see that

$$\frac{\partial \phi}{\partial y} = x^2z = x^2z + \frac{\partial}{\partial y}C(y, z)$$

This implies $\frac{\partial}{\partial y}C(y, z) = 0$ and thus $C(y, z) = C(z)$ is actually a function of z alone. Finally $\frac{\partial \phi}{\partial z} = x^2y$ and we conclude that

$$\frac{\partial \phi}{\partial z} = x^2y = x^2y + C'(z).$$

Thus $C'(z) = 0$ and necessarily $C(z) = C$, a constant. To recapitulate, we have found the desired potential function

$$\phi(x, y, z) = x^2yz + C.$$

Exercises for §27.1

Use your CAS to obtain graphical displays of the following vector fields. If your CAS does not have this capability, then simply graph a few representative vectors to simulate the graphical display of the vector field.

(1) $\vec{v} = x\vec{j}$.

(2) $\vec{v} = y\vec{i} - x\vec{j}$.

(3) $\vec{v} = -2x\vec{i} + \vec{j}$.

(4) $\vec{v} = x\vec{i} + y\vec{j} + z\vec{k}$.

(5) $\vec{v} = (z-y)\vec{i} + (x-z)\vec{j} + (y-x)\vec{k}$

(6) $\vec{v} = \vec{\nabla}\phi$ with $\phi(x,y) = y - x^2, y - x^3$, and $y - \sin(x)$.

(7) $\vec{v} = \vec{\nabla}\phi$ with $\phi(x,y,z) = x^2 + \frac{y^2}{4} + \frac{z^2}{9}$.

(8) $\vec{v} = \vec{\nabla}\phi$ with $\phi(x,y,z) = x - y + z$.

In the following exercises determine whether or not $\vec{v}$ is a gradient vector field and if it is find its potential function ϕ.

(9) $\vec{v} = y^2\vec{i} + 2xy\vec{j}$.

(10) $\vec{v} = 3x^2y\vec{i} + 2x^3\vec{j}$.

(11) $\vec{v} = 6x\vec{i} - 4y\vec{j}$.

(12) $\vec{v} = y\cos(xy)\vec{i} + x\cos(xy)\vec{j}$.

(13) $\vec{v} = y^2\vec{i} + 3xy\vec{j}$.

(14) $\vec{v} = \frac{y^2}{x}\vec{i} + 2y\ln(x)\vec{j}$.

(15) $\vec{v} = yz\vec{i} + 2xz\vec{j} + xy\vec{k}$.

(16) $\vec{v} = yz\vec{i} + xz\vec{j} + xy\vec{k}$.

(17) $\vec{v} = 2x\vec{i} + z\vec{j} + y\vec{k}$.

(18) $\vec{v} = (\cos(x))y^2z\vec{i} + 2(\sin(x))yz\vec{j} + (\sin(x))y^2\vec{k}$.

27.2 Line Integrals

Let $\vec{\gamma}(t) = \gamma_1(t)\vec{i} + \gamma_2(t)\vec{j} + \gamma_3(t)\vec{k}$ be a parameterization of a space curve. Geometrically the curve consists of the points

$$\gamma = \left\{(\gamma_1(t), \gamma_2(t), \gamma_3(t) \;\middle|\; t \in \mathbb{R}\right\}.$$

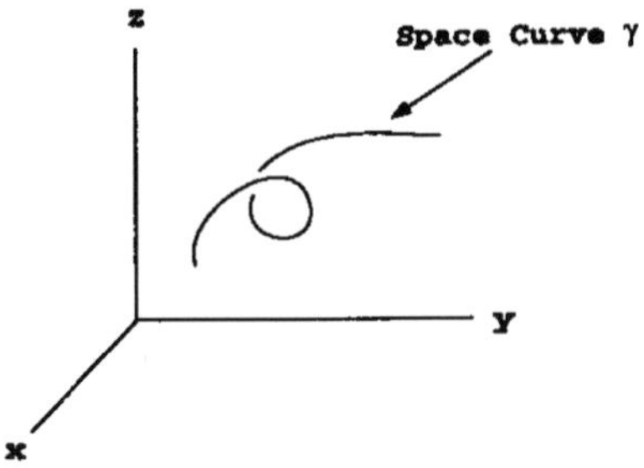

Consider now an arbitrary vector field $\vec{F}(x,y,z)$ restricted to our space curve γ, i.e., the set of vectors

$$\vec{F}(\gamma_1(t), \gamma_2(t), \gamma_3(t))$$

where t ranges over the real numbers. The graphical display of these vectors will resemble

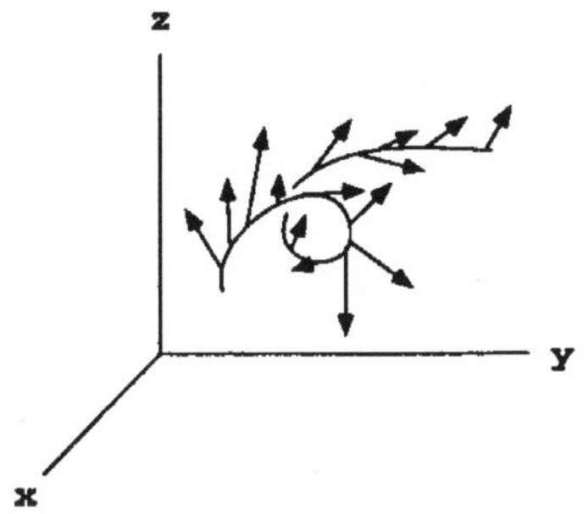

where the vectors which appear along the curve are induced by the vector field $\vec{F}$ (and are *not*, in general, tangent to the curve).

One good way to create this display is to first consider the graphical display of the vector field $\vec{F}$ itself,

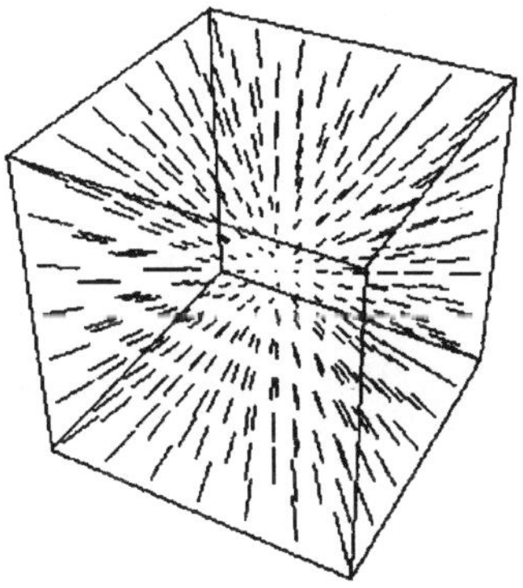

and then insert the space curve γ.

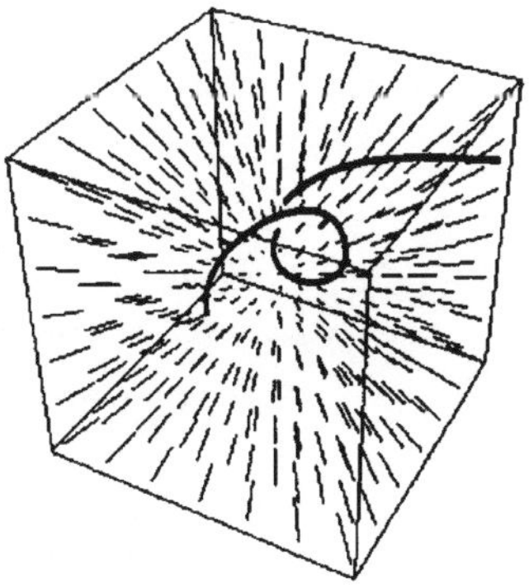

Example [27.11] (The wind) Let γ be the path of a small airplane as it flies through the air. We may think of the wind as a vector field which applies a force at each point in space. The restriction of the wind vector field to the curve γ gives a visualization of the wind forces on the airplane as it flies through the atmosphere.

Let

$$\vec{\mathbf{r}} = x\vec{\mathbf{i}} + y\vec{\mathbf{j}} + z\vec{\mathbf{k}}$$

be the position vector. If we restrict $\vec{\mathbf{r}}$ to a space curve

$$\gamma = \left\{ (\gamma_1(t), \gamma_2(t), \gamma_3(t)) \,\middle|\, t \in \mathbb{R} \right\}$$

we obtain

$$\vec{\mathbf{r}}(\gamma_1(t), \gamma_2(t), \gamma_3(t)) = \gamma_1(t)\vec{\mathbf{i}} + \gamma_2(t)\vec{\mathbf{j}} + \gamma_3(t)\vec{\mathbf{k}}.$$

In order to simplify our exposition we shall use the following notation:

$$\begin{aligned}
\vec{\mathbf{F}}(\gamma(t)) &= \vec{\mathbf{F}}(\gamma_1(t), \gamma_2(t), \gamma_3(t)) \\
\vec{\mathbf{r}}(\gamma(t)) &= \gamma_1(t)\vec{\mathbf{i}} + \gamma_2(t)\vec{\mathbf{j}} + \gamma_3(t)\vec{\mathbf{k}} \\
d\vec{\mathbf{r}} &= dx\vec{\mathbf{i}} + dy\vec{\mathbf{j}} + dz\vec{\mathbf{k}} \\
\frac{d\vec{\mathbf{r}}(\gamma(t))}{dt} &= \gamma_1'(t)\vec{\mathbf{i}} + \gamma_2'(t)\vec{\mathbf{j}} + \gamma_3'(t)\vec{\mathbf{k}} = \vec{\gamma}\,'(t).
\end{aligned} \tag{27.5}$$

Definition: *Let* $\vec{\mathbf{F}} = f_1(x, y, z)\vec{\mathbf{i}} + f_2(x, y, z)\vec{\mathbf{j}} + f_3(x, y, z)\vec{\mathbf{k}}$ *be a continuous vector field (i.e., the functions* f_1, f_2, *and* f_3 *are continuous) and let* $\gamma = (\gamma_1(t), \gamma_2(t), \gamma_3(t))$ *be a differentiable space curve on the interval* $a \le t \le b$. *The* ***line integral of*** $\vec{\mathbf{F}}$ ***over*** γ *is defined to be*

$$\int_\gamma \vec{\mathbf{F}} \cdot d\vec{\mathbf{r}} = \int_a^b \vec{\mathbf{F}}(\gamma(t)) \cdot \vec{\gamma}\,'(t)\, dt. \tag{27.6}$$

Remark: The physical concept of work allows us to translate the formal definition of the line integral into more intuitive terms. By definition, work is force times the distance moved. Thus if $\vec{\mathbf{F}}$ is a force field, the line integral

$$\int_\gamma \vec{\mathbf{F}} \cdot d\vec{\mathbf{r}}$$

represents the total work done as a particle is pushed along the curve γ by the forces of the vector field $\vec{\mathbf{F}}$.

To compute the line integral explicitly we assume that $\vec{\mathbf{F}}$ takes the form

$$\vec{\mathbf{F}} = f_1(x,y,z)\vec{\mathbf{i}} + f_2(x,y,z)\vec{\mathbf{j}} + f_3(x,y,z)\vec{\mathbf{k}},$$

and $\gamma = (\gamma_1(t), \gamma_2(t), \gamma_3(t))$. Then

$$\int_\gamma \vec{\mathbf{F}} \cdot d\vec{\mathbf{r}} = \int_\gamma \Big(f_1(x,y,z)dx + f_2(x.y.z)dy + f_3(x,y,z)dz\Big)$$

$$\begin{aligned} = \int_a^b \Big(& f_1(\gamma_1(t), \gamma_2(t), \gamma_3(t)\,)\,\gamma_1'(t) \\ & + f_2(\gamma_1(t), \gamma_2(t), \gamma_3(t)\,)\,\gamma_2'(t) \\ & + f_3(\gamma_1(t), \gamma_2(t), \gamma_3(t)\,)\,\gamma_3'(t)\Big)\; dt. \end{aligned} \tag{27.7}$$

Example [27.12] Let $\vec{\mathbf{F}} = xz\vec{\mathbf{i}} + yz\vec{\mathbf{j}} + x\vec{\mathbf{k}}$ be a vector field. Evaluate $\int_\gamma \vec{\mathbf{F}} \cdot d\vec{\mathbf{r}}$ along the space curve $\gamma = (1, t, t^2)$ for $0 \le t \le 1$.

In this example $\gamma_1(t) = 1$, $\gamma_2(t) = t$, and $\gamma_3(t) = t^2$. Hence $\gamma_1'(t) = 0$, $\gamma_2'(t) = 1$, $\gamma_3'(t) = 2t$, and

$$\begin{aligned} \int_\gamma \vec{\mathbf{F}} \cdot d\vec{\mathbf{r}} &= \int_0^1 \Big(1 \cdot t^2\gamma_1'(t) + t \cdot t^2\gamma_2'(t) + 1\gamma_3'(t)\Big) \\ &= \int_0^1 (t^3 + 2t)\; dt \;=\; \frac{5}{4}. \end{aligned}$$

Example [27.13] Evaluate the integral

$$\int_\gamma x^3yz\; dx + y^2x\; dy + 2xz\; dz$$

along the parabola γ which is given by the equation $y = x^2$ (in the $x\,y$–plane) on the interval $0 \le x \le 2$.

We attack this last example by first parameterizing γ,

$$\gamma(t) = (t, t^2, 0).$$

Since $x = \gamma_1(t) = t$, $y = \gamma_2(t) = t^2$, and $z = \gamma_3(t) = 0$ we see that $dx = \gamma_1'(t)\,dt = 1\,dt$, $dy = \gamma_2'(t)\,dt = 2t\,dt$, $dz = \gamma_3'(t)\,dt = 0$, and

$$\int_\gamma x^3yz\;dx + y^2x\;dy + 2xz\;dz = \int_0^2 t^5 \cdot 2t\;dt = \frac{256}{7}.$$

Exercises for §27.2

Use your CAS to evaluate the following line integrals $\int_\gamma \vec{\mathbf{F}}\cdot d\vec{\mathbf{r}}$. Graph the space curve γ in each case.

(1) $\vec{\mathbf{F}} = x^2y\vec{\imath} + 2xyz\vec{\mathbf{k}}$, $\gamma(t) = (-1, t, 0)$ for $0 \le t \le 1$.

(2) $\vec{\mathbf{F}} = x^2y\vec{\imath} + 2xyz\vec{\mathbf{k}}$, $\gamma(t)$ is the parabola $y = x^2$ on the plane $z = 1$ for $0 \le x \le 1$.

(3) $\vec{\mathbf{F}} = 3xz\vec{\imath} + x^2\vec{\jmath} + 2yz\vec{\mathbf{k}}$, $\gamma(t) = (t, t, t)$ for $1 \le t \le 2$.

(4) $\vec{\mathbf{F}} = 3xz\vec{\imath} + x^2\vec{\jmath} + 2yz\vec{\mathbf{k}}$, γ is the parabola $z = 2y^2$ on the plane $x = 3$ for $0 \le y \le 1$.

(5) $\vec{\mathbf{F}} = y\sqrt{x}\vec{\imath} - z\sqrt{y}\vec{\jmath} + x\vec{\mathbf{k}}$, $\gamma(t) = (1, t^3, 5)$ for $0 \le t \le 4$.

(6) $\vec{\mathbf{F}} = 3xz\vec{\imath} + x^2\vec{\jmath} + 2yz\vec{\mathbf{k}}$, γ is the boundary of a square which is in the $x\,y$–plane, centered at the origin, and whose sides have length 1.

(7) Evaluate $\int_\gamma (3x^2y\;dx + x^3\;dy)$ where γ is the line segment (in the $x\,y$–plane) which joins the points $(1,3)$ and $(-2,4)$.

(8) Evaluate $\int_\gamma (2xy\;dx + 3x^2y^3\;dy - xz\;dz)$ where γ is the line segment which joins the points $(1,-1,0)$ and $(0,2,-1)$.

(9) Let $\vec{\mathbf{F}} = -y\vec{\imath} + x\vec{\jmath}$. Graph the vector field $\vec{\mathbf{F}}$ and find the work done by the force field $\vec{\mathbf{F}}$ when moving a particle along the parabola $y = x^2$ on the interval $0 \le x \le 2$.

(10) Let $\vec{\mathbf{F}} = -y\vec{\imath} + x\vec{\jmath}$. Find the work done by the force field $\vec{\mathbf{F}}$ in moving a particle (clockwise or counterclockwise) around the unit circle in the $x\,y$–plane.

(11) Let $\vec{\mathbf{F}} = 2xy\vec{\imath} + x^2\vec{\jmath}$. Graph the vector field $\vec{\mathbf{F}}$ and find the work done by the force field $\vec{\mathbf{F}}$ when moving a particle exactly once around the unit circle in the $x\,y$–plane.

(12) Let $\vec{\mathbf{F}} = 2xyz\vec{\imath} + x^2z\vec{\jmath} + x^2y\vec{\mathbf{k}}$. Find the work done by the force field $\vec{\mathbf{F}}$ when moving a particle around the circle $(5, \cos(t), \sin(t))$ for $0 \le t \le 2\pi$.

27.3 Independence of Path

Let $\vec{F}$ be a continuous vector field and let $\vec{\mathbf{r}} = x\vec{\mathbf{i}} + y\vec{\mathbf{j}} + z\vec{\mathbf{k}}$ be the position vector field. Suppose

$$\gamma(t) = (\gamma_1(t), \gamma_2(t), \gamma_3(t))$$

is a smooth space curve on the interval $a \leq t \leq b$ which connects the points $A = \gamma(a)$ and $B = \gamma(b)$.

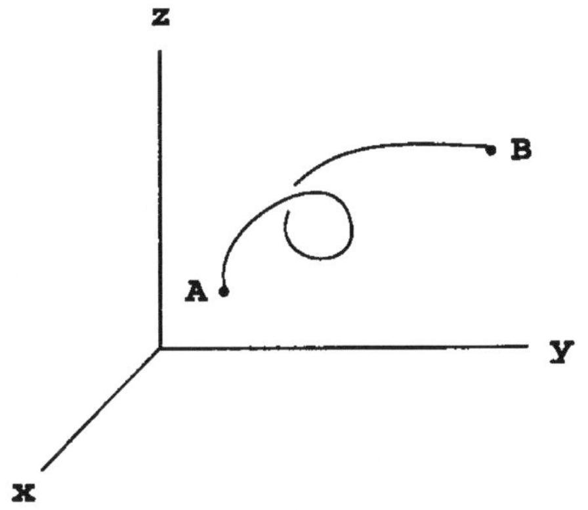

Since there are many possible space curves which connect A and B, for example,

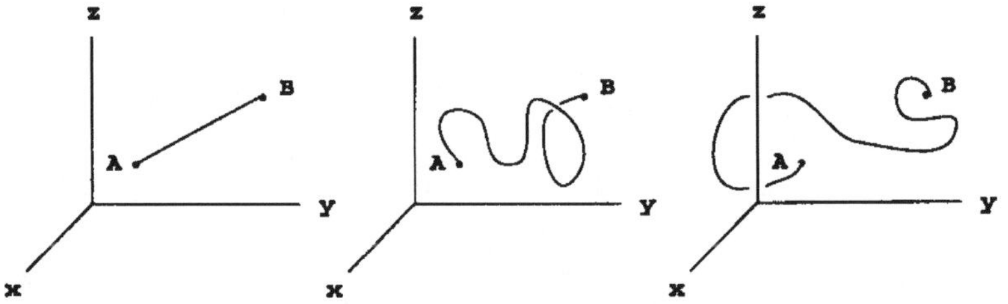

it is natural to ask the following question.

Question: Assuming the points A and B are fixed, does the line integral

$$\int_\gamma \vec{\mathbf{F}} \cdot d\vec{\mathbf{r}}$$

depend on the path γ which connects A and B?

A sensible approach to this question is to consider a particular vector field $\vec{\mathbf{F}}$ and a few specific paths.

Example [27.14] Let $\vec{\mathbf{F}}(x,y,z) = 2xy\vec{\mathbf{i}} + x^2\vec{\mathbf{j}} + \vec{\mathbf{k}}$. Compare the line integrals

$$\int_{\gamma_1} \vec{\mathbf{F}}\cdot d\vec{\mathbf{r}}, \qquad \int_{\gamma_2} \vec{\mathbf{F}}\cdot d\vec{\mathbf{r}}, \qquad \int_{\gamma_3} \vec{\mathbf{F}}\cdot d\vec{\mathbf{r}},$$

where γ_1 is the straight line between $(0,0,0)$ and $(1,1,1)$, $\gamma_2(t) = (t, t^3, t^2)$ for $0 \le t \le 1$, and $\gamma_3(t) = (t, \sin(\frac{5\pi}{2}t), t)$ where again $0 \le t \le 1$.

As can be seen in the figures below, all of the space curves defined above connect the points $(0,0,0)$ and $(1,1,1)$.

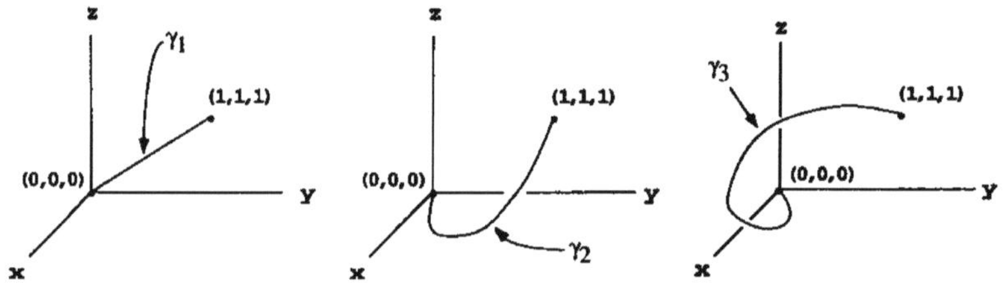

Now in the case of the first curve $x = t, y = t, z = t$, from which it follows that $dx = dy = dz = dt$, and

$$\begin{aligned}\int_{\gamma_1} \vec{\mathbf{F}}\cdot d\vec{\mathbf{r}} &= \int_{\gamma_1} \left(2xy\ dx + x^2\ dy + \ dz\right) \\ &= \int_0^1 \left(2t^2 + t^2 + 1\right)\ dt \\ &= \int_0^1 (1 + 3t^2)\ dt \ = \ 2.\end{aligned}$$

On γ_2, we have $x = t, y = t^3, z = t^2$, thus $dx = dt, dy = 3t^2\,dt, dz = 2t\,dt$, and

$$\begin{aligned}\int_{\gamma_2} \vec{\mathbf{F}}\cdot d\vec{\mathbf{r}} &= \int_{\gamma_2} 2xy\ dx + x^2\ dy + \ dz \\ &= \int_0^1 \left(2t^4 + 3t^4 + 2t\right)\ dt \\ &= \frac{2}{5} + \frac{3}{5} + 1 \ = \ 2.\end{aligned}$$

Finally when we integrate over γ_3, $x = t, y = \sin(\frac{5\pi}{2}t), z = t$. Hence we see that $dx = dz = dt,\ dy = \frac{5\pi}{2}\cos(\frac{5\pi}{2}t)\,dt$, and

$$\begin{aligned}\int_{\gamma_3} \vec{\mathbf{F}}\cdot d\vec{\mathbf{r}} &= \int_{\gamma_3} \left(2xy\,dx + x^2\,dy + dz\right) \\ &= \int_0^1 \left(2t\sin\left(\frac{5\pi}{2}t\right) + \frac{5\pi}{2}t^2\cos\left(\frac{5\pi}{2}t\right) + 1\right) dt \\ &= 2.\end{aligned}$$

This example suggests that the line integral $\int_\gamma \vec{\mathbf{F}}\cdot d\vec{\mathbf{r}}$ does *not* depend on the path γ. Is this true in general? The next example indicates that the question is more subtle.

Example [27.15] Let $\vec{\mathbf{F}}(x,y,z) = xy\vec{\mathbf{i}} + x^2\vec{\mathbf{j}} + \vec{\mathbf{k}}$. Compare the line integrals

$$\int_{\gamma_1} \vec{\mathbf{F}}\cdot d\vec{\mathbf{r}}, \qquad \int_{\gamma_2} \vec{\mathbf{F}}\cdot d\vec{\mathbf{r}},$$

where γ_1 and γ_2 are as in Example [27.14].

This example differs from the previous one only very slightly, yet upon computing the integrals

$$\begin{aligned}\int_{\gamma_1} \vec{\mathbf{F}}\cdot d\vec{\mathbf{r}} &= \int_0^1 (2t^2 + 1)\,dt = \frac{5}{3} \\ \int_{\gamma_2} \vec{\mathbf{F}}\cdot d\vec{\mathbf{r}} &= \int_0^1 (4t^4 + 2t)\,dt = \frac{9}{5}\end{aligned}$$

we see a very different behavior.

The discrepancy between these two seemingly similar examples is explained in the following theorem.

Theorem [27.16] *If* $\vec{\mathbf{F}}$ *is a gradient vector field then the line integral*

$$\int_\gamma \vec{\mathbf{F}}\cdot d\vec{\mathbf{r}}$$

is independent of the smooth path γ *which connects the points* A *and* B.

Before detailing the proof of this very strong theorem we remark that in Example [27.14] the vector field $\vec{\mathbf{F}}(x,y,z) = 2xy\vec{\mathbf{i}} + x^2\vec{\mathbf{j}} + \vec{\mathbf{k}}$ is a gradient vector field with potential function $\phi(x, y, z) = x^2y + z$. Thus we have, in working out this example, verified the conclusion of our theorem.

In order to prove Theorem [27.16] we will actually demonstrate an even more informative statement.

Theorem [27.17] *Let* $\vec{\mathbf{F}} = \vec{\nabla}\phi(x, y, z)$ *be a gradient vector field, and let* $\gamma(t)$ *be a smooth space curve on the interval* $a \leq t \leq b$. *Then*

$$\int_\gamma \vec{\mathbf{F}} \cdot d\vec{\mathbf{r}} = \phi(\gamma(b)) - \phi(\gamma(a)). \tag{27.8}$$

Proof: Remarkable as this theorem is it actually follows directly from our definitions and the multi–variable chain rule (see §23.5). In our notation $x = \gamma_1(t)$, $y = \gamma_2(t)$, $z = \gamma_3(t)$ and $dx = \gamma_1'(t)\ dt$, $dy = \gamma_2'(t)\ dt$, $dz = \gamma_3'(t)\ dt$. Thus

$$\int_\gamma \vec{\mathbf{F}} \cdot d\vec{\mathbf{r}} = \int_\gamma \vec{\nabla}\phi \cdot d\vec{\mathbf{r}}$$

$$= \int_\gamma \left(\frac{\partial \phi}{\partial x}\, dx + \frac{\partial \phi}{\partial y}\, dy + \frac{\partial \phi}{\partial z}\, dz \right)$$

$$= \int_a^b \left(\frac{\partial \phi}{\partial x}\frac{dx}{dt} + \frac{\partial \phi}{\partial y}\frac{dy}{dt} + \frac{\partial \phi}{\partial z}\frac{dz}{dt} \right) dt$$

$$= \int_a^b \left(\frac{d}{dt}\Big(\phi(\gamma_1(t), \gamma_2(t), \gamma_3(t)) \Big) \right) dt$$

$$= \phi(\gamma(b)) - \phi(\gamma(a)).$$

Example [27.18] Recompute the integrals in Example [27.14].

In this example $A = \gamma(0) = (0, 0, 0)$ and $B = \gamma(1) = (1, 1, 1)$. In addition it was already noted that

$$\vec{F}(x, y, z) = 2xy\vec{\mathbf{i}} + x^2\vec{\mathbf{j}} + \vec{\mathbf{k}} = \vec{\nabla}(x^2y + z),$$

i.e., $\phi(x, y, z) = x^2y + z$ is a potential function for $\vec{\mathbf{F}}$. Thus our theorem tells us that

$$\int_\gamma \vec{\mathbf{F}} \cdot d\vec{\mathbf{r}} = \phi(1,1,1) - \phi(0,0,0) = 2.$$

Remark: Since the integral which appears in Theorem [27.12] does not depend on the path joining the endpoints $\gamma(a)$, $\gamma(b)$ it can also be written as

$$\int_{\gamma(a)}^{\gamma(b)} \vec{\mathbf{F}} \cdot d\vec{\mathbf{r}}.$$

Example [27.19] Evaluate the integral

$$\int_{(0,1)}^{(1,\frac{\pi}{2})} 2x\sin(x^2y)\,dx + x^2\sin(x^2y)\,dy,$$

by finding a suitable potential function.

In this case the potential function is given by

$$\phi(x, y) = -\cos(x^2y),$$

since

$$\vec{\nabla}(\phi) = 2xy\sin(x^2y)\vec{\mathbf{i}} + x^2\sin(x^2y)\vec{\mathbf{j}}.$$

Applying Theorem [27.17] we see that

$$\int_{(0,1)}^{(1,\frac{\pi}{2})} 2x\sin(x^2y)\,dx + x^2\sin(x^2y)\,dy = -\cos\left(\frac{\pi}{2}\right) + \cos(0) = 1.$$

Closed Curves

If the curve $\gamma(t)$ on the interval $a \le t \le b$ is a **closed curve,** i.e., $\gamma(a) = \gamma(b)$, then its graphical display will be that of a (possibly twisted, punched, and stretched) closed loop which does not intersect itself.

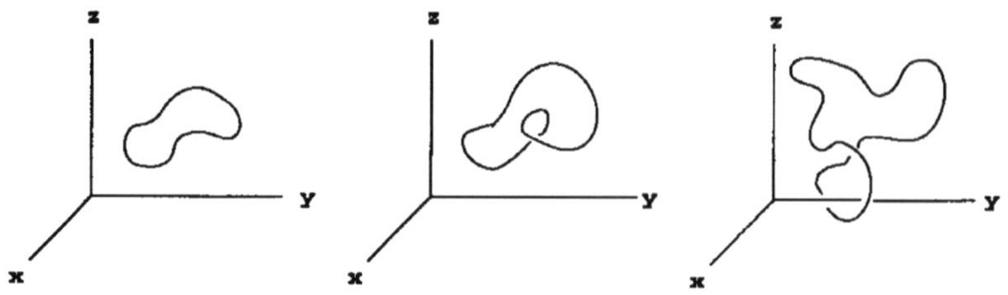

When these curves are traced out in stages, say on the interval $a \leq t \leq t_1$ for some choice of t_1,

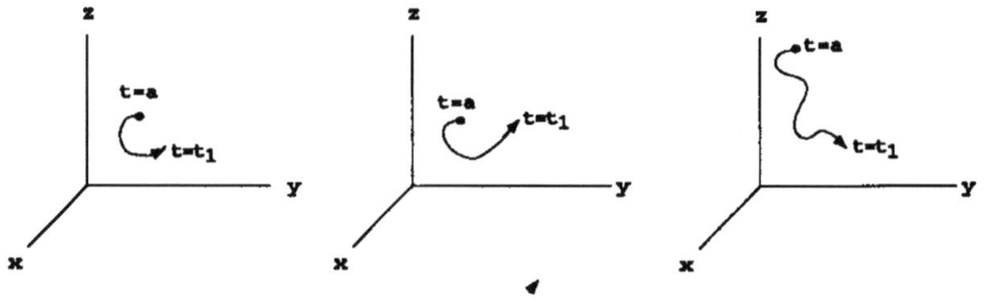

followed by the interval $t_1 \leq t \leq t_2$,

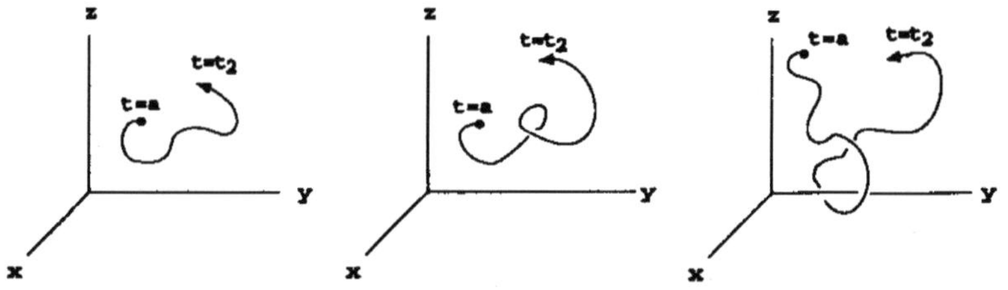

and finally on the interval $t_2 \leq t \leq b$ to complete the loop,

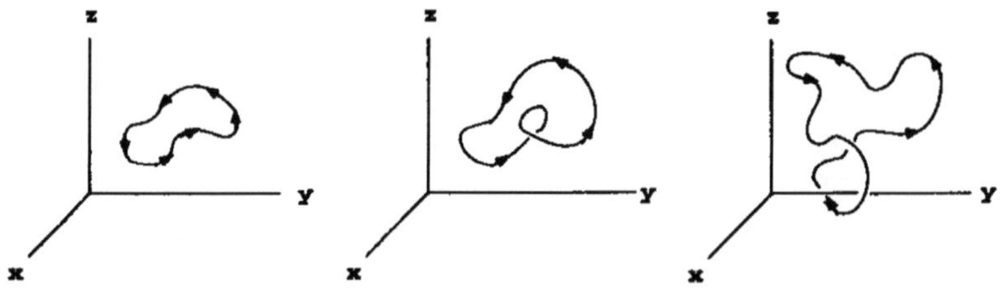

a natural counterclockwise motion appears. Now suppose we were to look at the closed curve from some far away point on the negative x–axis.

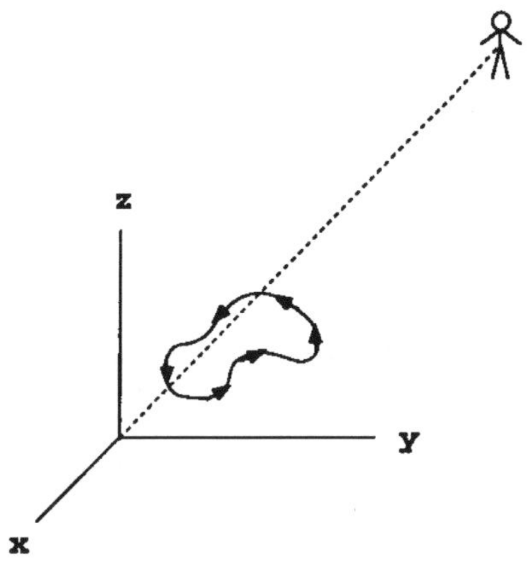

Upon tracing the curve the flow would look clockwise! Clearly the concepts of clockwise and counterclockwise rotation are not well defined in 3–dimensional space. We can however introduce the **reverse curve**

$$\gamma^-(t) = \gamma(a + b - t),$$

which satisfies the properties $\gamma^-(b) = \gamma(a)$ and $\gamma^-(a) = \gamma(b)$. When we fix a viewing perspective the reverse curve does have opposite orientation.

The definition of the reverse curve can be made for any path (closed or not), and it is easily demonstrated that given any continuous vector field and any smooth path γ,

$$\int_{\gamma^-} \vec{\mathbf{F}} \cdot d\vec{\mathbf{r}} = -\int_{\gamma} \vec{\mathbf{F}} \cdot d\vec{\mathbf{r}}.$$

Notation: When the curve γ is closed we shall denote the line integral of the vector field $\vec{F}$ on γ by

$$\oint_{\gamma} \vec{\mathbf{F}} \cdot d\vec{\mathbf{r}},$$

(the small circle in the middle of the integration sign being indicative of the fact that γ is closed).

We conclude this section with a striking result that is a direct consequence of Theorem [27.17].

Theorem [27.20] *If* $\vec{\mathbf{F}}$ *is a gradient vector field and* γ *is a smooth closed curve then*

$$\oint_{\gamma} \vec{\mathbf{F}} \cdot d\vec{\mathbf{r}} = 0. \tag{27.9}$$

Example [27.21] Let $\gamma(t) = (\cos(t), 3, \sin(t))$ for $0 \leq t \leq 2\pi$. Compute the reverse curve $\gamma^{-}(t)$ and evaluate

$$\oint_{\gamma^-} yz\,dx + xz\,dy + xy\,dz.$$

The curve γ is of course a circle with the indicated orientation (looked at from a point with positive x–component).

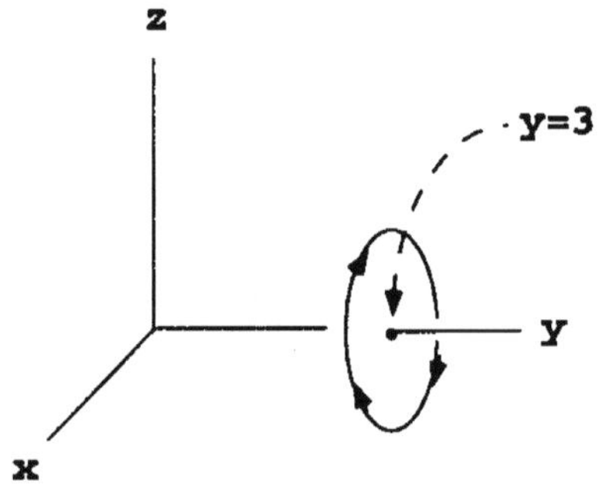

Since $a = 0$ and $b = 2\pi$ the reverse curve is given by

$$\begin{aligned}\gamma^{-}(t) &= \gamma(a+b-t)\\ &= \gamma(2\pi - t)\\ &= (\cos(2\pi - t), 3, \sin(2\pi - t)).\end{aligned}$$

The vector field in this example is in fact a gradient vector field

$$\vec{\mathbf{F}}(x,y,z) = yz\vec{\mathbf{i}} + xz\vec{\mathbf{j}} + xy\vec{\mathbf{k}} = \vec{\nabla}(xyz)$$

and thus

$$\oint_{\gamma^-} (yz\,dx + xz\,dy + xy\,dz) = 0.$$

Exercises for §27.3

Evaluate each of the following line integrals by finding a suitable potential function.

(1) $\int_{(0,0)}^{(1,1)} (y\,dx + (x + 3y^2)\,dy).$

(2) $\int_{(1,-1)}^{(0,3)} (2xy^3\,dx + 3x^2y^2\,dy).$

(3) $\int_{(0,0)}^{(1,\frac{\pi}{3})} (\sin(y)\,dx + x\cos(y)\,dy).$

(4) $\int_{(1,1)}^{(2,e)} (\ln(y)\,dx + \frac{x}{y}\,dy).$

(5) $\int_{(1,1)}^{(4,9)} (y \cdot \sqrt{\frac{y}{x}}\,dx + 3\sqrt{xy}\,dy).$

(6) $\int_{(0,0,0)}^{(1,1,1)} (2xyz\,dx + x^2z\,dy + x^2y\,dz).$

(7) $\int_{(0,1,1)}^{(\frac{\pi}{2},1,2)} (-(\sin(x))\,dx + z\,dy + y\,dz).$

(8) $\int_{(0,0,0)}^{(0,1,1)} (e^x\,dx + 2yz\,dy + y^2\,dz).$

(9) $\int_{(1,1,1)}^{(2,2,2)} (z\,dx + x\,dz).$

27.4 Green's Theorem in the Plane

Let $\gamma(t) = (\gamma_1(t), \gamma_2(t))$, with $a \le t \le b$, be a closed curve in the xy–plane which does not cross itself and is traversed only once as $a \le t \le b$.

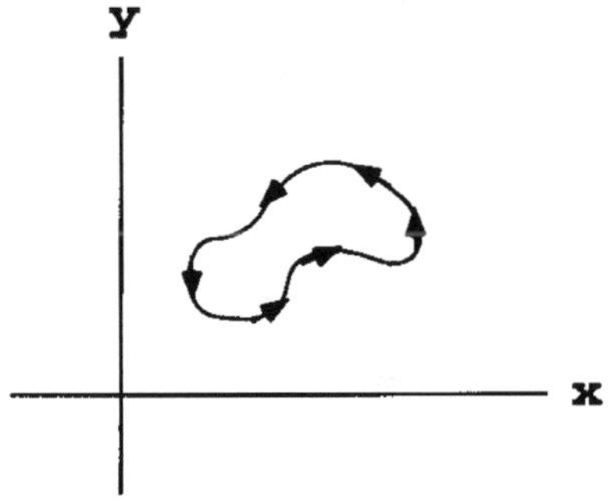

We claim that since γ lies in a 2–dimensional plane the concept of orientation can be made precise. If we imagine a 2–dimensional person (who spends her life in the 2–dimensional plane) traversing the curve $\gamma(t)$ as t increases from a to b

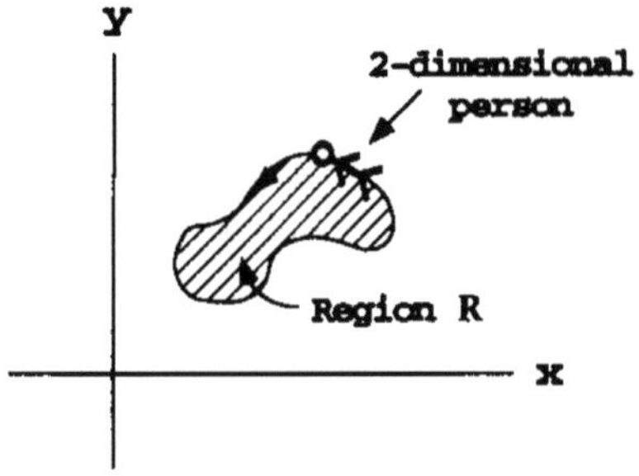

then the curve is said to have a **counterclockwise orientation** provided the region the curve surrounds, denoted $\mathcal{R}$, always lies to the left of the 2–dimensional person as she travels around the curve.

Example [27.22] Determine the orientation of $\gamma(t) = (\cos(t), -\sin(t))$ for the interval $0 \le t \le 2\pi$.

This closed curve is of course a circle. Let's plot the quarter–circle for the interval $0 \le t \le \frac{\pi}{2}$.

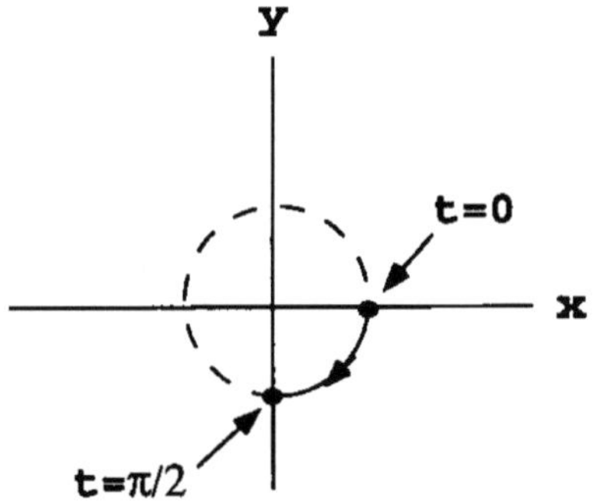

Clearly the orientation is clockwise.

Definition: *A curve $\gamma(t) = (\gamma_1(t), \gamma_2(t))$ for $a \le t \le b$ is said to be a* ***simple closed curve*** *if the following hold:*

(1) *$\gamma_1(t)$ and $\gamma_2(t)$ are continuous functions of t and are* ***piecewise differentiable****, i.e., there is a partition of the interval $a = t_0 \le t_1 \le \cdots \le t_n = b$ such that $\gamma_1(t)$ and $\gamma_2(t)$ are differentiable on each interval $t_i \le t \le t_{i+1}$.*

(2) *$\gamma(a) = \gamma(b)$ i.e., γ is closed.*

(3) *γ does not self–intersect.*

(4) *γ is traversed only once as t increases from a to b.*

We are now in a position to state a remarkable identity between line integrals around simple closed curves (with counterclockwise orientation) and double integrals over the regions which the curves enclose.

Theorem [27.23] (Green's Theorem) *Let $P(x, y)$ and $Q(x, y)$ be differentiable functions of x, y and let $\mathcal{R}$ be a region whose boundary, $\partial\mathcal{R}$, is a simple closed curve with counterclockwise orientation. Then*

$$\oint_{\partial\mathcal{R}} P\,dx + Q\,dy \;=\; \iint_{\mathcal{R}} \left(\frac{\partial Q}{\partial x} - \frac{\partial P}{\partial y}\right)\,dx\,dy. \tag{27.10}$$

We defer our proof briefly in order to illustrate this striking statement.

Example [27.24] Let $\vec{\mathbf{F}} = (x^2 - y^2)\vec{\mathbf{i}} + xy\vec{\mathbf{j}}$. Compute

$$\oint_{\partial\mathcal{R}} \vec{\mathbf{F}}\cdot d\vec{\mathbf{r}}$$

where $\partial\mathcal{R}$ is the boundary of the square pictured below.

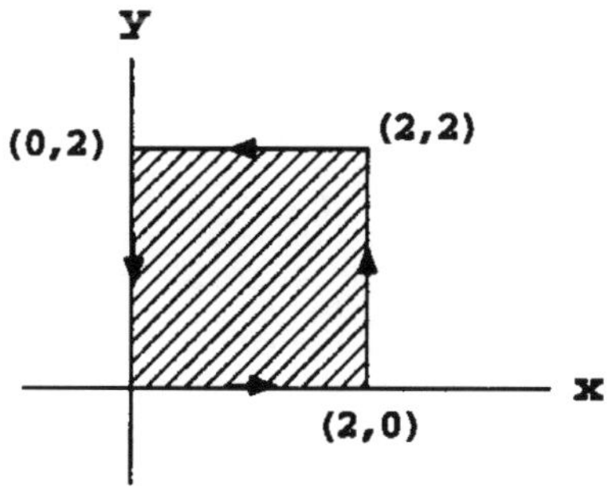

In this example $P(x,y) = x^2 - y^2$, $Q(x,y) = xy$, and Green's theorem tells us that

$$\oint_{\partial\mathcal{R}} \vec{\mathbf{F}}\cdot d\vec{\mathbf{r}} = \oint_{\partial\mathcal{R}} \Big((x^2 - y^2)\,dx + xy\,dy\Big)$$

$$= \iint_{\mathcal{R}} \left(\frac{\partial Q}{\partial x} - \frac{\partial P}{\partial y}\right) dx\,dy$$

$$= \int_0^2 \int_0^2 \Big(y - (-2y)\Big)\,dx\,dy$$

$$= 2\int_0^2 3y\,dy = 12.$$

Example [27.25] Given a region $\mathcal{R}$ whose boundary is a simple closed curve with counterclockwise orientation, show that

$$\text{Area}(\mathcal{R}) = \oint_{\partial\mathcal{R}} x\,dy. \qquad (27.11)$$

When we set $P = 0$ and $Q = x$ and use Green's theorem we immediately obtain the identity:

$$\oint_{\partial\mathcal{R}} x\,dy = \oint_{\partial\mathcal{R}} (P\,dx + Q\,dy)$$

$$= \iint_{\mathcal{R}} \left(\frac{\partial Q}{\partial x} - \frac{\partial P}{\partial y}\right) dx\,dy$$

$$= \iint_{\mathcal{R}} dx\,dy = \text{Area}(\mathcal{R}).$$

Example [27.26] Compute the area of the ellipse

$$\frac{x^2}{a^2} + \frac{y^2}{b^2} = 1$$

using Green's theorem.

To perform this computation we first parameterize the ellipse as a simple closed curve:

$$\gamma(t) = (a\cos(t), b\sin(t)),$$

for $0 \le t \le 2\pi$. Upon tracing this curve the orientation is quickly seen to be counterclockwise and thus by (27.11)

$$\text{Area(Ellipse)} = \oint_\gamma x\, dy.$$

To compute this line integral notice that $x = a\cos(t), y = b\sin(t), dy = b\cos(t)\, dt$ and hence

$$\text{Area(Ellipse)} = \int_0^{2\pi} a\cos(t) \cdot b\cos(t)\, dt$$

$$= ab\int_0^{2\pi} (\cos(t))^2\, dt$$

$$= \frac{ab}{2}\int_0^{2\pi} \Big(1 + \cos(2t)\Big)\, dt = ab\pi.$$

Before we can present the proof of Green's theorem (which turns out to be surprisingly elementary) we require the following,

Definition: *A region $\mathcal{R}$ is said to be* **simple** *if any horizontal or vertical line intersects its boundary in at most 2 points.*

For example, the following regions are simple,

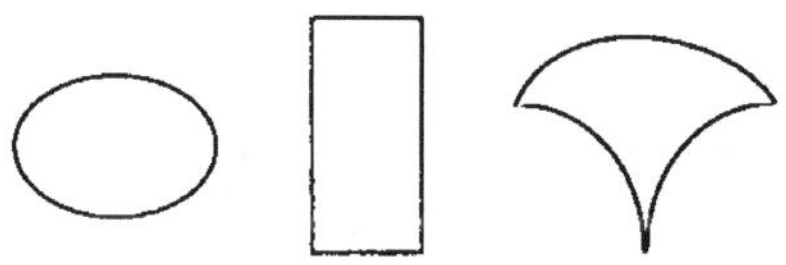

while the remainder are not.

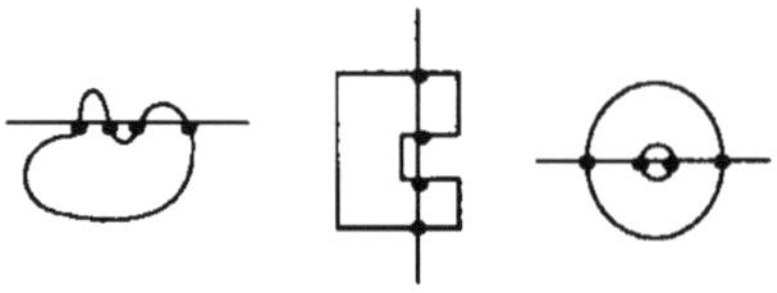

Proof of Green's Theorem:

We give the proof in the case $\mathcal{R}$ is a simple region whose boundary is again denoted $\partial\mathcal{R}$. The more general form of Green's theorem is obtained by decomposing a general region into simple regions such as in the diagram below.

Furthermore, we may assume that our simple region takes the form

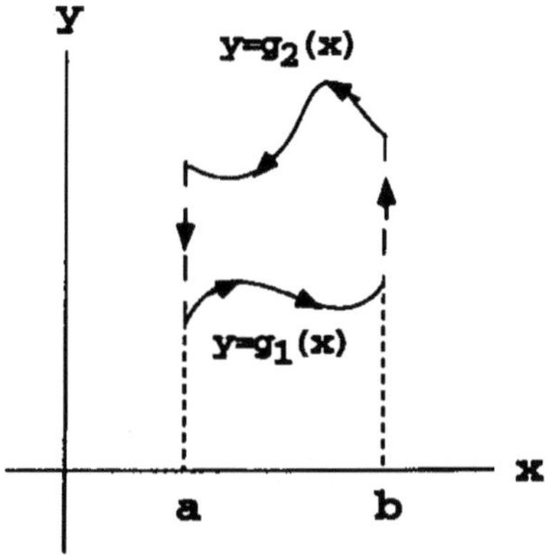

where the vertical lines in the above figure could have length 0.

To demonstrate the identity in Green's theorem it suffices to show that

$$\oint_{\partial\mathcal{R}} P\,dx = -\iint_{\mathcal{R}} \frac{\partial P}{\partial y}\,dx\,dy,$$

since the proof that

$$\oint_{\partial\mathcal{R}} Q\,dy = \iint_{\mathcal{R}} \frac{\partial Q}{\partial x}\,dx\,dy,$$

is entirely similar (and left to the reader).

Since $\partial\mathcal{R}$ is a union of four curves we parameterize each one separately. Notice that in the parameterization of the vertical segments, x is constant (either a or b). Thus $dx = 0$ on those segments and they do not contribute to the line integral $\oint_{\partial\mathcal{R}} P\,dx$. The remaining curves are parameterized by

$$x = x, \quad y = g_2(x), \qquad\qquad x = x, \quad y = g_1(x),$$

respectively. Hence,

$$\begin{aligned}
\oint_{\partial\mathcal{R}} P\,dx &= \int_a^b P(x, g_1(x))\,dx + \int_b^a P(x, g_2(x))\,dx \\
&= -\int_a^b \Big(P(x, g_2(x)) - P(x, g_1(x))\Big)\,dx \\
&= -\int_a^b \int_{g_1(x)}^{g_2(x)} \frac{\partial P}{\partial y}\,dx\,dy \\
&= -\iint_{\mathcal{R}} \frac{\partial P}{\partial y}\,dx\,dy.
\end{aligned}$$

Notice that the final identity is a consequence of the Fundamental Theorem of Calculus.

Exercises for §27.4

Determine the orientation of the following closed curves by graphing them with your CAS.

(1) $\gamma(t) = (\sin(t), -\cos(t))$ for $0 \leq t \leq 2\pi$.

(2) $\gamma(t) = \left(-\frac{\sin}{2}, -\frac{\cos(t)}{3}\right)$ for $0 \leq t \leq 2\pi$.

(3) $\gamma(t) = \left(\frac{\cos(t)}{2}, -\frac{\sin(t)}{5}\right)$ for $0 \leq t \leq 2\pi$.

(4) $\gamma(t) = (\cos(t) + \sin(t), \cos(t) - \sin(t))$ for $0 \leq t \leq 2\pi$.

Evaluate the following line integrals by using Green's Theorem.

(5) $\oint\limits_{\partial\mathcal{R}} ((x^2 + y^2)\,dx - xy\,dy)$, where $\partial\mathcal{R}$ is the boundary of the rectangle $\mathcal{R}$: $0 \leq x \leq 1$, $0 \leq y \leq 2$ with counterclockwise orientation.

(6) $\oint\limits_{\partial\mathcal{D}} ((3x^2y + y^3)\,dx - y^{10}\,dy)$, where $\partial\mathcal{D}$ is the boundary of the disk $\mathcal{D}$: $x^2 + y^2 \leq 4$ with counterclockwise orientation.

(7) $\oint\limits_{\partial\mathcal{T}} ((e^x + y)\,dx - (x - e^y)\,dy)$, where $\partial\mathcal{T}$ is the boundary of the triangle $\mathcal{T}$ whose vertices are located at $(0,0)$, $(0,-1)$, and $(1,0)$ with counterclockwise orientation.

(8) $\oint\limits_{\gamma}((y^2 + x)\,dx + (xy - y)\,dy)$, where γ is the ellipse $\left(\frac{\cos(t)}{3}, -\frac{\sin(t)}{2}\right)$ for $0 \leq t \leq 2\pi$.

(9) $\oint\limits_{\gamma}((y^2 + x^2)\,dx + (2x - y^2)\,dy)$, where γ is the circle $(-3\cos(t), 3\sin(t))$ for $0 \leq t \leq 2\pi$.

In the following exercises graph the given closed curve γ with your CAS. Compute the area of the region bounded by γ using Green's Theorem. Check your answer by evaluating the area by a double integral on your CAS.

(10) $\gamma = \left\{(x,y) \mid \frac{x^2}{4} + \frac{y^2}{9} = 1\right\}$.

(11) $\gamma = \left\{\left(-\frac{\cos(t)}{4}, \frac{\sin(t)}{25}\right) \mid 0 \leq t \leq 2\pi\right\}$.

(12) $\gamma = \left\{((\cos(t))^3, (\sin(t))^3) \mid 0 \leq t \leq 2\pi\right\}$.

(13) $\gamma = \left\{(\cos(t) + \sin(t), \cos(t) - \sin(t)) \mid 0 \leq t \leq 2\pi\right\}$.

(14) γ = boundary of the triangle whose vertices are located at the points $(0,0)$, $(1,2)$, and $(-1,4)$.

(15) γ = boundary of the quadrilateral whose vertices at located at the points $(0,0)$, $(1,0)$, $(2,1)$, and $(-1,5)$.

Additional exercises for Chapter XXVII

Use your CAS to obtain graphical displays of the following vector fields. If your CAS does not have this capability, then simply graph a few representative vectors to simulate the graphical display of the vector field.

(1) $\vec{\mathbf{v}} = x^2\vec{\mathbf{i}} + y^2\vec{\mathbf{j}} + (x^2 + y^2)\vec{\mathbf{k}}$.

(2) $\vec{\mathbf{v}} = -y\vec{\mathbf{i}} + x\vec{\mathbf{j}} + z\vec{\mathbf{k}}$.

(3) $\vec{\mathbf{v}} = \vec{\nabla}\phi$ with $\phi(x,y) = \cos(x) + \frac{1}{2}y^2$.

In the following exercises determine whether or not $\vec{\mathbf{v}}$ is a gradient vector field and if it is find its potential function ϕ.

(4) $\vec{\mathbf{v}} = x^2y\vec{\mathbf{i}} + xy^2\vec{\mathbf{j}}$.

(5) $\vec{v} = ye^{xy}\vec{i} + xe^{xy}\vec{j}$.

(6) $\vec{v} = 3x^2y^2z\vec{i} + 2x^3yz\vec{j} + x^3y^2\vec{k}$.

(7) $\vec{v} = (x^2 + yz)\vec{i} + (y^2 + zx)\vec{j} + (z^2 + xy)\vec{k}$.

In the following graph the force field $\vec{F}$ on your CAS and find the work done by the force field when moving a particle along the space curve γ. In each case graph γ.

(8) $\vec{F} = x^2\vec{i} + y^2\vec{j}$ and $\gamma(t) = (t, e^t)$ for $0 \le t \le 10$.

(9) $\vec{F} = xy\vec{i} + yz\vec{j} + zx\vec{k}$ and γ is the triangle in the plane $x + y + z = 2$ with vertices $(2,0,0), (0,2,0), (0,0,2)$ with a counter-clockwise orientation looking down the z–axis.

(10) $\vec{F} = 3x^2y^2z\vec{i} + 2x^3yz\vec{j} + x^3y^2\vec{k}$ and γ is the triangle in the plane $x + y + z = 2$ with vertices $(2,0,0), (0,2,0), (0,0,2)$ with a counter-clockwise orientation looking down the z–axis.

(11) $\vec{F} = (z - \sin(x))\vec{i} + (x + y^2)\vec{j} + (y + e^z)\vec{k}$ and γ is the intersection of the unit sphere and the cone $z = \sqrt{x^2 + y^2}$ with counter clockwise orientation looking down the z -axis.

(12) $\vec{F} = (x^2 + yz)\vec{i} + (y^2 + zx)\vec{j} + (z^2 + xy)\vec{k}$ and γ is the parabola $(t, t^2, 0)$ for $0 \le t \le 5$.

Evaluate each of the following line integrals by finding a suitable potential function.

(13) $\int_{(1,0)}^{(2,1/3)} (2xy^3dx + (1 - 3x^2y^2)dy)$.

(14) $\int_{(1,0)}^{(3,3)} \left((e^x \ln y - \frac{e^y}{x})dx + (\frac{e^x}{y} - e^y \ln(x))dy\right)$.

(15) Find a function $g(x)$ for which

$$\vec{v} = g(x)\,(x\sin(y) + y\cos(y))\,\vec{i} + g(x)\,(x\cos(y) - y\sin(y))\,\vec{j}$$

is a gradient vector field.

Evaluate the following line integrals by using Green's Theorem.

(16) $\oint (x^2 - y^2)\,dx + x\,dy$, where γ is the circle $(x-2)^2 + (y+3)^2 = 16$.

(17) $\oint \ln(1 + y)\,dx - \frac{xy}{1+y}\,dy$ where γ is the triangle with vertices $(0,0), (2,0)$, and $(0,5)$.

Compute the Areas of the following regions using Green's Theorem. Graph the region with your CAS.

(18) The region in the first quadrant enclosed by $y = x$, $y = 1/x$, and $y = x/10$.

(19) The region $\gamma(t) = \left(5\cos^3(t), \sin^3(t)\right)$ for $0 \le t \le 2\pi$.

(20) The region enclosed by the $y = 3$ and $y = x^2$.

Chapter XXVIII

Surface Integrals

28.1 Surface Integrals

Let $z = f(x, y)$ be a function of x, y which is differentiable on a bounded region $\mathcal{R}$ in the $x\,y$–plane.

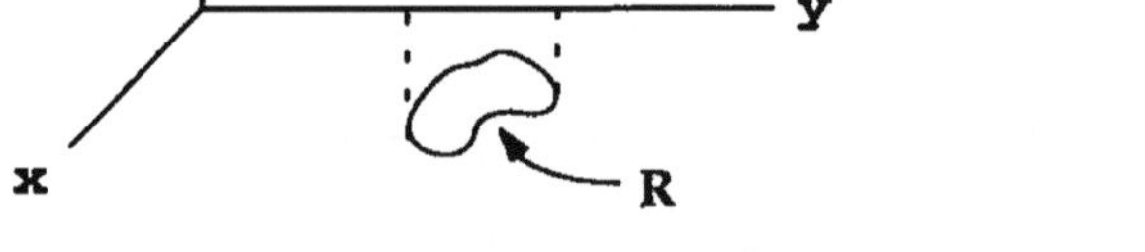

Denoting the piece of surface over $\mathcal{R}$ by

$$S \;=\; \{\; (x, y, z) \;\mid z = f(x, y) \text{ and } x, y \in \mathcal{R}\},$$

we claim that the area of S is given by the integral

$$\text{Area}(S) \;=\; \iint\limits_{\mathcal{R}} \sqrt{1 + \left(\frac{\partial f}{\partial x}\right)^2 + \left(\frac{\partial f}{\partial y}\right)^2}\; dx\, dy. \qquad (28.1)$$

This formula is reminiscent of the arc length formula we derived in §16.2 (see §22.2 which is also relevant). Before proving (28.1) we begin with some examples which demonstrate its utility.

Example [28.1] Find the area of the portion of the plane $x + 2y + 6z = 12$ which lies within the elliptic cylinder $\frac{x^2}{25} + \frac{y^2}{9} = 1$.

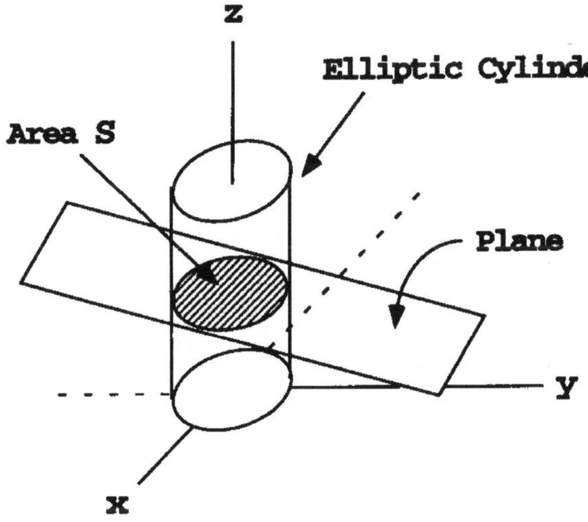

By definition

$$S = \left\{(x, y, z) \,\middle|\, x + 2y + 6z = 12 \text{ and } \frac{x^2}{25} + \frac{y^2}{9} \leq 1\right\}.$$

Thus $z = f(x, y) = \frac{1}{6}(12 - x - 2y)$ and $\frac{\partial f}{\partial x} = -\frac{1}{6}$, $\frac{\partial f}{\partial y} = -\frac{1}{3}$. Our formula now gives us the desired area:

$$\begin{aligned}
\text{Area}(S) &= \iint\limits_{\frac{x^2}{25}+\frac{y^2}{9}\leq 1} \sqrt{1 + \left(\frac{\partial f}{\partial x}\right)^2 + \left(\frac{\partial f}{\partial y}\right)^2}\, dx\, dy \\
&= \iint\limits_{\frac{x^2}{25}+\frac{y^2}{9}\leq 1} \sqrt{1 + \left(\frac{1}{36}\right) + \left(\frac{1}{9}\right)}\, dx\, dy \\
&= \frac{\sqrt{41}}{6} \iint\limits_{\frac{x^2}{25}+\frac{y^2}{9}=1} dx\, dy \\
&= \frac{\sqrt{41}}{6} \cdot 15\pi = \frac{5\sqrt{41}}{2}\pi.
\end{aligned}$$

Example [28.2] (Surface Area of a Sphere) Compute the surface area of a sphere of radius R.

As we have often observed, the points (x, y, z) on the surface of the sphere satisfy the equation $x^2 + y^2 + z^2 = R^2$, and the function

$$z = f(x,y) = \sqrt{R^2 - x^2 - y^2}$$

is well defined for $x^2 + y^2 \le R^2$. The graphical display is the familiar hemisphere of radius R.

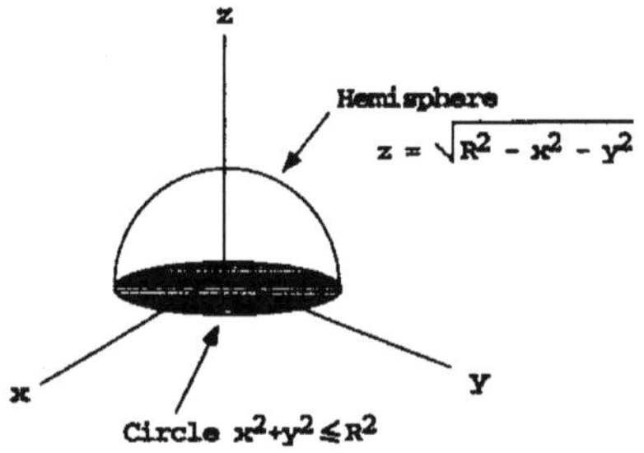

Now in this case the partial derivatives of $f(x, y)$ with respect to x and y are

$$\frac{\partial f}{\partial x} = \frac{1}{2}(R^2 - x^2 - y^2)^{-\frac{1}{2}} \cdot (-2x), \quad \frac{\partial f}{\partial y} = \frac{1}{2}(R^2 - x^2 - y^2)^{-\frac{1}{2}} \cdot (-2y),$$

respectively. Our formula dictates that the surface area of the hemisphere is given by the (improper) integral

$$\iint\limits_{x^2+y^2\le R^2} \sqrt{1 + \left(\frac{\partial f}{\partial x}\right)^2 + \left(\frac{\partial f}{\partial y}\right)^2}\, dx\, dy = \iint\limits_{x^2+y^2\le R^2} \sqrt{1 + \frac{x^2+y^2}{R^2 - x^2 - y^2}}\, dx\, dy$$

$$= R \iint\limits_{x^2+y^2\le R^2} \frac{dx\, dy}{\sqrt{R^2 - x^2 - y^2}}.$$

This integral lends itself naturally to polar coordinates in that the region we are integrating over can easily be described as a polar region. Thus we let

$$x = r\cos(\theta), \quad y = \sin(\theta), \quad dx\, dy = r\, dr\, d\theta,$$

and our integral becomes

$$R \iint\limits_{x^2+y^2\le R^2} \frac{dx\, dy}{\sqrt{R^2 - x^2 - y^2}} = R \int_0^{2\pi} \int_0^R \frac{r\, dr\, d\theta}{\sqrt{R^2 - r^2}}.$$

To complete this computation we use the simple substitution $u = R^2 - r^2$, $du = -2r\,dr$,

$$\int_0^R \frac{r\,dr\,d\theta}{\sqrt{R^2 - r^2}} = -\frac{1}{2}\int_{R^2}^0 \frac{du}{\sqrt{u}} = R,$$

implying

$$R\int_0^{2\pi}\int_0^R \frac{r\,dr\,d\theta}{\sqrt{R^2 - r^2}} = 2\pi R^2$$

is the surface area of the hemisphere, and $4\pi R^2$ is the surface area of the entire sphere.

Derivation of the Surface Area Formula

Recall (see §20.7) that if $\vec{v}$, $\vec{w}$ are non–zero vectors in 3–dimensional space which are not parallel then they span a parallelogram

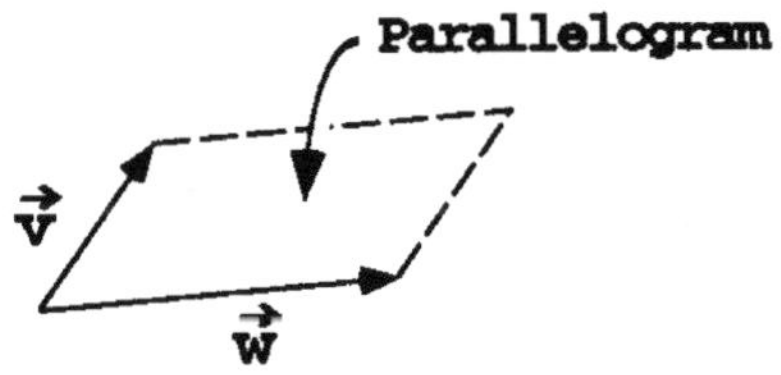

whose area is $|\vec{v} \times \vec{w}|$.

Now consider a small rectangle of area, $\Delta x\,\Delta y$ in the $x\,y$–plane. When we apply $z = f(x, y)$ to this small rectangle the result will be a small piece of the surface $z = f(x, y)$.

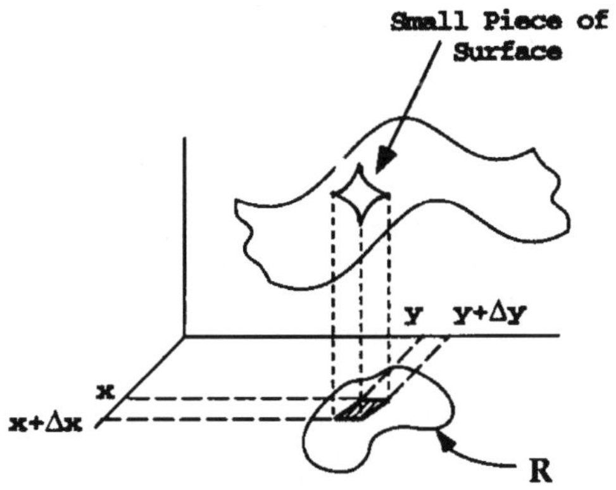

When we magnify this small piece of surface and compute its endpoints we obtain the following figure.

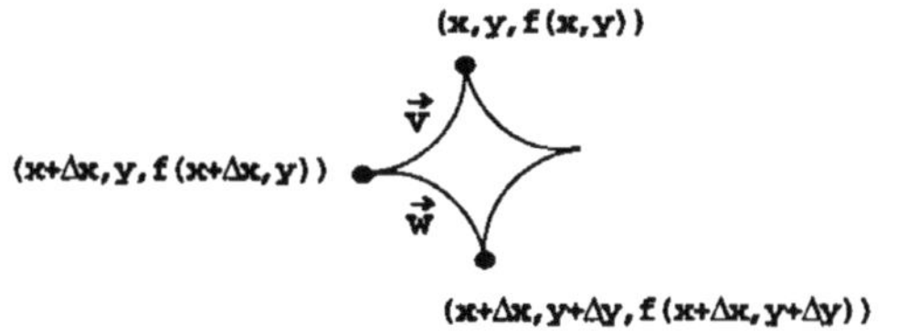

When we define

$$\begin{aligned}\vec{\mathbf{v}} &= \Delta x\vec{\mathbf{i}} + 0\vec{\mathbf{j}} + (f(x+\Delta x, y) - f(x))\vec{\mathbf{k}} \\ &\approx \Delta x\vec{\mathbf{i}} + \frac{\partial f}{\partial x}\Delta x\vec{\mathbf{k}}, \\ \vec{\mathbf{w}} &= 0\vec{\mathbf{i}} + \Delta y\vec{\mathbf{j}} + (f(x+\Delta x, y+\Delta y) - f(x+\Delta x, y))\vec{\mathbf{k}} \\ &\approx \Delta y\vec{\mathbf{j}} + \frac{\partial f}{\partial y}\Delta y\vec{\mathbf{k}},\end{aligned}$$

the area of our small piece of surface is approximately

$$\begin{aligned}|\vec{\mathbf{v}} \times \vec{\mathbf{w}}| &\approx \begin{vmatrix} \vec{\mathbf{i}} & \vec{\mathbf{j}} & \vec{\mathbf{k}} \\ \Delta x & 0 & \frac{\partial f}{\partial x}\Delta x \\ 0 & \Delta y & \frac{\partial f}{\partial y}\Delta y \end{vmatrix} \\ &= \left| -\frac{\partial f}{\partial x}\Delta x\,\Delta y\,\vec{\mathbf{i}} - \frac{\partial f}{\partial y}\Delta x\,\Delta y\,\vec{\mathbf{j}} + \Delta x\,\Delta y\,\vec{\mathbf{k}} \right| \\ &= \left(\sqrt{1 + \left(\frac{\partial f}{\partial x}\right)^2 + \left(\frac{\partial f}{\partial y}\right)^2}\right)\Delta x\,\Delta y.\end{aligned}$$

The total surface area in question is obtained by summing over all such small rectangles which lie in the region $\mathcal{R}$ and we thus obtain our formula:

$$\text{Area}(S) = \iint_{\mathcal{R}} \sqrt{1 + \left(\frac{\partial f}{\partial x}\right)^2 + \left(\frac{\partial f}{\partial y}\right)^2}\,dx\,dy.$$

Exercises for §28.1

Compute the surface area of the following surfaces S.

(1) S is the cone $z^2 = x^2 + y^2$ with $0 \le z \le 1$.

(2) S is the portion of the plane $2x - y + 3z = 4$ which lies within the elliptic cylinder $x^2 + \frac{y^2}{4} \le 2$.

(3) S is the surface $z = xy$ over a disk of radius 4 centered at the origin.

(4) S is the part of the unit sphere (centered at the origin, radius 1) above the plane $z = \frac{\sqrt{2}}{2}$.

(5) S is the ellipsoid $\frac{x^2}{a^2} + \frac{y^2}{b^2} + \frac{z^2}{c^2} = 1$.

(6) S is the portion of the sphere $x^2 + y^2 + z^2 = 4$ which lies within the elliptic cylinder $x^2 + \frac{y^2}{4} \le 1$.

28.2 Surface Integrals for Open Surfaces

When we consider surfaces we must separate those which are boundaries of solids, which are termed **closed surfaces**, and those which are not.

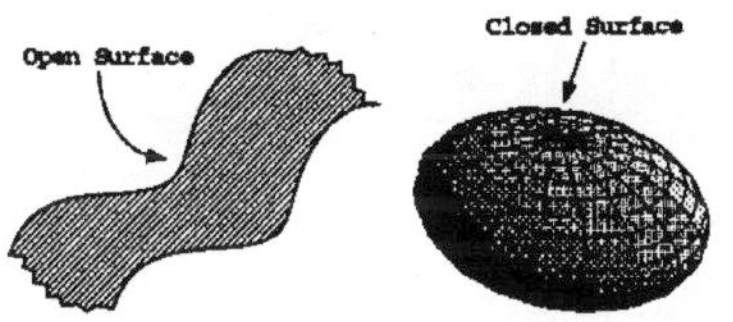

The latter type of surface, termed an **open surface**, will be the focus of the remainder of this section, while the case of closed surfaces is discussed in §28.3.

Let σ denote a small patch of 2–dimensional surface lying in 3–dimensional space.

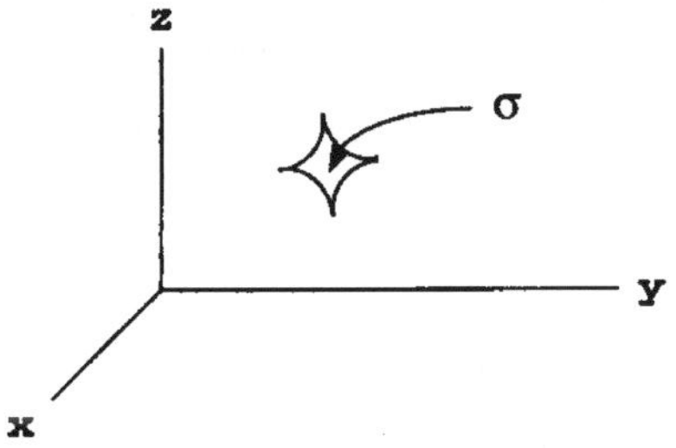

We shall assume that σ is actually *flat*, i.e., σ lies in a plane.

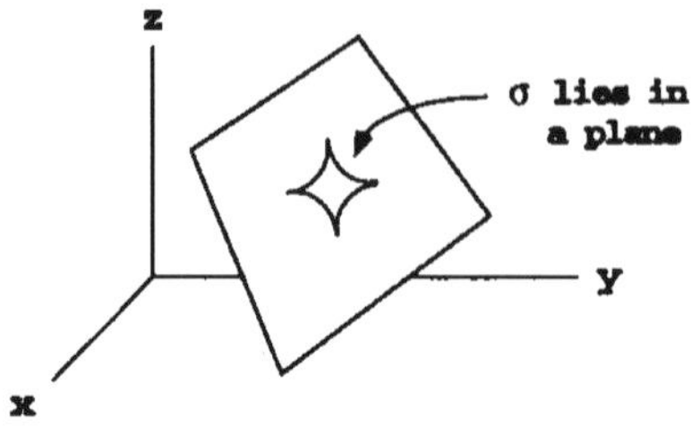

Definition: *A* ***surface vector****, denoted $\vec{\sigma}$, associated to σ is a vector perpendicular to σ which has length equal to the area of σ.*

The concept here is easily visualized: if Area(σ_1) is large then $\vec{\sigma}_1$ will be rather long, while $\vec{\sigma}_2$ will be a short vector when Area(σ_2) is itself small.

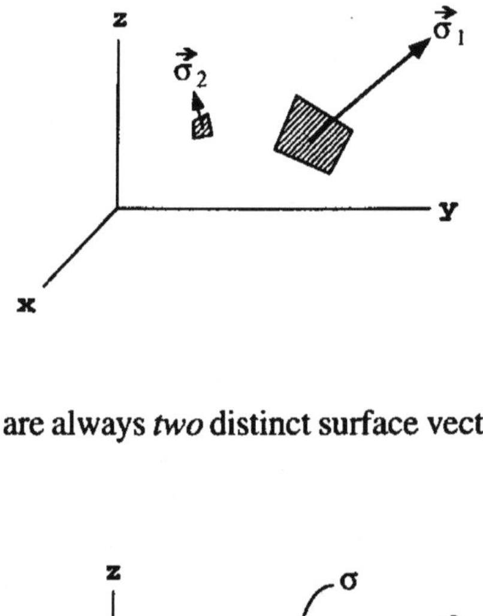

Notice further that there are always *two* distinct surface vectors which emanate from σ (since σ is flat).

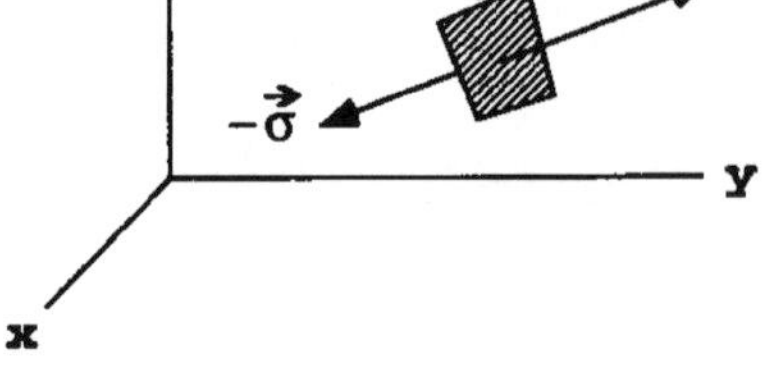

If we can express a surface vector in the form $\vec{\sigma} = a\vec{i} + b\vec{j} + c\vec{k}$ with $c \neq 0$, i.e., σ does not lie in a plane perpendicular to the xy–plane, then what distinguishes the two possibilities is the sign of the $\vec{\mathbf{k}}$–component, i.e., the sign of c. We introduce the following notation to keep this presentation clear.

Notation: Given a flat patch of 2–dimensional surface σ which does not lie in a plane perpendicular to the $x\,y$–plane, let $\vec{\sigma}^{+}$ denote the surface vector whose $\vec{\mathbf{k}}$–component is positive. We shall refer to $\vec{\sigma}^{+}$ as the $\vec{\mathbf{k}}$–positive surface vector. The $\vec{\mathbf{k}}$–negative surface vector, denoted by $\vec{\sigma}^{-}$ is defined to be

$$\vec{\sigma}^{-} = -\vec{\sigma}^{+}.$$

Observe that the $\vec{\mathbf{k}}$–component of $\vec{\sigma}^{-}$ is negative.

Remark: If σ does lie in a plane perpendicular to the $x\,y$–plane then the the $\vec{\mathbf{k}}$–component of σ is zero and we cannot distinguish between the two possible surface vectors via their $\vec{\mathbf{k}}$–components. The two surface vectors can, however, be distinguished by examining the sign of the $\vec{\mathbf{i}}$ or $\vec{\mathbf{j}}$–components. Similar definitions can be given in these cases. To simplify the discussion we now focus on surface vectors with non–zero $\vec{\mathbf{k}}$–component. This is the case for surfaces given in the form $z = f(x,y)$ where f is differentiable: a normal vector to such a surface is given by the gradient of $z - f(x,y)$ (see §23.6) i.e.,

$$\vec{\nabla}(z - f(x,y)) = -\frac{\partial f}{\partial x}\vec{\mathbf{i}} - \frac{\partial f}{\partial y}\vec{\mathbf{j}} + \vec{\mathbf{k}}.$$

Since the technical definition of a surface integral is somewhat involved we begin with a brief heuristic discussion. Given a vector field, $\vec{\mathbf{F}}$ (which can intuitively be thought of as a force or flow), and a small flat patch of surface, σ, with associated surface vectors $\vec{\sigma}^{+}$, $\vec{\sigma}^{-}$, the total $\vec{\mathbf{k}}$–positive force (flow or flux) across σ is approximately $\vec{\mathbf{F}}\cdot\vec{\sigma}^{+}$, while the total $\vec{\mathbf{k}}$–negative force across σ is approximately $\vec{\mathbf{F}}\cdot\vec{\sigma}^{-}$ where in both cases we are evaluating the vector field on the surface.

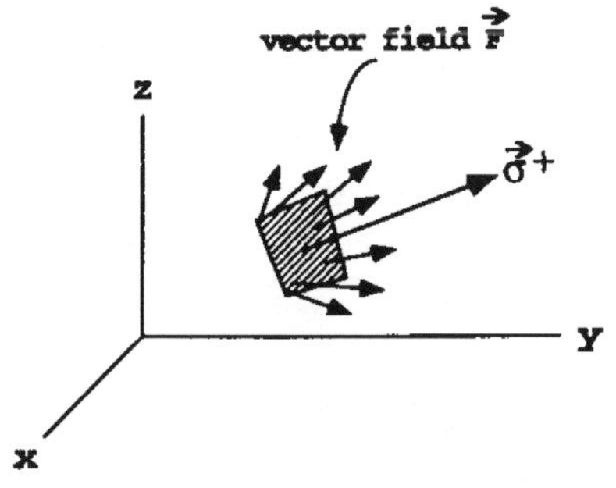

The total $\vec{\mathbf{k}}$–positive force (flow or flux) across a large surface S is obtained by first covering the surface with infinitely many infinitesimal patches, $d\sigma$ with associated surface vector $\overrightarrow{d\sigma}^{+}$, (making the surface resemble a patchwork quilt)

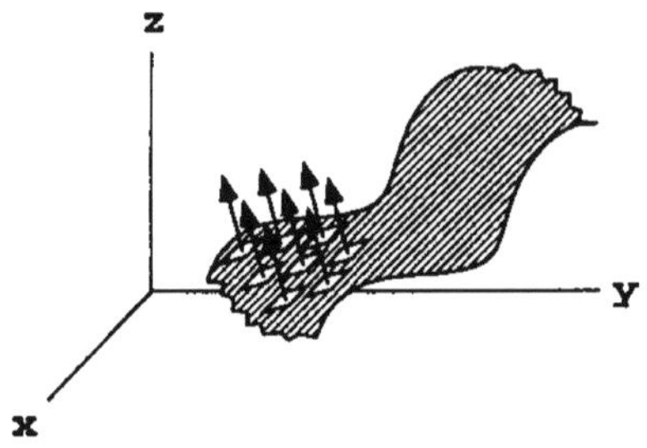

and then taking the limit of the sum of the fluxes of the infinitesimal flat patches. The limit of the sum is defined to be the surface integral $\iint_S \vec{\mathbf{F}}\cdot\overrightarrow{d\sigma}^{+}$. Similarly, $\iint_S \vec{\mathbf{F}}\cdot\overrightarrow{d\sigma}^{-}$ will denote the total $\vec{\mathbf{k}}$–negative flux across S.

In the above description we have tacitly assumed that when we choose the normal vectors along our surface (on our small patches), we were consistent in our choice of direction. This assumption is not as innocent as it may appear at first glance. While the displays above demonstrate no visible difficulty, problems do arise when twisted surfaces are considered. We begin with a vivid example. The **Möbius strip** is constructed by taking a long strip of paper,

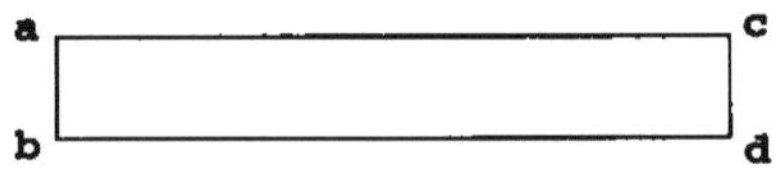

twisting it,

and then gluing together the opposite sides (i.e., in the figure above a is glued to d, and b is glued to c).

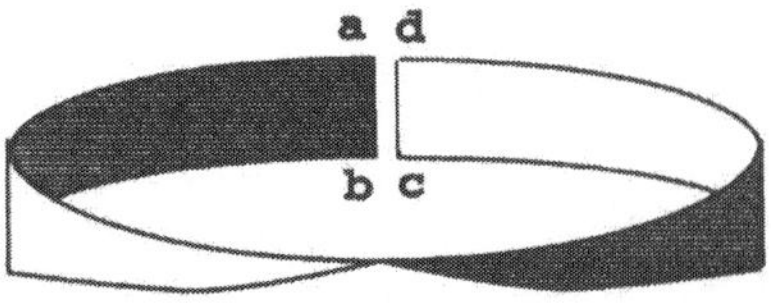

The result of these maneuvers is a Möbius strip (which we urge our readers to construct with paper and tape for themselves). What is unique (and troublesome) about this surface is that if you begin with a surface vector $\vec{\sigma}$ and travel along the surface, after one complete rotation you will land at the same point on the surface but the vector will point in the opposite direction. Thus it is not possible to consistently patch the Möbius strip with surface vectors which point in the same $\vec{\mathbf{k}}$–positive or $\vec{\mathbf{k}}$–negative direction. The Möbius strip is an example of what is called a non–orientable surface.

The general definition of orientability is motivated by this singular example. A surface S is said to be **orientable** provided that upon choosing a normal vector at a point P, when we travel along a closed path on S and return to P, the direction of the normal vector has not changed. Orientability allows us to patch our surface consistently. It should be noted that orientability is a common occurrence, for example the boundary of any simply connected three–dimensional solid is orientable, as is the graph of any differentiable function $z = f(x, y)$.

Let S be an orientable surface which is the graphical display of the differentiable function

$$z = f(x, y),$$

over a bounded region $\mathcal{R}$ in the $x\,y$–plane.

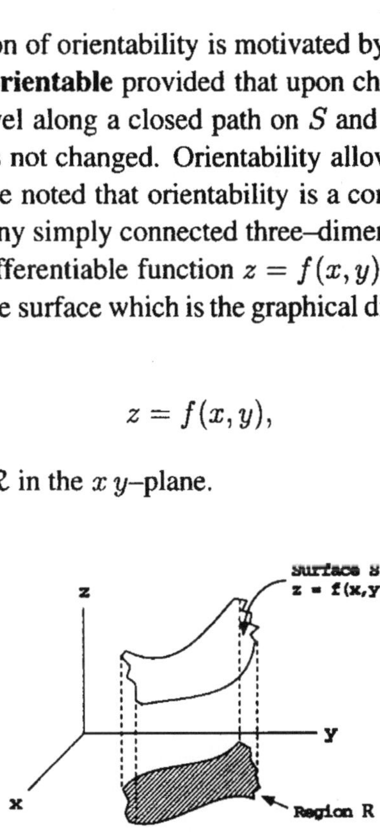

Let $d\sigma$ denote an infinitesimal patch of surface area on S whose projection onto the $x\,y$–plane is an infinitesimal rectangle of area $dx\,dy$.

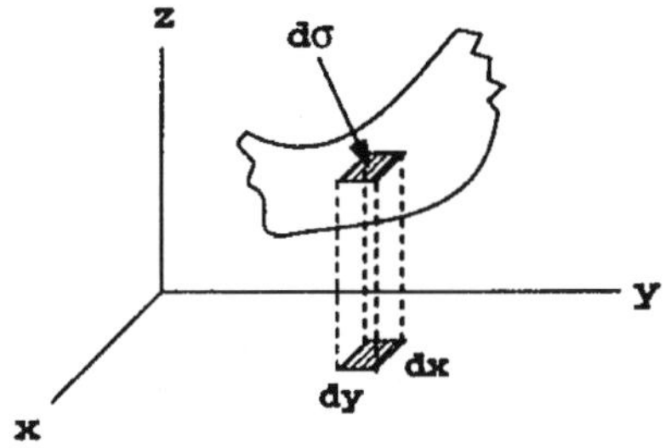

It was demonstrated in §28.1 that the area of $d\sigma$, which we shall denote by $d\sigma$, is given by the formula

$$d\sigma = \sqrt{1+\left(\frac{\partial f}{\partial x}\right)^2+\left(\frac{\partial f}{\partial y}\right)^2}\,dx\,dy. \tag{28.2}$$

Now recall (see §23.6) that a normal vector to the surface $z - f(x,y) = 0$ at the point (x,y,z) is given by the gradient of $z - f(x,y)$:

$$\vec{\nabla}(z - f(x,y)) = -\frac{\partial f}{\partial x}\vec{\mathbf{i}} - \frac{\partial f}{\partial y}\vec{\mathbf{j}} + \vec{\mathbf{k}}.$$

Note that this is a $\vec{\mathbf{k}}$–positive normal vector. The norm of this vector is the now familiar square root

$$\sqrt{1+\left(\frac{\partial f}{\partial x}\right)^2+\left(\frac{\partial f}{\partial y}\right)^2}.$$

(It is the fact that this square root appears in (28.2) that allows us to derive an elegant formula (28.5) for the surface integral.) Thus a unit $\vec{\mathbf{k}}$–positive normal vector $\vec{\mathbf{n}}^+$ is given by the quotient

$$\vec{\mathbf{n}}^+ = \frac{-\frac{\partial f}{\partial x}\vec{\mathbf{i}} - \frac{\partial f}{\partial y}\vec{\mathbf{j}} + \vec{\mathbf{k}}}{\sqrt{1+\left(\frac{\partial f}{\partial x}\right)^2+\left(\frac{\partial f}{\partial y}\right)^2}}, \tag{28.3}$$

while

$$\begin{aligned}\vec{\mathbf{n}}^- &= -\vec{\mathbf{n}}^+ \\ &= \frac{\frac{\partial f}{\partial x}\vec{\mathbf{i}} + \frac{\partial f}{\partial y}\vec{\mathbf{j}} - \vec{\mathbf{k}}}{\sqrt{1+\left(\frac{\partial f}{\partial x}\right)^2+\left(\frac{\partial f}{\partial y}\right)^2}},\end{aligned}$$

which is the unit $\vec{\mathbf{k}}$–negative normal vector to the surface $z - f(x,y) = 0$.

Associated to our infinitesimal surface patch there are two surface vectors: $\overrightarrow{d\sigma}^+$, and $\overrightarrow{d\sigma}^-$. By definition

$$\overrightarrow{d\sigma}^+ = \vec{\mathbf{n}}^+ \, d\sigma$$

$$= \frac{-\frac{\partial f}{\partial x}\vec{\mathbf{i}} - \frac{\partial f}{\partial y}\vec{\mathbf{j}} + \vec{\mathbf{k}}}{\sqrt{1 + \left(\frac{\partial f}{\partial x}\right)^2 + \left(\frac{\partial f}{\partial y}\right)^2}} \cdot \sqrt{1 + \left(\frac{\partial f}{\partial x}\right)^2 + \left(\frac{\partial f}{\partial y}\right)^2} \, dx \, dy \tag{28.4}$$

$$= \left(-\frac{\partial f}{\partial x}\vec{\mathbf{i}} - \frac{\partial f}{\partial y}\vec{\mathbf{j}} + \vec{\mathbf{k}} \right) dx \, dy,$$

and

$$\overrightarrow{d\sigma}^- = -\overrightarrow{d\sigma}^+ = \left(\frac{\partial f}{\partial x}\vec{\mathbf{i}} + \frac{\partial f}{\partial y}\vec{\mathbf{j}} - \vec{\mathbf{k}} \right) dx \, dy.$$

We are now in a position to solidify our heuristic discussion on surface integrals.

Definition: *Let* $\vec{\mathbf{F}} = P(x,y,z)\vec{\mathbf{i}} + Q(x,y,z)\vec{\mathbf{j}} + R(x,y,z)\vec{\mathbf{k}}$ *be a* ***differentiable vector field*** *(i.e., the* ***component functions*** *P, Q, and R are differentiable), and let S be an orientable surface given by the differentiable function* $z = f(x,y)$ *over a bounded region* $\mathcal{R}$ *in the* xy*–plane. The* $\vec{\mathbf{k}}$***–positive surface integral****, is defined to be*

$$\iint_S \vec{\mathbf{F}} \cdot \overrightarrow{d\sigma}^+ = \iint_{\mathcal{R}} \left(-\frac{\partial f}{\partial x}P - \frac{\partial f}{\partial y}Q + R \right) dx \, dy, \tag{28.5}$$

where P, Q, and R are evaluated on the surface, i.e., in (28.5) $P = P(x, y, f(x,y))$*, etc. This surface integral is also referred to as the total* $\vec{\mathbf{k}}$***–positive flux*** *across the surface S. The* $\vec{\mathbf{k}}$***–negative flux*** *across S is defined to be*

$$\iint_S \vec{\mathbf{F}} \cdot \overrightarrow{d\sigma}^- = \iint_{\mathcal{R}} \left(\frac{\partial f}{\partial x}P + \frac{\partial f}{\partial y}Q - R \right) dx \, dy, \tag{28.6}$$

where again P, Q, and R are evaluated on the surface.

Remark: The formulae (28.5) and (28.6) are obtained from (28.4) by simply taking the dot product of $\vec{\mathbf{F}}$ with $\overrightarrow{d\sigma}^+$ and $\overrightarrow{d\sigma}^-$ respectively.

Example [28.3] (Flux across a plane) Find the $\vec{\mathbf{k}}$–positive flux of the vector field $\vec{\mathbf{F}} = x^2\vec{\mathbf{i}} + xy\vec{\mathbf{j}} + \vec{\mathbf{k}}$ across part of the plane $ax + by + cz = d$ (where $c \neq 0$) where $0 \leq x \leq 1, 0 \leq y \leq 1$.

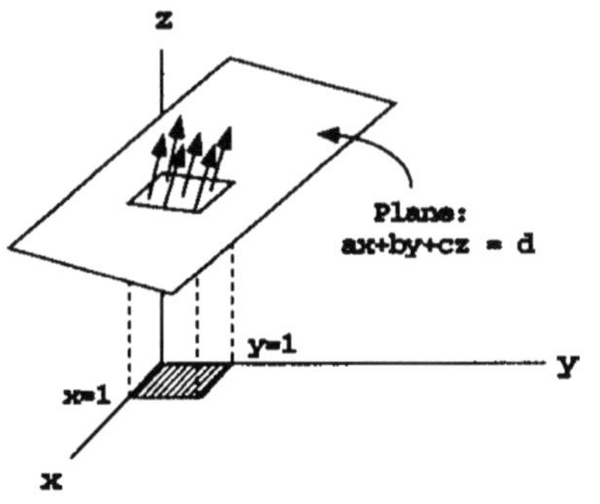

In order to compute this flux we begin by expressing the surface in the form $z = f(x,y)$ for $(x,y) \in \mathcal{R}$. This is easily accomplished: the plane is given by $z = \frac{1}{c}(d - ax - by)$ and the region $\mathcal{R}$ is simply $0 \leq x \leq 1, 0 \leq y \leq 1$. But then

$$\frac{\partial f}{\partial x} = -\frac{a}{c}, \quad \frac{\partial f}{\partial y} = -\frac{b}{c},$$

and by definition the $\vec{\mathbf{k}}$–positive flux is given by

$$\int_0^1 \int_0^1 \left(\frac{a}{c} \cdot x^2 + \frac{b}{c} \cdot xy + 1 \right) dx\, dy = \frac{1}{3}\frac{a}{c} + \frac{1}{4}\frac{b}{c} + 1.$$

Example [28.4] (Flux across a hemisphere) Let $\vec{\mathbf{F}} = x\vec{\mathbf{i}} + y\vec{\mathbf{j}}$. Evaluate the $\vec{\mathbf{k}}$–positive flux where S is the hemisphere $x^2 + y^2 + z^2 = 1$ with $z \geq 0$.

In this example $z = \sqrt{1 - x^2 - y^2} = f(x,y)$ is the equation of the hemisphere. Thus

$$\frac{\partial f}{\partial x} = \frac{1}{2}(1 - x^2 - y^2)^{-\frac{1}{2}} \cdot (-2x) = \frac{-x}{\sqrt{1 - x^2 - y^2}}$$

$$\frac{\partial f}{\partial y} = \frac{-y}{\sqrt{1 - x^2 - y^2}}.$$

The region $\mathcal{R}$ is simply the interior of the circle $x^2 + y^2 = 1$. Combining these pieces of information we see that

$$\iint_S \vec{\mathbf{F}} \cdot \overrightarrow{d\sigma}^{+} = \iint_{x^2+y^2 \leq 1} \left(\frac{x^2}{\sqrt{1 - x^2 - y^2}} + \frac{y^2}{\sqrt{1 - x^2 - y^2}} \right) dx\, dy.$$

This integral naturally lends itself to being evaluated by converting to polar coordinates, $x = r\cos(\theta)$, $y = r\sin(\theta)$:

$$\iint_S \vec{\mathbf{F}} \cdot \overrightarrow{d\sigma}^{+} = \int_0^{2\pi}\int_0^1 \frac{r^2}{\sqrt{1-r^2}} \cdot r\, dr\, d\theta$$

$$= 2\pi \int_0^1 \frac{r^3}{\sqrt{1-r^2}}\, dr \qquad \left(\text{set } u = 1 - r^2\right)$$

$$= \pi \int_0^1 \frac{1-u}{\sqrt{u}}\, du = \frac{4\pi}{3}.$$

Example [28.5] (Flux into a cone) The equation for the surface of the cone depicted below is given by $z = 1 - x^2 - y^2$. Find the total $\vec{\mathbf{k}}$–negative flux of the 2–dimensional position vector field $\vec{\mathbf{F}} = x\vec{\mathbf{i}} + y\vec{\mathbf{j}}$ going through the cone.

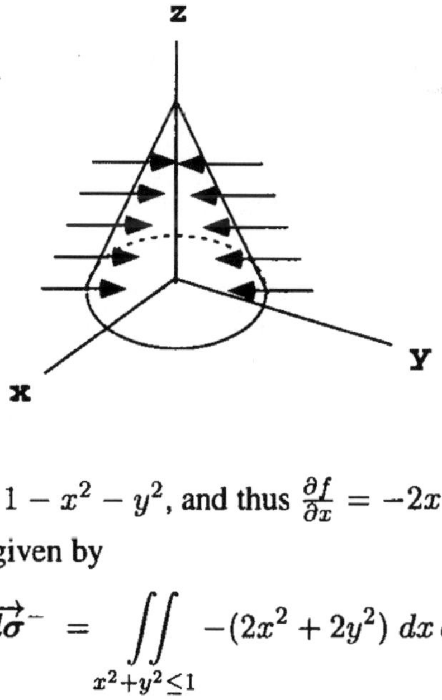

We have $z = f(x,y) = 1 - x^2 - y^2$, and thus $\frac{\partial f}{\partial x} = -2x$ and $\frac{\partial f}{\partial y} = -2y$, and the total $\vec{\mathbf{k}}$–negative flux is given by

$$\iint_S \vec{\mathbf{F}} \cdot \overrightarrow{d\sigma}^{-} = \iint_{x^2+y^2\le 1} -(2x^2 + 2y^2)\, dx\, dy$$

$$= -\int_0^{2\pi}\int_0^1 2r^2 \cdot r\, dr\, d\theta = -\frac{4\pi}{3},$$

after converting to polar coordinates.

Exercises for §28.2

Evaluate the surface integrals $\iint_S \vec{\mathbf{F}} \cdot \overrightarrow{d\sigma}^{+}$ and $\iint_S \vec{\mathbf{F}} \cdot \overrightarrow{d\sigma}^{-}$ for the following surfaces S and vector fields $\vec{\mathbf{F}}$.

(1) $\vec{\mathbf{F}} = xz\vec{\mathbf{i}} - \vec{\mathbf{j}} + y\vec{\mathbf{k}}$, where S is the portion of the plane $2x - y + 3z = 2$ for $0 \leq x \leq 1, 0 \leq y \leq 2$.

(2) $\vec{\mathbf{F}} = 2y\vec{\mathbf{i}} - 2x\vec{\mathbf{j}} + \vec{\mathbf{k}}$, where S is the surface of the parabolic cone $z = 4 - x^2 - y^2$ which lies above the $x\,y$–plane.

(3) $\vec{\mathbf{F}} = -x\vec{\mathbf{i}} + y\vec{\mathbf{j}} + x^2\vec{\mathbf{k}}$, where S is the portion of the surface $z = xy$ lying above the rectangle $-1 \leq x \leq 1, 0 \leq y \leq 2$.

(4) $\vec{\mathbf{F}} = x\vec{\mathbf{i}} + y\vec{\mathbf{j}} + z\vec{\mathbf{k}}$, where S is the surface of the cone $z = \sqrt{x^2 + y^2}$ for $0 \leq z \leq 1$.

(5) $\vec{\mathbf{F}} = x\vec{\mathbf{i}} + y\vec{\mathbf{j}} + z\vec{\mathbf{k}}$, where S is the portion of the surface $z = x^2 - xy + y^2$ lying above the rectangle $0 \leq x \leq 1, 1 \leq y \leq 2$.

(6) $\vec{\mathbf{F}} = xz\vec{\mathbf{i}} + yz\vec{\mathbf{j}} + x^2\vec{\mathbf{k}}$, where S is the lower hemisphere $x^2 + y^2 + z^2 = 1, z \leq 0$.

(7) $\vec{\mathbf{F}} = x^2\vec{\mathbf{i}} + y^2\vec{\mathbf{j}} + z\vec{\mathbf{k}}$, where S is the triangle whose vertices are located at the points $(1,0,0)$, $(0,1,0)$, and $(0,0,2)$.

(8) $\vec{\mathbf{F}} = 3xz\vec{\mathbf{i}} + \vec{\mathbf{j}} - y^2\vec{\mathbf{k}}$, where S is the triangle whose vertices are located at the points $(-1,1,0)$, $(1,0,1)$, and $(0,0,5)$.

28.3 Surface Integrals for Closed Surfaces

Closed surfaces (such as a sphere or an ellipsoid) are not expressible as the graphical display of a function $z = f(x,y)$. Observe that this is analogous to the fact that a closed curve in the $x\,y$–plane is never the graphical display of a function $y = g(x)$. To see this recall that by definition, given a function $z = f(x,y)$, for every point in the plane (x,y) there is a *unique* value $z = f(x,y)$. Thus the graphical display of the function cannot bend back as in the diagram below (and give a second z–value), to form a closed surface.

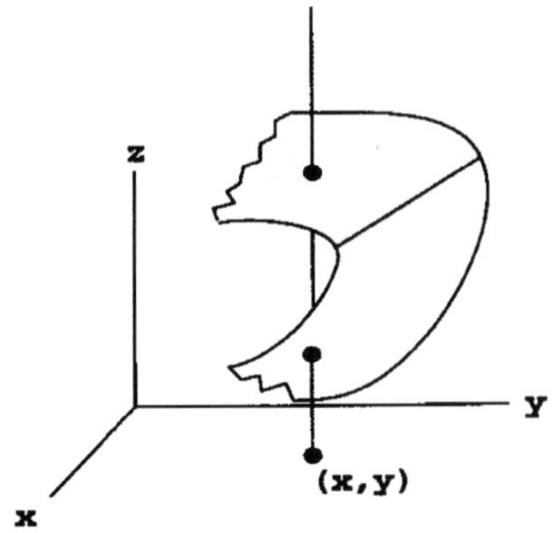

It is possible, however, to take a closed surface and express it as a union of several surfaces where each of the pieces is the graphical display of a function. The following examples illustrate this process.

Example [28.6] (Sphere) The function $z = \sqrt{R^2 - x^2 - y^2}$ has for its graphical display the upper hemisphere of radius R, denoted $S^{(\text{upper})}$, while the function $z = -\sqrt{R^2 - x^2 - y^2}$ gives the lower hemisphere, $S^{(\text{lower})}$. The entire sphere is simply the union $S = S^{(\text{lower})} \cup S^{(\text{upper})}$.

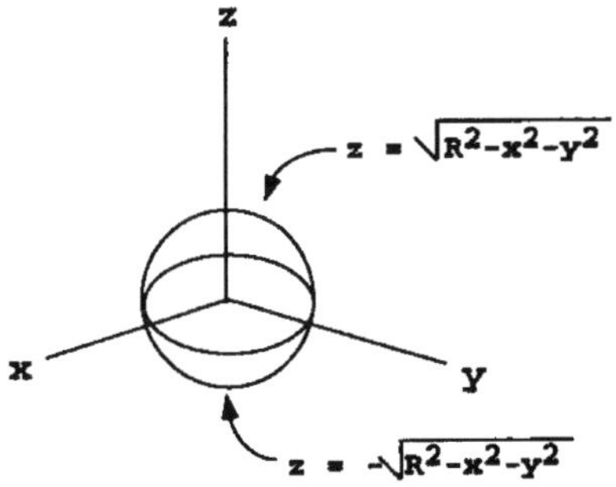

Example [28.7] (Cube) The unit cube is actually a union of 6 different open surfaces $S_1, S_2, S_3, S_4, S_5, S_6$ which are given by the 6 functions,

$$S_1 : \; z = 0, \qquad (0 \le x \le 1, \; 0 \le y \le 1)$$

$$S_2 : \; z = 1, \qquad (0 \le x \le 1, \; 0 \le y \le 1)$$

$$S_3 : \; y = 0, \qquad (0 \le x \le 1, \; 0 \le z \le 1)$$

$$S_4 : \; y = 1, \qquad (0 \le x \le 1, \; 0 \le z \le 1)$$

$$S_5 : \; x = 0, \qquad (0 \le y \le 1, \; 0 \le z \le 1)$$

$$S_6 : \; x = 1, \qquad (0 \le y \le 1, \; 0 \le z \le 1).$$

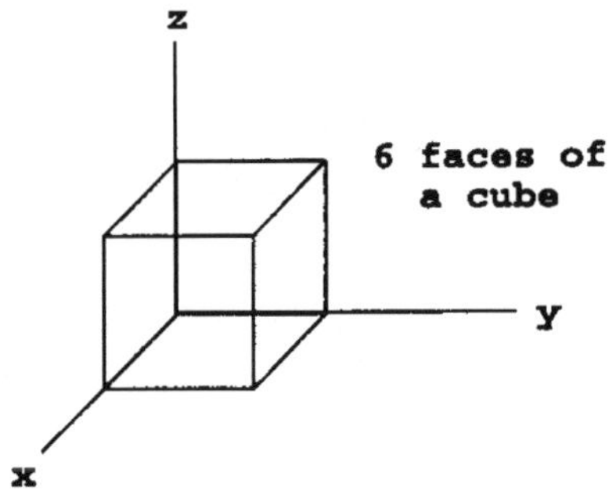

Once we have described our closed surface as the union of non–overlapping open surfaces

$$S = S_1 \cup S_2 \cup \cdots \cup S_m,$$

we are in a position to define the surface integral. Recall that to compute a surface integral surface vectors must be chosen, and there were always two ways to do this. It is a standard convention to choose the surface vectors pointing outward. (In making this convention we are tacitly assuming that our closed surface has a well–defined interior and exterior.) To accomplish this task we focus on each of the open surfaces S_i $(1 \le i \le m)$, and choose surface vectors $\vec{\sigma_i}$ which point away outward. Notice that in making these choices we will sometimes use the $\vec{\mathbf{k}}$–positive surface vector and sometimes the $\vec{\mathbf{k}}$–negative surface vector (as well as $\vec{\mathbf{j}}$ and $\vec{\mathbf{i}}$ positive and negative surface vectors). Now given a vector field, $\vec{\mathbf{F}}$, we define

$$\iint\limits_S \vec{\mathbf{F}} \cdot \overrightarrow{d\sigma} = \sum_{i=1}^{m} \iint\limits_{S_i} \vec{\mathbf{F}} \cdot \overrightarrow{d\sigma_i}. \tag{28.7}$$

Example [28.8] Let $\vec{\mathbf{F}} = x\vec{\mathbf{i}} + y\vec{\mathbf{j}}$. Evaluate the surface integral

$$\iint\limits_S \vec{\mathbf{F}} \cdot \overrightarrow{d\sigma}$$

where S is the surface of a sphere of radius 1.

In Example [28.6] we saw that $S = S^{\text{lower}} \cup S^{\text{upper}}$ where S^{upper} is given by the function $z = \sqrt{1 - x^2 - y^2}$ and S^{lower} is given by $z = -\sqrt{1 - x^2 - y^2}$. In this case the surface vectors of S^{upper} which point outward will be $\vec{\mathbf{k}}$–positive, while on S^{lower} the outward pointing surface vectors will be $\vec{\mathbf{k}}$–negative.

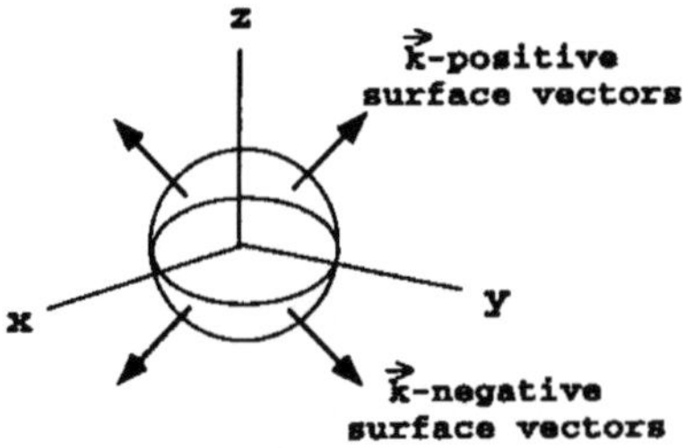

Recalling the computation in Example [28.4], we see that

$$\iint_S \vec{\mathbf{F}} \cdot \overrightarrow{d\sigma} = \iint_{S^{\text{upper}}} \vec{\mathbf{F}} \cdot \overrightarrow{d\sigma^+} + \iint_{S^{\text{lower}}} \vec{\mathbf{F}} \cdot \overrightarrow{d\sigma^-}$$

$$= 2 \iint_{x^2+y^2 \leq 1} \frac{x^2+y^2}{\sqrt{1-x^2-y^2}}\, dx\, dy$$

$$= \frac{8\pi}{3}.$$

Example [28.9] Let $\vec{\mathbf{F}} = x^2\vec{\mathbf{i}} - 2xz\vec{\mathbf{j}} + y\vec{\mathbf{k}}$. Evaluate the surface integral $\iint_S \vec{\mathbf{F}} \cdot \overrightarrow{d\sigma}$ where S is the surface of the unit cube given in Example [28.7].

In order to compute this surface integral we begin by computing the outward normal vectors to the various pieces S_1. On the surface S_1, $z = 0$ and $0 \leq x \leq 1$, $0 \leq y \leq 1$, hence $\vec{\mathbf{n}}_1 = -\vec{\mathbf{k}}$. Similarly for S_2, $\vec{\mathbf{n}}_2 = \vec{\mathbf{k}}$. The remaining sides of the cube are not actually graphical displays of functions of the form $z = f(x, y)$. The surface S_3 is in fact the graphical display of a function of the form $y = f(x, z)$, where $f(x, z) = 0$. Similarly S_4 is the graphical display of the function $y = f(x, z) = 1$. The normals in these cases are $\vec{\mathbf{n}}_3 = -\vec{\mathbf{j}}$ and $\vec{\mathbf{n}}_4 = \vec{\mathbf{j}}$. Finally, S_5 and S_6 are the graphical displays of the functions $x = f(y, z) = 0$ and $x = f(y, z) = 1$ respectively, and their normals are $\vec{\mathbf{n}}_5 = -\vec{\mathbf{i}}$ and $\vec{\mathbf{n}}_6 = \vec{\mathbf{i}}$.

Surface integrals of functions of the form $y = f(x, z)$ (or $x = f(y, z)$) can be evaluated in exactly the same manner as the case considered in §28.2 and §28.3 (notice that the concept of orientability is independent of any particular variable). Thus in this case the surface vectors are given by

$$\begin{aligned} \overrightarrow{d\sigma_1} &= \vec{\mathbf{n}}_1\, dx\, dy & \overrightarrow{d\sigma_2} &= \vec{\mathbf{n}}_2\, dx\, dy \\ \overrightarrow{d\sigma_3} &= \vec{\mathbf{n}}_3\, dx\, dz & \overrightarrow{d\sigma_4} &= \vec{\mathbf{n}}_4\, dx\, dz \\ \overrightarrow{d\sigma_5} &= \vec{\mathbf{n}}_5\, dy\, dz & \overrightarrow{d\sigma_6} &= \vec{\mathbf{n}}_6\, dy\, dz, \end{aligned}$$

and the surface integral is given by

$$\iint_S \vec{\mathbf{F}} \cdot \overrightarrow{d\sigma} = \int_0^1 \int_0^1 y \, dx \, dy + \int_0^1 \int_0^1 -y \, dx \, dy$$

$$+ \int_0^1 \int_0^1 2xz \, dx \, dz + \int_0^1 \int_0^1 -2xz \, dx \, dz$$

$$+ \int_0^1 \int_0^1 0 \, dy \, dz + \int_0^1 \int_0^1 dy \, dz$$

$$= 1.$$

Exercises for §28.3

Let $\vec{\mathbf{F}} = x\vec{\mathbf{i}} + y\vec{\mathbf{j}} + z\vec{\mathbf{k}}$ be the position vector field. In exercises (1)–(5), evaluate $\iint_S \vec{\mathbf{F}} \cdot \overrightarrow{d\sigma}$ for the given surface S.

(1) S is the surface of the unit cube $0 \le x \le 1, 0 \le y \le 1, 0 \le z \le 1$.

(2) S is the surface of the sphere $x^2 + y^2 + z^2 = 9$.

(3) S is the surface of the tetrahedron whose vertices are located at $(0,0,0)$, $(2,0,0)$, $(0,2,0)$, $(0,0,5)$.

(4) S is the surface of the cylinder $y^2 + z^2 \le 1, 0 \le x \le 2$.

(5) S is the surface of the cylinder $x^2 + z^2 \le 4, 0 \le y \le 2$.

(6) Let $\vec{\mathbf{F}} = -y\vec{\mathbf{i}} + x\vec{\mathbf{j}} + \vec{\mathbf{k}}$ be the spin vector field. Show that

$$\iint_S \vec{\mathbf{F}} \cdot \overrightarrow{d\sigma} = 0,$$

for each of the closed surfaces above.

28.4 The Divergence Theorem

Let $\vec{\mathbf{F}}(x,y,z) = P(x,y,z)\vec{\mathbf{i}} + Q(x,y,z)\vec{\mathbf{j}} + R(x,y,z)\vec{\mathbf{k}}$ be a differentiable vector field.

Definition: *The divergence of* $\vec{\mathbf{F}}$, *denoted* $\text{div}(\vec{\mathbf{F}})$, *is defined by the formula*

$$\begin{aligned}\text{div}(\vec{\mathbf{F}}) &= \vec{\nabla}\cdot\vec{\mathbf{F}} \\ &= \left(\frac{\partial}{\partial x}\vec{\mathbf{i}} + \frac{\partial}{\partial y}\vec{\mathbf{j}} + \frac{\partial}{\partial z}\vec{\mathbf{k}}\right)\cdot\vec{\mathbf{F}} \\ &= \frac{\partial P}{\partial x} + \frac{\partial Q}{\partial y} + \frac{\partial R}{\partial z}.\end{aligned}$$

Remark: The divergence of a vector field $\vec{\mathbf{F}}$ is a scalar valued function of the three variables x, y, z.

Example [28.10] Compute $\text{div}(\vec{\mathbf{F}})$ when $\vec{\mathbf{F}}(x,y,z) = 3x^2y\vec{\mathbf{i}} - 2x^3y^4z\vec{\mathbf{j}} + z^5\vec{\mathbf{k}}$.

Here we have

$$\begin{aligned}\text{div}(\vec{\mathbf{F}}) &= \frac{\partial}{\partial x}(3x^2y) + \frac{\partial}{\partial y}(-2x^3y^4z) + \frac{\partial}{\partial z}(z^5) \\ &= 6xy - 8x^3y^3z + 5z^4.\end{aligned}$$

Example [28.11] Compute $\text{div}(\vec{\mathbf{F}})$ when $\vec{\mathbf{F}}(x,y,z) = 2e^x\cos(yz)\vec{\mathbf{i}} + (\log y)\vec{\mathbf{j}} + \cos(xz)\vec{\mathbf{k}}$.

In this example

$$\text{div}(\vec{\mathbf{F}}) = 2e^x\cos(yz) + \frac{1}{y} - x\sin(xz).$$

We now consider a 3–dimensional solid body B and the surface enclosing B, i.e., the boundary of B which we denote ∂B. Clearly ∂B is a closed surface.

Definition: *A closed surface is termed **simple** provided:*

(1) *it is a finite union of pieces of surfaces which are specified by two variable differentiable functions,*

(2) *it is orientable,*

(3) *it does not intersect itself.*

We are now in a position to state the Divergence Theorem.

Theorem [28.12] (The Divergence Theorem) *Let B be a solid 3–dimensional body whose boundary ∂B is a simple closed surface. Given $\vec{\mathbf{F}}$ a differentiable vector field we have the identity*

$$\iiint_B \operatorname{div}(\vec{\mathbf{F}})\, dx\, dy\, dz = \iint_{\partial B} \vec{\mathbf{F}} \cdot \overrightarrow{d\sigma}.$$

Motivation for this theorem (which can be looked at as a generalization of Green's theorem), will be given in Chapter 29. For the moment we confine ourselves to illustrating this very useful theorem.

Example [28.13] Let $\vec{\mathbf{F}} = x\vec{\mathbf{i}} + y\vec{\mathbf{j}}$. Evaluate $\iint_{\partial B} \vec{\mathbf{F}} \cdot \overrightarrow{d\sigma}$ where B is the ball (i.e., solid sphere) of radius 1 centered at the origin.

The boundary of the solid sphere is simply the sphere of radius 1, and the reader will note we actually computed this integral in Example [28.8]. This example thus serves as a verification of the Divergence Theorem:

$$\operatorname{div}(\vec{\mathbf{F}}) = \vec{\nabla} \cdot \vec{\mathbf{F}} = 2,$$

and hence

$$\iint_{\partial S} \vec{\mathbf{F}} \cdot \overrightarrow{d\sigma} = \iint_S 2\, dx\, dy\, dz = \frac{8\pi}{3},$$

since the ball has volume $\frac{4\pi}{3}$.

Example [28.14] Let $\vec{\mathbf{F}} = x^2\vec{\mathbf{i}} - 2xz\vec{\mathbf{j}} + y\vec{\mathbf{k}}$. Evaluate

$$\iint_{\partial B} \vec{\mathbf{F}} \cdot \overrightarrow{d\sigma}$$

where B is the unit cube $B = \{(x, y, z) \mid 0 \le x \le 1,\ 0 \le y \le 1,\ 0 \le z \le 1\}$ and ∂B is its boundary.

The divergence in this example is

$$\operatorname{div}(\vec{\mathbf{F}}) = 2x,$$

and thus

$$\iint_{\partial B} \vec{\mathbf{F}} \cdot \overrightarrow{d\sigma} = \int_0^1 \int_0^1 \int_0^1 2x\, dx\, dy\, dz = 1.$$

Note that this agrees with our computation in Example [28.9].

Example [28.15] Let $\vec{\mathbf{F}} = (3x^2z^2 - 2x)\vec{\mathbf{i}} - 2y\vec{\mathbf{j}} - 2xz^3\vec{\mathbf{k}}$. Evaluate $\iint_{\partial E} \vec{\mathbf{F}} \cdot \overrightarrow{d\sigma}$ where ∂E is the boundary of the ellipsoid $E = \{(x, y, z) \mid x^2 + \frac{y^2}{4} + \frac{z^2}{9} \le 2\}$.

In this final example,

$$\begin{aligned} \operatorname{div}(\vec{\mathbf{F}}) &= \frac{\partial}{\partial x}(3x^2z^2 - 2x) + \frac{\partial}{\partial y}(2y) + \frac{\partial}{\partial z}(-2xz^3) \\ &= 6xz^2 - 2 + 2 - 6xz^2 \\ &= 0, \end{aligned}$$

and hence $\iint_{\partial E} \vec{\mathbf{F}} \cdot \overrightarrow{d\sigma} = 0$.

Exercises for §28.4

Find the divergence of he following vector fields.

(1) $\vec{\mathbf{v}} = 2x^3y\vec{\mathbf{i}} - z\vec{\mathbf{j}} + \vec{\mathbf{k}}$.

(2) $\vec{\mathbf{v}} = xz^2\vec{\mathbf{i}} + xyz\vec{\mathbf{j}} - 3xz^4\vec{\mathbf{k}}$.

(3) $\vec{\mathbf{v}} = xz\vec{\mathbf{i}} + \cos(y^2z)\vec{\mathbf{k}}$.

(4) $\vec{\mathbf{v}} = ye^x\vec{\mathbf{j}} - \ln(x^2y)\vec{\mathbf{k}}$.

(5) $\vec{\mathbf{v}} = e^{x+y}\vec{\mathbf{i}} - 2\cos(yz)\vec{\mathbf{j}} + z\vec{\mathbf{k}}$.

(6) $\vec{\mathbf{v}} = \frac{x}{y}\vec{\mathbf{i}} - \ln\left(\frac{x}{yz}\right)\vec{\mathbf{j}} + e^{z^2}\vec{\mathbf{k}}$.

Evaluate the following surface integrals $\iint\limits_{S} \vec{\mathbf{F}} \cdot \overrightarrow{d\sigma}$ by using the Divergence Theorem.

(7) $\vec{\mathbf{F}} = x^3\vec{\mathbf{i}} + y^3\vec{\mathbf{j}} + z^3\vec{\mathbf{k}}$, where S is the surface of a sphere of radius 1.

(8) $\vec{\mathbf{F}} = 3x^2y\vec{\mathbf{i}} - x\vec{\mathbf{j}} + z^2\vec{\mathbf{k}}$, where S is the surface of the unit cube $0 \le x \le 1$, $0 \le y \le 1, 0 \le z \le 1$.

(9) $\vec{\mathbf{F}} = x^4z\vec{\mathbf{i}} - 2xy^2z\vec{\mathbf{j}} + 3xz\vec{\mathbf{k}}$, where S is the surface of the rectangular box $0 \le x \le 1$, $0 \le y \le 2, -1 \le z \le 1$.

(10) $\vec{\mathbf{F}} = 3x^2y\vec{\mathbf{i}} - x\vec{\mathbf{j}} + z^2\vec{\mathbf{k}}$, where S is the surface of the tetrahedron whose vertices are located at $(0,0,0)$, $(1,0,0)$, $(0,1,0)$, and $(0,0,1)$.

(11) $\vec{\mathbf{F}} = \ln(xy)\vec{\mathbf{i}} + \cos(xz)\vec{\mathbf{j}} - \frac{z}{x}\vec{\mathbf{k}}$, where S is the surface of the sphere of radius 3 centered at $(5,5,0)$.

(12) $\vec{\mathbf{F}} = 3x\vec{\mathbf{i}} + 2\sin(x^3z)\vec{\mathbf{j}} - z\vec{\mathbf{k}}$, where S is the surface of the ellipsoid $x^2 + \frac{y^2}{4} + \frac{z^2}{9} \le 4$.

(13) $\vec{\mathbf{F}} = x\vec{\mathbf{i}} + y\vec{\mathbf{j}} + z\vec{\mathbf{k}}$, where S is the surface of the cylinder $x^2 + y^2 \le 9$ between the planes $z = 2$, $z = -2$.

(14) Given differentiable vector fields $\vec{\mathbf{F}}$ and $\vec{\mathbf{G}}$ and c a constant show that

$$\text{(i) } \operatorname{div}(\vec{\mathbf{F}} + \vec{\mathbf{G}}) = \operatorname{div}(\vec{\mathbf{F}}) + \operatorname{div}(\vec{\mathbf{G}}),$$

$$\text{(ii) } \operatorname{div}(c\vec{\mathbf{F}}) = c\operatorname{div}(\vec{\mathbf{F}}).$$

(15) Let $\vec{\mathbf{F}}$ be a differentiable vector field and $g(x,y,z)$ a differentiable function. Show that

$$\operatorname{div}(g \cdot \vec{\mathbf{F}}) = g \cdot \operatorname{div}(\vec{\mathbf{F}}) + \vec{\nabla} g \cdot \vec{\mathbf{F}}.$$

(16) Let f and g be differentiable functions. Show that

$$\operatorname{div}(\vec{\nabla} f \times \vec{\nabla} g) = 0.$$

(17) Given a twice differentiable function f we define the **Laplacian** of f to be

$$\nabla^2 f = \operatorname{div}(\vec{\nabla} f) = \frac{\partial^2 f}{\partial x^2} + \frac{\partial^2 f}{\partial y^2} + \frac{\partial^2 f}{\partial z^2}.$$

If f and g are twice differentiable functions show that

$$\nabla^2(fg) = f\nabla^2 g + g\nabla^2 f + 2\vec{\nabla} f \cdot \vec{\nabla} g.$$

(18) A function f is termed **harmonic** provided $\nabla^2 f = 0$. Show that the function $\frac{1}{\sqrt{x^2+y^2+z^2}}$ is harmonic everywhere except at the origin.

28.5 Curl and Spin

Imagine a vector field $\vec{\mathbf{F}} = P(x,y,z)\vec{\mathbf{i}} + Q(x,y,z)\vec{\mathbf{j}} + R(x,y,z)\vec{\mathbf{k}}$ as a field of forces, i.e., as a force at each point in space. If we place a disk in space, the following natural question arises: will the forces cause the disk to spin?

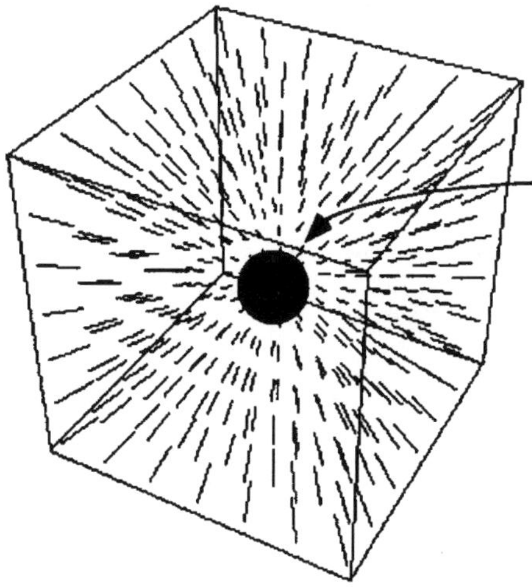

This question is not entirely esoteric: if we visualize the wind as a vector field acting on a moving airplane, the effect of spin will determine whether or not the flight is comfortable.

To answer this question we recall that

$$\text{curl}(\vec{\mathbf{F}}) = \begin{vmatrix} \vec{\mathbf{i}} & \vec{\mathbf{j}} & \vec{\mathbf{k}} \\ \frac{\partial}{\partial x} & \frac{\partial}{\partial y} & \frac{\partial}{\partial z} \\ P & Q & R \end{vmatrix}$$

$$= \left(\frac{\partial R}{\partial y} - \frac{\partial Q}{\partial z}\right)\vec{\mathbf{i}} + \left(\frac{\partial P}{\partial z} - \frac{\partial R}{\partial x}\right)\vec{\mathbf{j}} + \left(\frac{\partial Q}{\partial x} - \frac{\partial P}{\partial y}\right)\vec{\mathbf{k}}.$$

It can be shown, using basic principles of mechanics, that if $\vec{\mathbf{n}}$ is a unit normal vector to our disk, the disk will spin provided

$$\text{curl}(\vec{\mathbf{F}}) \cdot \vec{\mathbf{n}} \neq 0.$$

We define the **spin** of a small disk σ by the formula

$$\text{spin} = \text{curl}(\vec{\mathbf{F}}) \cdot \vec{\sigma},$$

where $\vec{\sigma}$ is a surface vector associated with our disk. We illustrate this concept with some simple examples.

Example [28.16] (Spin about an axis) Let $\vec{\mathbf{a}} = a_1\vec{\mathbf{i}} + a_2\vec{\mathbf{j}} + a_3\vec{\mathbf{k}}$ be a fixed vector, and let $\vec{\mathbf{r}} = x\vec{\mathbf{i}} + y\vec{\mathbf{j}} + z\vec{\mathbf{k}}$ be the position vector field (see §27.1). The graphical display of the vector field

$$\vec{\mathbf{F}} = \vec{\mathbf{a}} \times \vec{\mathbf{r}}$$

$$= \begin{vmatrix} \vec{\mathbf{i}} & \vec{\mathbf{j}} & \vec{\mathbf{k}} \\ a_1 & a_2 & a_3 \\ x & y & z \end{vmatrix}$$

$$= (a_2 z - a_3 y)\vec{\mathbf{i}} + (a_3 x - a_1 z)\vec{\mathbf{j}} + (a_1 y - a_2 x)\vec{\mathbf{k}}$$

consists of the vectors which spin about the fixed vector (which now resembles an axis), $\vec{\mathbf{a}}$.

That the vectors $\vec{\mathbf{v}} = \vec{\mathbf{a}} \times \vec{\mathbf{v}}$ spin about the axis $\vec{\mathbf{a}}$ is an immediate consequence of the right hand rule (see §20.6) for determining the cross product.

Example [28.17] Let $\vec{\mathbf{a}} = \vec{\mathbf{k}}$, in Example [28.16]. Knowing that the spin vector field is

$$\vec{\mathbf{F}} = \vec{\mathbf{a}} \times \vec{\mathbf{r}} = -y\vec{\mathbf{i}} + x\vec{\mathbf{j}}$$

compute the spin of a unit disk (disk or area 1) placed at the origin in the $x\,y$–plane.

Since a unit normal vector to the disk is simply $\vec{\mathbf{n}} = \vec{\mathbf{k}}$, the spin in question is given by

$$\text{spin} = \text{curl}(\vec{\mathbf{F}}) \cdot \vec{\mathbf{n}}$$

$$= \begin{vmatrix} \vec{\mathbf{i}} & \vec{\mathbf{j}} & \vec{\mathbf{k}} \\ \frac{\partial}{\partial x} & \frac{\partial}{\partial y} & \frac{\partial}{\partial z} \\ -y & x & 0 \end{vmatrix} \cdot \vec{\mathbf{k}}$$

$$= 2.$$

Example [28.18] (Position vector field) The position vector field, which is defined by $\vec{\mathbf{r}} = x\vec{\mathbf{i}} + y\vec{\mathbf{j}} + z\vec{\mathbf{k}}$, has no spin for any disk because

$$\operatorname{curl}(\vec{\mathbf{r}}) = 0.$$

This amounts to saying that the forces on a disk (placed anywhere) will cancel out.

Example [28.19] (Parallel vector field) Let $\vec{\mathbf{F}} = x\vec{\mathbf{j}}$. Visually $\vec{\mathbf{F}}$ consists of parallel vectors. Since $\operatorname{curl}(\vec{\mathbf{F}}) = \vec{\mathbf{k}} \neq 0$, we see that there will be spin on a disk place in the $x\,y$–plane.

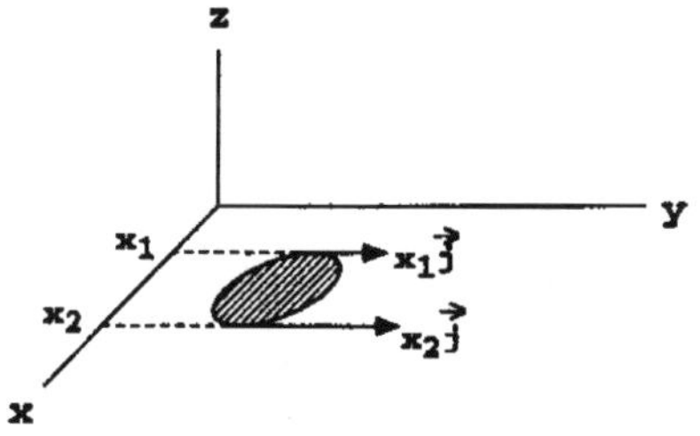

To ascertain the direction of the spin notice that if $x_2 > x_1$ the force of the vector $x_2\vec{\mathbf{j}}$ will be greater than the force of the vector $x_1\vec{\mathbf{j}}$. Referring to the figure above, the direction of spin will be counterclockwise.

Example [28.20] (Gradient Vector Field) Let

$$\vec{\mathbf{F}} = \vec{\nabla}\phi(x, y, z) = \frac{\partial \phi}{\partial x}\vec{\mathbf{i}} + \frac{\partial \phi}{\partial y}\vec{\mathbf{j}} + \frac{\partial \phi}{\partial z}\vec{\mathbf{k}}$$

be a gradient vector field with potential function ϕ. By definition, the $\operatorname{curl}(\vec{\mathbf{F}})$ is given by

$$\operatorname{curl}(\vec{\mathbf{F}}) = \left(\frac{\partial}{\partial y}\frac{\partial \phi}{\partial z} - \frac{\partial}{\partial z}\frac{\partial \phi}{\partial y}\right)\vec{\mathbf{i}} + \left(\frac{\partial}{\partial z}\frac{\partial \phi}{\partial x} - \frac{\partial}{\partial x}\frac{\partial \phi}{\partial z}\right)\vec{\mathbf{j}} + \left(\frac{\partial}{\partial x}\frac{\partial \phi}{\partial y} - \frac{\partial}{\partial y}\frac{\partial \phi}{\partial x}\right)\vec{\mathbf{k}}.$$

Recalling that when we reverse the order of partial differentiation the result remains the same (see §23.3), we see that

$$\frac{\partial}{\partial y}\frac{\partial \phi}{\partial z} = \frac{\partial}{\partial z}\frac{\partial \phi}{\partial y}, \quad \frac{\partial}{\partial z}\frac{\partial \phi}{\partial x} = \frac{\partial}{\partial x}\frac{\partial \phi}{\partial z}, \quad \frac{\partial}{\partial x}\frac{\partial \phi}{\partial y} = \frac{\partial}{\partial y}\frac{\partial \phi}{\partial x},$$

and hence $\operatorname{curl}(\vec{\mathbf{F}}) = 0$, i.e., *a gradient vector field has no spin.*

Exercises for §28.5

Using your CAS obtain the graphical display of the following vector fields $\vec{v}$. Compute for each vector field $\vec{v}$ the spin (up to sign) of a unit disk placed at the origin in the $x\,y$–plane, the $x\,z$–plane, and the $y\,z$–plane.

(1) $\vec{v} = z\vec{i} - x\vec{k}.$

(2) $\vec{v} = -y\vec{i} + z\vec{j} - y\vec{k}.$

(3) $\vec{v} = x\vec{i} + y\vec{j} - z\vec{k}.$

(4) $\vec{v} = 3\vec{i}.$

(5) $\vec{v} = 3y\vec{i}.$

(6) $\vec{v} = 3z\vec{i}.$

(7) $\vec{v} = (x^2 + y^2)\vec{k}.$

(8) $\vec{v} = 2xyz\vec{i} + x^2z\vec{j} + x^2y\vec{k}.$

(9) $\vec{v} = yz\cos(xyz)\vec{i} + xz\cos(xyz)\vec{j} + xy\cos(xyz)\vec{k}.$

(10) Given differentiable vector fields $\vec{F}$ and $\vec{G}$ and k a constant show that

$$\text{(i) } \operatorname{curl}(\vec{F} + \vec{G}) = \operatorname{curl}(\vec{F}) + \operatorname{curl}(\vec{G}),$$

$$\text{(ii) } \operatorname{curl}(k\vec{F}) = k\operatorname{curl}(\vec{F}).$$

(11) Given differentiable vector fields $\vec{F}$ and $\vec{G}$ show that

$$\operatorname{div}(\vec{F}\times\vec{G}) = \vec{G}\cdot\operatorname{curl}(\vec{F}) - \vec{F}\cdot\operatorname{curl}(\vec{G}).$$

(12) Given differentiable vector fields $\vec{F}$ and $\vec{G}$ show that

$$\vec{\nabla}(\vec{F}\cdot\vec{G}) = (\vec{F}\cdot\vec{\nabla})\vec{G} + (\vec{G}\cdot\vec{\nabla})\vec{F} + \vec{F}\times\operatorname{curl}(\vec{G}) + \vec{G}\times\operatorname{curl}(\vec{F}).$$

(13) Given a differentiable vector field $\vec{F}$ and a differentiable function $f(x, yz)$ demonstrate the identity

$$\operatorname{curl}(f\cdot\vec{F}) = f\cdot\operatorname{curl}(\vec{F}) + \vec{\nabla}f\times\vec{F}.$$

(14) Suppose that $\vec{F}$ has continuous second derivatives. Show that

$$\operatorname{div}(\operatorname{curl}\vec{F}) = 0.$$

(15) Given differentiable vector fields $\vec{F} = f_1(x, y, z)\vec{i} + f_2(x, y, z)\vec{j} + f_3(x, y, z)\vec{k}$ where each f_i is twice differentiable, define the **Laplacian** of $\vec{F}$ by

$$\nabla^2\vec{F} = \nabla^2 f_1\vec{i} + \nabla^2 f_2\vec{j} + \nabla^2 f_3\vec{k}$$

where $\nabla^2 f_i$ is the Laplacian defined in Exercises (17) of §28.4. Show that

$$\operatorname{curl}(\operatorname{curl}\vec{F}) = \vec{\nabla}(\operatorname{div}\vec{F}) - \nabla^2\vec{F}.$$

28.6 Stoke's Theorem

In this final section we again consider an open orientable surface S in 3–dimensional space which is the graphical display of a function $z = f(x, y)$ for $(x, y) \in \mathcal{R}$ where $\mathcal{R}$ is some bounded simply connected region in the $x\,y$–plane.

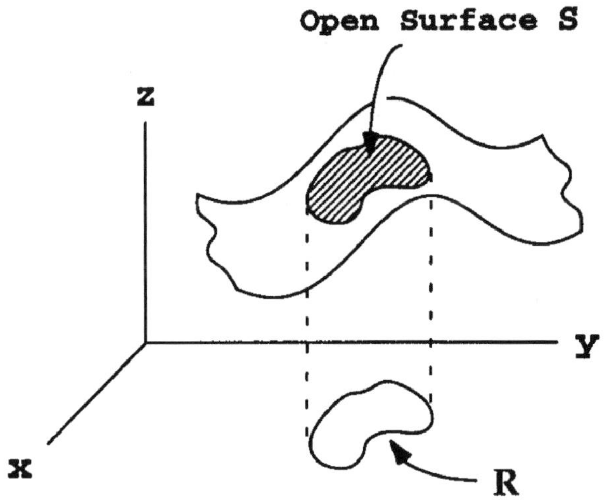

Given a differentiable vector field $\vec{\mathbf{F}}$ we wish to determine the total spin of $\vec{\mathbf{F}}$ on S. Following our general methodology we first patch S with infinitesimally small disks

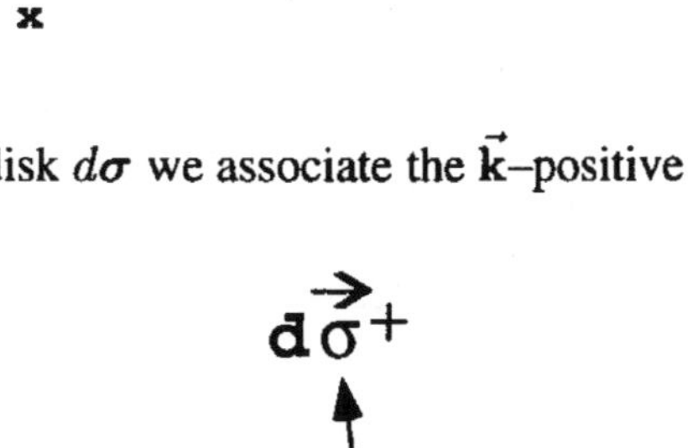

where with each small disk $d\sigma$ we associate the $\vec{\mathbf{k}}$–positive surface vector $\overrightarrow{d\sigma^+}$.

The spin on each tiny disk is given by curl($\vec{\mathbf{F}}$) $\cdot \overrightarrow{d\sigma^+}$, and the total spin across S is given by the limit of the sums of these spins, i.e., the total spin is the surface integral

$$\iint_S \text{curl}(\vec{\mathbf{F}}) \cdot \overrightarrow{d\sigma^+}.$$

Stoke's theorem is a remarkable identity which relates the integral above to the line integral around the curve bounding the surface S (which we shall denote ∂S). Before we can state the result we must discuss the orientation of the curve ∂S. Let $\vec{\mathbf{n}}$ be a $\vec{\mathbf{k}}$–positive normal vector to the surface S. By pointing the thumb of your right hand in the direction of $\vec{\mathbf{n}}$ and letting your hand wrap around the curve, an orientation of ∂S is induced.

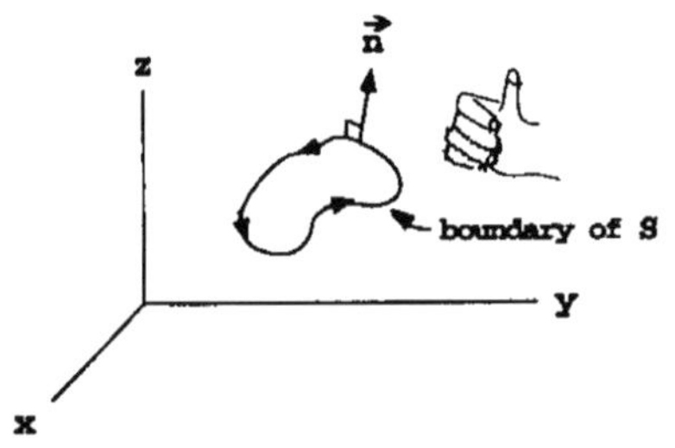

After a moment of observation it is clear that this orientation (as in the above diagram) is counterclockwise. **Stoke's Theorem** states that

$$\iint_S \text{curl}(\vec{\mathbf{F}}) \cdot \overrightarrow{d\sigma^+} = \oint_{\partial S} \vec{\mathbf{F}} \cdot d\vec{\mathbf{r}}.$$

Example [28.21] Verify Stoke's Theorem for the surface of the paraboloid $z = 1 - x^2 - y^2$ with $0 \le z \le 1$ and $\vec{\mathbf{F}} = z\vec{\mathbf{i}} + x\vec{\mathbf{j}} - y\vec{\mathbf{k}}$.

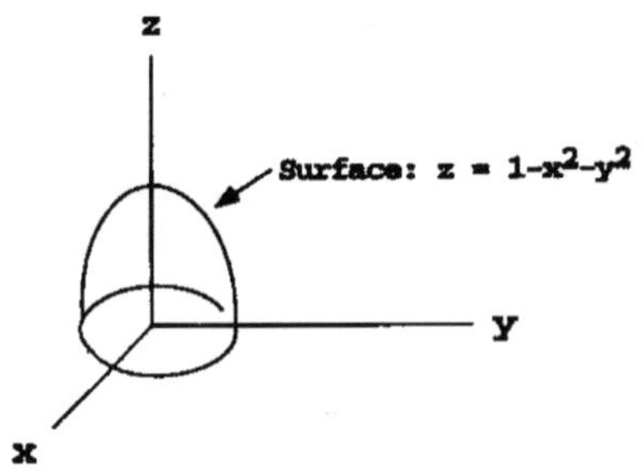

Denoting the surface here by S notice that ∂S is the unit circle in the $x\,y$–plane, i.e.,

$$\partial S = \Big\{(x, y, 0) \mid x^2 + y^2 = 1\Big\}.$$

In order to compute the line integral around ∂S we begin by parameterizing it:

$$x = \cos(\theta), \quad y = \sin(\theta), \quad z = 0 \quad (0 \le \theta \le 2\pi),$$

and thus

$$dx = -\sin(\theta), \quad dy = \cos(\theta), \quad dz = 0.$$

Then

$$\oint_{\partial S} \vec{\mathbf{F}} \cdot d\vec{\mathbf{r}} = \oint_{\partial S} (z\,dz + x\,dy - y\,dz)$$

$$= \int_0^{2\pi} (\cos(\theta))^2\, d\theta$$

$$= \pi.$$

On the other hand, when we express S in the form

$$x^2 + y^2 - 1 + z = 0,$$

we see that the $\vec{\mathbf{k}}$–positive surface vector is

$$\overrightarrow{d\sigma^+} = (2x\vec{\mathbf{i}} + 2y\vec{\mathbf{j}} + \vec{\mathbf{k}})\, dx\, dy.$$

Now the curl$(\vec{\mathbf{F}})$ is given by

$$\text{curl}(\vec{\mathbf{F}}) = \begin{vmatrix} \vec{\mathbf{i}} & \vec{\mathbf{j}} & \vec{\mathbf{k}} \\ \frac{\partial}{\partial x} & \frac{\partial}{\partial y} & \frac{\partial}{\partial z} \\ z & x & -y \end{vmatrix} = -\vec{\mathbf{i}} - \vec{\mathbf{j}} + \vec{\mathbf{k}},$$

thus

$$\iint_S \text{curl}(\vec{\mathbf{F}}) \cdot \overrightarrow{d\sigma^+} = \iint_{x^2+y^2 \le 1} (-2x - 2y + 1)\, dx\, dy$$

$$= \int_0^{2\pi} \int_0^1 (-2r\cos(\theta) - 2r\sin(\theta) + 1)\, r\, dr\, d\theta$$

$$= \pi,$$

and Stoke's Theorem is verified in this case.

Example [28.22] Consider the surface S of a cone given by $z^2 = x^2 + y^2$ with $0 \leq z \leq 1$. Verify Stoke's Theorem in the case $\vec{\mathbf{F}} = -y\vec{\mathbf{i}} + x\vec{\mathbf{j}} + \vec{\mathbf{k}}$.

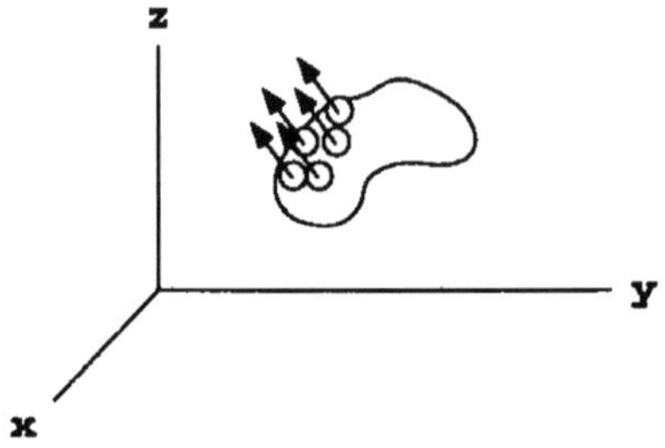

The cone in this example is the graphical display of the function $z = \sqrt{x^2 + y^2}$ with $0 \leq z \leq 1$. Now

$$\overrightarrow{d\sigma} = \frac{-x\vec{\mathbf{i}} - y\vec{\mathbf{j}}}{\sqrt{x^2 + y^2}} + z\vec{\mathbf{k}},$$

and

$$\text{curl}(\vec{\mathbf{F}}) = \begin{vmatrix} \vec{\mathbf{i}} & \vec{\mathbf{j}} & \vec{\mathbf{k}} \\ \frac{\partial}{\partial x} & \frac{\partial}{\partial y} & \frac{\partial}{\partial z} \\ -y & x & 1 \end{vmatrix} = 2\vec{\mathbf{k}}.$$

We conclude that

$$\iint_S \text{curl}(\vec{\mathbf{F}}) \cdot \overrightarrow{d\sigma} = \iint_{x^2+y^2 \leq 1} 2\, dx\, dy = 2\pi.$$

To complete the verification of Stoke's Theorem we begin by parameterizing ∂S, which in this case is the circle of radius 1 on the plane $z = 1$. The parameterization

$$x = \cos(\theta), \quad y = \sin(\theta), \quad z = 1,$$

allows us to quickly compute the line integral in question:

$$\oint_{\partial S} \vec{\mathbf{F}} \cdot d\vec{\mathbf{r}} = \oint_{\partial S} (-y\,dx + x\,dy + dz)$$

$$= \int_0^{2\pi} \left((\sin(\theta))^2 + (\cos(\theta))^2\right) d\theta$$

$$= 2\pi,$$

To conclude this chapter we present an easily demonstrated corollary of Stoke's Theorem.

Corollary [28.23] *If $\vec{\mathbf{F}}$ is a gradient vector field then for any orientable open surface which is the graphical display of a differentiable function $z = f(x, y)$ over a bounded region, we have*

$$\oint_{\partial S} \vec{\mathbf{F}} \cdot d\vec{\mathbf{r}} = 0$$

where ∂S denotes the boundary of S.

Proof: When we recall that $\text{curl}(\vec{\mathbf{F}}) = 0$ whenever $\vec{\mathbf{F}}$ is a gradient vector field (see Example [28.20]), Stoke's theorem immediately gives us our corollary:

$$\oint_{\partial S} \vec{\mathbf{F}} \cdot d\vec{\mathbf{r}} = \iint_S \text{curl}(\vec{\mathbf{F}}) \cdot \overrightarrow{d\sigma} = 0.$$

Exercises for §28.6

(1) Verify Stoke's theorem for the surface $z = 4 - x^2 - y^2$ with $0 \le z \le 4$ and $\vec{\mathbf{F}} = 2z\vec{\imath} - 2x\vec{\jmath} + 2y\vec{\mathbf{k}}$.

(2) Verify Stoke's theorem for the cone $z = \sqrt{x^2 + y^2}$ with $0 \le z \le 2$ and $\vec{\mathbf{F}} = y\vec{\imath} - x\vec{\jmath} + z\vec{\mathbf{k}}$.

(3) Verify Stoke's theorem for the surface which is a triangle whose vertices are located at $(1,0,0)$, $(0,1,0)$, and $(0,0,1)$ where $\vec{\mathbf{F}} = z^2\vec{\mathbf{i}} + y^2\vec{\mathbf{j}} + xy\vec{\mathbf{k}}$.

Use Stoke's theorem to evaluate the following surface integrals $\iint\limits_S \text{curl}(\vec{\mathbf{F}}) \cdot \overrightarrow{d\sigma}^{+}$

(4) S is the hemisphere $x^2 + y^2 + z^2 = 1$, $z \geq 0$ and $\vec{\mathbf{F}} = 2x^2z\vec{\mathbf{i}} - x\vec{\mathbf{j}} + y^2\vec{\mathbf{k}}$.

(5) S is the triangle whose vertices at located at $(1,0,0)$, $(0,-1,0)$, and $(0,0,1)$ where $\vec{\mathbf{F}} = -y\vec{\mathbf{i}} + z\vec{\mathbf{j}} - x\vec{\mathbf{k}}$.

Use Stoke's theorem to evaluate the following line integrals $\oint\limits_{\partial S} \vec{\mathbf{F}} \cdot d\vec{\mathbf{r}}$.

(6) $\vec{\mathbf{F}} = -y\vec{\mathbf{i}} + x\vec{\mathbf{j}} + z\vec{\mathbf{k}}$ and ∂S is the intersection of the cylinder $x^2 + y^2 = 1$ (z is arbitrary) and the plane $y + z = 3$ (x arbitrary).

(7) $\vec{\mathbf{F}} = x\vec{\mathbf{i}} + y\vec{\mathbf{j}} + z\vec{\mathbf{k}}$ and ∂S is a simple closed curve bounding an arbitrary simple surface S.

(8) Let S_1, S_2 be two surfaces with the same boundary curve, i.e., $C = \partial S_1 = \partial S_2$. Show that if S_1 and S_2 lie on opposite sides of C then

$$\iint\limits_{S_1} \text{curl}(\vec{\mathbf{F}}) \cdot \overrightarrow{d\sigma}^{+} = -\iint\limits_{S_2} \text{curl}(\vec{\mathbf{F}}) \cdot \overrightarrow{d\sigma}^{-}.$$

(9) Let C be a simple closed curve. Given differentiable functions f and g, show that

$$\oint\limits_C (f\vec{\nabla}g + g\vec{\nabla}f) \cdot d\vec{\mathbf{r}} = 0.$$

Additional exercises for Chapter XXVIII

Compute the surface area of the following surfaces S.

(1) S is the portion of the parabaloid $z = 2 - x^2 - y^2$ above the xy-axis.

(2) S is the portion of the surface $z = x + y^2$ above the triangle with vertices $(0,0)$, $(1,1)$, $(0,1)$.

(3) S is portion of the cone $z = x^2 + y^2$ where $\frac{1}{2} \leq z \leq 6$,

(4) S is the portion of the surface $y^2 = x$ between $z = 0$, $z = 4$, $y = 1$ and $y = 2$.

Evaluate the surface integrals $\iint\limits_S \text{curl}(\vec{\mathbf{F}}) \cdot \overrightarrow{d\sigma}^{+}$ and $\iint\limits_S \text{curl}(\vec{\mathbf{F}}) \cdot \overrightarrow{d\sigma}^{-}$ for the following surfaces S and the vector fields $\vec{F}$.

(5) $\vec{F} = x\vec{\mathbf{i}} + y^2\vec{\mathbf{j}} + z\vec{\mathbf{k}}$ and S is the surface in Problem 1.

(6) $\vec{F} = y\vec{\mathbf{j}} + z\vec{\mathbf{k}}$ and S is the portion of the parabaloid $z = x^2 + y^2$ below the plane $z = 4$.

(7) $\vec{F} = x^2\vec{\mathbf{i}} + y^2x\vec{\mathbf{j}} + z\vec{\mathbf{k}}$, where S is the surface $z = xy^2$ lying above the rectangle $0 \le x \le 2$, $-1 \le y \le 1$.

(8) $\vec{F} = x\vec{\mathbf{i}} + y\vec{\mathbf{j}} + z\vec{\mathbf{k}}$, and S is the portion of the cylinder $x^2 + z^2 = a^2$ in the first octant between the planes $x = 0$, $x = 9$, $z = y$, and $z = 2y$.

(9) $\vec{F} = (x+y)\vec{\mathbf{i}} + (y+z)\vec{\mathbf{j}} + (z+x)\vec{\mathbf{k}}$, where S is the portion of the plane $x + y + z = 1$ in the first octant.

Find the divergence of the following vector fields. Check your answers by computing with your CAS.

(10) $\vec{F} = e^{xy^2}\vec{\mathbf{i}} - \cos(y)\vec{\mathbf{j}} + \sin^2 x\vec{\mathbf{k}}$.

(11) $\vec{F} = \frac{1}{\sqrt{x^2+y^2+z^2}}\left(x\vec{\mathbf{i}} + y\vec{\mathbf{j}} + z\vec{\mathbf{k}}\right)$.

(12) $\vec{F} = \ln(x)\vec{\mathbf{i}} + e^{xyz}\vec{\mathbf{j}} + \tan^{-1}(\frac{x}{z})\vec{\mathbf{k}}$.

Evaluate the following surface integrals $\iint\limits_S \vec{F} \cdot \overrightarrow{d\sigma}$ by using the Divergence Theorem.

(13) $\vec{F} = e^{xy}\vec{\mathbf{i}} - \cos(y)\vec{\mathbf{j}} + \sin^2(x)\vec{\mathbf{k}}$, and S is the cube bounded by the coordinate planes $x = 1$, $y = 1$, and $z = 1$.

(14) $\vec{F} = z^3\vec{\mathbf{i}} - x^3\vec{\mathbf{j}} + y^3\vec{\mathbf{k}}$ and S is the sphere $x^2 + y^2 + z^2 = a^2$.

Compute for each vector field $\vec{\mathbf{v}}$ the spin (up to sign) of a unit disk placed at the origin in xy-plane, the zy-plane and the xz-plane. Use your CAS to obtain a graphical display of the vector field $\vec{\mathbf{v}}$.

(15) $\vec{\mathbf{v}} = \frac{1}{\sqrt{x^2+y^2+z^2}}\left(x\vec{\mathbf{i}} + y\vec{\mathbf{j}} + z\vec{\mathbf{k}}\right)$

(16) $\vec{\mathbf{v}} = e^{xy^2}\vec{\mathbf{i}} - \cos(y)\vec{\mathbf{j}} + \sin^2(x)x\vec{\mathbf{k}}$

(17) $\vec{\mathbf{v}} = \ln(x)\vec{\mathbf{i}} + e^{xyz}\vec{\mathbf{j}} + \tan^{-1}(\frac{x}{z})\vec{\mathbf{k}}$

Use Stoke's Theorem to evaluate the following surface integrals $\int_S \int \mathrm{curl}(\vec{F}) \cdot \overrightarrow{d\sigma}$.

(18) $\vec{F} = 3z\vec{\mathbf{i}} + 4xy\vec{\mathbf{j}} + 2y\vec{\mathbf{k}}$, where S is portion of the paraboloid $z = 4 - x^2 - y^2$ above the xy-plane.

(19) $\vec{F} = (z + \sin(x))\,\vec{\mathbf{i}} + (x+y)\vec{\mathbf{j}} + (y + e^z)\vec{\mathbf{k}}$ and S is the cone $z = \sqrt{x^2+y^2}$ beneath the plane $z = 3$.

Chapter XXIX

Differential Forms: An Overview

29.1 Differential Forms and the Wedge Product

The higher dimensional integration theorems we have encountered thus far, Green's Theorem, the Divergence Theorem, and Stoke's Theorem, appear, at first, rather distinct. One of the achievements of modern mathematics has been to develop a theory which demonstrates that these theorems are all special cases of a very general form of the Fundamental Theorem of Calculus. Recall that the essence of the Fundamental Theorem of Calculus is that differentiation and integration are inverse processes, i.e.,

$$\frac{d}{dx}(x^2 + C) = 2x, \qquad \int 2x\,dx \;=\; x^2 + C. \tag{29.1}$$

This foundational observation can be generalized to higher dimensions and in so doing we will arrive at the unifying theory known as the theory of differential forms .

To motivate this theory we begin by examining the various higher dimensional integrals which have arisen. The line integral takes the form

$$\int_C \Big(P(x,y,z), dx + Q(x,y,z)\,dy + R(x,y,z)\,dz\Big), \tag{29.2}$$

the most general surface integral is given by

$$\iint_S \Big(P(x,y,z)\,dx\,dy + Q(x,y,z)\,dy\,dz + R(x,y,z)\,dz\,dx\Big), \tag{29.3}$$

while triple integrals are given in the form

$$\iiint_B P(x,y,z)\,dx\,dy\,dz. \tag{29.4}$$

The expressions which appear within these integrals,

$$\begin{aligned}
&P(x,y,z)\,dx + Q(x,y,z)\,dy + R(x,y,z)\,dz\\
&\qquad P(x,y,z)\,dx\,dy + Q(x,y,z)\,dy\,dz + R(x,y,z)\,dz\,dx\\
&\qquad\qquad P(x,y,z)\,dx\,dy\,dz
\end{aligned} \tag{29.5}$$

are all termed **differential forms** (in dimensions 1, 2, and 3 respectively). The theory of differential forms will classify these expressions in a uniform way and will provide a powerful tool to view higher dimensional integration and differentiation as inverse processes.

Definition: *A* ***zero–form*** *is a differentiable function $f(x,y,z)$ in three variables x,y,z.*

The expressions dx, dy, dz are called differentials.

Definition: *A* ***one–form*** *is an expression of the form*

$$\omega = P(x,y,z)\,dx + Q(x,y,z)\,dy + R(x,y,z)\,dz \tag{29.6}$$

where P, Q, R are zero–forms.

Remark: Given two one–forms

$$\omega_i = P_i(x,y,z)\,dx + Q_i(x,y,z)\,dy + R_i(x,y,z)\,dz, \quad (i=1,2) \tag{29.7}$$

we define their sum (and difference) respectively by

$$\begin{aligned}
\omega_1 \pm \omega_2 &= (P_1(x,y,z) \pm P_2(x,y,z))\,dx\\
&+ (Q_1(x,y,z) \pm Q_2(x,y,z))\,dy\\
&+ (R_1(x,y,z) \pm R_2(x,y,z))\,dz.
\end{aligned} \tag{29.8}$$

Furthermore, if $f = f(x,y,z)$ is a zero–form and $\omega = P\,dx + Q\,dy + R\,dz$ is a one–form, we define

$$f\cdot\omega = f\cdot P\,dx + f\cdot Q\,dy + f\cdot R\,dz. \tag{29.9}$$

Having defined addition and subtraction of one–forms, and multiplication of a one–form by a zero–form, it is only natural to try to multiply two one–forms. The operation we will define is termed the **wedge product** and is denoted by $\wedge$ (which we pronounce *wedge*). The result of wedging two one–forms will be a **two–form**.

Consider the set which consists of sums and differences of the formal expressions

$$du \wedge dv \qquad \Big(du, dv \in \{dx, dy, dz\}\Big), \tag{29.10}$$

subject to the identities

$$dx \wedge dy = -dy \wedge dx, \;\; dy \wedge dz = -dz \wedge dy, \;\; dz \wedge dx = -dx \wedge dz, \tag{29.11}$$

and

$$dx \wedge dx = 0, \quad dy \wedge dy = 0, dz \wedge dz = 0. \tag{29.12}$$

We can extend the wedge product to arbitrary one–forms ω_1, ω_2 and consider **two–forms**

$$\omega_1 \wedge \omega_2, \tag{29.13}$$

once we specify that for any three one–forms $\omega_1, \omega_2, \omega$

$$\begin{aligned} (\omega_1 + \omega_2) \wedge \omega &= \omega_1 \wedge \omega + \omega_2 \wedge \omega, \\ \omega \wedge (\omega_1 + \omega_2) &= \omega \wedge \omega_1 + \omega \wedge \omega_2, \end{aligned} \tag{29.14}$$

and given a zero–form $f(x, y, z)$,

$$f \cdot \omega_1 \wedge \omega_2 = \omega_1 \wedge f \cdot \omega_2. \tag{29.15}$$

Example [29.1] Expand and simplify the wedge–product:

$$\Big((x^2 - 2y)\,dx - xy^2\,dz\Big) \wedge y\,dx.$$

To simplify this wedge product we apply (29.10) and(29.11):

$$\begin{aligned} \Big((x^2 - 2y)\,dx - xy^2\,dz\Big) \wedge y\,dx &= (x^2 - 2y)\,dx \wedge y\,dx + (-xy^2)\,dz \wedge y\,dx \\ &= y(x^2 - 2y)dx \wedge dx - xy^3\,dz \wedge dx \\ &= xy^3\,dx \wedge dz. \end{aligned}$$

Example [29.2] Expand and simplify the wedge–product:

$$(y\,dx - x\,dy) \wedge (x^2 y\,dx + y^2 x\,dz). \tag{29.16}$$

This example is similar to the previous one:

$$\begin{aligned}(y\,dx - x\,dy) &\wedge (x^2 y\,dx + y^2 x\,dz)\\ &= (y\,dx - x\,dy) \wedge x^2 y\,dx + (y\,dx - x\,dy) \wedge y^2 x\,dz\\ &= -x^3 y\,dy \wedge dx + y^3 x\,dx \wedge dz - y^2 x^2\,dy \wedge dz.\end{aligned} \tag{29.17}$$

The example above motivates the following definition,

Definition: *A **standard two–form** is an expression of the form*

$$P(x,y,z)\,dx \wedge dy \;+\; Q(x,y,z)\,dy \wedge dz \;+\; R(x,y,z)\,dz \wedge dx, \tag{29.18}$$

where P, Q, R are zero–forms.

Remarks:

(1) Any two–form can be transformed into a standard two–form using the identities of the wedge product. For example, the two–form

$$x\,dy \wedge dx - 3x^2yz\,dy \wedge dz + (xy^2 + z)\,dx \wedge dz \;=\; - x\,dx \wedge dy - 3x^2yz\,dy \wedge dz - (xy^2 + z)\,dz \wedge dx.$$

(2) It should be noted that the wedge product mimics the cross product of vectors in that were we to replace dx by $\vec{\mathbf{i}}$, dy by $\vec{\mathbf{j}}$, and dz by $\vec{\mathbf{k}}$, the identities the wedge satisfies transforms into identities the cross product satisfies.

The next objective in this program is to take the wedge product of a two–form and a one–form (keeping the identities of the wedge product in mind) and obtain a **three-form**. Since any two–form can be put into standard form we may restrict our attention to the wedge product

$$\Big(P\,dx \wedge dy \;+\; Q\,dy \wedge dz \;+\; R\,dz \wedge dx\Big) \wedge \Big(g_1\,dx + g_2\,dy + g_3\,dz\Big). \tag{29.19}$$

Using the identities (29.10) and (29.11), and in addition, specifying that

$$(\omega_1 \wedge \omega_2) \wedge \omega_3 = \omega_1 \wedge (\omega_2 \wedge \omega_3) \tag{29.20}$$

for all triples of one–forms $\omega_1, \omega_2, \omega_3$, we see that the only cases which do not amount to zero are of the form

$$f(x,y,z)\, du_1 \wedge du_2 \wedge du_3 \qquad \Big(du_i \in \{dx, dy, dz\},\ i = 1,2,3\Big) \tag{29.21}$$

where du_i are distinct. Since we can always permute the order of the du_i using (29.8) the following definition is appropriate.

Definition: *A standard three–form is an expression of the form*

$$f(x,y,z)\, dx \wedge dy \wedge dz, \tag{29.22}$$

where f is a zero–form.

Example [29.3] Put the wedge product

$$(y^2\, dx + x\, dy) \wedge (z\, dx \wedge dy - x\, dy \wedge dz)$$

into standard form.

Here again we use the identities of the wedge product:

$$\begin{aligned}
&(y^2\, dx + x\, dy) \wedge (z\, dx \wedge dy - x\, dy \wedge dz) \\
&\qquad = y^2\, dx \wedge (z\, dx \wedge dy - x\, dy \wedge dz) \\
&\qquad\quad + x\, dy \wedge (z\, dx \wedge dy - x\, dy \wedge dz) \\
&\qquad = y^2\, dx \wedge z\, dx \wedge dy - y^2\, dx \wedge x\, dy \wedge dz \\
&\qquad\quad + x\, dy \wedge z\, dx \wedge dy - x\, dy \wedge -x\, dy \wedge dz \\
&\qquad = 0 + -y^2\, dx \wedge x\, dy \wedge dz + 0 + 0 \\
&\qquad = -xy^2 dx \wedge dy \wedge dz.
\end{aligned} \tag{29.23}$$

Example [29.4] Put the three-form $x^3yz^2\, dz \wedge dy \wedge dx$ into standard form.

In this example we utilize (29.8) repeatedly:

$$\begin{aligned}
x^3yz^2\, dz \wedge dy \wedge dx &= x^3yz^2\, (dz \wedge dy) \wedge dx = -x^3yz^2\, (dy \wedge dz) \wedge dx \\
&= -x^3yz^2\, dy \wedge (dz \wedge dx) = x^3yz^2\, dy \wedge (dx \wedge dz) \\
&= x^3yz^2\, (dy \wedge dx) \wedge dz \\
&= -x^3yz^2\, dx \wedge dy \wedge dz
\end{aligned} \tag{29.24}$$

Exercises for §29.1

Expand and simplify the following wedge products. Express your final answers in standard form.

(1) $(dx + dy + dz) \wedge (dx + dy + dz)$.

(2) $(dx - x\,dy + z\,dz) \wedge dy$.

(3) $((x^2 - 2yz)\,dx + (x - 3y)\,dz) \wedge (dx \wedge dy)$.

(4) $(y^3\,dx + z\,dy) \wedge (x\,dx \wedge dy - 3y^2\,dy \wedge dx)$.

(5) $dx \wedge (y^2\,dx \wedge dy - 3x^2\,dx \wedge dz + 2\,dz \wedge dy)$.

(6) $(e^y\,dx) \wedge (x^3y\,dx - e^x\,dy + \cos(x^2yz)\,dz)$.

(7) $(e^{xy}\,dx - 3z^2\,dy + xy\,dz) \wedge (2\,dx - 3\,dy + e^y\,dz)$

(8) $(dx \wedge dz) \wedge (dy \wedge dx)$.

(9) $(dx \wedge x\,dy \wedge z\,dz) \wedge dx$.

(10) $3y^3\,dz \wedge (dx \wedge dy - dy \wedge dz + 2\,dx \wedge dz)$.

29.2 The d–Operator

In this section we will define the **d–operator** (denoted d) which will act as a derivative in higher dimensions. The d operator is defined on n–forms for $n = 0, 1, 2$ and when applied to an n–form gives an $(n+1)$–form.

Definition: *Let $f(x, y, z)$ be a zero–form. We define*

$$df = \frac{\partial f}{\partial x}\,dx + \frac{\partial f}{\partial y}\,dy + \frac{\partial f}{\partial z}\,dz. \tag{29.25}$$

Given a one–form $\rho = P\,dx + Q\,dy + R\,dz$ we let

$$d\rho = dP \wedge dx + dQ \wedge dy + dR \wedge dz. \tag{29.26}$$

Finally, when $\omega = P\,dx\wedge dy + Q\,dy\wedge dz + R\,dz\wedge dx$, the ***d–operator*** *is defined to be*

$$d\omega = dP \wedge (dx \wedge dy) + dQ \wedge (dy \wedge dz) + dR \wedge (dz \wedge dx). \tag{29.27}$$

Example [29.5] Let $f(x, y, z) = x^2yz^4$. Evaluate df.

We have

$$df = \frac{\partial f}{\partial x}\,dx + \frac{\partial f}{\partial y}\,dy + \frac{\partial f}{\partial z}\,dz = 2xyz^4\,dx + x^2z^4\,dy + 4x^2yz^3\,dz.$$

Example [29.6] Let $\rho = y\,dx - xz\,dy$. Evaluate $d\rho$.

In this case

$$\begin{aligned} d\rho &= dy \wedge dx - d(xz) \wedge dy \\ &= -dx \wedge dy - (z\,dx + x\,dz) \wedge dy \\ &= (-1 - z)\,dx \wedge dy + x\,dy \wedge dz. \end{aligned} \tag{29.28}$$

Example [29.7] Let $\omega = x^3yz^5dy \wedge dz$. Evaluate $d\omega$.

Following our definition once again we obtain

$$\begin{aligned} d\omega &= d(x^3yz^5) \wedge (dy \wedge dz) \\ &= (3x^2yz^5\,dx + x^3z^5\,dy + 5x^3yz^4\,dz) \wedge (dy \wedge dz) \\ &= 3x^2yz^5\,dx \wedge dy \wedge dz. \end{aligned} \tag{29.29}$$

Example [29.8] Let $\rho = P\,dx + Q\,dy + R\,dz$ be an arbitrary one–form. Evaluate $d\rho$.

Despite the generality of this example we can obtain a compact formula

$$\begin{aligned} d\rho &= dP \wedge dx + dQ \wedge dy + dR \wedge dz \\ &= \left(\frac{\partial P}{\partial x}\,dx + \frac{\partial P}{\partial y}\,dy + \frac{\partial P}{\partial z}\,dz\right) \wedge dx \\ &\quad + \left(\frac{\partial Q}{\partial x}\,dx + \frac{\partial Q}{\partial y}\,dy + \frac{\partial Q}{\partial z}\,dz\right) \wedge dy \\ &\quad + \left(\frac{\partial R}{\partial x}\,dx + \frac{\partial R}{\partial y}\,dy + \frac{\partial R}{\partial z}\,dz\right) \wedge dz \end{aligned}$$

$$= \left(\frac{\partial Q}{\partial x} - \frac{\partial P}{\partial y}\right) dx \wedge dy$$

$$+ \left(\frac{\partial R}{\partial y} - \frac{\partial Q}{\partial z}\right) dy \wedge dz$$

$$+ \left(\frac{\partial P}{\partial z} - \frac{\partial R}{\partial x}\right) dz \wedge dx.$$

Exercises for §29.2

Evaluate $d\omega$ for the following differential forms ω.

(1) $\omega = xyz$.

(2) $\omega = x^3yz^2$.

(3) $\omega = \cos(xz^2)$.

(4) $\omega = e^x$.

(5) $\omega = dx + dy$.

(6) $\omega = x^2y\,dx - y\,dy + dz$.

(7) $\omega = e^y\,dx - e^x\,dy + z\,dz$.

(8) $\omega = xyz\,dx - e^{x+y+z}\,dz$.

(9) $\omega = x^3y\,dx \wedge dz$.

(10) $\omega = \cos(xyz)\,(dy \wedge dz)$.

(11) $\omega = x\,dy \wedge dz - y\,dx \wedge dy$.

(12) $\omega = 3x^2y\,dx \wedge dy \wedge dz$.

(13) Let ω and η be k–forms for $k = 0, 1, 2$. Show that

(i) $d(\omega + \eta) = d\omega + d\eta$.

(ii) $d(d\omega) = 0$.

29.3 The Generalized Stoke's Theorem

Having defined the d–operator in §29.2 we are now in a position to state the general theorem from which Green's theorem, the Divergence theorem, and Stoke's theorem can be deduced. Let ω be a differential n–form (for $n = 1, 2, 3$) in standard form. Given an n–dimensional geometric body Ω (i.e., a curve, a surface, or a solid body), we now give meaning to the integral

$$\int_{\Omega} \omega. \tag{29.30}$$

Case (1) If $\omega = P\,dx + Q\,dy + R\,dz$ is a standard 1–form, and Ω is a curve, then

$$\int_{\Omega} \omega = \int_{\Omega} (P\,dx + Q\,dy + R\,dz), \tag{29.31}$$

is defined to be the line integral.

Case (2) If $P\,dx \wedge dy + Q\,dy \wedge dz + R\,dz \wedge dx$ is a 2–form and Ω is a 2–dimensional surface then

$$\int_{\Omega} \omega = \iint_{\Omega} \Big(P\,dx\,dy + Q\,dy\,dz + R\,dz\,dx\Big) \tag{29.32}$$

where we drop the wedge in the integral and evaluate the zero–forms P, Q, R on the surface Ω.

Case (3) If $\omega = f\,dx \wedge dy \wedge dz$ is a standard 3–form and Ω is a 3–dimensional solid, we define

$$\int_{\Omega} \omega = \iiint_{\Omega} f\,dx\,dy\,dz \tag{29.33}$$

to be the triple integral where here again the wedges are dropped.

The general form of the Fundamental Theorem of Calculus states that (under suitable hypothesis)

$$\int_{\Omega} d\omega = \int_{\partial\Omega} \omega, \tag{29.34}$$

where $\partial\Omega$ is the boundary of Ω.

Example [29.9] Let $\omega = P\,dx + Q\,dy + R\,dz$ and let S be a simple surface whose boundary ∂S is a closed curve which is smooth and does not self intersect. Deduce **Green's theorem** from (29.35).

Recalling equation (29.30)

$$d\omega = \left(\frac{\partial Q}{\partial x} - \frac{\partial P}{\partial y}\right) dx \wedge dy + \left(\frac{\partial R}{\partial y} - \frac{\partial Q}{\partial z}\right) dy \wedge dz + \left(\frac{\partial P}{\partial z} - \frac{\partial R}{\partial x}\right) dz \wedge dx. \tag{29.35}$$

we deduce that (29.35) takes the form

$$\iint_S \left(\frac{\partial Q}{\partial x} - \frac{\partial P}{\partial y}\right) dx\, dy + \left(\frac{\partial R}{\partial y} - \frac{\partial Q}{\partial z}\right) dy\, dz +$$

$$\left(\frac{\partial P}{\partial z} - \frac{\partial R}{\partial x}\right) dz\, dx = \int_{\partial S} \Big(P\, dx + Q\, dy + R\, dz\Big). \quad (29.36)$$

Equation (29.37) is actually a generalization of Green's theorem. To deduce the theorem in the form we encountered in §27.4 we consider a 1–form $\omega = P\,dx + Q\,dy$ where P, Q do not involve the variable z, i.e., P, Q are functions of x, y.

Example [29.10] Let $\omega = P\,dx \wedge dy + Q\,dy \wedge dz + R\,dz \wedge dx$ be a 2–form and let Ω be a 3–dimensional solid object whose boundary $\partial\Omega$ is a closed simple surface. Deduce the **Divergence theorem**.

Here again we begin by computing the effect of the d–operator on the differential form in question:

$$\begin{aligned} d\omega &= \frac{\partial P}{\partial z} dx \wedge dy \wedge dz + \frac{\partial Q}{\partial x} dx \wedge dy \wedge dz + \frac{\partial R}{\partial y} dx \wedge dy \wedge dz \\ &= \left(\frac{\partial P}{\partial z} + \frac{\partial Q}{\partial x} + \frac{\partial R}{\partial y}\right) dx \wedge dy \wedge dz. \end{aligned} \quad (29.37)$$

The general Fundamental theorem (29.23) now gives us

$$\iint_{\partial\Omega} \Big(P\,dx\,dy + Q\,dy\,dz + R\,dz\,dx\Big) = \iiint_{\Omega} \left(\frac{\partial P}{\partial z} + \frac{\partial Q}{\partial x} + \frac{\partial R}{\partial y}\right) dx\,dy\,dz,$$

which is a variation of the Divergence theorem.

Remark: The Fundamental Theorem of Calculus for functions of a single variable

$$\int_a^b f'(x)\,dx = f(b) - f(a),$$

can also be seen as a special case of (29.23) in the following way. If we look at $\omega = f(x)$ as a zero–form, then $d\omega = f'(x)\,dx$. Let Ω be the 1–dimensional line segment $a \le x \le b$. Then the boundary of Ω, which we still denote $\partial\Omega$, simply consists of the two endpoints a, b.

In this final case the generalized Fundamental Theorem will state that

$$\int_{\Omega} d\omega = \int_a^b f'(x)\,dx = \int_{\partial\Omega} \omega = f(b) - f(a).$$

We can be interpret these identities provided we specify that the integral of a 0–form is given by

$$\int_{\partial\Omega} \omega = f(b) - f(a). \tag{29.38}$$

Example [29.11] (Change of Coordinates) Consider the double integral $\iint\limits_{\mathcal{R}} f(x,y)\,dx\,dy$ over a two–dimensional region $\mathcal{R}$. Use wedge products to show that under a change of coordinates

$$x = g_1(u,v), \qquad y = g_2(u,v) \tag{29.39}$$

the double integral becomes

$$\iint\limits_{\mathcal{R}} f(x,y)\,dx\,dy = \iint\limits_{\mathcal{R}^*} f^*(u,v)\left(\frac{\partial g_1}{\partial u}\frac{\partial g_2}{\partial v} - \frac{\partial g_1}{\partial v}\frac{\partial g_2}{\partial u}\right) du\,dv, \tag{29.40}$$

where $f^*(u,v) = f(g_1(u,v), g_2(u,v))$ and R^* is the image under the transformation.

Notice here that we are actually re–verifying Proposition 25.14. The reader should observe that our original derivation of (29.40) was lengthy. The fact that we can easily obtain this result using differential forms (eliminating any need to memorize the formula) is a vivid demonstration of the power of this theory.

Since the transformation is given by $x = g_1(u,v),\ y = g_2(u,v)$ we have

$$dx = \frac{\partial g_1}{\partial u}\,du + \frac{\partial g_1}{\partial v}\,dv$$

$$dy = \frac{\partial g_2}{\partial u}\,du + \frac{\partial g_2}{\partial v}\,dv.$$

Thus we conclude that

$$dx \wedge dy = \left(\frac{\partial g_1}{\partial u}\,du + \frac{\partial g_1}{\partial v}\,dv\right) \wedge \left(\frac{\partial g_2}{\partial u}\,du + \frac{\partial g_2}{\partial v}\,dv\right)$$

$$= \left(\frac{\partial g_1}{\partial u}\frac{\partial g_2}{\partial v} - \frac{\partial g_1}{\partial v}\frac{\partial g_2}{\partial u}\right) du \wedge dv.$$

Additional exercises for Chapter XXIX

Expand and simplify the following wedge products. Express your final answers in standard form.

(1) $\left(x^2ydx + y^2xdy + z^ydz\right) \wedge \left(xy^2dx + yx^2dy + z^ydz\right)$.

(2) $\left(\ln(x^2)dx + e^{yx}dy + \cos(z^y)dz\right) \wedge \left(xy^2dx + yx^2dy\right)$.

(3) $\frac{1}{x^2+y^2+z^2}\left(x^2dx + dy + dz\right) \wedge \left(xdx + yx^2dy\right)$.

Evaluate $d\omega$ for the following differential forms ω.

(4) $\omega = \cos(y^2), dx - \sin(x^2)\, dy$.

(5) $\omega = 6zdx \wedge dy - xy\, dx \wedge dz$.

(6) $\omega = e^z \cos(y)\, dx - e^x \sin(x)\, dy + e^x \tan(x)\, dz$.

(7) $\omega = (y - z)dx \wedge dz - (z - x)dy \wedge dz + (x - y)dx \wedge dy$.

A differential form, ω is said to be closed if $d\omega \equiv 0$. Further ω is said to be exact if there is a differential form ψ such that $d\psi = \omega$. Determine which of the following are closed and which are exact or neither.

(8) $\omega = yz\, dx + xz\, dy + xy\, dz$.

(9) $\omega = x\, dx + x^2y^2\, dy + yz\, dz$.

(10) $\omega = 2xy^2\, dx \wedge dy + z\, dy \wedge dz$.

(11) $\omega = \cos(y^2)\, dx - \sin(x^2)\, dy$.

(12) $\omega = y\, dx + (x + 3y^2)\, dy$.

(13) $\omega = 2xy^3\, dx + (1 + x^2y^2)dy$.

Three diferential forms $\omega_1, \omega_2, \omega_3$ are said to be **linearly independent** if and only if $\omega_1 \wedge \omega_2 \wedge \omega_3 \neq 0$. In the following exercise determine if the differential forms are independent.

(14) $\omega_1 = dx, \omega_2 = dy, \omega_3 = dz$.

Chapter XXX

Fourier Series

30.1 Periodic Functions

Let $f(x)$ be a function whose domain is the set of real numbers $\mathbb{R}$. Fix a real number ω. We term $f(x)$ **periodic with period ω** provided

$$f(x+\omega) = f(x)$$

for all $x \in \mathbb{R}$. The study of periodic functions of period ω can immediately be reduced to the study of functions of period 1 (which we shall simply refer to as **periodic**) in the following manner. Given $f(x)$ of period ω define the new function

$$g(x) = f(\omega x).$$

Observe that $g(x)$ is periodic (i.e., of period 1), and any question we ask about $f(x)$ reduces to a question about $g(x)$.

The most intrinsic periodic functions which appear are the trigonometric functions

$$\cos(2\pi n x), \qquad \sin(2\pi n x) \qquad\qquad (n = 0, \pm 1, \pm 2, \dots). \tag{30.1}$$

It is a remarkable theorem (due to Fourier) that an arbitrary differentiable periodic function can be expressed as a linear combination (i.e., as a sum with coefficients) of these basic periodic functions (30.1).

Example [30.1] We may express the periodic function $\cos^2(2\pi x)$ as the linear combination

$$\cos^2(2\pi x) = \frac{1+\cos(4\pi x)}{2}.$$

Example [30.2] Express $\cos^3(2\pi x)$ as a linear combination of the basic periodic functions (30.1).

Here we apply **Euler's Formula** (see §19.7), which states

$$e^{2\pi ix} = \cos(2\pi x) + i\sin(2\pi x),$$

to see that

$$\begin{aligned}(e^{2\pi ix})^3 &= (\cos(2\pi x) + i\sin(2\pi x))^3 \\ &= \cos^3(2\pi x) + 3i\cos^2(2\pi x)\sin(2\pi x) \\ &\qquad - 3\cos(2\pi x)\sin^2(2\pi x) - i\sin^3(2\pi x).\end{aligned}$$

But

$$(e^{2\pi ix})^3 = e^{6\pi ix} = \cos(6\pi x) + i\sin(6\pi x).$$

Equating the real parts of the last two expressions we see that

$$\cos^3(2\pi x) - 3\cos(2\pi x)\sin^2(2\pi x) = \cos(6\pi x).$$

Since

$$\sin^2(2\pi x) = 1 - \cos^2(2\pi x)$$

we deduce that

$$4\cos^3(2\pi x) - 3\cos(2\pi x) = \cos(6\pi x),$$

i.e.,

$$\cos^3(2\pi x) = \frac{1}{4}(3\cos(2\pi x) + \cos(6\pi x)).$$

It can be shown that by performing various clever manipulations (similar to that which we used in Example [30.2]) any polynomial in the functions $\cos(2\pi x)$ or $\sin(2\pi x)$ can be expressed as a finite linear combination of the basic functions (30.1).

The theory of Fourier series is a generalization of the above observation to arbitrary periodic functions. In order to give a clear presentation (and avoid a mass of notation), we utilize the complex valued periodic function

$$e^{2\pi inx} = \cos(2\pi nx) + i\sin(2\pi nx) \qquad (n = 0, \pm1, \pm2, \dots). \tag{30.2}$$

Since

$$\cos(2\pi nx) = \frac{1}{2}\left(e^{2\pi inx} + e^{-2\pi inx}\right)$$

$$\sin(2\pi nx) = \frac{1}{2i}\left(e^{2\pi inx} - e^{-2\pi inx}\right)$$

the functions (30.2) can also be used to construct arbitrary periodic functions.

An infinite series of the type

$$\sum_{n=-\infty}^{\infty} a_n e^{2\pi inx} \tag{30.3}$$

is called a **Fourier series**. Observe that the identities

$$\begin{aligned} |e^{2\pi inx}| &= |\cos(2\pi nx) + i\sin(2\pi nx)| \\ &= \sqrt{\cos^2(2\pi nx) + \sin^2(2\pi nx)} = 1, \end{aligned}$$

imply that the Fourier series (30.3) converges absolutely provided

$$\sum_{n=-\infty}^{\infty} |a_n e^{2\pi inx}| = \sum_{n=-\infty}^{\infty} |a_n|$$

converges. In this case (30.3) converges to the function

$$f(x) = \sum_{n=-\infty}^{\infty} a_n e^{2\pi inx},$$

which must be periodic in that it is a linear combination of the basic periodic functions (30.2).

Proposition [30.3] *Let $\sum_{n=-\infty}^{\infty} a_n e^{2\pi inx}$ converge absolutely to a function $f(x)$. Then*

$$a_0 = \int_0^1 f(x)\,dx$$

is the average value of the function $f(x)$.

Proof: This proposition is based on the following computation:

$$\int_0^1 e^{2\pi inx}\,dx = \begin{cases} 1,\text{if } n = 0 \\ 0,\text{if } n = \pm 1, \pm 2, \ldots . \end{cases}$$

It immediately follows from this that

$$\begin{aligned}\int_0^1 f(x)\,dx &= \int_0^1 \sum_{n=-\infty}^{\infty} a_n e^{2\pi inx}\,dx \\ &= \sum_{n=-\infty}^{\infty} a_n \int_0^1 e^{2\pi inx}\,dx \\ &= a_0.\end{aligned}$$

Proposition [30.4] *Suppose the sum* $\sum_{n=-\infty}^{\infty} a_n e^{2\pi inx}$ *converges absolutely to an integrable function* $f(x)$. *Then*

$$a_n = \int_0^1 f(x)e^{-2\pi inx}\,dx.$$

Remark: Observe the analogy with Taylor series: Taylor coefficients are obtained by differentiation, Fourier coefficients are computed by integration.

Proof: Here we need only compute the integral in question:

$$\begin{aligned}\int_0^1 f(x)e^{-2\pi inx}\,dx &= \int_0^1 \left(\sum_{m=-\infty}^{\infty} a_m e^{2\pi imx}\right) e^{-2\pi inx}\,dx \\ &= \sum_{m=-\infty}^{\infty} a_m \int_0^1 e^{-2\pi i(m-n)x}\,dx \\ &= a_n\end{aligned}$$

since the last integral is 0 unless $m = n$.

Exercises for §30.1

(1) Express $(\sin 2\pi x)^3$ as a finite linear combination of the functions $\sin(2\pi nx)$ and/or $\cos(2\pi nx)$ with $n = 0, \pm 1, \pm 2, \ldots$

(2) Express the polynomial $2 - \sin(2\pi x) + 3\sin^2(2\pi x)) + 2\sin^3(2\pi x)$ as a finite linear combination of the functions $\sin(2\pi nx)$ and/or $\cos(2\pi nx)$ with $n = 0, \pm 1, \pm 2, \ldots$

(3) Express $\cos^5(2\pi x)$ as a finite linear combination of the functions $\sin(2\pi nx)$ and/or $\cos(2\pi nx)$ with $n = 0, \pm 1, \pm 2, \ldots$

(4) Show that for integers m, n,

$$\int_0^1 \cos(2\pi mx)\cos(2\pi nx)\,dx = 0$$

unless $m = n$.

(5) Show that for integers m, n,

$$\int_0^1 \cos(2\pi mx)\sin(2\pi nx)\,dx = 0.$$

(6) Show that for integers m, n,

$$\int_0^1 \sin(2\pi mx)\sin(2\pi nx)\,dx = 0$$

unless $m = n$.

(7) Evaluate the integrals

$$\int_0^1 (\cos(2\pi nx))^2\,dx, \qquad \int_0^1 (\sin(2\pi nx))^2\,dx$$

for $n = 0, \pm 1, \pm 2, \ldots$

30.2 Fourier Expansions of Periodic Functions

We showed in §30.1 that if the Fourier series

$$\sum_{n=-\infty}^{\infty} a_n e^{2\pi i n x}$$

converges then it must converge to a periodic function

$$f(x) = \sum_{n=-\infty}^{\infty} a_n e^{2\pi i n x}$$

Furthermore, if $f(x)$ is integrable then we must have

$$a_n = \int_0^1 f(x)e^{-2\pi inx}\,dx.$$

The question remains as to whether the converse is true.

Question: Can every integrable periodic function be expressed as a linear combination of the basic periodic functions $e^{2\pi inx}$?

The answer to this question was given in one fell swoop by Fourier in the following striking theorem.

Theorem [30.5] *Let $f(x)$ be a periodic function which is infinitely differentiable. Let*

$$\hat{f}(n) = \int_0^1 f(x)e^{-2\pi inx}\,dx.$$

Then the Fourier series

$$\sum_{n=-\infty}^{\infty} \hat{f}(n)e^{2\pi inx}$$

converges absolutely to the function $f(x)$.

Remark: The hypothesis in the above theorem, that f be infinitely differentiable, can be weakened considerably. For example, the theorem holds for functions f such that $\int_0^1 |f(x)|\,dx$ is defined and finite.

Proof: This proof is presented in four stages.

Step 1. We begin by showing that the series

$$\sum_{n=-\infty}^{\infty} \hat{f}(n)e^{2\pi inx}$$

does in fact converge absolutely to some function of x. The absolute convergence will follow immediately once we show that for all n,

$$|\hat{f}(n)| \leq \frac{c}{n^2}$$

for some constant $c > 0$. In order to verify this bound we begin by integrating the integral which defines $\hat{f}(n)$ by parts twice:

$$\hat{f}(n) = \int_0^1 f(x)e^{2\pi inx}\,dx$$

$$= -\int_0^1 f'(x)\frac{e^{2\pi inx}}{2\pi in}\,dx$$

$$= \int_0^1 f''(x)\frac{e^{2\pi inx}}{(2\pi in)^2}\,dx.$$

Upon taking absolute values of the latter equation, we see that

$$|\hat{f}(x)| \le \frac{c}{n^2}$$

where

$$c = \frac{1}{4\pi^2}\int_0^1 |f''(x)|\,dx,$$

(here we are tacitly using the easily verified fact that, in general, if $f(x)$ is an integrable function then $\left|\int_a^b f(x)\,dx\right| \le \int_a^b |f(x)|\,dx$).

Step 2. We now claim that it suffices to prove that

$$f(0) = \sum_{n=-\infty}^{\infty} \hat{f}(n) \tag{30.4}$$

for any function f which is periodic and infinitely differentiable. To see this fix a number x_0 in the interval $0 < x_0 < 1$, and define the function g by

$$g(x) = f(x + x_0).$$

The function g (which is termed the **translate of f by x_0**), is periodic and infinitely differentiable and thus if (30.4) holds for any function we have

$$g(0) = \sum_{n=-\infty}^{\infty} \hat{g}(n).$$

But $g(0) = f(x_0)$ and thus

$$\begin{aligned}
\hat{g}(n) &= \int_0^1 g(x)e^{-2\pi inx}\,dx \\
&= \int_0^1 f(x+x_0)e^{-2\pi inx}\,dx \qquad \text{(change variables: } x \to x - x_0\text{)} \\
&= \int_{x_0}^{x_0+1} f(x)e^{-2\pi in(x-x_0)}\,dx \\
&= e^{2\pi inx_0}\int_{x_0}^{x_0+1} f(x)e^{-2\pi inx}\,dx \\
&= e^{2\pi inx_0}\hat{f}(n).
\end{aligned}$$

We conclude that

$$f(x_0) = g(0) = \sum_{n=-\infty}^{\infty} \hat{f}(n)e^{2\pi inx_0}.$$

Remark: The final identity in the above argument is actually a special case of the following principle. If f is a periodic function and α is a real number then

$$\hat{f}(n) = \int_{\alpha}^{\alpha+1} f(x)e^{-2\pi inx}\,dx.$$

This can be verified by differentiating $I_n(\alpha) = \int_{\alpha}^{\alpha+1} f(x)e^{-2\pi inx}\,dx$ with respect to α via the Fundamental Theorem of Calculus,

$$\frac{d}{d\alpha}I_n(\alpha) = f(\alpha+1)e^{-2\pi in(\alpha+1)} - f(\alpha)e^{-2\pi in\alpha} = 0.$$

We conclude that $I_n(\alpha)$ is a constant independent of α (see Proposition [7.3]) and it must, therefore, be equal to $I_n(0) = \hat{f}(n)$ for any α.

Step 3. We may assume that $f(0) = 0$: if this is not the case, consider the new function $f(x) - f(0)$, which vanishes at 0, is periodic, and is infinitely differentiable.

Step 4. We are now reduced to considering a function $f(x)$ which is periodic, infinitely differentiable, and vanishes at 0. Recalling Step 2 of our proof, we need only show that

$$\sum_{n=-\infty}^{\infty} \hat{f}(n) = 0. \tag{30.5}$$

Since $f(0) = 0$ we may consider the new function $g(x)$ which is defined by the identity

$$g(x) = \frac{f(x)}{1 - e^{2\pi i x}},$$

where

$$g(0) = \lim_{x \to 0} \frac{f(x)}{1 - e^{2\pi i x}} = \lim_{x \to 0} \frac{f'(x)}{-2\pi i e^{2\pi i x}} = \frac{f'(0)}{-2\pi i},$$

by L'Hospital's rule (see §19.8). Observe that by definition, $g(x)$ is itself periodic and infinitely differentiable. Now,

$$\begin{aligned}
\hat{f}(n) &= \int_0^1 f(x)e^{-2\pi i n x}\,dx \\
&= \int_0^1 (1 - e^{2\pi i x}) \cdot g(x) \cdot e^{-2\pi i n x}\,dx \\
&= \int_0^1 g(x)e^{-2\pi i n x}\,dx - \int_0^1 g(x)e^{-2\pi i (n-1)x}\,dx \\
&= \hat{g}(n) - \hat{g}(n-1).
\end{aligned}$$

But then

$$\sum_{n=-\infty}^{\infty} \hat{f}(n) = \sum_{n=-\infty}^{\infty} \Big(\hat{g}(n) - \hat{g}(n-1)\Big) = 0,$$

since the latter sum is a telescoping sum where all the terms cancel:

$$\cdots \Big(\hat{g}(0) - \hat{g}(-1)\Big) + \Big(\hat{g}(1) - \hat{g}(0)\Big) + \Big(\hat{g}(2) - \hat{g}(1)\Big) + \cdots$$

This completes the proof.

30.3 Examples

Example [30.6] Graph the periodic square wave function $f(x)$ which is defined on the interval $0 \le x < 1$ by

$$f(x) = \begin{cases} \frac{1}{2}, & \text{if } 0 \le x < \frac{1}{2} \\ 1, & \text{if } \frac{1}{2} \le x < 1 \end{cases}$$

and otherwise satisfies the periodicity relation $f(x) = f(x+1)$. Expand $f(x)$ into a Fourier series.

The square wave function is graphed below.

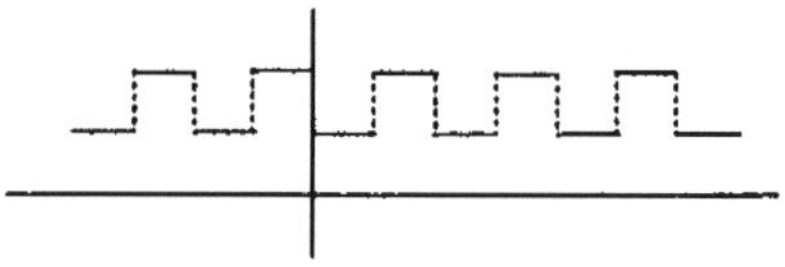

To compute the Fourier expansion

$$f(x) = \sum_{n=-\infty}^{\infty} \hat{f}(n) e^{2\pi i n x},$$

we first compute $\hat{f}(0)$:

$$\begin{aligned} \hat{f}(0) &= \int_0^1 f(x)\,dx \\ &= \int_0^{\frac{1}{2}} \frac{1}{2}\,dx + \int_{\frac{1}{2}}^1 dx \\ &= \frac{3}{4}. \end{aligned}$$

Next we compute $\hat{f}(n)$ when $n \neq 0$,

$$\begin{aligned} \hat{f}(n) &= \int_0^1 f(x) e^{-2\pi i n x}\,dx \\ &= \int_0^{\frac{1}{2}} \frac{1}{2} e^{-2\pi i n x}\,dx + \int_{\frac{1}{2}}^1 e^{-2\pi i n x}\,dx \\ &= \frac{e^{-\pi i n} - 1}{-4\pi i n} + \frac{1 - e^{-\pi i n}}{-2\pi i n} \\ &= \frac{e^{-\pi i n} - 1}{4\pi i n}, \end{aligned}$$

hence

$$\hat{f}(n) = \begin{cases} 0, & \text{if } n \text{ is even and } n \neq 0 \\ -\frac{1}{2\pi i n}, & \text{if } n \text{ is odd.} \end{cases}$$

We conclude that

$$f(x) = \frac{3}{4} - \sum_{n \text{ odd}} \frac{1}{2\pi i n} e^{2\pi i n x}.$$

Example [30.7] Graph the sawtooth function $f(x)$ which is defined on the interval $0 \leq x < 1$ by

$$f(x) = x$$

and otherwise satisfies the periodicity relation $f(x) = f(x+1)$. Expand $f(x)$ into a Fourier series.

The saw tooth function is graphed below.

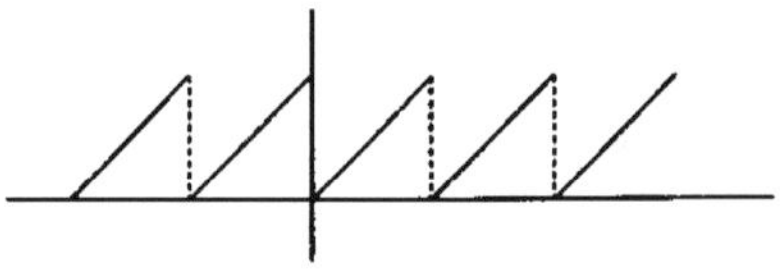

The Fourier coefficients are given by

$$\hat{f}(0) = \int_0^1 x\,dx = \frac{1}{2},$$

and for $n \neq 0$, integration by parts gives us

$$\begin{aligned} \hat{f}(n) &= \int_0^1 x e^{-2\pi i n x}\,dx \\ &= \left.\frac{x e^{-2\pi i n x}}{-2\pi i n}\right]_0^1 - \int_0^1 \frac{e^{-2\pi i n x}}{-2\pi i n}\,dx \\ &= \frac{1}{-2\pi i n}. \end{aligned}$$

Thus

$$f(x) = \frac{1}{2} - \sum_{n \neq 0} \frac{1}{2\pi i n} e^{2\pi i n x}.$$

Example [30.8] Graph the periodic triangular wave function

$$f(x) = \begin{cases} x, & 0 \leq x < \frac{1}{2} \\ 1 - x, & \frac{1}{2} \leq x < 1. \end{cases}$$

Approximate $f(x)$ with the finite Fourier series

$$f(x) \approx \hat{f}(0) + \hat{f}(1)e^{2\pi ix} + \hat{f}(-1)e^{-2\pi ix}$$
$$+ \hat{f}(2)e^{4\pi ix} + \hat{f}(-2)e^{-4\pi ix} + \hat{f}(3)e^{6\pi ix} + \hat{f}(-3)e^{-6\pi ix}.$$

Superimpose the graph of $f(x)$ with the graph of the above finite Fourier series and compare the displays.

The triangular wave function is graphed below.

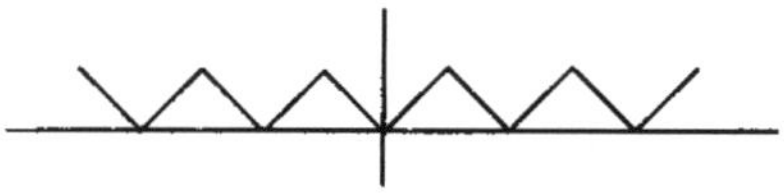

The Fourier coefficients of this function are given by

$$\begin{aligned}\hat{f}(0) &= \int_0^{\frac{1}{2}} x\,dx + \int_{\frac{1}{2}}^1 (1-x)\,dx \\ &= \frac{1}{8} + \frac{1}{8} = \frac{1}{4}\end{aligned}$$

and for $n \neq 0$,

$$\hat{f}(n) = \int_0^{\frac{1}{2}} xe^{-2\pi inx}\,dx + \int_{\frac{1}{2}}^1 (1-x)e^{-2\pi inx}\,dx.$$

The latter integral can be simplified with the substitution $x \mapsto 1-x$:

$$\begin{aligned}\int_{\frac{1}{2}}^1 (1-x)e^{-2\pi inx}\,dx &= \int_{\frac{1}{2}}^0 -xe^{2\pi inx}\,dx \\ &= \int_0^{\frac{1}{2}} xe^{2\pi inx}\,dx.\end{aligned}$$

Hence

$$\begin{aligned}\hat{f}(n) &= \int_0^{\frac{1}{2}} x(e^{-2\pi inx} + e^{2\pi inx})\,dx \\ &= 2\int_0^{\frac{1}{2}} x\cos(2\pi nx)\,dx \\ &= \left.\frac{2x\sin(2\pi nx)}{2\pi n}\right]_0^{\frac{1}{2}} - 2\int_0^{\frac{1}{2}} \frac{\sin(2\pi nx)}{2\pi n}\,dx \\ &= \frac{\cos(\pi n) - 1}{4\pi^2 n^2},\end{aligned}$$

and we conclude that

$$\hat{f}(n) = \begin{cases} 0, & \text{if } n \text{ is even and } n \neq 0 \\ -\frac{1}{2\pi^2 n^2}, & \text{if } n \text{ is odd.} \end{cases}$$

The finite Fourier series in question is thus given by

$$\begin{aligned} \sum_{-3 \leq n \leq 3} \hat{f}(n) e^{2\pi i n x} &= \frac{1}{4} - \frac{1}{2\pi^2}\left(e^{2\pi i x} + e^{-2\pi i x}\right) - \frac{1}{18\pi^2}\left(e^{6\pi i x} + e^{-6\pi i x}\right) \\ &= \frac{1}{4} - \frac{\cos(2\pi x)}{\pi^2} - \frac{\cos(6\pi x)}{9\pi^2}. \end{aligned} \qquad (30.6)$$

When we superimpose the graph of (30.6) with the graph of the triangular wave we obtain the following display.

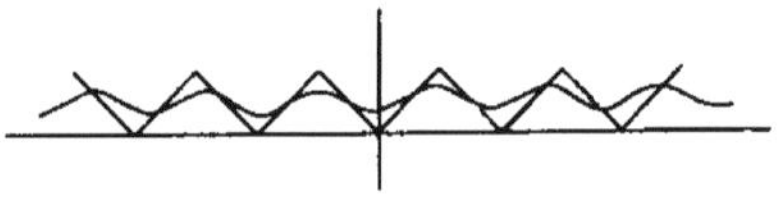

Example [30.9] (Plancherel's Identity) Show that if f is an infinitely differentiable periodic function then

$$\int_0^1 |f(x)|^2 \, dx = \sum_{n=-\infty}^{\infty} |\hat{f}(n)|^2.$$

Following our original definitions we see that

$$\begin{aligned} \int_0^1 |f(x)|^2 \, dx &= \int_0^1 \left(\sum_{n=-\infty}^{\infty} \hat{f}(n) e^{2\pi i n x} \right) \cdot \left(\sum_{m=-\infty}^{\infty} \overline{\hat{f}(m)} e^{-2\pi i m x} \right) dx \\ &= \sum_{n=-\infty}^{\infty} \sum_{m=-\infty}^{\infty} \hat{f}(n) \overline{\hat{f}(m)} \int_0^1 e^{2\pi i (n-m) x} \, dx \\ &= \sum_{n=-\infty}^{\infty} |\hat{f}(n)|^2, \end{aligned}$$

since the integral $\int_0^1 e^{2\pi i(n-m)x}\,dx$ equals 0 unless $m = n$, in which case it equals 1.

Example [30.10] (Euler) Establish the formula

$$\sum_{n=1}^{\infty} \frac{1}{n^2} = \frac{\pi^2}{6}.$$

Proof: Recall that we have already demonstrated (see Example [30.7]) that the Fourier coefficients for the sawtooth function $f(x) = x$ $(0 \le x < 1)$ are given by

$$\hat{f}(n) = \int_0^1 xe^{-2\pi inx}\,dx = \begin{cases} \frac{1}{2} & \text{if } n = 0 \\ -\frac{1}{2\pi in}, & \text{if } n \neq 0. \end{cases}$$

Therefore, by Plancherel's identity (see Example [30.9])

$$\begin{aligned} \int_0^1 |f(x)|^2\,dx &= \int_0^1 x^2\,dx \\ &= \frac{1}{3} \\ &= \frac{1}{4} + \frac{1}{4\pi^2}\sum_{n\neq 0}\frac{1}{n^2} \\ &= \frac{1}{4} + \frac{2}{2\pi^2}\sum_{n=1}^{\infty}\frac{1}{n^2} \end{aligned}$$

Solving for $\sum_{n=1}^{\infty} \frac{1}{n^2}$ the memorable formula appears,

$$\sum_{n=1}^{\infty} \frac{1}{n^2} = 2\pi^2\left(\frac{1}{3} - \frac{1}{4}\right) = \frac{\pi^2}{6}.$$

Exercises for §30.3

Graph the following periodic functions $f(x)$. Compute the Fourier coefficients $\hat{f}(n)$. Approximate $f(x)$ with the finite Fourier series

$$f(x) \approx \sum_{-4 \le n \le 4} \hat{f}(n)e^{2\pi inx}.$$

Superimpose the graph of $f(x)$ with the graph of the above finite Fourier series and compare the displays.

(1)

$$f(x) = \begin{cases} 1, & \text{if } 0 \le x < \frac{1}{3} \\ 3, & \text{if } \frac{1}{3} \le x < 1. \end{cases}$$

(2)

$$f(x) = \begin{cases} \frac{1}{2}, & \text{if } 0 \le x < \frac{1}{2} \\ x, & \text{if } \frac{1}{2} \le x < 1. \end{cases}$$

(3)

$$f(x) = \begin{cases} 4x, & \text{if } 0 \le x < \frac{1}{4} \\ \frac{4}{3}(1-x), & \text{if } \frac{1}{4} \le x < 1. \end{cases}$$

(4) $f(x) = x^2, \quad (0 \le x < 1).$

(5) $f(x) = e^{-x}, \quad (0 \le x < 1).$

(6)

$$f(x) = \begin{cases} 0, & \text{if } 0 \le x < \frac{1}{4} \\ x - \frac{1}{4}, & \text{if } \frac{1}{4} \le x < \frac{1}{2} \\ \frac{3}{4} - x, & \text{if } \frac{1}{2} \le x < \frac{3}{4} \\ 0, & \text{if } \frac{3}{4} \le x < 1. \end{cases}$$

(7) Show that

$$\sum_{n=1}^{\infty} \frac{1}{n^4} = \frac{\pi^4}{90}.$$

(8) Let f be an infinitely differentiable periodic function. Show that for every integer $k = 1, 2, 3, \ldots$ there exists a constant $c > 0$ such that

$$\hat{f}(n) < \frac{c}{|n|^k}$$

for all $n \neq 0$.

Hint: Integrate by parts k times.

Additional exercises for Chapter XXX

Graph the following periodic functions $f(x)$. Compute the Fourier coefficients $\hat{f}(n)$. Approximate $f(x)$ with a finite Fourier series and superimpose the graph of $f(x)$ with the the graph of the finite series for several n.

(1) $f(x) = x - x^2$, for $0 \leq x \leq 1$.

(2)

$$f(x) = \begin{cases} 4x^2, & \text{if } 0 \leq x \leq \frac{1}{2} \\ 4(1-x)^2, & \text{if } \frac{1}{2} \leq x \leq 1. \end{cases}$$

(3)

$$f(x) = \begin{cases} 0, & \text{if } 0 \leq x \leq \frac{1}{3c} \\ 1, & \text{if } \frac{1}{3c} \leq x \leq \frac{2}{3c} \\ 0, & \text{if } \frac{2}{3c} \leq x \leq 1 \end{cases}$$

for $c \geq 1$.

(4)

$$f(x) = \begin{cases} 0, & \text{if } \leq x \leq \frac{1}{c} \\ 1, & \text{if } \frac{1}{c} \leq x \leq \frac{c-1}{c} \\ 0, & \text{if } \frac{c-1}{c} \leq x \leq 1 \end{cases}$$

for $c > 1$.

(5) $f(x) = e^{2\pi x}$, $(0 \leq x \leq 1)$.

(6) $f(x) = e^{-ax}\cos(2\pi bx)$, $(0 \leq x \leq 1)$.

(7) $f(x) = e^{-ax}\sin(2\pi bx)$, $(0 \leq x \leq 1)$.

(8) $f(x) = \cosh(2\pi ax)$, $(0 \leq x \leq 1)$.

(9) $f(x) = \sinh(2\pi ax)$, $(0 \leq x \leq 1)$.

(10) Show that

$$\frac{\pi - x}{2} = \frac{\sin(x)}{1} + \frac{\sin(2x)}{2} + \frac{\sin(3x)}{3} + \cdots$$

for $0 < x < 2\pi$. (**Hint:** use the transformation $x = 2\pi u$ and find the Fourier series of the left side on the interval $0 < x < 2\pi$).

(11) Show that

$$\frac{3x^2 - 6\pi x + 2\pi^2}{12} = \sum_{n=1}^{\infty} \frac{\cos(nx)}{n^2},$$

for $0 \leq x \leq 2\pi$. **Hint:** Integrate the formula in problem (10).

(12) Show that

$$\frac{x^3 - 3\pi x^2 + 2\pi^2 x}{12} = \sum^{\infty} \frac{\sin(nx)}{n^3}$$

for $0 \leq x \leq 2\pi$.

APPENDIX

Elementary Functions and their Identities

A1. Binomials

$$(1+x)^2 = 1+2x+x^2$$

$$(1+x)^3 = 1+3x+3x^2+x^3$$

$$(1+x)^4 = 1+4x+6x^2+4^3+x^4$$

$$\vdots$$

$$(1+x)^n = \sum_{k=0}^{\infty} \binom{n}{k} x^k \quad \text{where } \binom{n}{k} = \frac{n(n-1)(n-2)\cdots(n-k+1)}{k!}, \quad \binom{n}{0} = 1.$$

$$(1+x)^{\frac{1}{2}} = 1+\tfrac{1}{2}x-\tfrac{1\cdot 1}{2\cdot 4}x^2+\tfrac{1\cdot 1\cdot 3}{2\cdot 4\cdot 6}x^3-\tfrac{1\cdot 1\cdot 3\cdot 5}{2\cdot 4\cdot 6\cdot 8}x^4+\cdots$$

$$\tfrac{1}{1-x} = 1+x+x^2+x^3+\cdots \quad (\text{for } |x|<1).$$

$$\tfrac{1}{1+x} = 1-x+x^2-x^3+\cdots \quad (\text{for } |x|<1).$$

A2. Trigonometric Functions

$$\sin(x) = \sum_{n=0}^{\infty} \frac{(-1)^n x^{2n+1}}{(2n+1)!}$$

$$\cos(x) = \sum_{n=0}^{\infty} \frac{(-1)^n x^{2n}}{(2n)!}$$

$$\tan\left(\tfrac{\pi x}{2}\right) = \tfrac{4x}{\pi} \sum_{n=1}^{\infty} \frac{1}{(2n-1)^2-x^2}$$

$$\sin(-x) = -\sin(x), \qquad \cos(-x) = \cos(x)$$

$$\sin(x) = \cos(x-\frac{\pi}{2}), \qquad \cos(x) = \sin(x+\frac{\pi}{2})$$

$$\tan(x) = \frac{\sin(x)}{\cos(x)}, \qquad \tan(-x) = -\tan(x)$$

$$\cos^2(x)+\sin^2(x) = 1$$

$$\sin(x \pm y) = \sin(x)\cos(y) \pm \sin(y)\cos(x)$$
$$\cos(x \pm y) = \cos(x)\cos(y) \mp \sin(x)\sin(x)$$
$$\tan(x \pm y) = \frac{\tan(x) \pm \tan(y)}{1 \mp \tan(x)\tan(y)}$$

$$\tan\left(\frac{x}{2}\right) = \frac{1-\cos(x)}{\sin(x)} = \frac{\sin(x)}{1+\cos(x)}$$
$$\sin(x) \pm \sin(y) = 2\sin\left(\frac{x \pm y}{2}\right)\cos\left(\frac{x \mp y}{2}\right)$$
$$\cos(x) + \cos(y) = 2\cos\left(\frac{x+y}{2}\right)\cos\left(\frac{x-y}{2}\right)$$
$$\cos(x) - \cos(y) = 2\sin\left(\frac{x+y}{2}\right)\sin\left(\frac{y-x}{2}\right)$$
$$\tan(x) \pm \tan(y) = \frac{\sin(x \pm y)}{\cos(x)\cos(y)}$$

$$\sin^2(x) = \frac{1-\cos(2x)}{2}, \qquad \cos^2(x) = \frac{1+\cos(2x)}{2}$$
$$\sin^3(x) = \frac{3\sin(x)-\sin(3x)}{4}, \qquad \cos^3(x) = \frac{3\cos(x)+\cos(3x)}{4}$$
$$\sin^4(x) = \frac{3-4\cos(2x)+\cos(4x)}{8}, \qquad \cos^4(x) = \frac{3+4\cos(2x)+\cos(4x)}{8}$$

$$\sin^2(x) - \sin^2(y) = \sin(x+y)\sin(x-y) = \cos^2(y) - \cos^2(x)$$
$$\cos^2(x) - \sin^2(y) = \cos(x+y)\cos(x-y) = \cos^2(y) - \sin^2(x)$$

$$\sin(2x) = 2\sin(x)\cos(x), \qquad \cos(2x) = 2\cos^2(x) - 1$$
$$\sin(3x) = 3\sin(x) - 4\sin^3(x), \qquad \cos(3x) = 4\cos^3(x) - 3\cos(x)$$
$$\sin(4x) = \cos(x)(4\sin(x) - 8\sin^3(x)), \qquad \cos(4x) = 8\cos^4(x) - 8\cos^2(x) + 1$$

$$\sin(x) = x\prod_{n=1}^{\infty}\left(1 - \frac{x^2}{\pi^2 n^2}\right)$$

$$\cos(x) = x\prod_{n=0}^{\infty}\left(1 - \frac{4x^2}{\pi^2(2n+1)^2}\right)$$

$$\frac{\sin(x)}{x} = \prod_{n=1}^{\infty}\left(1 - \frac{4}{3}\sin^2\left(\frac{x}{3^n}\right)\right)$$

A3. Exponential Functions

$$e^x = \sum_{n=0}^{\infty} \frac{x^n}{n!}, \qquad a^x = e^{x \cdot \ln(a)} = \sum_{n=0}^{\infty} \frac{(x \ln(a))^n}{n!}$$

$$a^{x+y} = a^x \cdot a^y, \qquad a^{x-y} = \frac{a^x}{a^y}$$

$$\sin(x) = \tfrac{1}{2}(e^{ix} - e^{-ix})$$

$$\cos(x) = \tfrac{1}{2}(e^{ix} + e^{-ix})$$

$$e^{ix} = \cos(x) + i\sin(x)$$

$$e^{inx} = (\cos(x) + i\sin(x))^n$$

$$e^{inx} = \cos(nx) + i\sin(nx) \ (n = \text{integer})$$

A4. Hyperbolic Functions

$$\sinh(x) = \sum_{n=0}^{\infty} \frac{x^{2n+1}}{(2n+1)!}, \qquad \cosh(x) = \sum_{n=0}^{\infty} \frac{x^{2n}}{(2n)!}$$

$$\sinh(x) = \frac{e^x - e^{-x}}{2}, \qquad \sinh(ix) = i\sin(x)$$

$$\cosh(x) = \frac{e^x + e^{-x}}{2}, \qquad \cosh(ix) = i\cos(x)$$

$$\tanh(x) = \frac{\sinh(x)}{\cosh(x)} = \frac{e^x - e^{-x}}{e^x + e^{-x}}, \qquad \tanh(ix) = i\tan(x)$$

$$\cosh^2(x) - \sinh^2(x) = 1$$

$$\sinh(x \pm y) = \sinh(x)\cosh(y) \pm \sinh(y)\cosh(x)$$

$$\cosh(x \pm y) = \cosh(x)\cosh(y) \pm \sinh(x)\sinh(y)$$

$$\sinh^2(x) = \frac{\cosh(2x) - 1}{2} \qquad \cosh^2(x) = \frac{\cosh(2x) + 1}{2}$$

$$\sinh(2x) = 2\sinh(x)\cosh(x) \qquad \cosh(2x) = 2\cosh^2(x) - 1$$

$$\sinh(x) = x\prod_{n=1}^{\infty}\left(1 + \frac{x^2}{\pi^2 n^2}\right) \qquad \cosh(x) = \prod_{n=1}^{\infty}\left(1 + \frac{4x^2}{\pi^2(2n+1)^2}\right)$$

A5. The Logarithm

$$\ln(1+x) = \sum_{n=1}^{\infty} \frac{(-1)^{n+1}\, x^n}{n} \quad (-1 < x \leq 1)$$

$$\ln(x) = \sum_{n=1}^{\infty} \frac{(-1)^{n+1}\,(x-1)^n}{n} \quad (0 < x \leq 2)$$

$$\ln(x) = \sum_{n=1}^{\infty} \frac{1}{n}\left(\frac{x-1}{x}\right)^n \quad (x \geq \tfrac{1}{2})$$

$$\ln\left(\frac{1+x}{1-x}\right) = 2\sum_{n=1}^{\infty} \frac{x^{2n-1}}{2n-1} \quad (|x| < 1)$$

$$\left(\frac{1-x}{x}\right)\cdot \ln\left(\frac{1}{1-x}\right) = 1 - \sum_{n=1}^{\infty} \frac{x^n}{n(n+1)} \quad (|x| < 1)$$

$$\ln(x\cdot y) = \ln(x) + \ln(y), \qquad \ln\left(\frac{x}{y}\right) = \ln(x) - \ln(y)$$

$$\ln\left(e^x\right) = e^{\ln(x)} = x$$

A6. Inverse Trigonometric Functions

$$-\frac{\pi}{2} \leq \arcsin(x) \leq \frac{\pi}{2} \quad (-1 \leq x \leq 1), \qquad \arcsin(\sin(x)) = x \quad (-\frac{\pi}{2} \leq x \leq \frac{\pi}{2})$$

$$0 \leq \arccos(x) \leq \pi \quad (-1 \leq x \leq 1), \qquad \arccos(\cos(x)) = x \quad (0 \leq x \leq \pi)$$

$$-\frac{\pi}{2} < \arctan(x) < \frac{\pi}{2} \quad (-\infty < x < \infty), \qquad \arctan(\tan(x)) = x \quad (-\frac{\pi}{2} < x < \frac{\pi}{2})$$

$$\arcsin(x) = \frac{\pi}{2} - \arccos(x) = \sum_{n=0}^{\infty} \frac{(2n)!}{(n!)^2(2n+1)!}\, 2^{-2n}\, x^{2n+1}$$

$$\arctan(x) = \sum_{n=0}^{\infty} \frac{(-1)^n x^{2n+1}}{2n+1} \quad (|x| \leq 1)$$

$$\arctan(x) = \frac{\pi}{2} - \frac{1}{x} + \frac{1}{3x^3} - \frac{1}{5x^5} + \frac{1}{7x^7} - \frac{1}{9x^9} + \cdots \quad (|x| \geq 1)$$

$$\arcsin(x) + \arccos(x) = \frac{\pi}{2}$$

$$\arcsin(x) = \arccos(\sqrt{1-x^2}) \quad (0 \leq x \leq 1)$$

$$= -\arccos(\sqrt{1-x^2}) \quad (-1 \leq x \leq 0)$$

$$\arccos(x) = \arcsin(\sqrt{1-x^2}) \quad (0 \leq x \leq 1)$$

$$= \pi - \arcsin(\sqrt{1-x^2}) \quad (-1 \leq x \leq 0)$$

$$\arctan(x) \;=\; \arcsin\left(\frac{x}{\sqrt{1+x^2}}\right)$$

Symbols

Index

Made in the USA
Monee, IL
01 March 2026

45210950R00356